中国地质大学（北京）庆祝建校七十周年文丛
A Series of Publications in Celebration of the 70th Anniversary of China University of Geosciences Beijing

雷涯邻自选集

Selected Academic Works of Lei Yalin

地质出版社
Geological Publishing House
· 北　京 ·
· Beijing ·

图书在版编目（CIP）数据

雷涯邻自选集 ：汉文、英文 / 雷涯邻著. -- 北京 ：地质出版社，2025. 1. --（中国地质大学（北京）庆祝建校七十周年文丛）. -- ISBN 978-7-116-14334-0

Ⅰ. P3-53

中国国家版本馆 CIP 数据核字第 2024GP7773 号

LEI YALIN ZIXUANJI

责任编辑：秦　切　祁向雷
责任校对：田建茹
出版发行：地质出版社
社址邮编：北京市海淀区学院路 31 号，100083
电　　话：（010）66554649（邮购部）；（010）66554469（编辑室）
网　　址：https：//www. gph. clmpg. com
印　　刷：天津画中画印刷有限公司
开　　本：787mm×1092mm　1/16
印　　张：25. 75
字　　数：645 千字
版　　次：2025 年 1 月北京第 1 版
印　　次：2025 年 1 月天津第 1 次印刷
定　　价：168. 00 元
书　　号：ISBN 978-7-116-14334-0

ISBN 978-7-116-14334-0
9 787116 143340

《中国地质大学（北京）庆祝建校七十周年文丛》

A Series of Publications in Celebration of the 70th Anniversary of China University of Geosciences Beijing

序

马俊杰

百年大计，教育为本。教育大计，教师为本。教师是教育发展的第一资源，是人类灵魂的工程师，是人类文明的传承者，是一所大学的立教之本、兴教之源。

在中国地质大学（北京）七十年的发展历程中，学校不仅向社会各界输送了近二十万名优秀人才，也造就了一批严谨治学、甘为人梯的名家，汇聚了一批蜚声中外、博古通今的大师。作为我国地球科学领域的先行者和领路人，他们艰苦创业，找矿立功，用实践始终回答着时代之问、人民之问；他们专业精深，潜心科研，上天入地下海登极，用行动不懈探索着地球的奥秘；他们坚守育人初心，传道授业，赓续薪火，做好学生为学、为事、为人的引路人；他们坚持“四个面向”，成为“向地球深部进军”的先行军、“两山论”的推动者与“共建地球生命共同体”的守护人。不论过去、现在还是将来，教师队伍永远是学校建设的重心和信心之所在，是学生成才的动力和助力之所在。

教学推动科研，科研促进教学，二者辩证统一，已经成为现代大学的基本标志。值此建校七十周年之际，学校决定编辑出版《中国地质大学（北京）庆祝建校七十周年文丛》，系统收录我校仍工作在一线的院士和知名学者具有代表性的学术成果，进一步传承他们的精品力作，弘扬他们的师道师风，推动地球科学创新发展。

文丛编辑工作自 2020 年 5 月启动，得到了专家学者的大力支持，文丛以地质、资源、环境为主要特色，辑录了多位相关领域执牛耳者的科研成果千余篇，许多文章发表时在国内外产生过重大影响，很多观点为地球科学领域弥补了空白。这些文章高深却不晦涩，博大但足够朴实，既可作为专业文献供晚学后辈查阅使用，又可作为地学科普让非专业人士开阔视野。从字里行间足以见得作者平生所见所学之丰富，更可以深切感受到对祖国寸土寸山之热爱。阅读这些优秀的学术作品，能够学习丰富的地学知识，体味深刻的家国情怀，感受一丝不苟的治学态度，追寻敢为人先的开拓精神。

对科学研究而言，只有站在巨人的肩膀上，才能推陈出新、继往开来。这套文丛既是学校数十年学术发展历程的缩影，也是学校接续血脉、进一步探索创新的起点。我们推出这套文丛的最终目的，是期待着广大读者能够从沉甸甸的文字中汲取向下扎根的养分，从厚重的文化积淀中获得向上生长的力量，在推动地球科学领域科技创新、助力生态文明建设、开展自然文化研究等方面催生新的理论、引领新的实践。

我们相信，《中国地质大学（北京）庆祝建校七十周年文丛》的出版，必将为中国地质大学（北京）建设成为地球科学领域世界一流大学的光辉事业注入崭新力量。

Foreword

Ma Junjie

From the perspective of history, education is the foundation of national development, and teachers are the key to educational undertakings. As the primary resources for the development of education, the engineers of the human soul, and the transmitters of human civilization, teachers lay the groundwork of education and are the guarantors of dynamic education in a university.

During its 70 years of development, the China University of Geosciences Beijing has not only sent nearly 200,000 talented graduates out into all circles of life, but also cultivated a cadre of scholars with rigorous academic attitudes and a willingness to pave the way for others, without concern for personal fame. It has gathered a group of masters renowned at home and abroad and well-versed in the learning of both ancient and modern times. As forerunners and leaders in the earth sciences in China, they have worked hard, making great contributions in mineral exploration and providing empirical answers to the great questions of the age. They are committed professionals, devoting themselves to scientific research and exploration of the Earth's mysteries. Adhering to the "Four Orientations", they have become the vanguard of the march into the inner depths of the planet, the promoters of the "Two Mountains" theory and the guardians of the shared future for life on the Earth. Whether in the past, at present or in the future, the teaching staff are always the centre of building an educational institution, and the driving force and source of help for students.

Teaching promotes scientific research and research in turn fosters better teaching. The reciprocity of teaching and research has become the basic code of modern universities. On the occasion of its 70th anniversary, the University decides to edit and issue *A Series of Publications in Celebration of the 70th Anniversary of the China University of*

Geosciences Beijing, which provides a representative sample of the academic achievements of academicians and leading scientists still working in the front line of education. This collection will pass on the legacy of their outstanding works, exemplify their pedagogy and professional ethics, and promote innovation in the earth sciences.

The editing of the series of publications started in May 2020 and has received strong support from experts and scholars. The collection touches on geology, natural resources and the environment, assembling more than a thousand articles published by eminent scientists. A number of the articles were of considerable influence both in China and abroad for their ideas filling in gaps in our understanding of the Earth. These articles are sophisticated but accessible, well designed to serve as professional literature for later generations, and can also be used to broaden the horizons of non-scientists. Between the lines, one can sense what the authors have seen and learned in their lives, and one can also feel their deep love for the motherland. Reading these excellent academic works, readers will acquire a wealth of geological knowledge and appreciate the profound ties of the authors to their homes and country, their meticulous attitude towards academic research, and their pioneering spirit in science.

In matters of scientific research, it is only by standing on the shoulder of giants that we can bring forth the new from the old, carrying on from the past and opening a way to the future. This compendium is a showcase for the academic development of the University over the decades, but also the starting point for the University as it trains worthy successors for further exploration and innovation. The ultimate purpose of this array of articles is to help readers draw knowledge and strength from these cultural deposits, so as to create new theories and lead new developments in the earth sciences, the construction of an ecological civilization, and the elaboration of the links between nature and culture.

We believe that the publication of this series will inject new energy into the glorious effort to build the China University of Geosciences Beijing into a world-class university in the earth sciences.

中国地质大学（北京）
庆祝建校七十周年文丛

作者简介

雷涯邻

湖北江陵人，1966年7月出生，二级教授，博士生导师，主要从事资源能源环境经济与政策领域的教学和研究工作。2005年6月至2020年5月，任中国地质大学（北京）党委常委、副校长（其间：2016年3月至2018年3月，挂职任吉林省长白山保护开发区党工委委员、管委会副主任）。2020年5月至2023年1月，任北京化工大学党委常委、副校长。2023年1月至今，任中国地质大学（北京）党委书记。现兼任教育部高等学校工商管理类专业教学指导委员会委员、自然资源部科学技术咨询委员会委员、中国高教学会碳中和与清洁能源教育专委会副理事长、中国现代产业学院协同创新平台副理事长及专家委员会副主任、国际期刊 *Resources Policy* 共同主编、中文《资源与产业》主编。入选北京市社科理论人才“百人工程”。先后获自然资源科技进步奖特等奖、二等奖，北京市高等教育教学成果二等奖等。先后主持国家社科基金重大专项、国家重点研发计划课题、国家软科学研究计划、国家重大科技专项专题、国家自然科学基金项目等国家级、省部级以及国际合作与企事业单位委托项目80余项。在《人民日报》《光明日报》等报刊和学术期刊发表论文200余篇，出版学术专著（译著、教材等）10余部。

A Series of Publications in Celebration of the 70th Anniversary of China University of Geosciences Beijing

Introduction to the author

Lei Yalin

Born in July 1966 in Hubei Province. She is a doctoral supervisor and professor of resources, energy, environmental economics and policy at the School of Economics and Management, China University of Geosciences, Beijing（CUGB）. She was a member of the Standing Committee of the Party Committee and vice president of CUGB from June 2005 to May 2020. During the period, she was a member of the Party Working Committee and deputy director of the Management Committee of Changbai Mountain Protection and Development Zone in Jilin Province from March 2016 to March 2018. She was appointed to the Standing Committee of the Party Committee and vice president of Beijing University of Chemical Technology from May 2020 to January 2023. She became the Secretary of the Party Committee of CUGB after January 2023. She is currently a member of the Business Administration Teaching Steering Committee of the Ministry of Education, a member of the Science and Technology Advisory Committee of the Ministry of Natural Resources, a vice chairman of the Carbon Neutral and Clean Energy Education Committee of the China Association of Higher Education, a vice chairman of the Cooperative Innovation Platform and a deputy director of the Expert Committee of the China Institute of Modern Industry, and a co-editor of *Resources Policy* and of an editor of *Resources and Industry*. She has been selected as the candidate of "100 People Project" from Beijing Social Science Theoretical Talents. She has won the special prize and the second prize of the Natural Resources Science and Technology Progress Award, and the second prize of Beijing Higher Education teaching achievements. She has presided over more than 80 projects funded by Major Project of National Social Science Fund, National Key Research and Development Project, National Soft Science Research Program, National Major Science and Technology Project, National Natural Science Foundation of China and other national, provincial and ministerial levels, as well as international cooperation and enterprises and institutions. She has published more than 200 papers in *People's Daily*, *Guangming Daily* and other newspapers and academic journals, and published more than 10 academic monographs（translated works, teaching materials, etc.）.

目　　录

Contents

能源经济

矿产资源经济

环境与健康

区域可持续发展

能源经济

能源、经济和环境（3E）系统研究综述

范凤岩 雷涯邻*

中国地质大学 人文经管学院，北京 100083

摘 要 随着工业化进程的加快以及社会经济的发展，人类对资源的索取持续增长，产生了一系列生态环境问题，使人类的生存发展面临严峻挑战。为了解决这些问题，资源、环境以及社会经济协调发展问题备受当今世界的关注。多年来，学术界对这一问题的研究方兴未艾。目前，国内外关于能源-经济-环境（3E）系统的研究成果主要集中在三者之间相关关系的探讨以及模型的构建上。本文在参阅国内外相关文献的基础上，就 3E 系统有代表性的研究进行了梳理和总结，并对国内外学者在该领域的研究进行了评述，为进一步开展相关问题的研究提供参考。

关键词 能源-经济-环境（3E）系统，可持续发展，关系，模型

An review on energy, economy and environment (3E) system

FAN Fengyan, LEI Yalin*

School of Humanities and Economic Management, *China University of Geosciences*, *Beijing* 100083, *China*

Abstract With the development of industrialization and social economy, human's demands for resources has sustained grown, which results in a series of ecological environment problems, the human's survival and development is facing severe challenges. In order to solve these problems, the coordinated development between resources, environment and social economy has attracted the world's attention. Over the years, the academic research on this problem is just unfolding. At present, the research about energy–economy–environment (3E) system in domestic and foreign countries are mainly focused on the relationship between the three and the model construction. Based on reviewing the relevant literature, this paper has combed and summarized the representative research and reviews the domestic and foreign scholars'

原文刊于《生态经济》，2013 年第 12 期。

基金项目 国家自然科学基金面上项目“基于动态 CGE 模型的燃料乙醇产业发展政策模拟与实证研究”（71173200）。

* 为通信作者。

research in this field, and providing reference for the further research of the related problems.

Keywords Energy-economy-environment (3E) system, Sustainable development, Relationship model

能源是社会发展的物质基础，环境是能源的载体，同时也是各类生物赖以生存的空间[1]。然而，伴随着工业化进程的不断加快和经济的迅猛发展，人类不合理地开发和使用自然资源，不顾后果地任意排放污染物，使世界面临着资源枯竭和退化、生态环境恶化和社会经济发展失衡的严峻挑战。国内外的历史经验表明，能源、环境与经济形成了一个相互关联的复杂有机统一体，寻求能源、经济和环境子系统协调发展不但是可持续发展理论的最本质要求，也是实现国民经济持续、稳步、高速发展的有效途径。如何协调好三者之间的关系，深入探究其变化规律，已成为政府和学术界亟待解决的重大课题。许多国家对此进行了大量的分析和研究工作，并取得了一些重要成果。这些成果对国际社会普遍关注的可持续发展理念、作用机制、战略构想及实现路径的研究，对构筑完善的政策法规体系具有重要的指导意义。

1 能源-经济-环境（3E）系统研究的历史沿革

经济增长是人类社会进步与发展最基本的事实。产业革命后很长一段时期内，人们单纯的追求物质财富的增长，把获取“经济利益”作为首要目标，并以国内生产总值（gross domestic product，简称 GDP）作为衡量发展的金标准。当时能源的稀缺性还未充分表现出来，因而其对经济增长的作用也没有引起学者的关注。大多数新古典经济学家在分析能源对经济增长的影响时，一般将能源视为资本、劳动、土地等主要生产要素产生的一个中间变量，并以柯布-道格拉斯（Cobb-Douglas）的资本和劳动力的双变量生产函数为其表现形式[2]。20 世纪 70 年代爆发的“石油危机”使能源的稀缺性因机械工业的迅速发展日渐显露。随着世界范围内能源需求的不断增加，矿物能源生产与消费引发的生态破坏和环境污染对人类生存构成了严重的威胁。罗马俱乐部《增长的极限》一书的出版，引发了人们对以往发展模式的思考，能源和环境在经济增长中的重要作用才被经济学家充分重视。各国研究人员开始利用经济学理论方法分别研究能源问题和环境问题，逐渐形成了以能源-经济、环境-经济二元系统为研究对象的理论体系[3]。20 世纪 80 年代以后，随着可持续发展观的提出并不断完善以及人类对能源、环境、经济认识程度的加深，人们逐渐意识到，只有把能源、环境、经济纳入一个整体中研究，才能更加全面、深入、系统地了解它们之间的作用机理。于是，国际上许多能源研究和环保机构开始联手和经济学家合作，构建能源（energy）-经济（economy）-环境（environment）3E 系统研究框架体系分析三者之间的发展规律与内在联系，展开对 3E 系统综合平衡和协调发展问题的研究[4]。

我国最初从苏联引入能源经济学的理论，对能源平衡利用的成本效益进行分析，并借鉴苏联经验对能源企业的经济管理问题展开研究[3]。改革开放以来，能源供应为适应经济的高速发展大大增加，成为经济发展的有力支撑，我国借鉴西方经济学的研究理论和方法，在能源系统领域开展了许多国家级、地区级或部门级的能源模型研究工作，在理论方法和实际应用上都取得了很大的进步。华中工学院国家能源模型（HNEM），华中工学院农村能源模型（HREM），能源科学工作者提出建立的国家能源供型体系（CNEM）以及国务院经济技术社会发展研究中心提出的能源供应系统规划与决策模型等都在建立能源模型系统方面进行了可喜的尝试[5]。然而，伴随着能源的大量消耗，能源消费造成的大气污染越来越严重，在环境问题受到全球广泛关注的大背景下，20 世纪 80 年代以来，我国逐渐在环境经济问题的研究方面投入更多的人力、物力和财力[3]。进入 90 年代，越来越多的学者开始将环境因素引入能源和经济二元系统中，从人口、能源结构、能源效率、技术、资源安全和环境等不同角度探讨能源、经济、环境三者之间的相互关系[3]。1984 年，国家计委和国务院能源办牵头组织了全国近 70 名专家学者启动了历时 5 年的中国“广义能源效率战略工程”项目[6]，试图探寻一种既有利于能源供应和环境保护，又不制约经济发展的新战略，拉开了“能源-经济-环境”协调发展问题研究的序幕。此后，中国学者根据中国国情，结合国内外有关 3E 系统的研究方法和手段，展开了一系列能源-经济-环境系统问题的研究，这些研究以中国经济社会的可持续发展和能源可持续利用为基本目标，通过定量的测量和评价，为能源环境战略和政策的科学制定提供了方法借鉴和理论准备[7]。

能源-经济-环境系统的研究主要是对社会发展系统中能源、经济、环境三个子系统之间交互作用程度测算方法和模型的研究，以实现各系统在发展演变过程中的合作、互补、和谐共生，其本质是子系统间的综合平衡与协调发展目前已成为世界各国决策层和学术界共同关注的热点话题。随着时间的推移，3E 系统研究的角度也越来越丰富，主要集中在对各子系统间相互关系的探讨、系统协调性评价、模型的构建等方面。

2 能源、环境与经济增长关系研究

随着人类对经济、能源、环境认识的加深，国际上开展了一系列平衡 3E 系统的综合研究，一些发达国家已将研究成果纳入国家的长期发展计划，国内关于该领域的研究成果也日渐丰富。以分析能源、经济、环境运行规律及两两之间相互关系为主要内容，实现经济社会协调可持续发展为目标的 3E 系统理论体系逐渐形成。

2.1 能源与经济增长的关系

国内外关于能源与经济增长的关系研究主要采用 Granger 因果检验或协整理论分析彼此间的因果关系。J. 克拉夫特（J. Kraft）和 A. 克拉夫特（A. Kraft）[8] 开拓性地对美国 1947—1974 年的数据进行因果检验，发现仅存在国民生产总值（gross national product，简称 GNP）到能源消费的单向因果关系，反之则不成立，指出能源保护政策不会影响 GNP 的增长。但能源消费与经济增长的关系在不同国家或地区有不同的结论。巴特都（Bhartendu）[9]、万坤（Wankeun）[10]、穆赫拉（Mehrara）[11] 等人的实证研究支持 J. 克拉夫特和 A. 克拉夫特的研究结论。埃达尔（Erdal）[12] 等对土耳其 1970—2006 年能源消费和 GNP 的因果关系进行考察则发现能源消费与 GNP 之间存在双向因果关系。阿瑟夫-埃达耶（Asafu-Adjaye）[13] 利用协整检验及误差修正模型对印度、印度尼西亚、菲律宾和泰国的能源消费与收入之间的关系进行研究发现，短期内印度和印度尼西亚存在从能源消费到收入的单向因果关系，泰国和菲律宾能源消费和收入之间存在双向因果关系。李健强（Lee Chien-Chiang）[14] 选取了18个发展中国家为研究对象，采用单位根检验、面板协整分析以及误差修正模型研究了其在1975—2001年之间能源消费和GDP之间的关系，研究发现长期和短期均存在从能源消费到GDP的单向因果关系。刘海莺和赵莹[15] 采用格兰杰因果关系检验及协整分析对1990—2009年中国经济增长与能源消费之间的动态关系进行了实证分析，结果发现国内生产总值是引起能源消费变化的Granger原因，但能源消费对经济增长的Granger因果关系显著性相对较差。可见，由于选取样本数据的观测时间段和对数据处理方法的不同，学者对不同国家能源与经济增长之间关系的研究并未达成一致。

除对能源消费和经济增长的因果关系进行探讨外，一些学者还采用不同方法考察了能源价格与经济增长的关系，尤其体现在能源价格波动对经济增长造成的影响。帕帕拜特罗（Papapetrou）[16] 应用Var模型考察了石油价格、股市收益、利率、实际经济活动与就业之间的关系，研究表明油价变动会对实际经济活动和就业产生影响。我国学者原鹏飞和吴吉林[17] 通过建立CGE模型研究了能源价格上涨情景下我国能源消费与经济增长的综合波动特征，研究表明：能源价格上涨使GDP、出口、就业水平下降，物价水平上涨，但对优化产业结构、降低能源强度产生了积极作用。张琳和何炼成[18] 基于1989—2010年的数据，就能源价格变动对经济增长与通货膨胀的影响进行了实证研究，结果表明能源价格、通货膨胀以及经济增长之间存在长期均衡关系，经济增长将在短期内增加能源消费需求，拉动能源价格上涨，进而对通货膨胀产生影响。

2.2 环境与经济增长的关系

关于环境与经济增长关系的研究较多，大多围绕环境库兹涅茨曲线（EKC）展开。国内外一些学者尝试利用各国截面数据、面板数据对环境库兹涅茨曲线进行实证分析以检验其存在性，即：环境恶化程度是否会随着经济发展呈现先上升后下降的倒U形曲线。国外学者塞尔登（Selden）[19]、帕纳约托（Panayotou）[20]、加里尔蒂（Galeotti）[21]、卡纳斯（Canas）[22]、迪茨（DietzS）[23] 及国内学者张娟[24]、李春米[25]、高宏霞、杨林[26-27] 等人的研究均证实了倒U形曲线的存在，也有部分研究对EKC的认识提出质疑。斯特恩（Stern）[28] 认为EKC的统计基础薄弱。丁达（Dinda）[29] 通过对以往文献的梳理指出仅有部分空气污染物符合EKC曲线，对EKC曲线的拐点并未达成一致。李善同（Bertinelli）和斯特罗布尔（Strobl）[30] 利用半参数回归分析对环境库兹涅茨曲线的存在性进行了重新检验，发现环境恶化与国家富裕度之间没有呈现出倒U形曲线。我国学者彭水军和包群[31] 采用省际面板数据对我国经济增长与包括水污染、大气污染在内的6类环境污染指标之间的关系进行了实证检验，结果表明经济增长与环境污染之间的关系有U形、倒U形和N形等多种形式，环境库兹涅茨倒U形曲线在很大程度上取决于污染指标以及估计方法的选取。

环境库兹涅茨曲线提出后，部分学者从生产规模、经济结构、科技水平、贸易等视角对 EKC 的形成动因进行了分析。一般而言，随着人均收入水平的提高，经济规模会随之增大，首先带来污染增加，环境状况恶化。但伴随经济水平的进一步提高，能源密集型产业会逐渐被技术和服务密集型产业代替，从而导致环境质量的改善[32]。科技进步则主要从两方面对环境质量的改善有所贡献，一是有效提高资源的使用效率，二是不断开发清洁技术取代传统技术[33]。从贸易的角度看，污染会通过国际贸易从富裕的国家转移到低收入国家，使发达国家环境质量提高，同时造成发展中国家环境质量进一步恶化[34]。格雷杰（Grainger）[35] 通过分析还发现，EKC 曲线下降段的出现得益于政府政策的正确制定和有效实施。

为了进一步分析经济增长过程中驱动工业化国家节能减排的因素，一些学者采用了更为细致的分解分析法研究污染变化贡献因素的特征以及这些因素的减排效应比较。刘兰翠（Liu Lancui）[36] 等人采用 LMDI 方法对中国 1998—2005 年间 36 个工业部分的终端能源消费的碳排放驱动因素进行分解分析，结果显示，工业结构变化对碳排放影响较大。何杰（He Jie）[37] 基于各省 13 个行业的数据，对中国工业 SO_2 排放量的驱动因素进行分析，结果表明经济规模和技术进步对能耗和排放影响最大，但两者作用方向相反。弗雷塔斯（Freitas）和金古町（Kaneko）[38] 对巴西能源消费产生的 CO_2 排放的驱动因素进行分解分析，发现经济活动和人口压力是污染排放的主要驱动因素，降低碳排放强度、能源结构向清洁能源方向发展对其减排作用较大。

2.3 能源、环境与经济增长的关系

系统可持续发展的目标之一就是在各个子系统之间建立一种稳定的协调发展状态[39]。因此，关于 3E 系统的协调度评价成为能源、环境、经济增长研究的主要内容，不同的学者从各自专业角度出发开展了大量相关工作。伦曾（Lenzen）和戴伊（Dey）[40] 选取能源消费、温室气体排放和一系列经济参数分析澳大利亚能源-环境-经济系统间的关系，结果表明：能源消费量的减少、温室气体减排要与增加就业和收入、减少进口等广泛的社会经济利益相协调。范中启和曹明[39] 在分析能源-经济-环境系统关系的基础上，设计了系统发展水平的指标体系，对我国 1990—2002 年的能源、经济与环境发展水平和系统协调度进行测度。赵芳[41] 拓宽了范中启和曹明的能源-经济-环境（3E）综合发展水平指标体系，借助主成分分析法和模糊数学隶属度数对中国 2000—2007 年间 3E 系统协调度进行测算，结果表明：3E 系统协调度处于弱协调状态；能源与经济处于轻度不协调等级，能源与环境属弱协调等级。胡绍雨[42] 通过对我国能源、经济与环境协调发展评价指标体系的模型构建，运用主成分分析法和因子分析法，测算了三个系统之间的协调发展系数，并进行了能源、经济与环境之间的综合发展评价。柯健[43]、薛乃川[1]、赵媛[44]、王霞[45]、史锦华[2]、李孝坤[46]、孙立成[47]、周荣敏和张燕[48]、李雨潼[49]、张正勇[50]、陈黎明和邓玲玲[51] 等学者也开展了相似研究。

随着能源、环境、经济研究的日渐深入，研究者更加注重采用统计学、运筹学以及系统工程学等方法对三者之间的关系进行更为精确的描述和定量分析。卡特（Carter）[52] 利用动态投入产出模型分析了环境污染排放量减少和新能源技术对于美国

未来10—15年经济增长率以及不同经济部门相对重要性的影响。奥里维拉（Oliveira）和安图尼斯（Antunes）[53] 建立了多目标线性规划模型，分析了能源政策、社会福利、经济增长和环境友好的可持续性，并为环境和经济发展提供政策途径。拉扎雷托（Lazzaretto）和托夫罗（Toffolo）[54] 基于多目标优化模型，采用系统分析和热力学方法对“3E”系统进行了分析，同时对能源消费过程中排放的工业废气所要付出的环境治理成本进行了估算。夏小勇[55] 用计量经济模型刻画能源消耗及环境条件对经济增长的影响。张阿玲等[56] 提出了改进的3E一体化模型，采用了部门活动水平分析、计量经济分析和线性规划分析方法，用于温室气体减排技术选择和减排对经济影响分析。崔和瑞和王娣[57] 通过建立我国能源-经济-环境的VAR模型对能源消费总量、GDP和二氧化硫排放量1995—2015年的数据进行了预测，并利用脉冲响应函数和方差分解对能源、经济和环境三者的动态关系进行了分析。

3 能源-环境-经济系统研究的模型选择

目前，能源安全和大气环境问题引起了国际社会的高度关注，随着能源、经济、环境研究的日渐深入，研究者更加注重采用模型对能源生产和消费过程中的资源浪费、环境污染等情况进行有效测算，3E系统模型正朝着复杂化、巨型化方向发展。主要包括内生经济增长模型、CGE模型、MARKAL模型、投入产出模型、多元线性规划模型等。

（1）内生经济增长模型。20世纪80年代中期以来，伴随内生增长理论的发展，以罗伯森（Robson）、休乌（Schou）、肖尔兹（Scholz）、鲍温伯格（Bovenberg）等为代表的经济学家通过将资源、环境或污染作为内生要素引入生产函数或效用函数，探讨其对经济长期增长的影响[58]，使内生经济增长模型成为研究资源消耗、环境恶化与经济可持续发展问题的主要理论分析工具。山姆（Sam）[59] 将能源消耗、环境质量纳入同一分析框架下，讨论经济的长期增长问题。他认为，能源消耗、国民经济总产出的增长与扩张、产品与服务的生产效果以及由于能源大量消耗导致的环境危机之间存在紧密的关系。国内学者王海建[60-63] 分别利用卢卡斯（Lucas）的人力资本积累、罗默（Romer）的“干中学”与R&D内生经济增长模型，建立了一系列耗竭型资源约束、环境污染的内生增长模型，研究资源耗竭、环境污染与经济增长之间的关系。于渤等[64] 将环境污染治理投入纳入R&D的内生增长模型，建立了一个同时考虑能源耗竭、环境污染及治理成本的可持续增长模型，运用最优控制方法探讨其平衡增长解，并在此基础上讨论经济增长与能源耗竭速度、污染治理投入之间满足的动态关系。

（2）MARKAL模型。MARKAL（market allocation）模型是一个基于单目标线性规划方法的能源系统分析工具，通常由4000~6000个变量及数量相当的方程式构建而成[65]，主要用于宏观能源规划和政策分析。吉伦（Gielen）和陈长红（Chen Changhong）[66] 应用线性MARKAL模型研究了2000—2020年上海能源系统SO_2、NO_x和CO_2减排的最优策略。拉森（Larson）等[67] 利用MARKAL模型，建立了包含能源转换和终端使用技术、能代表所有经济部门的中国能源系统模型，研究了在维持经济社会发展、确保国家能源供应安全和未来50年环境可持续发展的前提下，不同情境下的中国未来的能源技术战略。在中国，陈文颖等[68-69] 在MARKAL模型基础上，建立了能源-环境-经济耦合的非线性动态规划模型——中国MARKAL-MACRO模型，分析碳减排对能源系统以及GDP的潜在影响。清华大学佟庆等[70] 将MARKAL模型应用于北京市中远期能源系统建设研究，根据北京市未来能源、经济、环境协调发展的要求，进行了几种可能的未来能源系统发展情景分析。

（3）可计算一般均衡模型。可计算一般均衡（computable general equilibrium，CGE）模型通常由涵盖一个经济体系中生产、需求、贸易、收入等各种经济要素的一组庞大的联立方程式构成，在投入产出分析的基础上用以描述经济体中各产业部门间与最终需求部门的关联性[3]。20世纪70年代以后，CGE模型开始延伸至能源、环境领域，模拟能源、环境与经济之间的互动影响，现已在能源贸易及税收政策等方面广泛使用。如哥廷吉尔（Gottinger）[71] 通过构建能源-经济-环境的可计算一般均衡模型模拟欧盟单边和多边政策工具对温室气体排放的影响。结果表明排放标准、可拍卖许可证和温室气体排放税都能达到减排目标，但对GDP、福利指数以及再分配效应影响有所不同。魏涛远等[72]、王灿等[73] 利用可计算一般均衡模型定量分析了征收碳税对中国经济和温室气体排放的影响。结果表明：征收碳税使中国的经济状态恶化，但有助于碳减排政策的实现。黄英娜等[74] 采用相同方法对征收从价能源环境税进行模拟后也指出，征收能源环境税有助于减少污染排放，但不能从根本上促进能源利用效率的提高，并对宏观经济造成负面影响。

（4）投入产出模型。投入产出模型通过编制能够体现国民经济各部门生产与分配使用之间的平衡关系投入产出表，运用线性代数的计算方法揭示国民经济各部门、再生产各环节之间的内在联系，以此作为经济分析、预测的依据。投入产出法使用十分广泛，3E系统分析只是它一个方面的用途。哈德纳（Hawdona）和皮尔森伯（Pearsonb）[75] 通过建立英国10部门的投入产出模型，反映能源、环境以及经济效应三者之间的相互影响关系。伦岑（Lenzen）和戴伊（Dey）[76] 考虑到能源消费、温室气体排放和经济参数对澳大利亚经济的影响，基于投入-产出原理，对环境引致的支出选择进行了分析，结果表明：能源消费与温室气体排放的减少与广泛的社会经济利益相吻合。雷明[77,78] 利用投入产出核算进行了一系列的分析。通过计算资源使用、污染物排放和资源恢复、污染治理数据，设计出相对完整的资源-能源-环境-经济综合投入产出表，建立资源-环境-经济的绿色国民经济核算体系，对中国的能源经济环境状况进行了分析和预测。部分学者还将它与多目标规划模型相结合，用于总体经济分析及经济政策仿真。李林红等[79] 通过建立多目标投入产出模型对滇池流域经济与环境协调发展的问题作了分析。姜涛等[80] 建立了基于动态投入产出原理的可持续发展多目标最优规划模型，分析人口、资源、环境、经济相互之间的作用和影响，进而分析并模拟中国中长期可持续发展状况。

除以上模型外，国内外一些研究还通过其他模型对能源生产和消费过程中的经济、环境、技术等问题进行评价和测算。布里亚（Boudria）[81] 采用RAINS-ASIA模型对中国和印度在1990—2020年间可再生能源使用潜力和污染物减排的成本效益进行了分析。桑里（Syri）[82] 应用PRIMES和RAINS模型比较2010年前欧盟三种不同情景下的能源需求状况，并分析了它们对碳排放、酸性污染物以及地表臭氧的影响。希纳（Shina）[83] 使用LEAP模型分析了韩国使用填埋的废物气体发电对能源市场、发电成本和温室气体排放的影响。张阿玲和李继峰[56] 通过构造混合式3E模型TH-3EM（清华能源-经济-环境混合评价模型），对中国的节能战略选择进行分析和研究。杜慧滨和顾培亮[84] 通过建立能源-环境-经济系统演化模型，探讨系统与外部环境之间、子系统之间和子系统与外部环境之间的相互关系以及内部协调发展机制。

4 研究评述与未来发展方向

随着能源和环境在人们日常生产和生活过程中扮演着越来越重要的角色，能源、环境、经济的平衡可持续发展在世界范围内引起了普遍关注。国内外众多学者一方面采用时间序列数据对能源消费与经济增长关系进行协整和因果关系检验的研究，另一方面围绕环境库兹涅茨曲线探讨经济收入对环境的影响，同时将能源和环境纳入模型中研究能源、环境与经济增长关系的成果则相对有限。国外对三者之间关系的研究起步较早，多采用生产函数、协整分析等方法，国内学者的研究起步较晚，更多地侧重于对3E系统协调性评价与分析。从现有文献的分析中可以看出，计量经济学软件和计算机辅助决策支持系统在3E系统可持续发展的研究中应用广泛，CGE模型、投入-产出模型、定性和定量相结合的系统集成模型的建立与使用，使区域能源-经济-环境系统协调发展的研究方法不断更新和完善，但由于不同的研究人员对数据的处理方式以及评价方法的不同，研究结论还存在着较大的分歧。

面对社会经济系统对能源的需求与生态系统对能源供给之间日益增大的供求矛盾[85]，能源环境问题已经纳入资源、环境、经济的综合分析系统中。但总体看来，现有的研究还存在一些不足，未来仍需在某些方面进行加强。

（1）模型的进一步研究。当前，不同的学者构建众多模型用于能源环境领域的分析。然而国外的分析方法与评价体系是基于特定国家或地区的历史背景进行的，参数的设定和基线情景的各种假设通常与中国实际情况有所差异，影响其应用的可信度[86]。由于研究问题的不同，应加强对模型参数的修正和检验，着手对不同模型进行比较，以适应新领域、新问题的研究需要。

（2）加强对数据的处理分析。3E系统研究对数据要求较高，分析时离不开大量的连续数据。而多数发展中国家统计系统建设和环境保护工作起步较晚，使得一些长期连续的能源、环境和经济数据难以获得[3]，且不同地区由于统计口径的差异，可比性较差。未来研究在使用模型时，应先对数据的差异进行分析处理，减少国家或地区间的不同造成的限制。

（3）注重区域间差异的比较分析。不同区域自然条件禀赋不同，经济发展阶段、产业结构以及经济增长方式也各不相同，因此可能会削弱统计分析结论的显著性[87]。中国地区间的空间差异显著，经济发展很不平衡，各地区对能源消费的需求、污染治理的投入不尽相同，采用全国总量数据很难反映 3E 系统间的真实关系，因此，在对我国能源、环境、经济协调性进行分析时，应按照经济发展水平对不同地区的实际情况和不同层次的需要分别展开研究，才能更好地反映中国的能源环境实际状况。

（4）评价指标体系设立的系统性。评价指标体系的建立多见于对 3E 系统协调性评价的文献中，但建立的指标体系没有跳出传统的可持续发展指标体系的框架，缺乏一个能够全面、科学地表征环境破坏和资源损耗整体水平的一般性环境指标体系[88]。有的指标与模型还是多层次、多指标的罗列，多数情况下只是采用单一或各种具体指标代表能源、环境或经济整体发展水平的高低。因而，加强指标体系设立的确定性、系统性，尤其加强可再生资源和生态破坏评价方面的研究对 3E 系统可持续发展战略选择具有重要意义。

（5）关于环境成本核算问题的研究。目前，环境资源价值的核算尚处在探索阶段，通常采取的做法是将社会效益和生态效益转化为经济效益，但由于数据获取不易、信息受限等不确定因素的存在以及无法排除认为因素对生态环境效益价值量化结果的影响，对于污染物产生的环境代价的损失还很难进行准确估量[89]。然而，随着环境保护活动支出占国内生产总值比重的逐渐攀升，外部资源要素投入效益评估的缺失或错误将直接导致资源配置的不合理[90]。因而，客观有效地衡量环境成本价值成为协调能源、环境、经济发展的重要途径，也是未来亟待解决的重要课题。

参考文献

[1] 薛乃川，贡璐，安尼瓦尔·阿木提. 新疆能源、经济与环境复合系统协调度综合评价［J］. 干旱区资源与环境，2009（5）：56-60.

[2] 史锦华. 民族地区能源、经济与环境系统协调评价研究［J］. 广西民族研究，2009（3）：178-183.

[3] 邓玉勇，杜铭华，雷仲敏. 基于能源-经济-环境（3E）系统的模型方法研究综述［J］. 甘肃社会科学，2006（3）：209-212.

[4] 王丽丽. 浙江能源-经济-环境协调发展研究——基于能源环境公平性视角［D］. 杭州：浙江理工大学，2010.

[5] 张丽峰. 中国能源供求预测模型及发展对策研究［D］. 北京：首都经济贸易大学，2006.

[6] 严良，张春梅，康淑娟. 青海省能源与经济、环境协调发展分析及其优化［J］. 价格理论与实践，2007（4）：21-22.

[7] 陈军. 中国非可再生能源战略评价模型与实证研究［D］. 武汉：中国地质大学，2008.

[8] Kraft J，Kraft A. On the Relationship between Energy and GNP［J］. Energy Dvelopment，1978（3）：401-403.

[9] Bhartendu S，Cohen S J. Impact of CO_2-induced Climate Change on Residential Heating and Cooling Energy Requirements in Ontario，Canada［J］. Energy and Buildings，1987（10）：99-108.

[10] Oh W，Lee K. Causal Relationship between Energy Consumption and GDP Revisited：the Case of Korea 1970—1999［J］. Energy Economics，2004，26：51-59.

[11] Mehrara M. Energy Consumption and Economic Growth：the Case of Oil Exporting Countries［J］. Energy policy，2007，35（5）：2939-2945.

[12] Erdal G，Erdal H，Esengün K. The Causality between Energy Consumption and Economic Growth in Turkey［J］. Energy Policy，2008，36（10）：3838-3842.

[13] Asafu-Adjaye J. The Relationship between Energy Consumption, Energy Prices and Economic Growth: Time Series Evidence from Asian Developing Countries [J]. Energy Economics, 2000, 22 (6): 615-625.

[14] Lee C C. Energy Consumption and GDP in Developing Countries: A Cointegrated Panel Analysis [J]. Energy Economics, 2005, 27 (3): 415-427.

[15] 刘海莺, 赵莹. 能源消费与中国经济增长关系的实证分析 [J]. 统计与决策, 2011 (3): 128-129.

[16] Papapetrou E. Oil Price Shocks, Stock Market, Economic Activity and Employment in Greece [J]. Energy Economics, 2001, 23 (5): 511-532.

[17] 原鹏飞, 吴吉林. 能源价格上涨情景下能源消费与经济波动的综合特征 [J]. 统计研究, 2011, 28 (9): 57-65.

[18] 张琳, 何炼成. 能源价格、通货膨胀与经济增长: 互动机理与实证检验 [J]. 江海学刊, 2012 (1): 91-97, 238.

[19] Selden T M, Song D. Environmental Quality and Development: Is There a Kuznets Curve for Air Pollution Emissions? [J]. Journal of Environmental Economics and Management, 1994, 27 (2): 147-162.

[20] Panayotou T. Demystifying the Enviromental Kuznets Curve: Turning a Black Box into a Policy Tool [J]. Evironment and development economics, 1997, 2 (4): 465-484.

[21] Galeotti M, Lanza A. Desperately Seeking Environmental Kuznets [J]. Environmental Modelling & Software, 2005, 20 (11): 1379-1388.

[22] Canas Â, Ferrão P, Conceição P. A New Environmental Kuznets Curve? Relationship between Direct Material Input and Income Per Capita: Evidence from Industrialised Countries [J]. Ecological Economics, 2003, 46 (2): 217-229.

[23] Dietz S, AdgerbW N. Economic Growth, Biodiversity Loss and Conservation Effort [J]. Journal of Environmental Management, 2003, 68 (1): 23-35.

[24] 张娟. 经济增长与工业污染: 基于中国城市面板数据的实证研究 [J]. 贵州财经学院学报, 2012 (4): 32-36.

[25] 李春米. 经济增长、环境规制与产业结构-基于陕西省环境库兹涅茨曲线的分析 [J]. 兰州大学学报: 社会科学版, 2010, 38 (5): 96-102.

[26] 高宏霞, 杨林, 王节. 经济增长与环境污染关系的研究-基于环境库兹涅茨曲线的实证分析 [J]. 云南财经大学学报, 2012 (2): 70-77.

[27] 高宏霞, 杨林, 付海东. 中国各省经济增长与环境污染关系的研究与预测——基于环境库兹涅茨曲线的实证分析 [J]. 经济学动态, 2012 (1): 52-57.

[28] Stern D I. The Rise and Fall of the Environmental Kuznets Curve [J]. World Development, 2004, 32 (8): 1419-1439.

[29] Dinda S. Environmental Kuznets Curve Hypothesis: A Survey [J]. Ecological Economics, 2004, 49 (4): 431-455.

[30] Bertinelli L, Strobl E. The Environmental Kuznets Curve Semi-Parametrically Revisited [J]. Economics Letters, 2005, 88 (3): 350-357.

[31] 彭水军, 包群. 经济增长与环境污染-环境库兹涅茨曲线假说的中国检验 [J]. 财经问题研究, 2006 (8): 3-17.

[32] 钟茂初, 张学刚. 环境库兹涅茨曲线理论及研究的批评综论 [J]. 中国人口·资源与环境, 2010, 20 (2): 62-67.

[33] Pasche M. Technical Progress, Structural Change, and the Environmental Kuznets Curve [J]. Ecological Economics, 2002, 42 (3): 381-389.

[34] Taylor M S, Copeland B R. Trade, Growth, and the Environment [J]. Journal of Economic Literature, 2004, 42: 7-71.

[35] Grainger A. The Forest Transition: An Alternative Approach [J]. Area, 1995, 27 (3): 242-251.

[36] Liu L C, Fan Y, Wu G, et al. Using LMDI Method to Analyze the Change of China's Industrial CO_2 Emissions from Final Fuel Use: An Empirical Analysis [J]. Energy Policy, 2007, 35 (11): 5892-5900.

[37] He J. What is the Role of Openness for China's Aggregate Industrial SO_2 Emission?: A Structural Analysis Based on the DivisiaDecomposition Method [J]. Ecological Economics, 2010, 69 (4): 868-886.

[38] Freitas L C, Kaneko S. Decomposition of CO_2 Emissions Change from Energy Consumption in Brazil: Challenges and Policy Implications [J]. Energy Policy, 2011, 39 (3): 1495-1504.

[39] 范中启，曹明. 能源-经济-环境系统可持续发展协调状态的测度与评价 [J]. 预测，2006 (4): 68-72.

[40] Lenzen M, Dey C J. Economic, Energy and Greenhouse Emissions Impacts of some Consumer Choice, Technology and Government Outlay Options [J]. Energy Economics, 2002, 24 (4): 377-403.

[41] 赵芳. 中国能源-经济-环境 (3E) 协调发展状态的实证研究 [J]. 经济学家，2009 (12): 35-41.

[42] 胡绍雨. 我国能源、经济与环境协调发展分析 [J]. 技术经济与管理研究，2013 (4): 78-82.

[43] 柯健，李超. 基于DEA聚类分析的中国各地区资源、环境与经济协调发展研究 [J]. 中国软科学，2005 (2): 144-148.

[44] 赵媛，梁中，袁林旺. 能源与社会经济环境协调发展的多目标决策——以江苏省为例 [J]. 地理科学，2001, 21 (2): 164-169.

[45] 王霞. 能源-经济-环境复杂系统持续协调发展评价指标体系与方法研究——以山东省为例 [J]. 特区经济，2006 (9): 369-370.

[46] 李孝坤，韦杰. 重庆都市区环境与经济协调发展演进分析及对策探讨 [J]. 经济地理，2005, 25 (3): 387-390.

[47] 孙立成，梅强，周德群. 区域3E系统协调发展水平PLS-SEM测度模型及应用研究 [J]. 运筹与管理，2012, 21 (3): 119-128.

[48] 周荣敏，张燕. 能源-经济-环境系统综合发展水平实证研究 [J]. 商业时代，2011 (22): 130-131.

[49] 李雨潼. 基于3E协调度分析的黑龙江省低碳经济发展路径研究 [J]. 求是学刊，2013, 40 (1): 73-77.

[50] 张正勇，刘琳，唐湘玲. 乌鲁木齐市3E系统协调发展评价研究 [J]. 安徽农业科学，2011, 39 (7): 4379-4380, 4385.

[51] 陈黎明，邓玲玲. 我国3E系统协调度测度及其影响因素分析 [J]. 统计与决策，2012 (16): 126-129.

[52] Carter A P. Energy Environment, and Economic Growth [J]. The Bell Joumal of Economics and Management Science, 1974. 5 (2): 578-592.

[53] Oliveira C, Antunes C H. A Multiple Objective Model to Deal with Economy-Energy-Environment Interactions [J]. European Journal of Operational Research, 2004, 153 (2): 370-385.

[54] Lazzaretto A, Toffolo A, Reini M, et al. Four Approaches Compared on the TADEUS (Thermoeconomic Approach to the Diagnosis of Energy Utility Systems) Test Case [J]. Energy, 2006, 31 (10-11): 1586-1613.

[55] 夏小勇，王苏楠. 能源、环境与经济增长的计量分析 [J]. 湖北农机化，2008 (3): 38-39.

[56] 张阿玲，郑淮，何建坤. 适合中国国情的经济、能源、环境 (3E) 模型 [J]. 清华大学学报：自

然科学版，2002（12）：53-37.
[57] 崔和瑞，王娣. 基于 VAR 模型的我国能源-经济-环境（3E）系统研究［J］. 北京理工大学学报：社会科学版，2010（1）：25-30.
[58] 王彦彭. 我国能源环境与经济可持续发展［D］. 北京：首都经济贸易大学，2010：26-29.
[59] Sam H S. Energy Economic Growth and the Environment［M］. Baltimore and London：Johns HoPkins University Press，1972.
[60] 王海建. 资源约束、环境污染与内生经济增长［J］. 复旦学报：社会科学版，2000（1）：76-80.
[61] 王海建. 耗竭性资源管理与人力资本积累内生经济增长［J］. 管理工程学报，2000（3）：17-19，2.
[62] 王海建. 资源环境约束之下的一类内生经济增长模型［J］. 预测，1999（4）：37-39.
[63] 王海建. 耗竭性资源、R&D 与内生经济增长模型［J］. 系统工程理论方法应用，1999（3）：41-45.
[64] 于渤，黎永亮，迟春洁. 考虑能源耗竭、污染治理的经济持续增长内生模型［J］. 管理科学学报，2006（4）：16-21.
[65] 冯等田. 中国区域能源经济与环境规制［D］. 兰州：兰州大学，2007.
[66] Gielen D，Changhong C. The CO_2 Emission Reduction benefits of Chinese Energy Policies and Environmental Policies：A Case Study for Shanghai，Period 1995—2020［J］. Ecological Economics，2001，39（2）：257-270.
[67] Larson E D，Zongxin W，DeLaquil P，et al. Future Implications of China's Energy-Technology Choices［J］. Energy Policy，2003，31（12）：1189-1204.
[68] 陈文颖，高鹏飞，何建坤. 用 MARKAL-MACRO 模型研究碳减排对中国能源系统的影响［J］. 清华大学学报：自然科学版，2004（3）：55-60.
[69] 陈文颖，高鹏飞，何建坤. 二氧化碳减排对中国未来 GDP 增长的影响［J］. 清华大学学报：自然科学版，2004（6）：24-27.
[70] 佟庆，白泉，刘滨，等. MARKAL 模型在北京中远期能源发展研究中的应用［J］. 中国能源，2004（6）：37-41.
[71] Gottinger H W. Greenhouse Gas Economics and Computable General Equilibrium［J］. Journal of Policy Modeling，1998，20（5）：537-580.
[72] 魏涛远，格罗姆斯洛德. 征收碳税对中国经济与温室气体排放的影响［J］. 世界经济与政治，2002（8）：48-50.
[73] 王灿，陈吉宁，邹骥. 基于 CGE 模型的 CO_2 减排对中国经济的影响［J］. 清华大学学报：自然科学版，2005（12）：39-42.
[74] 黄英娜，郭振仁，张天柱. 应用 CGE 模型量化分析中国实施能源环境税政策的可行性［J］. 城市环境与城市生态，2005（2）：21-23.
[75] Hawdona D，Pearsonb P. Input-Output Simulations of Energy，Environment，Economy Interactions in the UK［J］. Energy Economics，1995，17（1）：73-86.
[76] Lenzen M，Dey C J. Economic，Energy and Greenhouse Emissions Impacts of some Consumer Choice，Technology and Government Outlay Options［J］. Energy Economics，2002，24（4）：377-403.
[77] 雷明. 资源-经济-环境投入产出核算应用研究——中国能源-资源-经济-环境综合分析（1992~2020）［J］. 数量经济基数经济研究，1998（11）：61-65.
[78] 雷明. 绿色国内生产总值（GDP）核算［J］. 自然资源学报，1998（4）：33-39.
[79] 李林红，介俊，吴莉明. 昆明市环境保护投入产出表的多目标规划模型［J］. 昆明理工大学学报：自然科学版，2001（1）：105-107，112.
[80] 姜涛，袁建华，何林，等. 人口-资源-环境-经济系统分析模型体系［J］. 系统工程理论与实践，2002（12）：68-73.

[81] Boudri J C, Hordijk L, Kroeze C, et al. The Potential Contribution of Renewable Energy in Air Pollution Abatement in China and India [J]. Energy Policy, 2002. 30 (5): 409-424.
[82] Syri S, Amann M, Capros P, et al. Low CO_2 Energy Pathways and Regional Air Pollution in Europe [J]. Energy Policy, 2001, 29 (11): 871-884.
[83] Shin H C, Park J W, Kim H S, et al., Environmental and Economic Assessment of Landfill Gas Electricity Generation in Korea Using LEAP Model [J]. Energy Policy, 2005, 33 (10): 1261-1270.
[84] 杜慧滨，顾培亮. 区域发展中的能源-经济-环境复杂系统 [J]. 天津大学学报：社会科学版，2005 (5): 44-47.
[85] 潘新华. 陕北地区能源开发、环境保护与经济协调发展的思考 [J]. 企业经济，2005 (9): 129-131.
[86] 蒋金荷，姚愉芳. 气候变化政策研究中经济-能源系统模型的构建 [J]. 数量经济基数经济研究，2002 (7): 41-45.
[87] 丁继红，年艳. 经济增长与环境污染关系剖析——以江苏省为例 [J]. 南开经济研究，2010 (2): 64-79.
[88] 王长征，刘毅. 经济与环境协调研究综述 [J]. 中国人口·资源与环境，2002，12 (3): 32-36.
[89] 赵芳. 能源-经济-环境非协调发展原因的经济学解释 [J]. 中国人口·资源与环境，2008，18 (4): 67-72.
[90] 王德发，阮大成，王海霞. 工业部门绿色 GDP 核算研究——2000 年上海市能源-环境-经济投入产出分析 [J]. 财经研究，2005 (2): 66-75.

Decomposition analysis of energy-related carbon emissions from the transportation sector in Beijing

FAN Fengyan[1,2], LEI Yalin[1,2,*]

1. *School of Humanities and Economic Management, China University of Geosciences, Beijing* 100083, *China*
2. *Key Laboratory of Carrying Capacity Assessment for Resource and Environment, Ministry of Land and Resource, Beijing* 100083, *China*

Abstract In the process of rapid development and urbanization in Beijing, identifying the potential factors of carbon emissions in the transportation sector is an important prerequisite to controlling carbon emissions. Based on the expanded Kaya identity, we built a multivariate generalized Fisher index (GFI) decomposition model to measure the influence of the energy structure, energy intensity, energy intensity, output value of per unit traffic turnover, transportation intensity, economic growth and population size on carbon emissions from 1995 to 2012 in the transportation sector of Beijing. Compared to most methods used in previous studies, the GFI model possesses the advantage of eliminating decomposition residuals, which enables it to display better decomposition characteristics (Ang et al., 2004). The results show: (1) The primary positive drivers of carbon emissions in the transportation sector include the economic growth, energy intensity and population size. The cumulative contribution of economic growth to transportation carbon emissions reaches 334.5%. (2) The negative drivers are the transportation intensity and energy structure, while the transportation intensity is the main factor that restrains transportation carbon emissions. The energy structure displays a certain inhibition effect, but its inhibition is not obvious. (3) The contribution rate of the output value of per unit traffic turnover on transportation carbonemissions appears as a flat "M". To suppress the growth of carbon emissions in transportation further, the government of Beijing should take the measures of promoting the development of new energy vehicles, limiting private vehicles' increase and promoting public transportation, evacuating non-core functions of Beijing and continuingly controlling population size.

Keywords Beijing, Transportation sector, Carbon emissions, Decomposition analysis, Policy suggestions

1 Introduction

In the 21st century, global warming and the energy crisis pose a serious challenge to sustainable human development. Low carbon development has become the global consensus for

原文刊于 *Transportation Research Part D*，2016 年第 46 卷。原文发表时为彩色。

* 为通信作者。

growth. However, with the rapid development of the economy and society, cities have become a major source of carbon emissions, and urban transportation is an important contributor to urban carbon emissions. According to UN statistics, urban carbon emissions account for 75% of the total global carbon emissions, in which transportation accounting for 17.5% (Li and Liu, 2014). Our calculation reveals that the transportation carbon emissions of Beijing in 2012 accounted for approximately 22.28% of the total carbon emissions. In the worldwide, transportation has already become the second highest carbon dioxide (CO_2) emission sector, trailing the electricity and heat generation (IEA, 2014).

Increased transportation carbon emissions have become an important factor restricting the sustainable development of the global economy. Low-carbon transportation and related issues have thus become a substantial topic in the community. Multiple studies have investigated the factors influencing carbon emissions in the transportation sector. Lakshmanan (1997) developed a decomposition scheme to identify the magnitude and the relative effects of the various factors in the U.S. transportation energy use and CO_2 emissions between 1970 and 1991. The results revealed that the growth in the propensity to travel, population, and gross domestic product (GDP) were the three most important factors driving U.S. transportation energy use and CO_2 emissions. Data covering 146 urbanized areas in the United States indicated that the population density, transit share, freeway lane miles per capita, private vehicle occupancy, and average travel time had a statistically significant explanatory effect on passenger travel-related CO_2 emissions (Mishalani et al., 2014). Lu et al. (2007) adopted the Divisia index approach to explore the effects of five factors on the total carbon dioxide emissions from highway vehicles in Germany, Japan, South Korea and Taiwan during 1990—2002. Their results suggested that the rapid growth in the economy and of vehicle ownership were the most important factors for the increased CO_2 emissions, whereas population intensity contributed significantly to decreasing the emissions. Timilsina and Shrestha (2009) analyzed the potential factors influencing the growth of transportation sector CO_2 emissions in selected Asian countries during 1980—2005 and identified changes in per capita CDP, population growth and transportation energy intensity as the main factors driving transportation sector CO_2 emission growth. Schipper et al. (1997) performed a decomposition analysis on the changes in freight energy use of 10 industrialized countries from 1973 to 1992 to identify the relative contribution of the activity, modal structure, and energy intensity to increases in the energy use observed in each country. The major findings were as follows: domestic freight volumes increased with trucks carrying the majority of the increased freight. Freight energy use and the associated carbon emissions increased markedly and were rising when compared to the emissions associated with passenger travel. The energy used for freight will continue to increase unless substantial reductions occur in the energy intensities of truck freight. Andreoni and Galmarini (2012) also used a decomposition analysis to investigate the main factors influencing CO_2 emissions from the water and aviation transportation sectors in Europe during 2001—2008. The results indicated that economic growth was the main factor behind CO_2 emissions. Kwon (2005) investigated the key factors in the change in CO_2 emissions from car travel in Great Britain over the last 30 years using the IPAT identity. The results revealed that car driving distance per person was a dominant force for the growth of emissions. Technology factors, such as fuel efficiency and substitution of diesel fuel, partly cancelled out the growth effects, while the contribution of the other factors was relatively small. Mazzarino (2000) adopted a comparative static approach to analyze the main factors by determining the variation of transportation sector CO_2 emissions in Italy during 1980—1995. The results indicated that the main driving force in the variation of CO_2 emissions was the growth of the GDP. Timilsina and Shrestha (2009) determined the factors responsible for the growth of transportation sector CO_2 emissions in 20 Latin American and Caribbean (LAC) countries during 1980—2005 by decomposing the emission growth into components associated with changes in the fuel mix, modal shift, economic growth, emission coefficients and transportation energy intensity. The key finding was that economic growth and the changes in transportation energy intensity were the principal factors driving transportation sector CO_2 emissions growth in the countries considered.

Currently, research on the factors influencing transportation carbon emissions in China remains limited. Wang et al. (2011) applied the logarithmic mean Divisia index (LMDI) method to determine the factors that influence changes in transportation sector CO_2 emissions in China over the period 1985—2009. The results showed that highway transportation was the most substantial source of CO_2 emissions. The per capita economic activity effect and transportation modal shifting effect were primarily responsible for driving the growth in the transportation sector CO_2 emissions. The transportation intensity effect and transportation services share effect were the main drivers of the reduction of CO_2 emissions in China. However, the emission coefficient effect played a minor role over the study period. Through a daily activity survey conducted in Beijing from 2000 to 2011, Wang and Liu (2014) examined the effects of individual travel behavior on carbon emissions from urban transportation in Beijing. The results showed that the vehicle-use intensity, disposable income per capita and population size were the main drivers for the increase of household daily travel carbon emissions. Both the transportation intensity and emission coefficient had significant effects on the reduction of carbon emissions. However, the transportation mode share played a minor role. Based on data covering Beijing, Tianjin, Shanghai and Chongqing, Su et al. (2011) empirically analyzed the factors influencing urban transportation carbon emissions during 1995—2009. The results showed that the city population size and vehicle ownership had important effects on urban transportation carbon emissions by influencing passenger and freight turnover. Urban passenger and freight turnover had significant positive effects on urban transportation carbon emissions, whereas the proportion of buses had a significant negative effect. Yang et al. (2014) and Wu et al. (2012) applied the LMDI decomposition method to analyze transportation sector carbon emissions in Jiangsu and Shanghai, respectively, and yielded consistent conclusions: the positive driving factors were the economic output, population size and industrial structure, whereas the negative drivers were the transportation energy structure and energy intensity. Shen and Chi (2012) analyzed the driving factors of CO_2 emissions from transportation sector in China from 1991 to 2009. The results showed that the main positive drivers of CO_2 emissions from the transportation sector were the urbanization level, proportion of tertiary industries and total population, and that the main negative drivers were the traffic turnover per unit of GDP, energy consumption per unit of traffic turnover, secondary and tertiary industries value per unit of population, and the contribution of tertiary industry to GDP. Based on the STIRPAT model and the energy-related carbon emissions from transportation in Jilin Province from 1999—2011, Gao et al. (2013) revealed that the population, per capita GDP, transportation, city investment rate, and number of private cars displayed positive driving effects on carbon emissions, and that energy consumption per GDP, in contrast, had an inhibitory effect on carbon emissions. The main influencing factors of the relevant literature are summarized in Table 1.

Table 1 Literature on the main factors influencing transportation carbon emissions

Author	Region	Year	The main influencing factors								
			P	EO	EI	ES	MS	FM	TI	IS	CI
Mishalani et al.	America	2014	√								
Lu et al.	4 countries	2007	√	√							
Timilsina and Shrestha	Asian countries	2009	√	√	√			√			
Lakshmanan	America	1997	√	√			√				
Schipper et al.	10 countries	1997			√		√	√			√
Andreoni and Galmarini	Europe	2012		√	√					√	
Kwon	England	2005	√								√
Mazzarino	Italy	2000	√	√	√		√	√	√		
Timilsina and Shrestha	20 countries	2009		√	√			√			
Wang et al.	China	2011	√	√			√		√		
Wang and Liu	Beijing	2014	√				√		√		
Su et al.	4 provinces of China	2011	√								
Yang et al.	Jiangsu	2014	√	√	√	√				√	
Wu et al.	Shanghai	2012	√	√	√	√					
Shen and Chi	China	2012	√						√	√	
Gao et al.	Jilin	2013	√	√	√						

Note: P, EO, EI, ES, MS, FM, TI, IS, CI respectively represent population, economic output, energy intensity, energy structure, modal structure, fuel mix, transportation intensity, industrial structure and carbon intensity.

Increasing the energy-savings and emission-reductions in transportation has become an inevitable goal to relax the tension of energy supply, to ease the pressures on the environmental capacity and to build a low-carbon ecological city. Beijing is an international metropolis, and the transportation carbon emissions continue to grow along with the increasing expansion of urban space and increased car ownership. From 1995—2012, the average annual growth rate of transportation carbon emissions in Beijing reached 11.81% and is expected to grow continuously. It has become a primary issue to analyze the key factors affecting urban transportation carbon emissions and their influence degree in establishing low-carbon urban transportation policy and controlling urban transportation carbon emissions. Previous studies have adopted different countries or regions as research areas, analyzing the influencing factors of transportation carbon emissions. However, we only found Wang and Liu (2014) discussed the effects of individual behavior on carbon emissions from urban transportation in Beijing. A systematic analysis of the factors affecting transportation sector CO_2 emissions in Beijing has not been involved. Moreover, industrial characteristics in the factor decomposition analysis have not been included in most previous studies. This study aims to draw a comprehensive picture of the driving forces of the changing carbon emissions related to transportation sector in Beijing by building a multivariate generalized Fisher index (GFI) decomposition model based on the expanded Kaya identity using the data from 1995 to 2012 in Beijing. Policy suggestions will be proposed following the research results.

2 Methodology and data

We use the expanded generalized Fisher index method to analyze different factors on transportation carbon emissions in Beijing based on the expanded Kaya identity. The main reason is that the most commonly used index decomposition methods of the Laspeyres and Divisia index methods cannot handle the residuals in the decomposition process and the explanation of all the changes in carbon emissions is inhibited. The generalized Fisher index has better decomposition characteristics than the Laspeyres and Divisia index methods. Ang et al. (2004) compared the generalized Fisher index method with five widely employed index decomposition methods on a basis of the factor-reversal, time-reversal, proportionality, aggregation, zero-value robust and negative-value robust tests [see Table 1 in Ang et al. (2004) for details]. The generalized Fisher index passed all of the tests except the aggregate test, and achieved complete factor decomposition. This suggests the generalized Fisher index should be a better way to perform decomposition.

2.1 The expanded Kaya identity

The Kaya identity was proposed by Japanese scholar Yoichi Kaya (Kaya, 1990) for the first time at an IPCC workshop; it is a useful tool for analyzing the factors of carbon emissions (Albrecht et al., 2002). The expression is:

$$C = (C/E)(E/\mathrm{GDP})(\mathrm{GDP}/P)P \tag{1}$$

The Kaya identity relates energy-related carbon emissions with energy (E), economic growth level (GDP) and population (P). Its structure is relatively simple and only explores the quantitative relationship among the carbon emissions, energy intensity, economic growth and population at a national level without considering the influence of structural factor. Therefore, we added the energy structure as a factor to the Kaya identity because of the available data, and we decomposed carbon emissions into energy structure, energy efficiency, economic growth and population. The expansion of the equation is as follows:

$$C = \sum_i C_i = \sum_i \frac{C_i}{E_i}\frac{E_i}{E}\frac{E}{\mathrm{GDP}}\frac{\mathrm{GDP}}{P}P \tag{2}$$

where C represents the carbon emissions; C_i is the carbon emissions of the i-th fuel type; E_i and E represent the i-th fuel consumption and the total primary energy consumption, respectively; GDP is the gross domestic product; and P is population.

As a type of induced demand of a national economy, energy-related carbon emissions in the transportation sector are influenced by regional economic development level, energy intensity of the sector and other factors. Therefore, Beijing's transportation carbon emissions are expanded as the following according to Eq. (2) and the characteristics of the transportation sector:

$$C = \sum_i C_i = \sum_i \frac{C_i}{E_i}\frac{E_i}{E}\frac{E}{\mathrm{GDP}_{\mathrm{tr}}}\frac{\mathrm{GDP}_{\mathrm{tr}}}{V}\frac{V}{\mathrm{GDP}}\frac{\mathrm{GDP}}{P}P \tag{3}$$

where $\mathrm{GDP}_{\mathrm{tr}}$ is the GDP in the transportation sector; V represents the traffic turnover; C_i/E_i is the carbon emissions of the ith fuel type, namely, the carbon emission coefficient; E_i/E represents the total energy consumption share of the ith fuel type in the transportation sector, namely, the energy structure; $E/\mathrm{GDP}_{\mathrm{tr}}$ is the energy consumption per unit of GDP in the transportation sector, namely the energy intensity; $\mathrm{GDP}_{\mathrm{tr}}/V$ represents the value of the transportation GDP brings by per unit transportation turnover; V/GDP represents the traffic turnover per unit of GDP, defined as the transportation intensity; and Y/P represents the per capita GDP, namely, the economic growth. P represents population size.

2.2 Generalized Fisher index decomposition

The Fisher index decomposition method was first proposed by Fisher in 1922. The model is expressed as follows:

$$V = \sum_i V_i = \sum_i X_{1i}X_{2i} \tag{4}$$

where V represents an aggregate indicator, i denotes the sub-category of the aggregate, and X_{1i} and X_{2i} indicate the decomposed variables. The effect associated with changes in X_1 and X_2 from year 0 to year T is given in Eq. (5):

$$D = \frac{V^T}{V^0} = \frac{\sum_i X_{1i}^T X_{2i}^T}{\sum_i X_{1i}^0 X_{2i}^0} = D_{X_1} \cdot D_{X_2} \tag{5}$$

According to the Fisher index formula, we have

$$D_{X_1} = \left[\frac{\sum_i X_{1i}^T X_{2i}^0}{\sum_i X_{1i}^0 X_{2i}^0} \cdot \frac{\sum_i X_{1i}^T X_{2i}^T}{\sum_i X_{1i}^0 X_{2i}^T}\right]^{\frac{1}{2}} \tag{6}$$

$$D_{X_2} = \left[\frac{\sum_i X_{1i}^0 X_{2i}^T}{\sum_i X_{1i}^0 X_{2i}^0} \cdot \frac{\sum_i X_{1i}^T X_{2i}^T}{\sum_i X_{1i}^T X_{2i}^0} \right]^{\frac{1}{2}} \tag{7}$$

Some scholars have used the traditional Fisher model to study energy and relative gas emissions (Liu and Ang, 2003; Boyd and Roop, 2004), but in empirical studies, two factors are insufficient for meeting the needs of solving this problem. Ang et al. (2004) extended the two factors to multiple factors and proposed the generalized Fisher index. The specific steps are as follows:

V is the total index is composed of n elements:

$$V = \sum_i X_1 X_2 \cdots X_n \tag{8}$$

Define the set $N = \{1, 2, \ldots, n\}$ in which the cardinality of N is n. S is a subset of N, for which the cardinality is s'. Define afunction $V(S) = \sum (\Pi_{l \in S} X_l^{\ T} \Pi_{m \in N/S} X_m^{\ 0})$ and $V(\Phi) = \sum (\Pi_{m \in N} X_m^{\ 0})$ where Φ is a null set and the superscripts denote the current year T and the base year 0. According to the "geometric average" principle, V^T/V^0 is divided into n parts, and the decomposition results of factor X_j $(j = 1, 2, \ldots, n)$ are given by the following:

$$D_{X_j} = \prod_{\substack{S \subset N \\ J \in S}} \left[\frac{V(S)}{V(S/\{j\})} \right]^{\frac{1}{n} \frac{1}{\binom{n-1}{s'-1}}} = \prod_{\substack{S \subset N \\ J \in S}} \left[\frac{V(S)}{V(S/\{j\})} \right]^{\frac{(s'-1)!\,(n-s')!}{n!}} \tag{9}$$

D_{X_j} $(j = 1, 2, \ldots, n)$ is the decomposition factor of the generalized Fisher index method.

According to the above method, Eq. (3) is defined for the energy-related carbon emission coefficient $X_{1i} = C_i/E_i$, energy structure $X_{2i} = E_i/E$, energy intensity $X_3 = E/\mathrm{GDP}_{\mathrm{tr}}$, output value of per unit traffic turnover $X_4 = \mathrm{GDP}_{\mathrm{tr}}/V$, transportation intensity $X_5 = V/\mathrm{GDP}$, economic growth $X_6 = \mathrm{GDP}/P$, and population size $X_7 = P$. Thus, the carbonemission formula canbe written as follows:

$$C = \sum_i C_i = \sum_i \frac{C_i}{E_i} \frac{E_i}{E} \frac{E}{\mathrm{GDP}_{\mathrm{tr}}} \frac{\mathrm{GDP}_{\mathrm{tr}}}{V} \frac{V}{\mathrm{GDP}} \frac{\mathrm{GDP}}{P} P = \sum_i X_{1i} X_{2i} X_3 X_4 X_5 X_6 X_7 \tag{10}$$

Generally, the carbon emission coefficients of all types of fossil fuels X_{1i} are fixed; therefore, the factors affecting transportation carbon emissions in Beijing are mainly the energy structure, energy intensity, output value of per unit traffic turnover, transportation intensity, economic growth and population size. The change in emissions can be decomposed into the following:

$$C^T / C^0 = D_{X_1} D_{X_2} D_{X_3} D_{X_4} D_{X_5} D_{X_6} \tag{11}$$

In Eq. (11), C^T is the carbon emissions in year T; C^0 is the carbon emissions in the base year; D_{X_1} is the energy structure effect where X_1 is the product of the energy consumptions structure (E_i/E) and corresponding carbon emission coefficients (C_i/E_i), D_{X_2} is the energy intensity effect, D_{X_3} is the output value of per unit traffic turnover effect, D_{X_4} is the transportation intensity effect, D_{X_5} is the economic growth effect, and D_{X_6} is the population size effect. Among these values, the following can be written:

$$D_{X_1}=\left[\frac{\sum_i X_{1i}^T X_2^0 X_3^0 X_4^0 X_5^0 X_6^0}{\sum_i X_{1i}^0 X_2^0 X_3^0 X_4^0 X_5^0 X_6^0}\right]^{\frac{1}{6}}\left[\begin{array}{l}\frac{\sum_i X_{1i}^T X_2^T X_3^0 X_4^0 X_5^0 X_6^0}{\sum_i X_{1i}^0 X_2^T X_3^0 X_4^0 X_5^0 X_6^0}\cdot\frac{\sum_i X_{1i}^T X_2^0 X_3^T X_4^0 X_5^0 X_6^0}{\sum_i X_{1i}^0 X_2^0 X_3^T X_4^0 X_5^0 X_6^0}\cdot\\ \frac{\sum_i X_{1i}^T X_2^0 X_3^0 X_4^T X_5^0 X_6^0}{\sum_i X_{1i}^0 X_2^0 X_3^0 X_4^T X_5^0 X_6^0}\cdot\frac{\sum_i X_{1i}^T X_2^0 X_3^0 X_4^0 X_5^T X_6^0}{\sum_i X_{1i}^0 X_2^0 X_3^0 X_4^0 X_5^T X_6^0}\cdot\\ \frac{\sum_i X_{1i}^T X_2^0 X_3^0 X_4^0 X_5^0 X_6^T}{\sum_i X_{1i}^0 X_2^0 X_3^0 X_4^0 X_5^0 X_6^T}\end{array}\right]^{\frac{1}{30}}\cdot$$

$$\left[\begin{array}{l}\frac{\sum_i X_{1i}^T X_2^T X_3^T X_4^0 X_5^0 X_6^0}{\sum_i X_{1i}^0 X_2^T X_3^T X_4^0 X_5^0 X_6^0}\cdot\frac{\sum_i X_{1i}^T X_2^T X_3^0 X_4^T X_5^0 X_6^0}{\sum_i X_{1i}^0 X_2^T X_3^0 X_4^T X_5^0 X_6^0}\cdot\frac{\sum_i X_{1i}^T X_2^T X_3^0 X_4^0 X_5^T X_6^0}{\sum_i X_{1i}^0 X_2^T X_3^0 X_4^0 X_5^T X_6^0}\cdot\\ \frac{\sum_i X_{1i}^T X_2^T X_3^0 X_4^0 X_5^0 X_6^T}{\sum_i X_{1i}^0 X_2^T X_3^0 X_4^0 X_5^0 X_6^T}\cdot\frac{\sum_i X_{1i}^T X_2^0 X_3^T X_4^T X_5^0 X_6^0}{\sum_i X_{1i}^0 X_2^0 X_3^T X_4^T X_5^0 X_6^0}\cdot\frac{\sum_i X_{1i}^T X_2^0 X_3^T X_4^0 X_5^T X_6^0}{\sum_i X_{1i}^0 X_2^0 X_3^T X_4^0 X_5^T X_6^0}\cdot\\ \frac{\sum_i X_{1i}^T X_2^0 X_3^T X_4^0 X_5^0 X_6^T}{\sum_i X_{1i}^0 X_2^0 X_3^T X_4^0 X_5^0 X_6^T}\cdot\frac{\sum_i X_{1i}^T X_2^0 X_3^0 X_4^T X_5^T X_6^0}{\sum_i X_{1i}^0 X_2^0 X_3^0 X_4^T X_5^T X_6^0}\cdot\frac{\sum_i X_{1i}^T X_2^0 X_3^0 X_4^T X_5^0 X_6^T}{\sum_i X_{1i}^0 X_2^0 X_3^0 X_4^T X_5^0 X_6^T}\cdot\\ \frac{\sum_i X_{1i}^T X_2^0 X_3^0 X_4^0 X_5^T X_6^T}{\sum_i X_{1i}^0 X_2^0 X_3^0 X_4^0 X_5^T X_6^T}\end{array}\right]^{\frac{1}{60}}\cdot$$

$$\left[\begin{array}{l}\frac{\sum_i X_{1i}^T X_2^T X_3^T X_4^T X_5^0 X_6^0}{\sum_i X_{1i}^0 X_2^T X_3^T X_4^T X_5^0 X_6^0}\cdot\frac{\sum_i X_{1i}^T X_2^T X_3^T X_4^0 X_5^T X_6^0}{\sum_i X_{1i}^0 X_2^T X_3^T X_4^0 X_5^T X_6^0}\cdot\frac{\sum_i X_{1i}^T X_2^T X_3^T X_4^0 X_5^0 X_6^T}{\sum_i X_{1i}^0 X_2^T X_3^T X_4^0 X_5^0 X_6^T}\cdot\\ \frac{\sum_i X_{1i}^T X_2^T X_3^0 X_4^T X_5^T X_6^0}{\sum_i X_{1i}^0 X_2^T X_3^0 X_4^T X_5^T X_6^0}\cdot\frac{\sum_i X_{1i}^T X_2^T X_3^0 X_4^T X_5^0 X_6^T}{\sum_i X_{1i}^0 X_2^T X_3^0 X_4^T X_5^0 X_6^T}\cdot\frac{\sum_i X_{1i}^T X_2^T X_3^0 X_4^0 X_5^T X_6^T}{\sum_i X_{1i}^0 X_2^T X_3^0 X_4^0 X_5^T X_6^T}\cdot\\ \frac{\sum_i X_{1i}^T X_2^0 X_3^T X_4^T X_5^T X_6^0}{\sum_i X_{1i}^0 X_2^0 X_3^T X_4^T X_5^T X_6^0}\cdot\frac{\sum_i X_{1i}^T X_2^0 X_3^T X_4^T X_5^0 X_6^T}{\sum_i X_{1i}^0 X_2^0 X_3^T X_4^T X_5^0 X_6^T}\cdot\frac{\sum_i X_{1i}^T X_2^0 X_3^T X_4^0 X_5^T X_6^T}{\sum_i X_{1i}^0 X_2^0 X_3^T X_4^0 X_5^T X_6^T}\cdot\\ \frac{\sum_i X_{1i}^T X_2^0 X_3^0 X_4^T X_5^T X_6^T}{\sum_i X_{1i}^0 X_2^0 X_3^0 X_4^T X_5^T X_6^T}\end{array}\right]^{\frac{1}{60}}\cdot$$

$$\left[\frac{\sum_i X_{1i}^T X_2^T X_3^T X_4^T X_5^T X_6^0}{\sum_i X_{1i}^0 X_2^T X_3^T X_4^T X_5^T X_6^0}\cdot\frac{\sum_i X_{1i}^T X_2^T X_3^T X_4^T X_5^0 X_6^T}{\sum_i X_{1i}^0 X_2^T X_3^T X_4^T X_5^0 X_6^T}\cdot\frac{\sum_i X_{1i}^T X_2^T X_3^T X_4^0 X_5^T X_6^T}{\sum_i X_{1i}^0 X_2^T X_3^T X_4^0 X_5^T X_6^T}\cdot\frac{\sum_i X_{1i}^T X_2^T X_3^0 X_4^T X_5^T X_6^T}{\sum_i X_{1i}^0 X_2^T X_3^0 X_4^T X_5^T X_6^T}\cdot\frac{\sum_i X_{1i}^T X_2^0 X_3^T X_4^T X_5^T X_6^T}{\sum_i X_{1i}^0 X_2^0 X_3^T X_4^T X_5^T X_6^T}\right]^{\frac{1}{30}}\left[\frac{\sum_i X_{1i}^T X_2^T X_3^T X_4^T X_5^T X_6^T}{\sum_i X_{1i}^0 X_2^T X_3^T X_4^T X_5^T X_6^T}\right]^{\frac{1}{6}} \quad (12)$$

D_{X_2}, D_{X_3}, D_{X_4}, D_{X_5}, and D_{X_6} can be obtained by Eq. (9).

2.3 Data sources

Energy consumption data in the transportation sector are adopted from energy balance tables for Beijing in previous China Energy Statistical Yearbook. The fuel types include coal, gasoline, kerosene, diesel, fuel oil, liquefied petroleum gas, natural gas, heat and electricity. Consistent with most studies (Yang et al., 2014a; Shen and Chi, 2012), we assume that heat and electricity do not directly generate carbon emissions. To ensure comparability among the data, energy sources were converted into standard coal equivalent according to standard coal coefficients from appendix 4 of the 2013 China Energy Statistical Yearbook. Carbon emission coefficients upon the standard coal equivalent are referred to Zhao (2009).

In the China Energy Statistical Yearbooks, transportation, storage and postal services are identified as one single industry sector. The proportion of energy consumed by the storage and postal services is small [accounting for approximately 7.6% in 2007, according to the National Bureau (He, 2012)]; therefore, the total energy consumed by this category is approximately the value consumed by transportation. Economic and traffic turnover data originate from past Beijing Statistical Yearbooks. The sample data spans the period from 1995 to 2012, and the statistical data are reported at current prices. To eliminate the effect of price fluctuations, GDP and GDP_{tr} have been converted into the 1995 value according to the corresponding value.

3 Results

The decomposition results for the factors influencing transportation carbon emissions in the above formulas are shown in Table 2 and in Fig. 1.

Table 2 Decomposition effect of carbon emissions from the transportation sector in Beijing

Year	D_{X_1}	D_{X_2}	D_{X_3}	D_{X_4}	D_{X_5}	D_{X_6}
1995—1996	0.995	0.998	1.220	0.894	1.031	1.007
1996—1997	1.000	0.922	1.128	0.908	1.106	0.985

Continued

Year	D_{X_1}	D_{X_2}	D_{X_3}	D_{X_4}	D_{X_5}	D_{X_6}
1997—1998	1. 021	0. 981	1. 187	0. 847	1. 101	1. 005
1998—1999	0. 994	1. 104	1. 060	0. 927	1. 101	1. 009
1999—2000	0. 993	0. 993	1. 040	0. 954	1. 068	1. 084
2000—2001	1. 007	1. 163	0. 970	0. 958	1. 065	1. 016
2001—2002	1. 001	1. 042	0. 979	0. 954	1. 091	1. 028
2002—2003	0. 985	1. 021	0. 999	0. 937	1. 083	1. 023
2003—2004	1. 000	1. 187	0. 929	1. 026	1. 114	1. 025
2004—2005	1. 002	0. 987	0. 896	1. 054	1. 091	1. 030
2005—2006	0. 982	1. 190	1. 081	0. 881	1. 091	1. 041
2006—2007	0. 979	1. 103	1. 109	0. 851	1. 097	1. 047
2007—2008	0. 977	1. 122	1. 001	0. 953	1. 037	1. 057
2008—2009	1. 012	0. 996	1. 038	0. 900	1. 046	1. 050
2009—2010	1. 001	0. 961	0. 956	1. 060	1. 048	1. 055
2010—2011	0. 990	1. 008	0. 909	1. 080	1. 038	1. 029
2011—2012	0. 990	0. 965	1. 028	0. 966	1. 049	1. 025
1995—2012	0. 930	1. 967	1. 582	0. 400	3. 345	1. 654

Note: D_{X_1}, D_{X_2}, D_{X_3}, D_{X_4}, D_{X_5}, D_{X_6} and V represent energy structure effect, energy intensity effect, output value of per unit traffic turnover effect, transportation intensity effect, economic growth effect, population size effect.

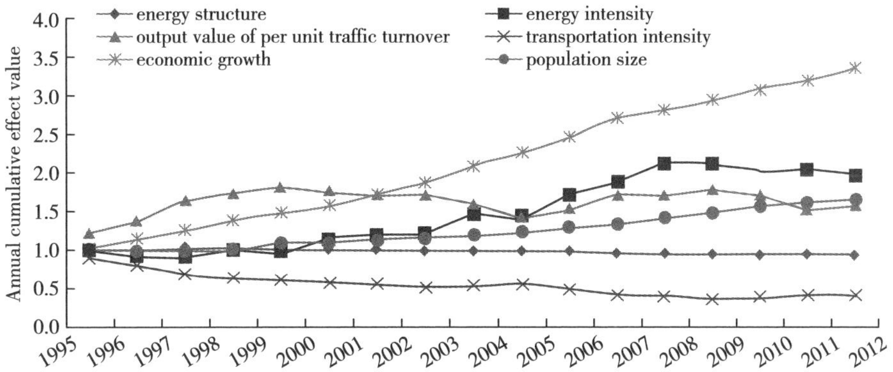

Fig. 1 Influencing factors of transportation carbon emissions in Beijing from 1995 to 2012

(1) Economic growth, energy intensity, population size and output value of per unit traffic turnover play positive roles in increasing transportation carbon emissions, with cumulative contributions of 334. 5%, 196. 7%, 165. 4% and 158. 2%, respectively. Among them, economic growth contributes the most in increasing transportation carbon emissions, Energy intensity, population size and output value of per unit traffic turnover generally display positive effects on transportation carbon emissions, but the effects are weaker than that of economic growth.

(2) The effect of energy intensity on transportation carbon emissions shows a fluctuating trend over time, alternating performance of inhibition and promotion. During 1998—1999, 2000—2004, 2005—2008, and 2010—2011, energy intensity displays a promotion role, while an inhibition role in the other years. The annual cumulative effect of output value of per unit traffic turnover on transportation carbon emissions appears as a flat "M", where the effect on transportation carbon emissions is negative in 2000—2005 and 2009—2011 but positive in the other years.

(3) Transportation intensity and energy structure inhibit transportation carbon emissions. The decrease in transportation intensity has a great contribution to the reduction of transportation carbon emissions. The cumulative effect value decreased from 0.894 in 1995 to 0.400 in 2012, which indicates a strengthening trend of inhibition.

(4) The annual effect of transportation intensity on transportation carbon emissions fluctuates in different periods. A promoting effect appears in 2003—2005 and 2009—2011 but an inhibitory effect emerges in 1995—2003, 2005—2009, and 2011—2012. The effect of energy structure on transportation carbon emissions is similar to energy intensity with inhibition and promotion effects appearing alternatively. Although its contribution is small, energy structure cumulatively plays an inhibiting role in transportation carbon emissions.

4 Discussion

The above results indicate that the contribution of each factor to the changes of transportation carbon emissions in Beijing is different. The economic growth, energy intensity, population size and output value of per unit traffic turnover play promotion roles in increasing transportation carbon emissions, while the transportation intensity and energy structure generally inhibit the emissions. The effects of energy intensity, output value of per unit traffic turnover, transportation intensity and energy structure fluctuate in some way at different stages. Effects of these six decomposition factors are discussed one by one in the following sections.

4.1 Energy structure

Energy structure shows an inhibitory effect on transportation carbon emissions in Beijing, which is consistent with a general theoretical expectation. However, a significant inhibitory effect has not been observed in this study. The result here is consistent with the previous observations on Jiangsu and Shanghai in China (Yang et al., 2014a; Wu et al., 2012). In fact, the energy structure adjustment has an indispensable role in controlling transportation carbon emissions resulting from energy consumption. Since 2004, Beijing has adopted a series of measures, such as substituting coal with oil and gas, to optimize the energy structure and promote natural gas, heat, electricity and other clean energy sources, making natural gas, heat and electricity accounted for 14.6% of energy consumption in 2012 against 9.1% in 2004, thereby slowing down the increase of transportation carbon emissions in Beijing. However, the situation is difficult to change in the short term that the energy consumption in the transportation sector is dominated by oil. Therefore, the demand for fossil fuels such as oil and other fossil fuels makes the inhibition of carbon emissions from energy consumption inapparent.

As shown in Table 2, the annual effect of energy structure on the growth of transportation carbon emissions in Beijing shows some volatility. This is mainly due to a lack of substantial technology revolution in transportation during the studied period, The inhibitory effect by advancing in energy structure thus is not strong enough to overcome the promotion effects brought by other factors, e. g. , economic growth, energy intensity, etc.

4.2 Energy intensity

Energy intensity is calculated by the energy consumed per unit of transportation sector GDP, and thus reflects the overall energy efficiency of the transportation sector. Concerning the effect of energy intensity on transportation carbon emissions, previous studies have not achieved consistent results yet because different regions are chosen in different studies. Lakshmanan (1997) analyzed the factors influencing transportation carbon emissions in the United States and pointed out that, the decreased energy intensity in passenger vehicles and light trucks is the main factor to reduce transportation carbon emissions. Timilsina and Shrestha (2009) analyzed the potential factors influencing the transportation sector CO_2 emissions in selected Asian countries during 1980—2005 and found that the decline in energy intensity corresponds to the decreasing CO_2 emissions in Mongolia, but the increasing emissions in Bangladesh, the Philippines and Vietnam. Timilsina and Shrestha (2009) determined the factors responsible for the growth in transportation sector CO_2 emissions in 20 Latin American and Caribbean countries during 1980—2005. Energy intensity is the main driving force in Bolivia, Caribbean, Cuba, Ecuador, Guatemala, Honduras, Panama, Paraguay, and the other Latin America in increasing transportation carbon emissions except in Cuba due to energy shortages during 1989—1994.

The results of this study show that energy intensity is an important factor causing the increase of transportation-related carbon emissions in Beijing. During 1995—2012, energy intensity in the transportation sector of Beijing increased from 1.982 standard coal equivalent per 10^4 yuan to 3.469 standard coal equivalent per 10^4 yuan (in 1995 prices), and the development over time is shown in Fig. 2. This observation may be due to the fast growing of private cars and a large number of state-owned vehicles[1]. For these parts of vehicles, energy consumption has been included while GDP cannot be calculated. As counted, the private cars grew from 494143 in 2000 to 2483494 in 2008, which caused rapid growth of gasoline, kerosene, diesel oil consumption. Thus, the contribution of energy intensity to transportation

---

[1] State-owned cars refers to the official vehicles purchased by public funds from all levels of the administrative organs, social groups, institutions, state-owned enterprises and other units.

sector carbon emissions in Beijing reflects the growth of private cars on transportation carbon emissions in some way. Inhibition role of energy intensity appears mainly post 2008 after the introduction of the fourth phase of national emission standards for motor vehicles in Beijing and usage limitation rule for motor vehicles based on license numbers.

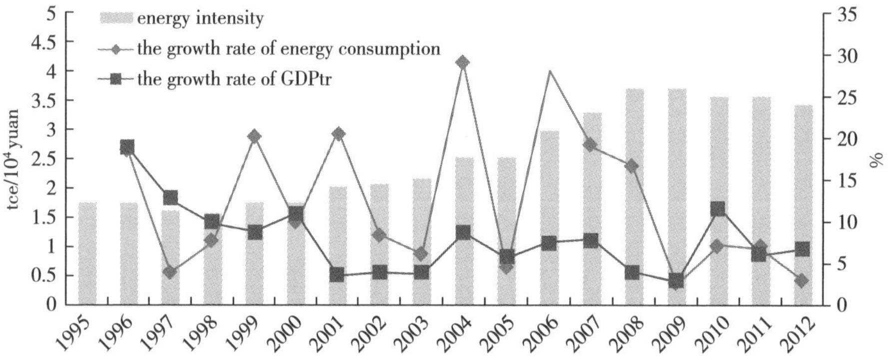

Fig. 2 The development over time of transportation energy intensity in Beijing

4.3 Output value of per unit traffic turnover

The effect of output value of per unit traffic turnover on transportation carbon emissions was not discussed in previous studies. Only Liu (2013) discussed the effect of output value of per unit traffic turnover on the transportation added value. The index used here is inverse to traffic turnover per unit transportation output value, V/GDP_{tr}, and could reflect the level of transportation efficiencyto a certain extent. For example, a bus is more efficient than a car and its output value of per unit traffic turnover GDP_{tr}/V is lower. The output value of per unit traffic turnover on transportation carbon emissions appears as a flat "M" for Beijing. Total freight turnover displayed a negative trend against the cumulative effect of output value of per unit traffic turnover, while GDP_{tr} of Beijing keep steadily growing during the studied period (Figs. 1 and 3). Therefore, the fluctuation of GDP_{tr}/V in Beijing corresponds to the change of total freight turnover over time. Total freight turnover reflects the change of the economic cycle. From 2000 to 2005, Chinese accession to the World Trade Organization (WTO) increased the demand for cargo transportation. After 2006, due to the subprime mortgage crisis in the United States, logistics and freight transportation sector of Beijing, which plays an important role in international (and internal) production and trade, was depressed to a certain degree. After 2009, attributing to economic recovery of Beijing that promoted the rapid development of business and expanded the scale of commercial circulation, total freight traffic turnover has also increased.

4.4 Transportation intensity

The transportation intensity effect is a decisive factor contributing to the reduction of transportation carbon emissions, with a cumulative effect of 0.400. This is closely related to the change of transportation intensity. From 1995—2012, the transportation turnover per 10^4 Yuan GDP in Beijing declined by 60.02% (in 1995 prices). As shown in Fig. 4, transportation intensity declining owes to industrial adjustment policy in Beijing, which restricts high energy consumption, high pollution and high emission industries and encourages the development of high value - added high - technical industries. This is consistent with the analysis of the influencing factors of transportation sector CO_2 emissions in China by Wang et al. (2011). Results of this study further shows that transportation intensity inhibited transportation carbon emissions except 2003—2005, 2009—2011. In 2003, the growth rate of passenger turnover in Beijing fell slightly because of Severe Acute Respiratory Syndrome (SARS). After the SARS crisis, passenger turnover rebound rapidly, and the growth rate increased significantly. In 2006, Beijing implemented credit card system, part of the passenger turnover may not be calculated, resulting in the decreased passenger turnover in 2006. After 2009, when the US subprime mortgage crisis undermined and the economy of Beijing was recovered, traffic turnover grew rapidly. Therefore, during 2003—2005 and 2009—2011, the growth rate of traffic turn-over exceeded that of GDP, and transportation intensity promote the transportation carbon emissions in Beijing. Accumulatively, the transportation intensity plays the main inhibition role on transportation carbon emissions in Beijing.

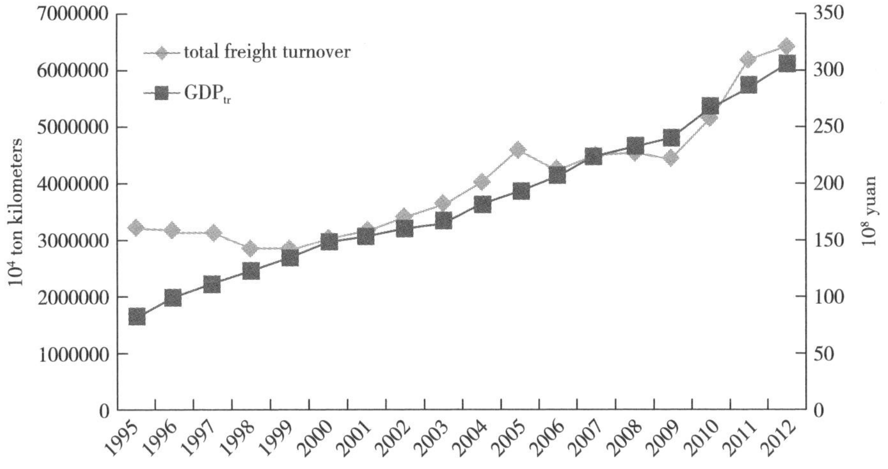

Fig. 3 Evolution of total freight turnover

4.5 Economic growth

Expanding the scale of economic output is the decisive factor in the increasing transportation carbon emissions in Beijing. This is confirmed in most studies (Lu et al., 2007; Andreoni and Galmarini, 2012; Mazzarino, 2000). Our calculation shows that the per capita GDP rose from 12690 yuan in 1995 to 42443 yuan in 2012, equivalent to an increase of 234% (in 1995 prices). The corresponding transportation carbon emissions increased from 811000 tons to 5405000 tons, yielding an increase of 566%. The increased living standards of residents have

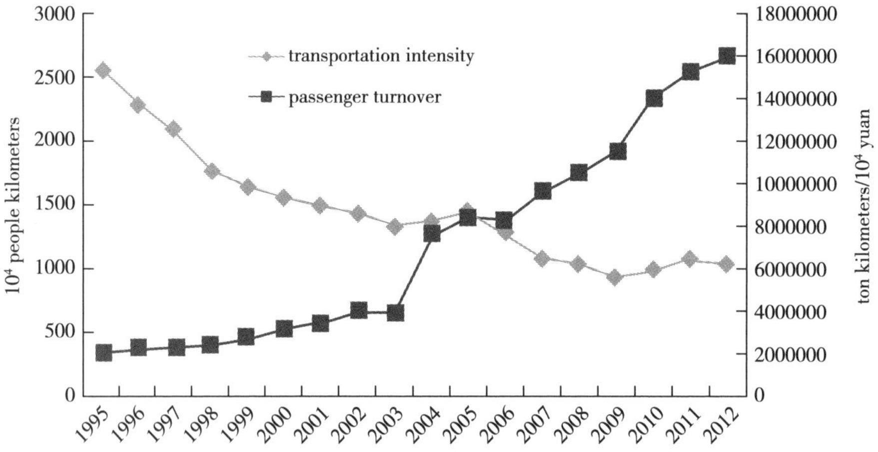

Fig. 4 Evolution of transportation intensity in Beijing

increased both tourism and the demand for private cars. The rapid development of the economic system establishes a new challenge to efficient logistics, and thus leads to substantially increasing of the energy consumption for transportation carbon emissions. Lin and Jiang (2009) noted that the process of urbanization and industrialization will basically end in 2020. Thus, we infer that Beijing's economy is not slowing down until 2020. The rigid expansion of energy consumption and high dependence on fossil energy will lead to successive growth in transportation carbon emissions. It is expected that economic growth will still be the main driving force of transportation carbon emissions in the next long period.

4.6 Population size

As for the effect of population size on transportation carbon emissions, Timilsina and Shrestha (2009), Lakshmanan (1997), and Wang and Liu (2014) noted that population size is an important factor promoting the growth of transportation carbon emissions based on their studies of different countries and regions. Our research found that the population size is a factor that increased carbon emissions, with a cumulative effect of 165.4%. Wang et al. (2011) noted that, assuming an increase of one permanent population, day trips in Beijing will increase 2.64 person-times. Li (2010) suggested that the number of cars will plus one every 2.5 people move to Beijing. With Beijing's continuing urbanization in recent years, the registered permanent population reached 21.14 million in 2013, resulting in a substantial pressure on urban transportation and promoting the increase of transportation carbon emissions.

5 Conclusions and suggestions

5.1 Conclusions

Based on the data for energy consumption and economic growth in the transportation sector in Beijing from 1995 to 2012, we established a transportation carbon emissions decomposition model of energy-related carbon emissions with the expanded Kaya identity and generalized Fisher index method. The effects have been measured of the energy structure, energy intensity, output value of per unit traffic turnover, transportation intensity, economic growth and population size on transportation carbon emissions from energy consumption in Beijing. The following conclusions can be drawn:

(1) The energy intensity, output value of per unit traffic turnover, economic growth and population size display positive effects on transportation carbon emissions in Beijing, with cumulative contributions of 334.5%, 196.7%, 165.4% and 158.2%, respectively. The successive growth of economic output is the dominant factor of transportation carbon emissions in Beijing. The energy intensity, population size and output value of per unit traffic turnover generally display weaker effects than economic growth.

(2) The energy structure and transportation intensity have inhibitory effects on the transportation carbon emissions in Beijing. The transportation intensity is the main contributor to the decline in transportation carbon emissions. The inhibitory effect of the energy structure is weaker than that for the transportation intensity. The inhibition of both combined is limited and cannot overcome the increased effect of carbon emissions caused by other factors such as population size and economic growth.

5.2 Suggestions

According to the above conclusions and considering the actual demand of transportation in Beijing, we suggest the following major measures to save energy and reduce carbon emissions.

(1) Promote the development of new energy vehicles. The study shows that improvements in the energy structure inhibit transportation carbon emissions because of "low carbon" fuel adjustment. In order to improve the energy consumption structure of cars, Beijing may further promote new energy vehicles, e. g. electric ones. Taking into account of the problems that new energy vehicles face such as high acquisition and use costs, immature technology level, insufficient charging stations and other issues, recommendations are raised as follows: Power recharging stations of new energy automobiles should be greatly expanded both in station number and in location number. Further, a comprehensive taxation, finance, government procurement and other supporting policies may be put into action to encourage the vehicle companies to develop relevant key technology, thus to reduce the purchase and use costs of new energy vehicles and increase actual utilization rate of new energy vehicles.

(2) Limit private vehicles' increase and implement the reform of state-owned cars to promote public transportation. The results show that the energy consumption by rapid increase in private vehicles and a large number of state-owned vehicles cannot be compensated by the reduction of energy intensity based on technological progress, thus becoming important driving

factors in the growth of transportation carbon emissions. The driving restriction policy and vehicle license allocation using a lottery did not reduce fuel consumption as much as expected (Wang et al., 2014; Yang et al., 2014b). Beijing has a large number of state-owned cars. Measures as the current state-owned cars reform, including auctioning state-owned cars, setting limits for the state-owned cars will not only help alleviate the urban traffic burden and improve public transport efficiency but will also save energy and reduce emissions. As for private cars, in addition to the existing policies, we may learn more experience from developed countries to further increase the cost of private cars by collecting fees for car purchase, license fees, gasoline taxes and other expenses in order to control their use. To encourage public transportation, the government needs to build a better public transportation system, such as increasing supplies according to the flow of people and develop rapid transit lines (Zhuo, 2015), as well as trying to provide multi-level public transport services in the form of customized buses. Meanwhile, the travel costs from public transportation should be further reduced.

(3) Evacuate non-core functions of Beijing and adjust industrial structure. The results show that economic growth plays a significant role in promoting transportation carbon emissions and optimization of industrial structure is conductive to the reduction of transportation intensity in Beijing. Reductions in transportation carbon emissions may be realized by evacuating non-core functions from Beijing to achieve a better industrial structure as well as to keep a suitable speed of development of GDP. As the capital of the country, Beijing may retain the core functions of the national political, cultural, international exchange, science and technology innovation center and eliminate high pollution and high energy-consuming industries such as iron and steel, nonferrous metals, building materials, chemical, textile, machinery, printing, and papermaking. As these industries have higher requirements for the convenient and efficient regional transportation system and contribute a lot to carbon emissions from transportation.

(4) Continuingly control the population size and improve population quality. Urban economic growth results in a larger urban population which is an important source of increasing transportation carbon emissions of the city. By considering energy conservation and emission reduction, we fully support the coordinating development strategy among Beijing, Tianjin and Hebei. Projects, money may be transferred to Tianjin City and Hebei Province to promote the flow and evacuation of the population. Anyway, population control policy has to be kept to avoid excessive population growth by increasing the standard of migrants entering the city and continuing to optimize the structure and quality of the population as the core of population development.

Acknowledgments The authors express sincere thanks for the financial support from the National Natural Science Foundation of China under Grant No. 71173200 and the support from the Development and Research Center of China Geological Survey under Grant No. 12120114056601. Editorial handling by

Dr. H. Oliver Gao is highly appreciated. We also would like to thank the two anonymous referees for their helpful comments that have improved our manuscript a lot.

References

Andreoni, V., Galmarini, S., 2012. European CO_2 emission trends: a decomposition analysis for water and aviation transport sectors. Energy 45 (1), 595-602.

Albrecht, J., François, D., Schoors, K., 2002. A Shapley decomposition of carbon emissions without residuals. Energy Policy 30 (9), 727-736.

Ang, B. W., Liu, F. L., Chung, H. S., 2004. A generalized Fisher index approach to energy decomposition analysis. Energy Econ. 26 (5), 757-763.

Boyd, G. A., Roop, J. M., 2004. A note on the Fisher ideal index decomposition for structural change in energy intensity. Energy J. 25 (1), 87-101.

Gao, B., Xu, Q. T., Li, Y. B., et al., 2013. Calculation and driving factors analysis of carbon emissions from traffic and transportation energy consumption in Jilin Province. Econ. Geogr. 33 (9), 25 - 30 (in Chinese).

He, C. H., 2012. Research on Low Carbon Traffic Development of Shanghai based on the LMDI Model. Thesis of Hefei University of Technology, Hefei, China (in Chinese).

IEA, 2014. CO_2 Emissions from Fuel Combustion Highlights. Office of Management and Administration, Paris, France.

Kaya, Y., 1990. Impact of carbon dioxide emission control on GNP growth: Interpretation of proposed scenarios. Paper presented at the IPCC Energy and Industry Subgroup, Response Strategies Working Group, Paris, France.

Kwon, T. H., 2005. Decomposition of factors determining the trend of CO_2 emissions from car travel in Great Britain (1970—2000). Ecol. Econ. 53 (2), 261-275.

Lakshmanan, T. R., 1997. Factors underlying transportation CO_2 emissions in the U. S. A.: a decomposition analysis. Transport. Res. Part D: Transport Environ. 2 (1), 1-15.

Li, J., Liu, H. Y., 2014. Research on the development of low carbon of urban transportation: taking Tianjin as an example. Ecol. Econ. 30 (2), 154-157 (in Chinese).

Li, X. J., 2010. Solve traffic congestion in Beijing with zero population growth. Transport Bus. China 11, 42 (in Chinese).

Lin, B. Q., Jiang, Z. J., 2009. Environmental Kuznets curve prediction and impact factors analysis on carbon dioxide emissions in China. Manage. World 4, 27-36 (in Chinese).

Liu, F. L., Ang, B. W., 2003. Eight methods for decomposing the aggregate energy intensity of industry. Appl. Energy 76 (1-3), 15-23.

Liu, X. Y., 2013. Factor Decomposition and Reduction Potential of CO_2 Emissions from the Transport Industry. Thesis of Chang'an University, Xi'an, China (in Chinese).

Lu, I. J., Lin, S. J., Lewis, C., 2007. Decomposition and decoupling effects of carbon dioxide emission from highway transportation in Taiwan, Germany, Japan and South Korea. Energy Policy 35 (6), 3226-3235.

Mazzarino, M., 2000. The economics of the greenhouse effect: evaluating the climate change impact due to the transport sector in Italy. Energy Policy 28 (13), 957-966.

Mishalani, R. G. et al., 2014. Modeling the relationships among urban passenger travel carbon dioxide emissions, transportation demand and supply, population density, and proxy policy variables. Transport. Res. Part D: Transport Environ. 33 (12), 146-154.

Su, T. Y., Zhang, J. H., Li, L., et al., 2011. Influencing factors of urban traffic carbon emission: an empirical study with panel data of big four city of China. Ind. Eng. Manage. 16 (5), 134-138 (in Chinese).

Schipper, L., Scholl, L., Price, L., 1997. Energy use and carbon emissions from freight in 10 industrialized countries: an analysis of trends from 1973 to 1992. Transport. Res. Part D: Transport Environ. 2 (1), 57-76.

Shen, M. H., Chi, X. W., 2012. Analysis of the driving factors on traffic carbon emissions growth in China's transportation sector. Jianghuai Forum 1, 31-38 (in Chinese).

Timilsina, G. R., Shrestha, A., 2009a. Factors affecting transport sector CO_2 emissions growth in Latin American and Caribbean countries: an LMDI decomposition analysis. Int. J. Energy Res. 33 (4), 396-414.

Timilsina, G. R., Shrestha, A., 2009b. Transport sector CO_2 emissions growth in Asia: underlying factors and policy options. Energy Policy 37 (11), 4523-4539.

Wang, L. L., Xu, J. T., Qin, P., 2014. Will a driving restriction policy reduce car trips? —The case study of Beijing, China. Transport. Res. Part A: Pol. 67, 279-290.

Wang, S. G., Duan, T. T., Cun, X. B., et al., 2011a. Research on traffic congestion in Beijing from the perspective of economics. Foreign Investment China 253, 186-187 (in Chinese).

Wang, W. W., Zhang, M., Zhou, M., 2011b. Using LMDI method to analyze transport sector CO_2 emissions in China. Energy 36 (10), 5909-5915.

Wang, Z., Liu, W., 2014. The impacts of individual behavior on household daily travel carbon emissions in Beijing. China Energy Proc. 61, 1318-1322.

Wu, K. Y., He, C. H., Wang, G. X., et al., 2012. Measurement and decomposition analysis on carbon emissions of transportation industry in Shanghai. Econ. Geogr. 32 (11), 45-51, 37 (in Chinese).

Yang, L. J., Wu, W., Su, Q., et al., 2014a. Carbon emissions of transportation energy consumption and its decoupling analysis in Jiangsu Province. Resour Environ. Yangtze Basin 23 (10), 1383 - 1390 (in Chinese).

Yang J., Liu Y., Qin P, et al., 2014b. A review of Beijing's vehicle registration lottery: Short-term effects on vehicle growth and fuel consumption, 75, 157-166.

Zhao, M., 2009. Carbon emissions from energy consumption in Shanghai. Res. Environ. Sci. 8, 112-117 (in Chinese).

Zhuo, J. C., 2015. Analysis on license plate restriction policy and purchase restriction in Beijing. Mod. Econ. Inform. 5, 35 (in Chinese).

The impacts of renewable energy and technological innovation on environment-energy-growth nexus: New evidence from a panel quantile regression

CHEN Wenhui[1,2], LEI Yalin[1,2,*]

1. *School of Humanities and Economic Management, China University of Geosciences, Beijing* 100083, *China*
2. *Key Laboratory of Carrying Capacity Assessment for Resource and Environment, Ministry of Land and Resources, Beijing* 100083, *China*

Abstract To mitigate climate change, many studies have been conducted to identify the determinants of CO_2 emissions. However, a consensus has not been reached yet on the issue because past work has often not considered the unobserved individual heterogeneity across countries. Therefore, this study revisits the environment-energy-growth nexus by employing a panel quantile regression to incorporate the effects of renewable energy consumption and technological innovation within the research background of global 30 countries over the period 1980—2014. The advantage of this method is considering the distributional heterogeneity to provide a detailed description of linkage between the CO_2 emissions and driving factors at different emissions levels. The results show that the effects of determinants on CO_2 emissions are heterogeneous. For high-emissions countries, the function of renewable energy consumption is limited in reducing CO_2 emissions due to the smaller proportion of renewable energy use. Moreover, technological innovation greatly affects countries with relatively higher CO_2 emissions. Therefore, one option is to financially support and apply technological innovations to generate renewable energy at lower costs as well as increase energy efficiency. Moreover, transforming the economic growth mode is helpful to transfer from non-renewable to renewable sources of energy to meet energy demand.

Keywords Carbon dioxide emissions, Renewable energy consumption, Technological innovation, Panel quantile regression

1 Introduction

With rapid economic development and technological progress, the demand for energy in developed and developing countries has increased. Although energy consumption can lead to economic growth, it is the principal reason for environmental degradation[1]. Over the past few decade, environmental pollution has become one of the greatest global issues because of large

---

原文刊于 *Renewable Energy*, 2018 年第 123 卷。原文发表时为彩色。

* 为通信作者。

increases in green-house gas (GHG) emissions[2]. With the development of industry, GHG emissions, carbon dioxide (CO_2) emissions in particular, have sharply increased[3]. Moreover, the IPCC (2007) predicted that energy-related CO_2 emissions in 2030 will have increased by 40%-110%. Therefore, it is necessary to investigate the energy consumption-related factors that influence changes in CO_2 emissions to realize energy-saving emission reduction and achieve low carbon economy development.

Regarding the main factors driving CO_2 emissions, the majority of studies have focused on economic growth and aggregate energy consumption. However, according to [4] and [5], aggregate energy consumption and economic growth alone may fail to explain changes in CO_2 emissions. In fact, in order to attain sustainable economic development, countries have turned their attention to the development and utilization of renewable energy. We will rely on renewable energy in the future[6]. In 2014, renewable energy accounted for 19.2% of global final energy consumption. The International Energy Agency (IEA) predicted that renewable energy will be the fastest growing component of global energy demand. For example, renewable energy technologies are the fastest growing sources of electricity in the US. Furthermore, environmentalist[7,8] have increased interest in the distribution of renewable energy resources in the power system because renewable energy is a virtually carbon-free energy source that can serve as a potential solution to climate change problems. However, according to [9], a much smaller part of work in the literature investigated the linkage between economic growth and energy consumption by energy source (i.e., renewable energy consumption and non-renewable energy consumption). Therefore, we need to pay attention to the impact of renewable energy consumption on CO_2 emissions.

Nomenclature

Abbreviation	
CO_2	Total carbon dioxide emissions
NRE	Non-renewable energy consumption
RE	Renewable energy consumption
PGDP	Real per capita gross domestic product
TECH	Technological innovation
FDI	Foreign direct investment
NRE	International Energy Agency
FINA	Financial development
POP	Population size
Symbols	
L	Natural logarithms

Because technological innovation is crucial in improving energy efficiency[10-12], it is a factor worthy of our attention when we explore the determinants of CO_2 emissions. High technology enables the economy to produce a level of output consuming a lower level of energy[13]. Moreover, technological innovation could lead to quicker adoption of renewable energy to meet energy demands and change the energy consumption structure. Even though previous researchers have extensively analyzed the relationship between disaggregated energy consumption, CO_2 emissions and economic growth, they did not include the effect of technological innovation in the environment-energy-growth nexus. Therefore, this paper discusses the influence of technological innovation on CO_2 emissions to expand the research target.

In terms of research method, most of the existing literature[14-19] adopted conventional panel estimation techniques that failed to consider unobserved individual and distributional heterogeneities across countries in the panel. In addition, some studies rely on a multivariate framework without including some control variables which lead to the omitted variables bias❶. As a result, existing studies did not come to the consensus conclusion concerning the effects of driving factors on CO_2 emissions. To fill these gaps, we examine the determinants of CO_2 emissions using panel quantile regression, adding relevant variables as control variables.

This paper makes several contributions to the existing literature. First, as shown in Table 1 in the appendix section, most existing studies used aggregate energy consumption as variable in their models and thus failed to estimate the effects of energy consumption by source. Instead, we include renewable and non-renewable energy sources separately in this models, and as such, it provides us some new finding about their disaggregated impacts on CO_2 emissions. Second, it employs the panel quantile regression method to empirically investigate the influence of driving factors on CO_2 emissions at different levels of CO_2 emissions in global 30 countries, which could provide more complete results compared to conventional panel regression and demonstrate the possible heterogeneity. Third, in order to avoid the omitted-variable bias faced by previous studies, several related control variables are included in our model. Indeed, against the background of an apparent lack of consensus in the literature on the environment-energy-growth nexus issue, the results of this study could provide scientific information for policy-makers to target policies to meet the CO_2 emissions level of a specific country rather than on the average of the entire group.

The outline of this paper is organized as follows: Section 2 reviews the related literature on the environment-energy-growth issues. Section 3 explores the panel quantile regression model and data, and Section 4 presents the results and a detailed discussion. The last section concludes the paper and presents related policy implications.

❶ 1 Omitted variables bias occurs when some variables are correlated with the dependent variable and one or more other independent variables but they are omitted in a regression model.

2 Literature review

There are three main research groups studying the environment-energy-growth nexus in the environmental economics literature. First, scholars[20-24] focuses on examining the validity of the Environmental Kuznets Curve (EKC) hypothesis. In fact, in the most EKC framework, one limitation of this group is that they identified the relation between environment and income in a bivariate framework and hence suffers from omitted variable bias[25-29]. Therefore, in order to circumvent omitted variable bias in EKC framework[30-32], introduced energy consumption into the link between CO_2 emissions and economic growth (Appendix Table 1) and found that the environmental EKC hypothesis was invalid. Some scholars[25,33-36] confirmed the existence of U-shaped relationship between CO_2 emissions and economic growth by building on the Environmental Kuznets Curve (EKC) framework. While there was also some evidence that the EKC hypothesis was an N-shaped relationship[37]. Indeed, existing studies also failed to find evidence of a consensus conclusion about the effects of economic growth on CO_2 emissions.

Actually, the influence of energy consumption in CO_2 emissions is also controversial. Some scholars concluded that total energy consumption has a positive effect on CO_2 emissions[25,32,38-40]. Some researchers[25,32] thought that there existed bidirectional causality between CO_2 emissions and economic growth. While, according to [39, 41], there was unidirectional causality from energy consumption to CO_2 emissions. Some results of Granger causality tests indicated the existence of a positive causality from CO_2 emissions to energy consumption[6].

Second, the majority of the environment-energy-growth nexus comes from aggregate energy consumption and conventional fossil energy consumption. Indeed, a much smaller body of work in the literature discussed the relationships between non-renewable energy consumption, renewable energy use, economic growth and CO_2 emissions (Appendix Table 2). All previous studies showed that the non-renewable energy consumption stimulates the CO_2 emissions, but the findings varied regarding the effects of renewable energy consumption.

Renewable energy consumption has contributed to CO_2 emissions reduction and improvement in technical and economic efficiency[42,43]. concluded the unidirectional causality exists between renewable energy use and economic growth. While, some researchers[44,45] revealed a bidirectional causality between them. And, some scholars[18,23,24,46-48] found that renewable energy mitigated CO_2 emissions. While[15,49,50], supported the renewable energy consumption has positive effect on CO_2 emissions. In addition[51], thought that GDP and renewable energy consumption contributed to CO_2 for lower threshold, while renewable energy use mitigated CO_2 for upper threshold. In conclude, empirical findings regarding the nexus of economic growth, energy consumption and CO_2 emissions are mixed. Overall, the majority of previous researchers reported incongruent empirical findings towards this puzzled relationship.

Table 1 Variable definitions and data sources

Variable	Definition	Units	Source
CO_2	Total CO_2 emissions from the consumption of energy	Thousand metric ton	EIA
NRE	Total primary energy (coal, oil and gas) consumption	Thousand tonnes oil equivalent	BP-statistical review
RE	Total renewable energy consumption	Thousand tonnes oil equivalent	BP-statistical review
TECH	Total patent applications	people	Database of the World Bank
PGDP	real per capita GDP	constant 2010 US $	Database of the World Bank
FINA	Financial development, domestic credit to private sector (% of GDP)	Percent	Database of the World Bank
TR	Trade (% of GDP)	Percent	Database of the World Bank
FDI	Foreign direct investment, net inflows (% of GDP)	Percent	Database of the World Bank
POP	Population size	people	Database of the World Bank

Table 2 The classification of selected countries

Continents	Number	Country Name
Europe & Central Asia	13	Netherlands (NLD), Norway (NOR), Finland (FIN), Germany (DEU), Greece (GRC), Ireland (IRL), Denmark (DNK), Portugal (PRT), Spain (ESP), Sweden (SWE), Turkey (TUR), France (FRA), United Kingdom (GBR)
South Asia	2	Bangladesh (BGD), India (IND)
North America	2	Canada (CAN), United States (USA)
Latin America & Caribbean	3	Chile (CHL), Mexico (MEX), Peru (PER)
East Asia & Pacific	6	China (CHN), Indonesia (IDN), Japan (JPN), New Zealand (NZL) Philippines (PHL), Thailand (THA)
Middle East & North Africa	3	Iran, Islamic Rep. (IRN), Egypt (EGY), Israel (ISR)
Sub-Saharan Africa	1	South Africa (ZAF)

Studies in the third strand address the importance of technological innovation on this issue (Appendix Table 3)[52]. concluded a unidirectional causality from technological innovation to renewable energy use. Besides[53], also obtained that technological innovation is significant in decreasing the use of fossil fuels based on ARDL approach. Some scholars[16,45] found that research and development (R&D) expenditures has positive influence in economic growth. Investments in R&D and change in technology could reduce the CO_2 emissions[54]. Therefore, technological innovation played a major role in reducing energy consumption and improving energy efficiency[13].

In conclude, it has not yet recognized a consensus conclusion on this environment-energy-growth issue. Some reasons may explain the different conclusions reached in previous works: The samples used for research are different; the models and methods adopted to study relationships vary; and the control variables included in the model are diverse[55,56][57]. argued that the shortcoming of previous studies was that the results suffered from deviation due to a neglect of distributional heterogeneity. In addition, because of the only two or three control variables in previous studies, there was m omitted-variable bias.

Therefore, this study revisits the environment-energy-growth nexus by employing a panel quantile regression to incorporate the effects of renewable energy consumption and technological innovation. The advantage of this method is that the distributional heterogeneity may be included to provide a detailed description of linkage between the CO_2 emissions and driving factors at different emissions levels. Besides, in order to eliminate the omitted-variable bias faced by previous studies, several related control variables are included in our model.

Table 3 Summary of descriptive statistics

Variable	Mean	Maximum	Minimum	Std. Dev.	Skewness	Kurtosis	Jarque-Bera	Probability
LCO_2	11. 99463	16. 06905	8. 857537	1. 488117	0. 54761	2. 670323	52. 98187	0. 0000
LNRE	11. 1318	14. 90477	8. 023526	1. 423147	0. 546366	2. 749299	50. 9051	0. 0000
LRE	8. 139872	12. 59104	-0. 792857	2. 054402	-0. 966409	4. 876657	293. 9338	0. 0000
LTECH	8. 493805	13. 74098	4. 532599	1. 833962	0. 557294	3. 230372	52. 46277	0. 0000
LPGDP	9. 361507	11. 42512	5. 863168	1. 410646	-0. 742941	2. 436558	102. 2751	0. 0000
FINA	2. 168107	32. 30119	-28. 62426	3. 103832	1. 223639	25. 59758	20923. 91	0. 0000
TR	58. 13706	209. 6567	12. 00868	28. 41316	1. 536372	6. 978022	1023. 29	0. 0000
FDI	74. 71746	221. 2885	5. 771342	48. 43095	0. 759819	2. 835242	94. 62593	0. 0000
LPOP	12. 96174	16. 59913	10. 2621	1. 330705	0. 63191	2. 825279	65. 92466	0. 0000

Note: "L" means the variable in natural logarithms.

3 Methodology and data

3.1 Panel quantile regression model

In this study, a panel quantile regression model is built to investigate the impacts of economic growth, technological innovation, and renewable and non-renewable energy consumption on CO_2 emissions. Based on the results of Hausman tests, we employed a fixed effect model for regression. By adopting a fixed effect panel quantile regression method, we are able to depict the determinants of CO_2 emissions throughout the conditional distribution.

Quantile regression, which was proposed by Ref. [58], is helpful in investigating asymmetric features of variable distributions. This method can obtain the estimated coefficients across the different quantiles. The conditional quantile of y_i given x_i, is as follows:

$$Qy_i(\tau \mid x_i) = x_i^T\beta_\tau \tag{1}$$

Quantile regression is robust to outliers and heavy distributions. However, these methods do not take into account the unobserved heterogeneity of a country[57]. Therefore, some scholars[59-64] studied the econometric theory of applying quantile regressions to panel data. In this paper, we employ a panel quantile method with fixed effects, which makes it possible to estimate the conditional heterogeneous covariance effects of carbon emissions drivers, thus controlling for unobserved individual heterogeneity. Consider the following fixed effect panel quantile regression model:

$$Qy_{it}(\tau_k \mid \alpha_i,\ x_{it}) = \alpha_i + x_{it}^T\beta(\tau_k);\ i = 1,\ \cdots,\ N;\ t = 1,\ \cdots,\ T \tag{2}$$

In Eq. (2), α_i have a pure location shift effect on the conditional quantiles of the response. The effects of the covariates x_{it} are permitted to depend upon the quantile τ of interest. i is the index of individual and t is the index of time. N is the number of observations on the individual i. T is the number of observations on the time t. [59] treated unobservable fixed effect as parameters to be jointly estimated with the covariate effects for different quantiles. A unique characteristic of this method is the introduction of a penalty term in the minimization to address the computational problem of specifically estimating a mass of parameters; the parameter estimate is calculated as follows:

$$\min_{(\alpha,\ \beta)} = \sum_{k=1}^{K}\sum_{t=1}^{T}\sum_{n=1}^{N} w_k\rho_{\tau_k}(y_{it} - \alpha_i - x_{it}^T\beta(\tau_k)) + \lambda\sum_{i}^{N}\mid \alpha_i \mid,$$
$$i = 1,\ \cdots,\ N;\ t = 1,\ \cdots,\ T \tag{3}$$

Where K is the index of quantiles, x is the matrix of explanatory variables, ρ_{τ_k} is the quantile loss function, and w_k is the relative weight given to the k-th quantile, which controls for the contribution of the k-th quantile on the estimation of the fixed effect.

Furthermore, we estimate the effect of economic growth (PGDP), technological innovation (TECH), renewable energy consumption (RE) and non-renewable energy consumption (NRE) on CO_2 emissions by modifying the specifications of previous studies. In this paper, we employ equally weighted quantiles $w_k = 1/K$[65,66] and set $\lambda = 1$[67]. We specify the conditional quantiles function for quantil etas follows

$$Qy_{it}(\tau | \alpha_i, \xi_t, x_{it}) = \alpha_i + \xi_t + \beta_{1\tau}LNRE_{it} + \beta_{2\tau}LRE_{it} + \beta_{3\tau}LTECH_{it} + \beta_{4\tau}LPGDP_{it} \quad (4)$$

Where the countries are indexed by i and time by t. y_{it} is the CO_2 emissions indicator.

This paper employs a panel quantile regression model that considers the unobserved individual and distributional heterogeneity. Moreover, to avoid an omitted-variable bias, based on pre-viously published studies[29,56,68-72], population (*POP*), trade openness (*TR*), foreign direct investment (*FDI*), and financial development (*FINA*) are selected as control variables in the model (Equation (5)).

$$Qy_{it}(\tau | \alpha_i, \xi_t, x_{it}) = \alpha_i + \xi_t + \beta_{1\tau}LNRE_{it} + \beta_{2\tau}LRE_{it} + \beta_{3\tau}LTECH_{it} + \beta_{4\tau}LPGDP_{it} + \beta_{5\tau}FINA_{it} + \beta_{6\tau}TR_{it} + \beta_{7\tau}FDI_{it} + \beta_{8\tau}LPOP \quad (5)$$

3.2 Data and descriptive statistic

3.2.1 Data source

In this paper, the dependent variable is CO_2 emissions, and the explanatory variables are the real per capita GDP (constant 2010 US $), technological innovation ([53, 73] use patent data as a proxy for technological innovation), renewable energy consumption and non-renewable energy consumption (fossil fuel energy consumption). The data cover the period of 1980—2014 which was chosen due to the availability of the data. All variables are transformed with natural logarithms to help eliminate heteroscedasticity. In order to overcome omitted-variable bias, we consider relevant variables as control variables. According to previous researches[29,56,68-72], we choose trade openness, foreign direct investment, the population size, and financial development as control variables. The variables descriptions, data definitions and data sources are shown in Table 1.

The dataset comprises a panel of 30 countries during the 1980—2014 period and includes data on CO_2 emissions, economic growth, renewable and non-renewable energy consumption, technological innovation, per capita GDP, foreign direct investment, financial development, trade openness and population size. The variables data are chosen to cover as many countries as possible. Map 1❶ lists the 30 countries included in the model, which are distributed across the seven continents (Table 2).

3.2.2 Data description

The statistical descriptions of all variables in the model are presented in Table 3. Recall that the skewness coefficient equals 0 for a symmetric distribution, while the kurtosis coefficient

---

❶ 图 1 见原文，刊于 *Renewable Enevgy*，2018 年第 123 卷。原图发表时为彩色。

equals 3 for the normal distribution. The results show that the distributions of variables are skewed and more concentrated than the normal distribution with longer tails. Therefore, the OLS regression approach may yield biased results. Besides, the Jarque－Bera statistical test results indicate the non－normality of the unconditional distribution of variables. This is an additional finding that supports our use of the quantile regression approach to describe the complete picture of the heterogeneous impacts of the influences factors on CO_2 emissions in this paper.

According to the scatter plots (Fig. 1) which describe the link between CO_2 emissions and its driving factors, it visualizes a lot of linear and nonelinear relationships. Therefore, adopting the quantile regression method to estimate the impacts of the driving forces on CO_2 emissions is reasonable and applicable.

4 Results

To study the heterogeneity of the effect of influence factors on different levels of CO_2 emissions and avoid omitted－variable bias, we develop our quantile regression model with relevant control variables. In this section, first, we conduct the stationarity test. Then, we implement the panel cointegration test between CO_2 emissions and its driving factors. Last, we present the results of the quantile regression model.

4.1 Panel unit root test

Before using the panel quantile regression model, the panel unit root test is used to test the stationarity of variables. LLC (Levine－Line－Chu) test and Breitung test are first kind of panel data unit root tests which assume that the different cross－section sequences in panel data have a common unit root process. The second type is the unit root test with different roots including IPS (Im－Pesarane－Skin) test, Fishere－ADF test and Fishere－PP test. Results of the panel unit root tests is presented in Table 4. These results show that we could not be rejected the null hypothesis of the existence of a unit root at the 5% level. Since all variables are non－stationary, a differencing method is used to eliminate the non－stationary trend. The results show that the null hypothesis could almost be completely rejected at the 5% level. Thus, it can be concluded that all the variables are stationary at first－order difference.

4.2 Panel co－integration results

As indicated by the results of the panel unit root test, all variables are stationary at their first differences. Thus, Johansen Fisher panel cointegration test is used to examine the long－run relationship among variables. The results obtained from the trace test and maxeigen test are reported in Table 5. The results suggest that there are at least two cointegration relationships between CO_2 emissions and all the explanatory variables.

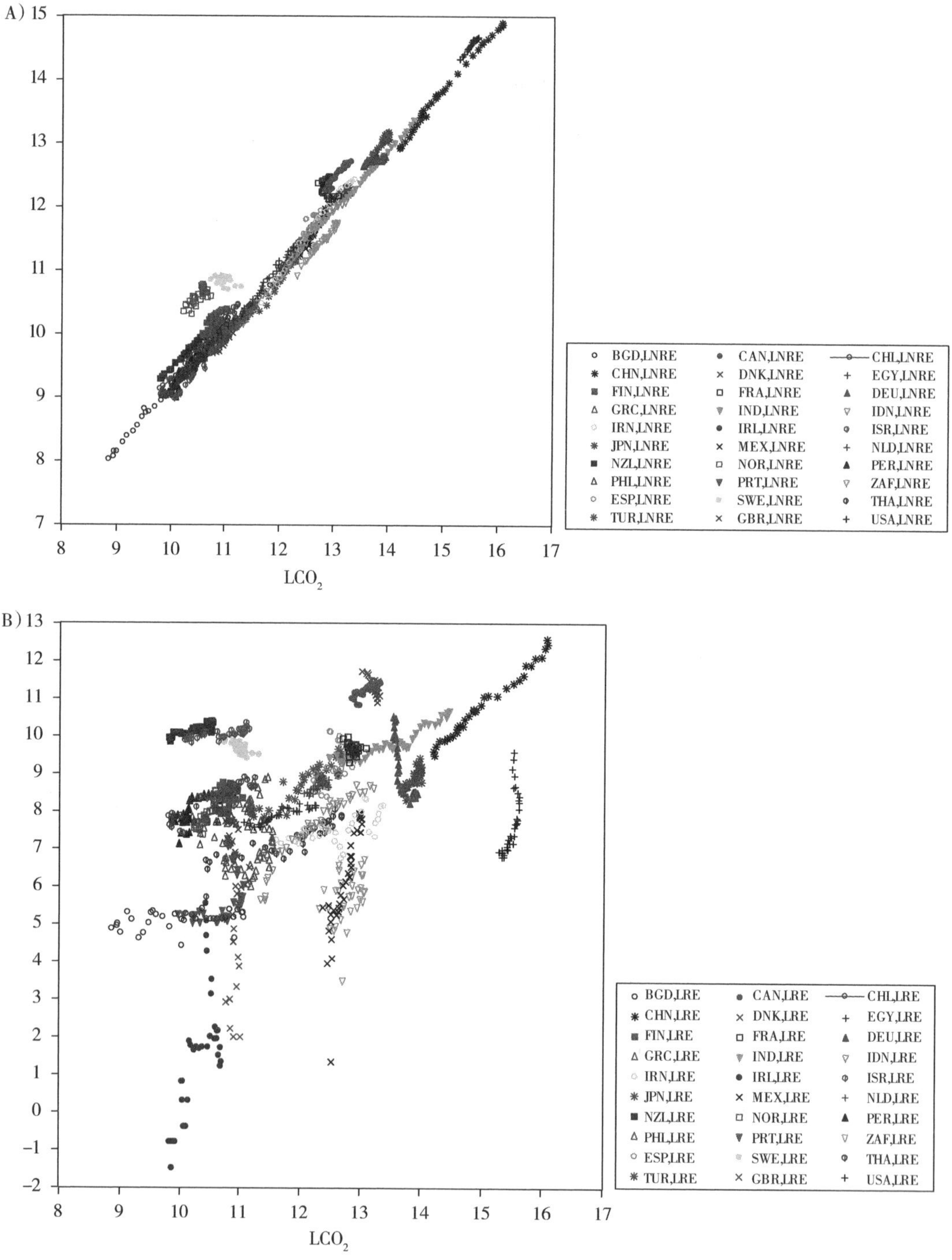

Fig. 1 Scatter chart of the links between the effect factors and CO_2 emissions

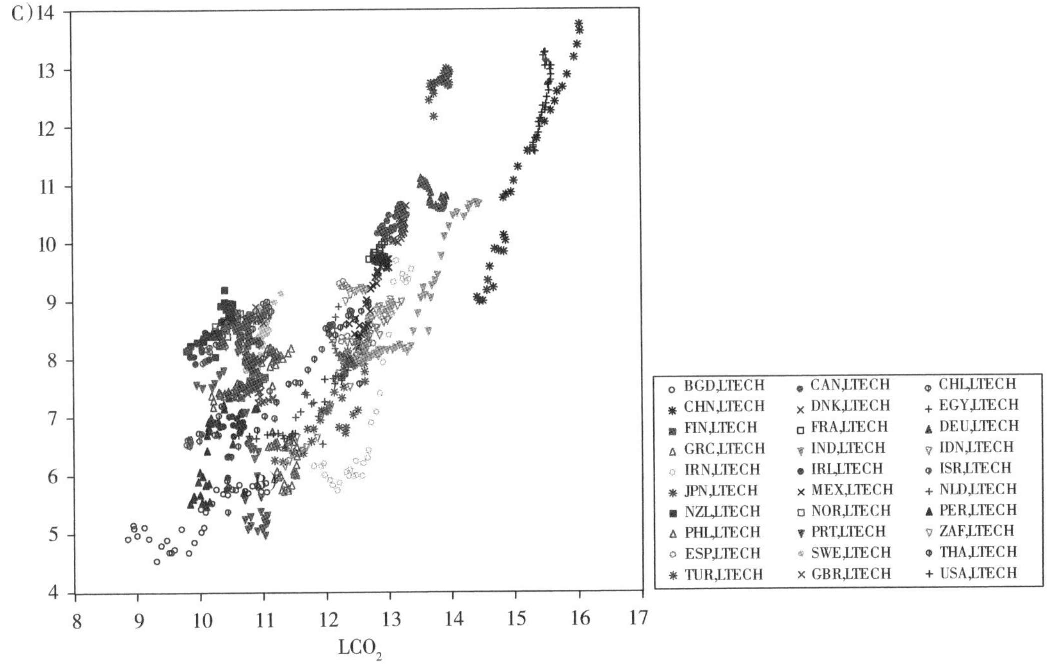

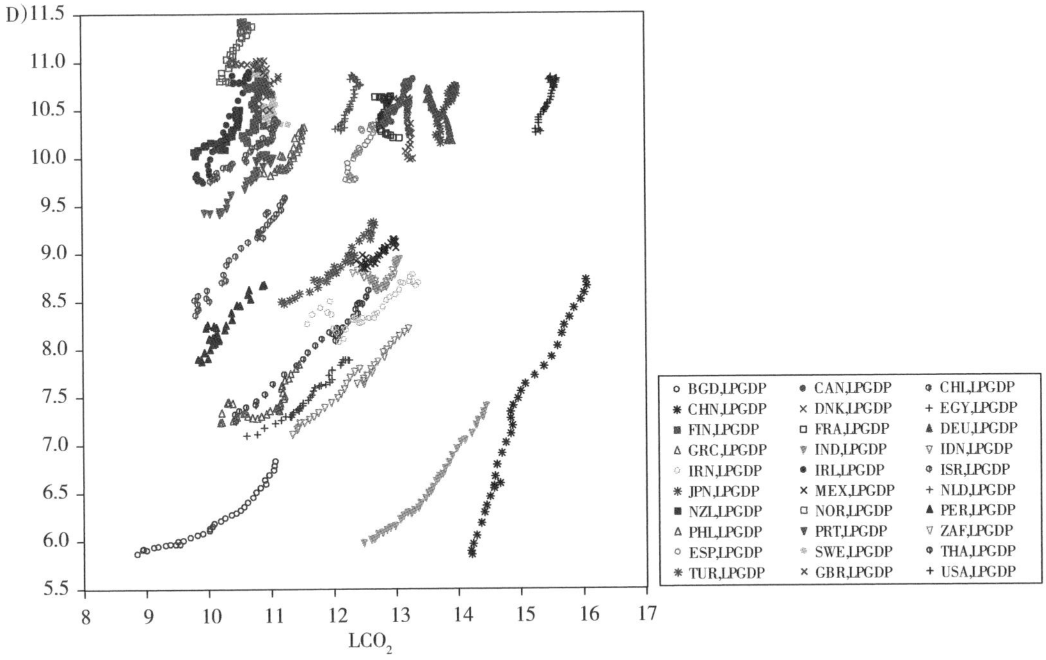

Fig. 1 (continued)

Table 4 Panel unit root test

Variable	Statistic	Levels					First difference				
		LLC	Breitung	IPS	Fisher-ADF	Fisher-PP	LLC	Breitung	IPS	Fisher-ADF	Fisher-PP
LCO_2	value	-1. 527	0. 432	-1. 850	16. 842	47. 070	-6. 231	-3. 644	-5. 492	45. 220	45. 307
	Prob. **	0. 063	0. 667	0. 032	0. 078	0. 000	0. 000	0. 000	0. 000	0. 000	0. 000
LPGDP	value	-1. 896	0. 772	-0. 927	17. 050	9. 614	-5. 588	-4. 384	-5. 003	42. 148	35. 417
	Prob. **	0. 029	0. 780	0. 177	0. 073	0. 475	0. 000	0. 000	0. 000	0. 000	0. 000
LNREC	value	-0. 522	0. 502	0. 434	8. 524	50. 880	-7. 635	-5. 976	-6. 443	52. 184	51. 805
	Prob. **	0. 301	0. 692	0. 668	0. 578	0. 000	0. 000	0. 000	0. 000	0. 000	0. 000
LREC	value	-1. 793	2. 253	-2. 606	39. 788	8. 063	-1. 646	-3. 358	-3. 638	31. 650	37. 206
	Prob. **	0. 037	0. 988	0. 005	0. 000	0. 623	0. 050	0. 000	0. 000	0. 001	0. 000
LTP	value	-1. 107	-0. 878	-1. 484	14. 760	38. 396	-5. 638	-3. 485	-6. 499	53. 135	82. 932
	Prob. **	0. 134	0. 190	0. 069	0. 141	0. 000	0. 000	0. 000	0. 000	0. 000	0. 000
TR	value	2. 500	-1. 225	-1. 746	18. 340	21. 473	-4. 313	-3. 632	-3. 622	34. 712	109. 785
	Prob. **	0. 994	0. 110	0. 040	0. 050	0. 018	0. 000	0. 000	0. 000	0. 000	0. 000
IS	value	-1. 262	-0. 324	0. 145	10. 648	4. 245	-6. 401	-6. 187	-6. 612	53. 917	63. 020
	Prob. **	0. 103	0. 373	0. 558	0. 386	0. 936	0. 000	0. 000	0. 000	0. 000	0. 000
FDI	value	-1. 208	-2. 337	-1. 787	20. 304	25. 470	-6. 675	-6. 378	-7. 935	65. 122	308. 830
	Prob. **	0. 113	0. 010	0. 037	0. 027	0. 005	0. 000	0. 000	0. 000	0. 000	0. 000
Fina	value	-0. 235	0. 117	0. 738	7. 218	11. 323	-4. 166	-4. 613	-3. 590	33. 085	91. 216
	Prob. **	0. 407	0. 547	0. 770	0. 705	0. 333	0. 000	0. 000	0. 000	0. 000	0. 000

Table 5 Johansen Fisher panel cointegration test

Hypothesized No. Of CE (s)	Stat. [a] from trace test	Prob.	Stat. [a] from maxeigen test	Prob.
None	372.2	0.000	252.2	0.000
At most 1	176.2	0.000	118.1	0.000
At most 2	96.39	0.002	65.47	0.293
At most 3	72.43	0.130	60.14	0.471
At most 4	80.97	0.037	80.97	0.037

[a] Probabilities are computed using asymptotic Chi-square distribution.

4.3 Panel quantile regression results

To illustrate the advantages of panel quantile regression model, OLS estimation results are also listed in the second columns of Table 6. The estimated results of the panel quantile regression are provided in Table 6 and Figs. 2-5. The results indicate that the coefficients of non-renewable energy consumption and per capita GDP on CO_2 emissions are positive and significant in the OLS regression. The impact of renewable energy consumption on CO_2 emissions is negative. However, the coefficient of technological innovation is insignificant in the OLS regression model. Additionally, the results (Fig. 2) provide evidence that the OLS regression results provide only an incomplete description of the effects of determinants on CO_2 emissions compared with the panel quantile regressions. A comparison of the OLS method and quantile regression results indicates that effects of the driving factors on emission are heterogeneous among countries with different levels of CO_2 emissions. Therefore, it is appropriate to use the panel quantile regression method to investigate the relationships among variables.

Table 6 Quantile regression results of driving factors on CO_2 emissions

Variables	OLS	Quantile regressions								
		10th	20th	30th	40th	50th	60th	70th	80th	90th
LNRE	1.02639*	0.96930*	1.03544*	0.98822*	0.95406*	0.95882*	0.94865*	1.04490*	1.11623*	1.12541*
LRE	-0.03425*	-0.06910*	-0.05123*	-0.04355*	-0.03177*	-0.0322*	-0.02112*	-0.02279*	-0.01138*	-0.01622*
LTECH	-0.00158	0.00436	0.00386	-0.00803*	-0.0080*	-0.01225*	-0.02240*	-0.03750*	-0.04836*	-0.04732*
LPGDP	0.13875*	0.22627*	0.15909*	0.12646*	0.11678*	0.09731*	0.09376*	0.02926*	0.01709*	0.00390*
FINA	-0.00038	-0.00683*	0.00112*	0.00168*	-0.0005*	0.00063*	0.00016	0.00163	0.00324*	-0.00187*
TR	0.00043	0.00205*	0.00138*	0.00079*	0.00077*	0.00078*	0.00106*	0.00043*	-0.00051*	-0.00064*
FDI	-0.00024*	0.00029*	0.00072*	0.00048*	0.00044*	-0.00006*	-0.00027*	-0.00039*	-0.00025*	-0.00006*
LPOP	0.12175*	0.04671*	0.03363*	0.10208*	0.13124*	0.12977*	0.15366*	0.09468*	0.13427*	0.15321*

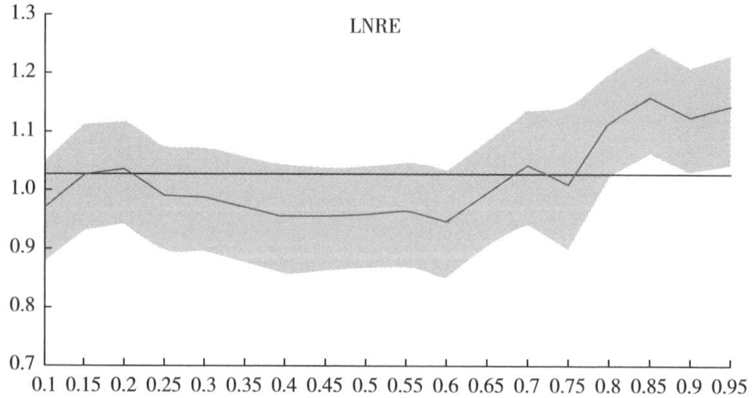

Fig. 2 Panel quantile regressions coefficients of non-renewable energy consumption

Note: The black solid line represents the corresponding OLS estimate coefficient

The quantile regression results based on Equation (5) are shown in Table 6, and they highlight several important findings. Regarding the impacts of energy consumption by source, increases in non-renewable energy consumption increase the level of CO_2 emissions. Moreover, compared with other driving factors, the non-renewable energy consumption has the greatest positive effect on CO_2 emissions across quantiles. These results are consistent with the findings of [74] for EU countries[23], for 27 advanced countries[15], for MENA countries[47,75], for OECD countries and [76] for 19 developed and developing countries. In addition, there are significant differences in the effect of non-renewable energy consumption on CO_2 emissions across lower and higher quantiles in the conditional distribution of CO_2 emissions. The influential intensity of non-renewable energy consumption on CO_2 emissions increases from 0.9486 to 1.125 and shows a rising trend (Fig. 2). For the high-emissions countries (such as China, United States, Canada and India), the non-renewable energy consumption has been increasing significantly in the past decades, and result in the increase of CO_2 emissions. While, at lower quantiles which refer to those lower CO_2 emission countries (such as Norway, Ireland, New Zealand, Chile, Denmark and Sweden), the growth rate of non-renewable energy consumption is quite slow over time during the sample period, thus the effect of non-renewable energy consumption on CO_2 emissions becomes relatively small.

Based on Fig. 3, the renewable energy consumption have a negative impact on CO_2 emission. This result is line with the those empirical reported by Ref. [50] for EU countries[23], for 27 advanced countries[18,47,75], for OECD countries[77], for Turkey and [24] for Kenya. However, this finding contrasts with the positive relationship between renewable energy consumption and CO_2 emissions found by Ref. [74] for EU countries[15], for MENA Region and [76] for a group of 19 developed and developing countries. This result is different from the findings of [78], who concluded that the use of renewable energy failed to mitigate CO_2 emissions in US. Moreover, the effect of renewable energy consumption on CO_2 emissions is heterogeneous: the impacts of renewable energy consumption on CO_2 emissions in the upper quantiles are smaller than those in the lower quantiles. For the high-CO_2-emissions countries, non-renewable energy consumption controls carbon emissions, whereas the function of renewable energy consumption is limited in reducing carbon emissions. This is mainly because fossil energy is still the main source of energy in these high-emissions countries, and the use and proportion of renewable energy is insufficient.

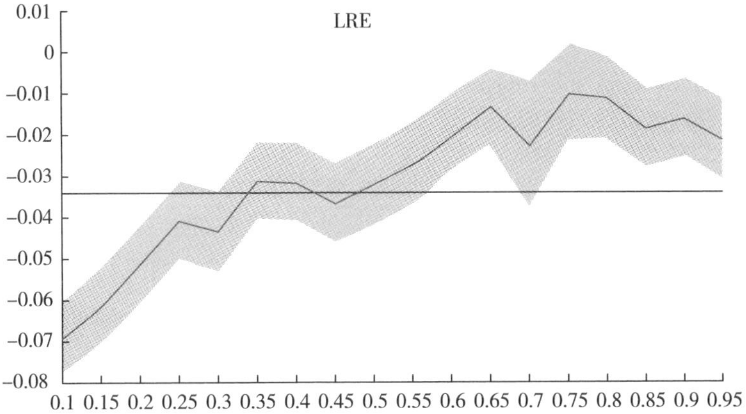

Fig. 3 Change in panel quantile regressions coefficients of renewable energy consumption

In 2014, the largest CO_2 producers were the United States Russia, India and China which is the biggest emitter. For instance, according to the EIA, the total energy-related CO_2 emissions in China reached 9376 million tons in 2014, accounting for 27.79% of the world's total emissions. Thus, Chinese government has taken measures to change the energy structure. The renewable energy use in China has a growth trend as it an average annual growth rate of 8.99%. Although China ranks at the top in the world in terms of renewable energy use with the increasing adoption of renewable energy resources, the share of non-renewable energy in total primary energy consumption has been more than 88% over the last two decades. In the US case[78], pointed out that the share of renewables in total energy consumption increased only marginally. However, regarding to the lower CO_2 emission countries, such as, Israel and New Zealand, the proportion of the use of renewable energy is more than 50%. In fact[79], found that the threshold level of proportion of renewable energy in the total energy use is at least 8.3889% which it can mitigate CO_2 emissions largely. Therefore, according to the effects of energy consumption on CO_2 emissions are heterogeneous in different countries, to effectively reduce CO_2 emissions, the high-CO_2-emissions countries strive to not only increase the renewable energy consumption but also enhance the share of clean and renewable energy, while the low-emissions countries should further rely on the renewable energy to meet energy demand.

The other finding is that the PGDP coefficients are highly significant and have a positive sign in various quantiles (Table 6), this finding is similar to the results of [30, 31, 36, 39, 80-

82]. However, this finding contrasts with the negative relationship between economic growth and CO_2 emissions found by Ref. [83] for Italy. Moreover, the effects of economic growth on CO_2 emissions continuously decrease from the lower-quantile countries to the upper-quantile countries (Fig. 4). It indicates that with more CO_2 emissions, economic growth will exert a smaller influence. Because, when the economic develops, the contradiction between and environmental pollution and economic growth also escalate gradually. Consequently, the rapid growth of economic has promoted the rapid increase in carbon emissions. However, for the high-emissions countries which the economy stay at the high stage of development, the environmental regulations of government would be improved. In these high-emissions countries, although CO_2 emissions have steadily increased, the energy intensity has decreased. For instance, China is the world's largest carbon emissions country, the change of China's economic growth mode makes the economic growth rate higher than the energy consumption. Therefore, the energy intensity of China has been decreasing from 10.9 (tce/yuan) in 1985 to 4.15 (tce/yuan) in 2014, which decreases the energy intensity and inhibits the increasing rate of carbon emissions. Similarly [84], obtained that income does not Granger cause the carbon emissions in the US. Therefore, the economic growth is not the main cause of growing CO_2 emissions for these high-emissions countries.

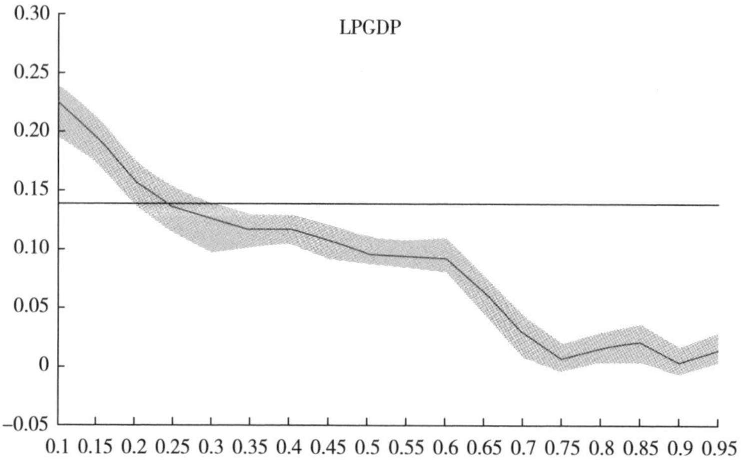

Fig. 4 Change in panel quantile regressions coefficients of per capita GDP

Additionally, the effect of technological innovation on CO_2 emissions is negative. Technological innovation has become a crucial element in the promotion of energy conservation, the reduction of CO_2 emissions and the substitution of fossil fuels with renewable energy. In fact[85,86], emphasized the importance of technology innovation in the development of renewable energy sector. Moreover, technological innovation is crucial for improving energy efficiency and reducing energy consumption[12,13,53,85,87,88]. Moreover, for the countries that have relatively higher CO_2 emissions, the effect of technological innovation has a greater negative impact on CO_2

emissions than in low-emissions countries (Fig. 5). For high-emissions countries, rapid economic growth consumes more energy than that in low-emissions countries, which implies that there is a further improvement for energy technology to improve energy efficiency and increase the proportion of renewables in these high-emissions and high-income countries. These high-emissions countries have invested heavily in advanced energy technology. As a result, the technological progress is the main factor for reducing carbon emissions. The IEA has reported that investments in renewable energy technologies in the power sector will increase from $270 billion in 2015 to $400 billion in 2030. Among the high-emissions countries, China made the largest renewable energy investments in 2014 which is $83.3 billion, up 39% from 2013. And, the US invested a total of $38.3 billion, up 7% in one year. Japan was third at $35.7 billion, which was 10% higher than in 2013. Therefore, in order to effectively reduce carbon emissions, the high-emissions countries should increase investment in advanced energy-saving technology and encourage the development of the technology innovation.

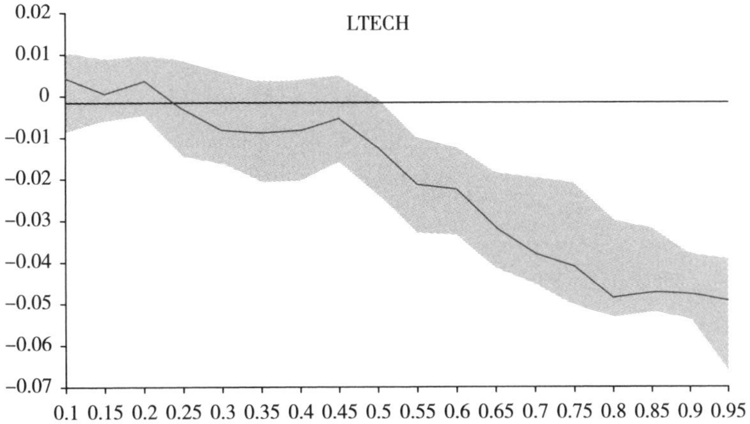

Fig. 5 Change in panel quantile regressions coefficients of technological innovation

4.4 The results of control variables

The panel quantile regression diagrams of control variables included in the model are provided in Fig. 6. First, there is a clear positive relationship between population and CO_2 emissions, as findings in the literature have indicated previously. Second, the financial development coefficient is positive and significant in the low and high quantiles except for the 10th, 40th and 90th. Financial development helps the economic growth, which results in more energy use and increasing CO_2 emissions. The similar result is supported by Refs. [33, 39, 89]. However, financial sector development also boosts investment in energy efficient technologies, thereby reducing CO_2 emissions. This result is in line with the finding by Refs. [90-93].

Regarding the effects of FDI and trade openness on CO_2 emissions are clearly significant and positive at low quantiles. The results are consistent with those reported by Refs. [25, 79, 94].

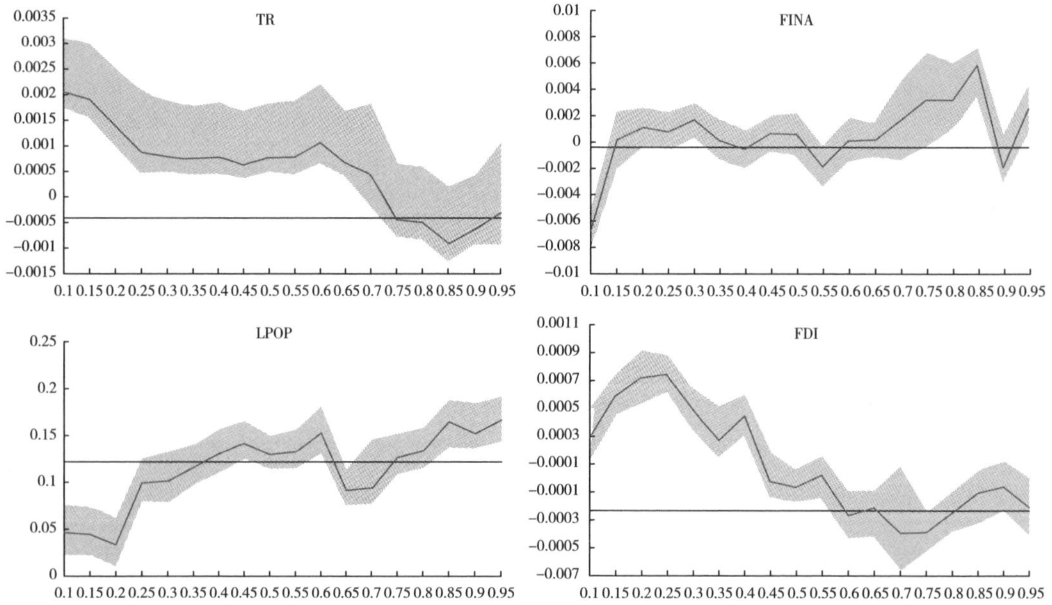

Fig. 6 Change in panel quantile regressions coefficients of control variables

For the low-emissions countries, increases in trade openness and more FDI can lead to higher GDP, higher energy consumption and higher levels of CO_2 emissions. Moreover, FDI could largely promote CO_2 emissions with relaxed environmental standards[30]. Therefore, the low-emissions countries should evaluate the environmental impact of FDI before introducing foreign investors into the country. However, for the high-emissions countries, FDI and trade openness both have negative impacts on CO_2 emissions. The results are similar to the findings of [2, 46, 48, 95-98]. Because increases in FDI increases energy efficiency through the adoption of new technologies, leading to lower emissions. Additionally, due to the technology spillover through trade flows, and these high-emissions and economically developed countries have made progress in developing new technologies and adopting environmentally friendly technologies over the last few decades. Thus, the impact of trade openness on the CO_2 emissions is negative. Therefore, the high-emissions countries could improve the level of FDI to mitigate the environment pollution to some extent.

5 Conclusion and suggestions

In this paper, a panel quantile regression model is developed to investigate the effects of economic growth, technological innovation, and renewable and non-renewable energy consumption on CO_2 emissions for global 30 countries during 1980—2014. This analysis extends the traditional panel regression method to consider the unobserved individual and distributional heterogeneities. Additionally, in order to overcome omitted-variable bias, related control variables are selected in the model. Therefore, this method could describe the relationships between CO_2 emissions and its influencing factors at different emissions levels. Based on the empirical results, several important conclusions are drawn, as follows.

5.1 Conclusions

(1) The results show that the impacts of driving factors on CO_2 emissions are heterogeneous. A comparison between conventional panel regression results and panel quantile regression with fixed effect results has emphasized the importance of considering the distributional heterogeneity of CO_2 emissions. This finding shows that the results of conventional panel regression under-or over-estimate the effect of driving factors on CO_2 emissions. In this paper, distributional analysis results could provide the scientific basis for policy-maker to target policies to specific country with different level of CO_2 emissions rather than to the entire group.

(2) Regarding the energy consumption by source, the non-renewable energy use has the greatest effect on carbon emissions across the quantiles. These results are consistent with the findings of [15, 23, 46, 47, 74, 76]. The renewable energy consumption have a negative impact on CO_2 emission. This result is line with those empirical reported by Refs. [14, 18, 23, 47, 75, 77]. However, this finding contrasts with those of [15, 17, 76]. Moreover, for the high emissions countries, the effect strength of non-renewable energy consumption on CO_2 emissions is stronger than in low-emissions countries. On the contrary, the influence strength of renewable energy consumption on CO_2 emissions is larger when the level of CO_2 emissions is relatively lower. Because non-renewable energy are still dominant in energy consumption structure in these high-emissions countries. Moreover, despite the increased level of renewable energy consumption, the mitigating effect on CO_2 emissions is limited due to the smaller proportion of renewable energy use in energy mix.

(3) The impact of economic growth on CO_2 emissions is positive. This result is similar to the findings of [30, 31, 36, 39, 80-82]. However, this finding contrasts with the negative relation-ship between economic growth and CO_2 emissions found by Ref. [83] for Italy. In addition, the effect strength of economic growth on CO_2 emissions continuously decreases from the low-emissions countries to the high-emissions countries, implying that with more CO_2 emissions, economic growth will exert a smaller influence. There are two main reasons, first, for the high-emissions countries which the economy stay at the high stage of development, the environmental regulations of government would be improved. Second, economic growth requires energy consumption as well as adds investments in renewable energy technologies. Therefore, low-carbon economy model could alleviate contradiction between environment pollution and economic growth.

(4) The effect of technological innovation has a greater negative effect on CO_2 emissions in the countries with relatively higher CO_2 emissions than in the low-emissions countries. This

finding is observed because technological innovation has become a crucial element in improving energy efficiency and reducing energy consumption[12,13,53,85,87,88]. These high-emissions countries have invested heavily in advanced energy technology. As a result, the technological progress is the main factor for reducing carbon emissions and the substitution of non-renewable energy with renewable energy for the countries that have relatively higher CO_2 emissions.

(5) For the low-emissions countries, increases in trade and more FDI can lead to economic growth and increasing energy consumption, ultimately, the higher levels of CO_2 emissions. However, the trade development and greater FDI mitigate CO_2 emissions in the high-CO_2-emissions countries because there is technology spillover through trade flows and because increases in FDI help to increase energy efficiency through the adoption of new technologies that lead to lower emissions.

In future work, we focus on the relationship between technological innovation and the renewable energy consumption in different country. How is it possible to stimulate technological innovation and raise the use of renewable energy to achieve emission-reducing and low carbon economy objectives?

5.2 Suggestions

The demand for non-renewable energy resulting from increased economic growth will accelerate the generation of CO_2 emissions, especially the high-emissions countries. Thus, the high-CO_2-emissions countries strive to reduce consumption and reliance on fossil fuels and add renewable sources to their energy mixes to cut down on pollution. Because, the renewable energy consumption could effectively mitigate carbon emissions in low-carbon countries, thus, they should improve the proportion of renewable energy sources to meet energy needs. Another option to control CO_2 emissions is increasing investment in the technological sector to provide opportunities to import new technologies. Thus, technological innovation could result in consuming from renewable energy at lower costs and increasing energy efficiency. Moreover, transforming the economic growth mode is helpful to transfer from non-renewable to renewable sources of energy to meet energy demand and effectively reduce CO_2 emissions. In addition, according to the effects of FDI on CO_2 emissions are heterogeneous in different countries, the high-emissions countries could improve the level of FDI to mitigate the environment pollution to some extent, while the low-emissions countries should evaluate the environmental impact of FDI before introducing foreign investors into the country.

Acknowledgement The authors express their sincere thanks for the support from the National Natural Science Foundation of China under Grant No. 71173200, National Research Council of Science and Technology Major Project under Grant No. 2016ZX05016005-003, the Development and Research Center of China Geological Survey under Grant No. 12120114056601 and No. 12120113093200, the Fundamental Research Funds for the Central Universities under Grant No. 5953200859632.

Appendix

Table 1 Studies focusing on total energy consumption in environment-energy-growth literature

Author	Country	Year	Variables	Methods	Results	EKC Hypothesis
[30]	Russia	1990—2007	CO_2, EC, GDP, GDP^2	Johansen cointegration, OLS, VECM	EC contributes to CO_2; EC ↔ GDP; GDP ↔ CO_2, EC ↔ CO_2	No
[31]	China	1995—2009	CO_2, EC, GDP, GDP^2, URB, TI	OLS with FE, GMM	TI, GDP → CO_2	No
[32]	USA	1960—2010	CO_2, EC, GDP, GDP^2, TR, URB, FD	ARDL, VECM causality	EC and URB contribute to CO_2, TR mitigates CO_2, CO_2 ↔ GDP; EC ↔ CO_2; TR ↔ GDP; CO_2 ↛ TR; GDP → EC	No
[6]	Turkey	1960—2000	CO_2, EC, GDP	Granger causality	CO_2 → EC; GDP ↛ CO_2	Not investigated
[99]	Malaysia	1971—1999	CO_2, EC, GDP	VECM causality	CO_2 → GDP; GDP ↔ EC	Not investigated
[100]	China	1990—2012	CO_2, EC, GDP	VECM	EC and GDP contribute to CO_2	Not investigated
[25]	EU Countries	1992—2010	CO_2, EC, GDP, GDP^2, TR, URB	FMOLS, Granger causality	EC, TR and URB contribute to CO_2; GDP ↔ CO_2, EC ↔ CO_2, TR ↔ CO_2, GDP ↔ EC	Yes
[39]	Turkey	1974—2010	CO_2, EC, GDP, GDP^2, FD	ARDL, VECM causality	EC and FD contribute to CO_2; EC, GDP → CO_2; FD ↔ CO_2	Yes
[33]	BRIC countries	1992—2007	CO_2, EC, GDP, GDP^2, FD	OLS	FD and EC contribute to CO_2; EC ↔ GDP; GDP ↔ CO_2	Yes
[38]	Turkey	1974—2010	CO_2, EC, GDP, FD	ARDL	EC and FD contribute to CO_2	Yes
[36]	China	1996—2012	CO_2, EC, GDP, GDP^2, TR, URB	GMM	EC, TR and URB contributes to CO_2	Yes
[37]	Canada	1948—2004	CO_2, GDP	PLS	positive correlation	Yes

Table 2 Studies about energy consumption by sources in environment-energy-growth literature

Author	Country	Year	Variables	Methods	Results
[17]	EU countries	1990—2008	CO_2, GDP, GDP^2, REC, NREC	OLS with FE	REC and NREC contribute to CO_2
[50]	EU countries	1996—2010	CO_2, GDP, GDP^2, GDP^3, REC	OLS with FE, RE	REC mitigates CO_2
[23]	Panel of 27 advanced countries	1990—2012	CO_2, GDP, GDP^2, REC, NREC, TR, URB, P	FMOLS	REC mitigates CO_2, NREC contributes to CO_2
[15]	MENA countries	1980—2009	CO_2, GDP, GDP^2, REC, NREC	FMOLS, DOLS	REC and NREC contribute to CO_2
[101]	OECD countries	1977—2010	CO_2, GDP, GDP^2, REC	FMOLS, DOLS	REC mitigates CO_2
[75]	OECD countries	1980—2010	CO_2, GDP, GDP^2, REC, NREC, TR	FMOLS, DOLS	REC and TR mitigate CO_2; NREC contributes to CO_2
[102]	OECD countries	1980—2011	CO_2, GDP, GDP^2, REC, NREC, POP	GMM, AMG	REC mitigates CO_2; NREC contribute to CO_2
[77]	Turkey	1961—2010	CO_2, GDP, GDP^2, REC	ARDL	REC mitigates CO_2
[24]	Kenya	1980—2012	CO_2, Y, REC, NREC, TR, FD, UB	ARDL	REC and FD mitigates CO_2 NREC, TR and URB contribute to CO_2
[48]	Malaysia	1980—2009	CO_2, GDP, GDP^2, REC, TR	ARDL	TR and REC mitigate CO_2
[79]	OECD countries	1996—2005	CO_2, GDP, CPI, REC	Threshold effect	GDP and REC contribute to CO_2 for lower threshold; REC mitigates CO_2 for upper threshold.
[103]	Panel of developed and developing countries	1984—2007	CO_2, GDP, Nuclear, REC	Cointegration test	Nuclear mitigates CO_2, REC and CO_2 contribute to CO_2
[104]	USA, Japan, France, Korea, Spain, Canada	1970—2007	CO_2, GDP, Nuclear	Johansen cointegration	Nuclear mitigates CO_2 in all countries, GDP mitigates CO_2 in USA, Canada and France
[105]	Italy	1960—2011	CO_2, GDP, REC, NREC, TR	Gregory-Hansen cointegration, ARDL model	REC and GDP mitigate CO_2, NREC contributes to CO_2
[19]	South America	1980—2010	CO_2, GDP, P, REC	FMOLS, Granger causality	GDP $\leftrightarrow$REC; $CO_2 \leftrightarrow$GDP; $CO_2 \leftrightarrow$REC;
[78]	USA	1960—2007	CO_2, GDP, Nuclear, REC	Toda-Yamamoto causality	Nuclear $\rightarrow CO_2$, $CO_2 \rightarrow$REC, GDP $\rightarrow$REC, $CO_2 \leftrightarrow$GDP

(continued)

Author	Country	Year	Variables	Methods	Results
[106]	USA	1965—2012	NUC, RE, CO_2 and GDP	VECM Granger causality	REC → CO_2
[107]	BRICS countries	1971—2010	REC, GDP, TR, CO_2	ARDL	REC ↔ GDP; CO_2→REC
[22]	7 Central American countries	1980—2010	REC, GDP, COAL PRICE, OIL PRICE	FMOLS and Regime-wise Granger-causality Tests	REC ↔ GDP REC ↔ CO_2

Table 3 Studies about the effect of technological innovation on environment-energy-growth nexus

Author	Country	Year	Variables	Methods	Results
[16]	China	1978—2008	GDP, REC, K, L, R&D	OLS	REC, K and R&D increase GDP
[45]	OECD countries	1990—2010	GDP, REC, K, L, R&D	OLS with FE	REC, K, L, R&D increase GDP
[52]	4 Nordic countries	1975—2012	GDP, R&D, CO_2, REC	VAR	R&D→REC, GDP →REC
[53]	Malaysia	1970—2009	TI, GDP, CO_2, CE	ARDL	TI mitigate NREC
[73]	New Zealand and Norway	1971—2010	TI, GDP, CO_2, CE	ARDL	TI affects clean energy-growth nexus
[13]	Malaysia	1980—2012	TI, EC, GDP, TR	ARDL	TI reduce EC

Note: CE (clean energy), EC (energy consumption), REC (renewable energy consumption), NREC (non-renewable energy consumption); R&D (Research and Development); OLS (Ordinary Least Squares), FMOLS (Fully Modified Ordinary Least Squares), DOLS (Dynamic Ordinary Least Squares), FE (Fixed effects), RE (Random effects), GDP (economic growth), CPI (consumer price index), P (population), TR (trade openness), FD (financial development), TI (technological innovation) URB (Urbanization), K (capital), L (labor); ↔ (bidirectional Granger causality), → (unidirectional Granger causality), ↛ (no Granger causality).

References

[1] K. Saidi, M. Ben Mbarek, Nuclear energy, renewable energy, CO_2 emissions, and economic growth for nine developed countries: evidence from panel Granger causality tests, Prog. Nucl. Energy 88(2016)364-374.

[2] E. Dogan, F. Seker, The influence of real output, renewable and non-renewable energy, trade and financial development on carbon emissions in the top renewable energy countries, Renew. Sustain. Energy Rev. 60 (2016) 1074-1085.

[3] P. Y. Chen, S. T. Chen, C. S. Hsu, C. C. Chen, Modeling the global relationships among economic growth, energy consumption and CO_2 emissions, Renew. Sustain. Energy Rev. 65 (2016) 420-431.

[4] Y. J. Zhang, The impact of financial development on carbon emissions: an empirical analysis in China, Energy Pol. 39 (4)(2010) 2197-2203.

[5] I. Ozturk, A. Acaravci, The long-run and causal analysis of energy, growth, openness and financial development on carbon emissions in Turkey, Energy Econ. 36 (2013) 262-267.

[6] P. Sadorsky, Renewable energy consumption, CO_2 emissions and oil prices in the G7 countries, Energy Econ. 31 (3)(2009) 456-462.

[7] M. H. Amini, B. Nabi, M. R. Haghifam, Load management using multi-agent systems in smart distribution network, in: Power and Energy Society General Meeting, 2013, pp. 1-5.

[8] K. G. Boroojeni, M. H. Amini, A. Nejadpak, S. S. Iyengar, B. Hoseinzadeh, C. L. Bak, A theoretical bilevel control scheme for power networks with large-scale penetration of distributed renewable resources, in: IEEE International Conference on Electro Information Technology, 2016, pp. 0510-0515.

[9] E. Dogan, Analyzing the linkage between renewable and non-renewable energy consumption and economic growth by considering structural break in time-series data, Renew. Energy 99 (2016) 1126-1136.

[10] K. Fisher-Vanden, G. H. Jefferson, H. Liu, Q. Tao, What is driving China's decline in energy intensity? Resourc. Energy Economic. 26 (1)(2004) 77-97.

[11] L. Hang, M. Tu, The impacts of energy prices on energy intensity: evidence from China, Energy Pol. 35 (5)(2007) 2978-2988.

[12] Z. Nan, M. D. Levine, L. Price, Overview of current energy-efficiency policies in China, Energy Pol. 38 (11)(2010) 6439-6452.

[13] K. Sohag, R. A. Begum, S. M. S. Abdullah, M. Jaafar, Dynamics of energy use, technological innovation, economic growth and trade openness in Malaysia, Energy 90(2015) 1497-1507.

[14] A. Jesus Lopez-Menendez, R. Perez, B. Moreno, Environmental costs and renewable energy: Re-visiting the environmental kuznets curve, J. Environ. Manag. 145 (2014) 368-373.

[15] S. Farhani, J. Ben Rejeb, Energy consumption, economic growth and CO_2 emissions: evidence from panel data for MENA region, Int. J. Energy Econ. Pol. 2 (2)(2012) 71-81.

[16] Y. Fang, Economic welfare impacts from renewable energy consumption: the China experience, Renew. Sustain. Energy Rev. 15 (9)(2011) 5120-5128.

[17] G. Bölük, M. Mert, Fossil & renewable energy consumption, GHGs (green-house gases) and economic growth: evidence from a panel of EU (European Union) countries, Energy 74 (2014) 439-446.

[18] F. Bilgili, E. Kocak, U. Bulut, The dynamic impact of renewable energy consumption on CO_2 emissions: a revisited Environmental Kuznets Curve approach, Renew. Sustain. Energy Rev. 54(2016) 838-845.

[19] N. Apergis, J. E. Payne, Renewable energy, output, carbon dioxide emissions, and oil prices: evidence from South America, Energy Sources B Energy Econ. Plann. 10 (3)(2015) 281-287.

[20] S. Dinda, D. Coondoo, Income and emission: a panel data-based cointegration analysis, Ecol. Econ. 57 (2)(2006) 167-181.

[21] U. Al-Mulali, B. Saboori, I. Ozturk, Investigating the environmental Kuznets curve hypothesis in Vietnam, Energy Pol. 76 (2015) 123-131.

[22] I. Ozturk, U. Al-Mulali, Investigating the validity of the environmental Kuznets curve hypothesis in Cambodia, Ecol. Indicat. 57 (2015) 324-330.

[23] U. Al-Mulali, I. Ozturk, The investigation of environmental Kuznets curve hypothesis in the advanced economies: the role of energy prices, Renew. Sustain. Energy Rev. 54 (2016) 1622-1631.

[24] U. Al-Mulali, S. A. Solarin, I. Ozturk, Investigating the presence of the environmental Kuznets curve (EKC) hypothesis in Kenya: an autoregressive distributed lag (ARDL) approach, Natural Hazards 80 (3)(2016) 1729-1747.

[25] A. Kasman, Y. S. Duman, CO_2 emissions, economic growth, energy consumption, trade and urbanization in new EU member and candidate countries: a panel data analysis, Econ. Modell. 44 (2015) 97-103.

[26] M. Sebri, O. Ben - Salha, On the causal dynamics between economic growth, renewable energy consumption, CO_2 emissions and trade openness: fresh evidence from BRICS countries, Renew. Sustain. Energy Rev. 39 (2014) 14-23.

[27] B. Lin, M. Moubarak, Renewable energy consumption - economic growth nexus for China, Renew. Sustain. Energy Rev. 40 (C)(2014) 111-117.

[28] A. Moheeldeen, Income-carbon emissions nexus for Middle East and North Africa countries: a Semi-parametric approach, Int. J. Energy Econ. Pol. 7 (2)(2017) 152-159.

[29] B. Saboori, A. Soleymani, CO_2 emissions, economic growth and energy consumption in Iran: a co-integration approach, Int. J. Environ. Sci. (2011).

[30] H. T. Pao, C. M. Tsai, Modeling and forecasting the CO_2 emissions, energy consumption, and economic growth in Brazil, Energy 36 (5) (2011) 2450-2458.

[31] L. Du, C. Wei, S. Cai, Economic development and carbon dioxide emissions in China: provincial panel data analysis, China Econ. Rev. 23 (2) (2012) 371-384.

[32] E. Dogan, B. Turkekul, CO_2 emissions, real output, energy consumption, trade, urbanization and financial development: testing the EKC hypothesis for the USA, Environ. Sci. Pollut. Res. 23 (2) (2016) 1203-1213.

[33] H. T. Pao, C. M. Tsai, Multivariate Granger causality between CO_2 emissions, energy consumption, FDI (foreign direct investment) and GDP (gross domestic product): evidence from a panel of BRIC (Brazil, Russian Federation, India, and China) countries, Energy 36 (1)(2011) 685-693.

[34] F. Seker, H. M. Ertugrul, M. Cetin, The impact of foreign direct investment on environmental quality: a bounds testing and causality analysis for Turkey, Renew. Sustain. Energy Rev. 52 (2015) 347-356.

[35] K. Gokmenoglu, N. Ozatac, B. M. Eren, Relationship between industrial production, financial development and carbon emissions: the case of Turkey, Procedia Economics & Finance 25(2015)463-470.

[36] T. Li, Y. Wang, D. Zhao, Environmental Kuznets Curve in China: new evidence from dynamic panel analysis, Energy Pol. 91 (2016) 138-147.

[37] J. He, P. Richard, Environmental kuznets curve for CO_2 in Canada, Ecological Economics 69 (5) (2010) 1083-1093.

[38] K. Gokmenoglu, N. Ozatac, B. M. Eren, Relationship between industrial production, financial development and carbon emissions: the case of Turkey, Procedia Economics and Finance 25(2015)463-470.

[39] F. Seker, H. M. Ertugrul, M. Cetin, The impact of foreign direct investment on environmental quality: a bounds testing and causality analysis for Turkey, Renew. Sustain. Energy Rev. 52 (2015) 347-356.

[40] S. Wang, Q. Li, C. Fang, C. Zhou, The relationship between economic growth, energy consumption, and CO_2 emissions: empirical evidence from China, Sci. Total Environ. 542 (Pt A)(2016) 360-371.

[41] A. Omri, CO_2 emissions, energy consumption and economic growth nexus in MENA countries: evidence from simultaneous equations models, Energy Econ. 40 (2013) 657-664.

[42] C. T. Tugcu, A. K. Tiwari, Does renewable and/or non-renewable energy consumption matter for total factor productivity (TFP) growth? Evidence from the BRICS, Renew. Sustain. Energy Rev. 65 (2016) 610-616.

[43] F. Kula, The long-run relationship between renewable electricity consumption and GDP: evidence from panel data, Energy Sources Part B 9 (2)(2014) 156-160.

[44] N. Apergis, J. E. Payne, Renewable energy, output, carbon dioxide emissions, and oil prices: evidence from South America, Energy Sources Part B 10(3) (2015) 281-287.

[45] R. Inglesi - Lotz, The impact of renewable energy consumption to economic growth: a panel data Application, Energy Econ. 53 (2015) 58-63.

[46] M. Ben Jebli, S. Ben Youssef, I. Ozturk, Testing environmental Kuznets curve hypothesis: the role of

renewable and non - renewable energy consumption and trade in OECD countries, Ecol. Indicat. 60 (2016) 824-831.

[47] S. Shafiei, R. A. Salim, Non-renewable and renewable energy consumption and CO_2 emissions in OECD countries: a comparative analysis, Energy Pol. 66 (2014) 547-556.

[48] J. Sulaiman, A. Azman, B. Saboori, The potentialof renewable energy: using the environmental kuznets curve model, Am. J. Environ. Sci. 9 (2) (2013) 103-112.

[49] N. Apergis, J. E. Payne, Renewable energy consumption and economic growth: evidence from a panel of OECD countries, Energy Pol. 38 (1)(2010) 656-660.

[50] A. J. L′opez-Men′endez, R. P′erez, B. Moreno, Environmental costs and renewable energy: re-visiting the Environmental Kuznets Curve, J. Environ. Manag. 145 (1)(2014) 368.

[51] C. L. Chiu, T. H. Chang, What proportion of renewable energy supplies is needed to initially mitigate CO_2 emissions in OECD member countries? Renew. Sustain. Energy Rev. 13 (6-7)(2009) 1669-1674.

[52] M. Irandoust, The renewable energy-growth nexus with carbon emissions and technological innovation: evidence from the Nordic countries,, Ecol. Indicat. 69 (2016) 118-125.

[53] C. F. Tang, E. C. Tan, Exploring the nexus of electricity consumption, economic growth, energy prices and technology innovation in Malaysia, Appl. Energy 104 (4)(2013) 297-305.

[54] A. Jones, An environmental assessment of food supply chains: a case study on dessert Apples, Environ. Manag. 30 (4)(2002) 560-576.

[55] P. K. Narayan, R. Smyth, Multivariate granger causality between electricity consumption, exports and GDP: evidence from a panel of Middle Eastern countries, Energy Pol. 37 (1)(2009) 229-236.

[56] S. Rafiq, R. Salim, I. Nielsen, Urbanization, openness, emissions, and energy intensity: a study of increasingly urbanized emerging economies, Energy Econ. 56 (2016) 20-28.

[57] H. M. Zhu, L. J. Duan, Y. W. Guo, K. M. Yu, The effects of FDI, economic growth and energy consumption on carbon emissions in ASEAN-5: evidence from panel quantile regression, Econ. Modell. 58 (2016) 237-248.

[58] R. Koenker, G. Bassett, Tests of linear hypotheses and l1 estimation, Econometrica 50 (6) (1982) 1577-1583.

[59] R. Koenker, Quantile regression for longitudinal data, J. Multivariate Anal. 91 (1)(2004) 74-89.

[60] C. Lamarche, Robust penalized quantile regression estimation for panel data, J. Econom. 157 (2)(2010) 396-408.

[61] A. F. Galvao Jr., Quantile regression for dynamic panel data with fixed effects, J. Econom. 164 (1) (2011) 142-157.

[62] I. A. Canay, A simple approach to quantile regression for panel data, Econom. J. 14 (3)(2011) 368-386.

[63] Y. J. Zhang, Y. L. Jin, J. Chevallier, B. Shen, The effect of corruption on carbon dioxide emissions in APEC countries: a panel quantile regression analysis, Technol. Forecast. Soc. Change 112 (2016) 220-227.

[64] U. Al-mulali, H. G. Fereidouni, J. Y. M. Lee, Electricity consumption from renewable and non-renewable sources and economic growth: evidence from Latin American countries, Renew. Sustain. Energy Rev. 30 (2014) 290-298.

[65] J. B. Ang, CO_2 emissions, energy consumption, and output in France, Energy Pol. 35 (10) (2007) 4772-4778.

[66] N. Apergis, J. E. Payne, The electricity consumption-growth nexus: renewable versus non-renewable electricity in Central America, Energy Sources Part B 7 (4)(2012) 423-431.

[67] O. Damette, P. Delacote, On the economic factors of deforestation: what can we learn from quantile analysis? Econ. Modell. 29 (6)(2012) 2427-2434.

[68] M. A. Boutabba, The impact of financial development, income, energy and trade on carbon emissions: evidence from the Indian economy, Econ. Modell. 40 (C)(2014) 33-41.

[69] A. Kasman, Y. S. Duman, CO_2 emissions, economic growth, energy consumption, trade and urbanization in new EU member and candidate countries: a panel data analysis, Econ. Modell. 44(44)(2015)97-103.

[70] M. Sebri, O. Ben-Salha, On the causal dynamics between economic growth, renewable energy consumption, CO_2 emissions and trade openness: fresh evidence from BRICS countries, Renew. Sustain. Energy Rev. 39 (39)(2014)14-23.

[71] W. H. You, H. M. Zhu, K. Yu, C. Peng, Democracy, Financial Openness, Global Carbon, Dioxide emissions: heterogeneity across existing emission levels, World Dev. 66 (2015) 189-207.

[72] H. Zhu, L. Duan, Y. Guo, K. Yu, The effects of FDI, economic growth and energy consumption on carbon emissions in ASEAN-5: evidence from panel quantile regression, Econ. Modell. 58 (2016) 237-248.

[73] Q. Fei, R. Rasiah, J. S. Leow, The impacts of energy prices and technological innovation on the fossil fuel-related electricity-growth nexus: an assessment of four net energy exporting countries, J. Energy South Afr. 25 (3)(2014) 37-46.

[74] G. Boluk, M. Mert, Fossil & renewable energy consumption, GHGs (green-house gases) and economic growth: evidence from a panel of EU (European Union) countries, Energy 74 (2014) 439-446.

[75] M. B. Jebli, S. B. Youssef, I. Ozturk, Testing environmental Kuznets curve hypothesis: the role of renewable and non-renewable energy consumption and trade in OECD countries, Ecol. Indicat. 60(2016)824-831.

[76] N. Apergis, J. E. Payne, K. Menyah, Y. Wolde-Rufael, On the causal dynamics between emissions, nuclear energy, renewable energy, and economic growth, Ecol. Econ. 69 (11)(2010) 2255-2260.

[77] G. Bölük, M. Mert, The renewable energy, growth and environmental Kuznets curve in Turkey: an ARDL approach,, Renew. Sustain. Energy Rev. 52 (2015) 587-595.

[78] K. Menyah, Y. Wolderufael, CO_2 emissions, nuclear energy, renewable energy and economic growth in the US, Energy Pol. 38 (6)(2010) 2911-2915.

[79] C. L. Chiu, T. H. Chang, What proportion of renewable energy supplies is needed to initially mitigate CO_2 emissions in OECD member countries? Renew. Sustain. Energy Rev. 13 (6-7)(2009) 1669-1674.

[80] M. Ben Jebli, S. Ben, Youssef, the environmental Kuznets curve, economic growth, renewable and non-renewable energy, and trade in Tunisia, Renew. Sustain. Energy Rev. 47 (2015) 173-185.

[81] V. G. R. Chandran, C. F. Tang, The impacts of transport energy consumption, foreign direct investment and income on CO_2 emissions in ASEAN-5 economies, Renew. Sustain. Energy Rev. 24 (2013) 445-453.

[82] B. Ozcan, The nexus between carbon emissions, energy consumption and economic growth in Middle East countries: a panel data analysis, Energy Pol. 62 (2013) 1138-1147.

[83] J. P. C. Sento, V. Moutinho, CO_2 emissions, non-renewable and renewable electricity production, economic growth, and international trade in Italy, Renew. Sustain. Energy Rev. 55 (2016) 142-155.

[84] U. Soytas, R. Sari, B. T. Ewing, Energy consumption, income, and carbon emissions in the United States, Ecol. Econ. 62 (3-4)(2007) 482-489.

[85] C. D. V. Costa, E. L. Rovere, D. Assmann, Technological innovation policies to promote renewable energies: lessons from the European experience for the Brazilian case, Renew. Sustain. Energy Rev. 12 (1)(2008) 65-90.

[86] S. Silva, I. Soares, O. Afonso, Economic and environmental effects under resource scarcity and substitution between renewable and non-renewable resources, Energy Pol. 54 (3)(2013) 113-124.

[87] F. Jamil, E. Ahmad, Income and price elasticities of electricity demand: aggregate and sector-wise analyses, Energy Pol. 39 (9)(2011) 5519-5527.

[88] J. Pan, G. Zhuang, Y. Zheng, Clarification of the concept of low-carbon economy and analysis of its core

elements, Int. Econ. Rev. (2010).

[89] S. Farhani, I. Ozturk, Causal relationship between CO_2 emissions, real GDP, energy consumption, financial development, trade openness, and urbanization in Tunisia, Environ. Sci. Pollut. Res. 22 (20)(2015) 15663-15676.

[90] A. Tamazian, J. P. Chousa, K. C. Vadlamannati, Does higher economic and financial development lead to environmental degradation: evidence from BRIC countries, Energy Pol. 37 (1)(2009) 246-253.

[91] U. Al-mulali, C. Weng-Wai, L. Sheau-Ting, A. H. Mohammed, Investigating the environmental Kuznets curve (EKC) hypothesis by utilizing the ecological footprint as an indicator of environmental degradation, Ecol. Indicat. 48 (2015) 315-323.

[92] A. Jalil, M. Feridun, The impact of growth, energy and financial development on the environment in China: a cointegration analysis, Energy Econ. 33 (2)(2011) 284-291.

[93] M. Shahbaz, S. Nasreen, F. Abbas, O. Anis, Does foreign direct investment impede environmental quality in high-, middle-, and low-income countries? Energy Econ. 51 (2015) 275-287.

[94] M. Shahbaz, N. Khraief, G. S. Uddin, I. Ozturk, Environmental Kuznets curve in an open economy: a bounds testing and causality analysis for Tunisia, Renew. Sustain. Energy Rev. 34 (2014) 325-336.

[95] E. Dogan, F. Seker, S. Bulbul, Investigating the impacts of energy consumption, real GDP, tourism and trade on CO_2 emissions by accounting for cross-sectional dependence: a panel study of OECD countries, Curr. Issues Tourism (2015).

[96] D. E, T. B, CO_2 emissions, real output, energy consumption, trade, urbanization and financial development: testing the EKC hypothesis for the USA, Environ. Sci. Pollut. Res. Int. 23(2)(2015) 1203-1213.

[97] M. S. Hossain, Panel estimation for CO_2 emissions, energy consumption, economic growth, trade openness and urbanization of newly industrialized countries, Energy Pol. 39(11)(2011)6991-6999.

[98] M. Shahbaz, Q. M. A. Hye, A. K. Tiwari, N. C. Leitao, Economic growth, energy consumption, fInancial development, international trade and CO_2 emissions in Indonesia, Renew. Sustain. Energy Rev. 25(2013) 109-121.

[99] J. B. Ang, Economic development, pollutant emissions and energy consumption in Malaysia, J. Pol. Model. 30(2)(2008)271-278.

[100] S. Wang, Q. Li, C. Fang, C. Zhou, The relationship between economic growth, energy consumption, and CO_2 emissions: empirical evidence from China, Sci. Total Environ. 542 (2016) 360-371.

[101] F. Bilgili, E. Koçak, ü. Bulut, The dynamic impact of renewable energy consumption on CO_2 emissions: a revisited Environmental Kuznets Curve approach, Renew. Sustain. Energy Rev. 54 (2016) 838-845.

[102] S. Shafiei, R. A. Salim, Non-renewable and renewable energy consumption and CO_2 emissions in OECD countries: a comparative analysis, Energy Pol. 66 (3)(2014) 547-556.

[103] N. Apergis, J. E. Payne, K. Menyah, Y. Wolde-Rufael, On the causal dynamics between emissions, nuclear energy, renewable energy, and economic growth, Ecol. Econ. 69 (11)(2010) 2255-2260.

[104] J. Baek, D. Pride, On the incomeenuclear energye CO_2 emissions nexus revisited, Energy Econ. 43 (2)(2014) 6-10.

[105] J. P. C. Bento, V. Moutinho, CO_2 emissions, non-renewable and renewable electricity production, economic growth, and international trade in Italy, Renew. Sustain. Energy Rev. 55(March)(2016)142-155.

[106] N. Apergis, J. E. Payne, Renewable and non-renewable electricity consumption-growth nexus: evidence from emerging market economies, Appl. Energy 88(12)(2011)5226-5230.

[107] N. B. Behmiri, J. R. Pires Manso, Does Portuguese economy support crude oil conservation hypothesis? Energy Pol. 45 (2012) 628-634.

A new approach for crude oil price prediction based on stream learning

GAO Shuang, LEI Yalin *

School of Humanities and Economic Management, China University of Geosciences, Beijing 100083, *China*

Abstract Crude oil is the world's leading fuel, and its prices have a big impact on the global environment, economy as well as oil exploration and exploitation activities. Oil price forecasts are very useful to industries, governments and individuals. Although many methods have been developed for predicting oil prices, it remains one of the most challenging forecasting problems due to the high volatility of oil prices. In this paper, we propose a novel approach for crude oil price prediction based on a new machine learning paradigm called stream learning. The main advantage of our stream learning approach is that the prediction model can capture the changing pattern of oil prices since the model is continuously updated whenever new oil price data are available, with very small constant overhead. To evaluate the forecasting ability of our stream learning model, we compare it with three other popular oil price prediction models. The experiment results show that our stream learning model achieves the highest accuracy in terms of both mean squared prediction error and directional accuracy ratio over a variety of forecast time horizons.

Keywords Crude oil, Economic geology, Prediction model, Machine learning, Stream learning

1 Introduction

Crude oil is a natural liquid fossil fuel found in geological formations beneath the earth's surface. It has mostly been extracted by oil drilling, which comes after the studies of structural geology, sedimentary basin analysis, and reservoir characterization (Guerriero et al., 2012). Crude oil is one of the most important energy resources on earth. So far, it remains the world's leading fuel, with nearly one-third of global energy consumption.

Crude oil prices are determined by many factors and have a big impact on the global environment and economy. Although crude oil prices were firm in early 2014, they fell sharply from mid 2014. In January 2016, the U. S. refiner acquisition cost for crude oil imports, as a proxy for world oil price, is only $28.81 per barrel on average, and the West Texas Intermediate (WTI) crude oil spot price, as the benchmark oil price in North America, is only $31.68 per barrel on

原文刊于 *Geoscienca Frontiers*，2017 年第 8 卷。原文发表时为彩色。

* 为通信作者。

average (EIA, 2016). The prices have dropped by more than seventy percent since June 2014.

The world's environment is affected by the oil price falling. With the drop of oil prices, the fuel bills are lowered. As a result, consumers are very likely to use more oil and thus increase the carbon emission. In addition, there is less incentive to develop renewable and clean energy resources. On the other hand, sustained low oil prices could lead to a drop in global oil and gas exploration and exploitation activities.

Fluctuating oil prices also play an important role in the global economy (Husain et al., 2015). The fallin oil prices would result in a modest boost to global economic activity, although the owners of oil sectors suffer income losses. Recent research from the World Bank shows that for every 30% decline of oil prices, the global GDP (Gross Domestic Product) would be increased by 0.5%. At the same time, the drop of oil prices would reduce the cost of living, and hence the inflation rate would fall.

There is no doubt that crude oil price forecasts are very useful to industries, governments as well as individuals. Thus, forecasting crude oil prices has been the subject of research by both academia and industry. Many methods and approaches have been developed for predicting oil prices. However, due to the high volatility of oil prices (Regnier, 2007), it remains one of the most challenging forecasting problems.

In recent years, machine learning techniques have been used in many applications in geosciences (Alavi et al., 2016). Machine learning provides powerful computational tools and algorithms that can learn from and make predictions on data. In this paper, we propose a novel approach for crude oil price prediction based on a new machine learning paradigm called stream learning (Gama et al., 2009; Bifet et al., 2010). The main advantage of our stream learning approach is that the prediction model can capture the changing pattern of oil prices since the model is continuously updated whenever new oil price data are available, with very small constant overhead. We compare our stream learning model with three other popular oil price prediction models for predicting two types of oil prices (the U. S. refiner acquisition cost for crude oil imports and the WTI crude oil spot price). The experiment results show that our stream learning model achieves the highest accuracy in terms of both mean squared prediction error and directional accuracy ratio over a variety of forecast time horizons.

2 Methods

2.1 Literature review

We divide crude oil price forecasting approaches into three categories: (1) heuristic approaches; (2) econometric models; and (3) machine learning techniques.

Heuristic approaches for oil price prediction include professional and survey forecasts, which are mainly based on professional knowledge, judgments, opinion and intuition. Another heuristic approach, the so-called no-change forecast, uses the current price of oil as the best prediction of future oil prices. Despite its simplicity, the no-change forecast appeared to be a good baseline approach for oil price prediction and was better than other heuristic judgmental approaches (Alquist et al., 2013).

Econometric models are the most widely used approaches for oil price prediction, which include autoregressive moving average (ARMA) models and vector autoregressive (VAR) models, with possibly different input variables (Pindyck, 1999; Frey et al., 2009). These econometric models provide more accurate prediction than the no-change model at least at some horizons (Alquist et al., 2013; Baumeister and Kilian, 2015). Recently, a forecast combination approach was proposed by Baumeister and Kilian (2015), which combines 6 different oil price prediction models including both econometric models (such as the VAR model) and the no-change model. It should be noted that most of the econometric models are linear models and are not be able to capture the nonlinearity of oil prices.

Several machine learning techniques were proposed for oil price prediction, such as artificial neural networks (ANN) (Yu et al., 2008; Kulkarni and Haidar, 2009), and support vector machine (SVM) (Xie et al., 2006). These are nonlinear models which may produce more accurate predictions if the oil price data are strongly nonlinear (Behmiri and Pires Manso, 2013). However, these machine learning techniques, like other traditional machine learning techniques, rely on a fixed set of training data to train a machine learning model and then apply the model to a test set. Such an approach works well if the training data and the test data are generated from a stationary process, but may not be effective for non-stationary time series data such as oil price data.

2.2 A new stream learning approach

In this paper, we propose a new stream learning approach for oil price prediction. Unlike traditional machine learning algorithms that use "one-shot" data analysis and focus on homogeneous and stationary data, stream learning algorithms have been developed to handle applications where continuous data streams are generated from non-stationary processes (Gama et al., 2009; Bifet et al., 2010).

Under the traditional supervised machine learning framework, one typically splits labeled data examples into a training set and a test set. The training set is used to train a machine learning model using a machine learning technique such as ANN or SVM. The test set then is used to test the performance of the machine learning model (note that sometimes there is also a development set which is used to tune the parameters of the machine learning model). For such an approach to be useful, there is an underlying assumption that the data examples in the training set and in the test set are homogeneous (e.g., generated from a stationary process), so that the trained machine learning model can capture the pattern of the data examples in the test set and produce accurate predictions for them. However, oil price time series data are not stationary, and thus traditional machine learning approaches may not produce accurate predictions.

In recent years, a new machine learning paradigm called stream learning has emerged to handle real-world applications where (1) there is a continuous flow of data as opposed to a fixed sample of independent and identically distributed (i. i. d.) examples, and (2) the data are generated by a non-stationary process instead of a stationary process (potentially at very high speed). Examples of stream learning applications include social networks, web mining, scientific data, financial data, etc.

Suppose we start with a set of initial training data which include a sequence of historical oil prices and the associated input data vectors, denoted by $D = \{(x_{-m+1}, y_{-m+1}), (x_{-m+2}, y_{-m+2}), \cdots, (x_{-1}, y_{-1}), (x_0, y_0)\}$, whereytis the oil price at timeslot t, and x_t is a vector of input data variables for predicting y_t. In this paper, we use the oil prices of the previous o timeslots to predict the oil price of timeslot t, i. e. , $x_t = (y_{t-o}, \cdots, y_{t-2}, y_{t-1})$, where o ranges from 1 to 10. Our framework is applicable to general time slot unit (also called forecast time horizon), which could be daily, weekly, monthly, quarterly, etc.

Suppose we want to predict the oil prices for the next n time slots, denoted by $y_1, y_2, \cdots, y_n$ and again letxtbe input data vector for predicting y_t, $t=1, \cdots, n$. We propose the following stream learning procedure for predicting oil prices.

Stream learning procedure

1. Use the data in the initial training set D to train a machine learning model, denoted by M_1, and use M_1 to predict the oil price for time slot 1.

2. For time slot t, $t = 2, \cdots, n$:

Add (x_{t-1}, y_{t-1}) to the training set D and update the machine learning model, denoted by M_t, and use M_t to predict the oil price for time slot t.

Our stream learning approach is a supervised machine learning method which uses a set of labeled training data to train an initial model. The main features and advantages of using stream learning for oil price prediction include:

(1) The machine learning model will be updated whenever new oil price data are available, so the model continuously evolves over time, and can capture the changing pattern of oil prices.

(2) For non-stationary time series data such as oil prices, a forgetting mechanism (e. g. , sliding windows, fading factors) will be deployed when updating the machine learning model.

(3) Updating the model requires only a small constant time per new data example, as opposed to re-training the model using the entire training data set.

Massive Online Analysis (MOA) is an open-source framework software that allows to build and run experiments of machine learning or data mining on evolving data streams (Bifet et al. ,

2010, 2012). We use MOA to develop stream learning models for oil price prediction. We have tried several machine learning models, including linear regression, perceptron and regression trees. To tune the parameters of a specific model in searching for the optimal solution, we split the initial training data into 90% as the training set and 10% as the development set. The development set is used to tune the model parameters, which is a standard approach in machine learning. We find that peceptron achieves the best accuracy for predicting oil prices and hence we use it as the core machine learning method within MOA.

2.3 Data transformation

A stationary process is a stochastic process whose joint probability distribution does not change when shifted in time (Priestley, 1988). Therefore, parameters of a stationary process such as the mean and variance do not change over time and do not follow any trends. A weaker form of stationarity is known as weak-sense stationarity, or wide-sense stationarity, which only requires that the mean and autocovariance do not vary with respect to time.

Under the traditional machine learning framework, to build a statistically sound machine learning model, the data time series need to be stationary or at least weakly stationary over the whole training and evaluation period. However, the oil price time series are not stationary. Under our stream learning approach, we no longer require that the oil price time series be stationary over the whole training and evaluation period, because the machine learning model will be continuously updated over time to capture the changing pattern of oil price time series. Nevertheless, when we update the machine learning model and apply the model to predict the oil price for the next time slot, there is still an underlying assumption that oil prices are stationary over the relatively shorter time period covering the current training data time slots and the next timeslot.

In statistics, data transformation techniques are usually applied so that the data are more likely to meet the assumptions of a statistical inference procedure. For time series data, it is common to difference the data to improve stationarity (Priestley, 1988; Witten et al., 2011). Therefore, we apply the following data transformation technique for the oil price time series:

$$y'_t = y_t - y_{t-1}$$

and we find that it improves the accuracy of oil price prediction models.

3 Results and discussion

In this section, we evaluate the performance of our stream learning model and compare it with three popular oil price prediction models.

3.1 Data and other prediction models

We use two types of oil price data to evaluate the accuracy of different oil price prediction models. The first one is the U. S. refiner acquisition cost for crude oil imports, which is the weighted average cost of all oil imported into the U. S. It can be viewed as a proxy for world oil price. The second one is the WTI crude oil spot price, which is used as a benchmark in oil pricing in North America, and is commonly cited in the press. Both data can be obtained from the U. S. 6 Energy Information Administration (EIA) website (EIA, 2016).

We compare our stream learning model with three other oil price prediction models. The first one is the no-change model, which is a commonly used heuristic model for oil price prediction. The second one is a model trained with artificial neural networks (ANN), which is a classical machine learning model for oil price prediction (Yu et al., 2008; Kulkarni and Haidar, 2009). The third one is the forecast combination model proposed by Baumeister and Kilian (2015), which is a state-of-the-art econometric oil price prediction model.

To have a robust evaluation, we apply the same long evaluation period from January 1992 to September 2012 as in Baumeister and Kilian (2015). For a fair comparison, we use the same sets of training data and test data for evaluating our stream learning method and the other two state-of-the-art supervised learning methods (the ANN model and the forecast combination model). Specifically, for predicting WTI crude oil price, we have used oil prices from January 1986 to December 1991 as training data, and oil prices from January 1992 to September 2012 as test data. For predicting U. S. refiner acquisition cost for crude oil imports, we have used oil prices from January 1974 to December 1991 as training data, and oil prices from January 1992 to September 2012 as test data.

The forecast time horizons include 1-month, 3-month, 6-month, 9-month and 12-month. Note that monthly oil prices are calculated by EIA based on daily data by taking an unweighted average of the daily closing spot prices for a given product over the specified time period. In Fig. 1 we plot monthly WTI crude oil spot price (both original oil price time series and time series after difference transformation) from January 1986 to January 2016.

3.2 Performance metrics

We use two standard performance metrics in the oil price prediction literature for comparing different oil price prediction models. The first metric is Mean Squared Prediction Error (MSPE). MSPE of a prediction model measures the average of the squares of the prediction errors. The prediction error is the difference between the true value and the predicted value. Let y_1, y_2, $\cdots$, y_n be the true oli prices and $\hat{y}_1$, $\hat{y}_2$, $\cdots$, $\hat{y}_n$ be the predicted oil prices under an oil price prediction model, then the MSPE of that model is:

$$\text{MSPE} = \frac{1}{n}\sum_{i=1}^{n}(\hat{y}_i - y_i)^2$$

For comparison purposes, we use the no-change model as the baseline model and express the MSPE of another model as a ratio relative to the MSPE of the no-change model. If the MSPE ratio of a model is less than 1, then the model is more accurate than the no-change model in terms of MSPE.

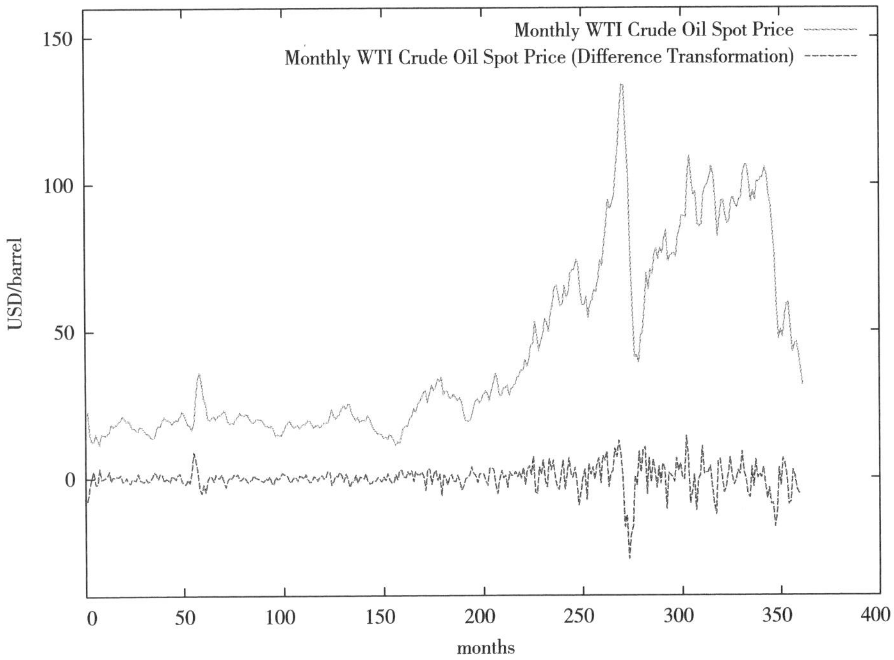

Fig. 1 Monthly WTI crude oil spot price in U. S. dollar per barrel (original oil price time series and time series after difference transformation) from January 1986 to January 2016

The second metric is Directional Accuracy Ratio (DAR), which measures the accuracy of predicting the direction of oil price change (i. e., whether oil price increases or decreases in the next time slot). It can be computed as follows:

$$\mathrm{DAR} = \frac{1}{n}\sum_{i=1}^{n} \mathrm{d}t$$

where $\mathrm{d}t = 1$ if $(\hat{y}_t - y_{t-1})(y_t - y_{t-1}) > 0$ and $\mathrm{d}t = 0$ otherwise. Note that if we do a random guess of the oil price direction by tossing a fair coin, the DAR would be 0. 5. Thus, if the DAR of a model is greater than 0. 5, then the model is better than a random guess.

3.3 Evaluation

Tables 1 and 2 summarize the results for predicting the U. S. refiner acquisition cost for crude oil imports. Our stream learning model achieves the lowest MSPE and the highest DAR among all prediction models for the 1-month, 6-month, 9-month and 12-month forecast time horizons. For example, for the 1-month time horizon, the MSPE of our stream learning model is 11. 87, with an error reduction of 31. 1% compared with the no-change model (17. 23), an error reduction of 13. 7% compared with the ANN model (13. 75), and an error reduction of 25. 3% compared with the forecast combination model in Baumeister and Kilian (2015) (15. 89). The DAR of our stream learning model is 0. 594, which is higher than both ANN model (0. 582) and forecast combination model (0. 570). For the 3-month horizon, although our model has a lower accuracy than the forecast combination model, it is more accurate than the no-change model, and more accurate than the ANN model in terms of MSPE.

Table 1 MSPE of different oil price prediction models: U. S. refiner Acquisition Cost for Crude Oil Imports

Time horizon		No-change	ANN	Forecast combination	Stream learning
1-month	MSPE	17.23	13.75	15.89	11.87
	Ratio	1	0.798	0.922	0.689
3-month	MSPE	97.13	96.92	88.00	95.37
	Ratio	1	0.998	0.906	0.981
6-month	MSPE	221.75	205.21	212.21	156.07
	Ratio	1	0.925	0.957	0.704
9-month	MSPE	277.09	276.06	262.68	183.79
	Ratio	1	0.996	0.948	0.663
12-month	MSPE	307.72	303.23	280.64	176.02
	Ratio	1	0.985	0.912	0.572

Notes: MSPE stands for Mean Squared Prediction Error, lower is more accurate. Boldface indicates the best performance score among all four prediction models for a given forecast time horizon.

Table 2 DAR of different oil price prediction models: U. S. refiner Acquisition Cost for Crude Oil Imports

Time horizon		Random guess	ANN	Forecast combination	Stream learning
1-month	DAR	0.5	0.582	0.570	0.594
3-month	DAR	0.5	0.586	0.592	0.562
6-month	DAR	0.5	0.510	0.556	0.598
9-month	DAR	0.5	0.594	0.575	0.602
12-month	DAR	0.5	0.643	0.627	0.667

Notes: DAR stands for Directional Accuracy Ratio, higher is more accurate. Boldface indicates the best performance score among all four prediction models for a given forecast time horizon.

Tables 3 and 4 summarize the results for predicting the WTI crude oil spot price. Again, our stream learning model achieves the highest accuracy among all prediction models for all forecast time horizons except for the 3-month time horizon. For example, for the 1-month time horizon, the MSPE of our stream learning model is 17.16, with an error reduction of 16.2% compared with the no-change model (20.48), an error reduction of 8.7% compared with the ANN model (18.79), and an error reduction of 8.0% compared with the forecast combination model (18.66). The DAR of our stream learning model is 0.570, which is also more accurate than both the ANN model (0.542) and the forecast combination model (0.521).

Table 3 MSPE of different oil price prediction models: West Texas Intermediate (WTI) Crude Oil Price

Time horizon		No-change	ANN	Forecast combination	Stream learning
1-month	MSPE	20. 48	18. 79	18. 66	17. 16
	Ratio	1	0. 917	0. 911	0. 838
3-month	MSPE	101. 81	101. 77	92. 24	99. 58
	Ratio	1	0. 999	0. 906	0. 978
6-month	MSPE	233. 03	226. 61	224. 17	181. 47
	Ratio	1	0. 972	0. 962	0. 778
9-month	MSPE	293. 54	290. 25	279. 45	214. 34
	Ratio	1	0. 989	0. 952	0. 730
12-month	MSPE	328. 31	307. 62	300. 73	234. 98
	Ratio	1	0. 937	0. 916	0. 716

Notes: MSPE stands for Mean Squared Prediction Error, lower is more accurate. Boldface indicates the best performance score among all four prediction models for a given forecast time horizon.

Table 4 DAR of different oil price prediction models: West Texas Intermediate (WTI) Crude Oil Price

Time horizon		Random guess	ANN	Forecast combination	Stream learning
1-month	DAR	0. 5	0. 542	0. 521	0. 570
3-month	DAR	0. 5	0. 586	0. 576	0. 546
6-month	DAR	0. 5	0. 469	0. 543	0. 602
9-month	DAR	0. 5	0. 502	0. 571	0. 606
12-month	DAR	0. 5	0. 526	0. 605	0. 610

Notes: DAR stands for Directional Accuracy Ratio, higher is more accurate. Boldface indicates the best performance score among all four prediction models for a given forecast time horizon.

4 Conclusion

Forecasting crude oil prices is a very challenging problem due to the high volatility of oil prices. In this paper, we developed a new oil price prediction approach using ideas and tools from stream learning, a machine learning paradigm for analysis and inference of continuous flow of non-stationary data. Our stream learning model will be updated whenever new oil price data are available, so the model continuously evolves over time, and can capture the changing pattern of oil prices. In addition, updating the model requires only a small constant time per new data example, as opposed to re-training the model using the entire training data set. The experiment results show that our stream learning model outperformed three other popular oil price prediction models over a variety of forecast time horizons.

Acknowledgments This research was supported by grants from the National Natural Science Foundation of

China (Grant No. 71173200), and the Strategic Research Center for Oil and Gas Resources, Ministry of Land and Resources of the People's Republic of China (Grant No. 2014BJYQ03).

References

Alavi, A. H., Gandomi, A. H., Lary, D. J., 2016. Progress of machine learning in geosciences: preface. Geoscience Frontiers 7(1), 1-2.

Alquist, R., Kilian, L., Vigfusson, R. J., 2013. Forecasting the price of oil. In: Elliott, G., Timmermann, A. (Eds.), Handbook of Economic Forecasting. North Holland, Amsterdam, pp. 427-507.

Baumeister, C., Kilian, L., 2015. Forecasting the real price of oil in a changing worlde a forecast combination approach. Journal of Business and Economic Statistics 33 (3), 338-351.

Behmiri, N. B., Pires Manso, J. R., 2013. Crude oil price forecasting techniques: a comprehensive review of literature. Alternative Investment Analyst Review 3, 18-29.

Bifet, A., Holmes, G., Kirkby, R., Pfahringer, B., 2010. MOA: massive online analysis. The Journal of Machine Learning Research 11, 1601-1604.

Bifet, A., Kirkby, R., Kranen, P., Reutemann, P., 2012. Massive Online Analysis Manual. Centre for Open Software Innovation. Available: http://ufpr. dl. sourceforge. net/project/moa - datastream/documentation/Manual. pdf.

EIA, 2016. U. S. Energy Information Administration [Internet]. [cited 2016 Mar 12]. Available: http: //www. eia. gov/petroleum/data. cfm#prices.

Frey, G., Manera, M., Markandya, A., Scarpa, E., 2009. Econometric models for oil price forecasting: a critical survey. CESifo Forum 1, 29-44.

Gama, J., Sebastiao, R., Rodrigues, P. P., 2009. Issues in Evaluation of Stream Learning Algorithms. Proceedings of the 15th ACM SIGKDD International Conference on Knowledge Discovery and Data Mining, Jun 28eJul 1. France, Paris, pp. 329-338.

Guerriero, V., et al., 2012. A permeability model for naturally fractured carbonate reservoirs. Marine and Petroleum Geology 40 (2), 115-134.

Husain, A. M., et al., 2015. Global implications of lower oil prices. IMF Survey Magazine 1-41.

Kulkarni, S., Haidar, I., 2009. Forecasting model for crude oil price using artificial neural networks and commodity future prices. International Journal of Computer Science and Information Security 2 (1).

Pindyck, R. S., 1999. The long-run evolution of energy prices. The Energy Journal 1-27.

Priestley, M. B., 1988. Non-linear and Non-stationary Time Series Analysis. Academic Press.

Regnier, E., 2007. Oil and energy price volatility. Energy Economics 29 (3), 405-427.

Witten, I. H., Frank, E., Hall, M. A., 2011. Data Mining: Practical Machine Learning Tools and Techniques, third ed. Morgan Kaufmann.

Xie, W., Yu, L., Xu, S., Wang, S., 2006. A New Method for Crude Oil Price Forecasting Based on Support Vector Machines. Proceedings of the 6th International Conference on Computational Science (ICCS), May 28-31, Reading, UK, pp. 444-451.

Yu, L., Wang, S. Y., Lai, K. K., 2008. Forecasting crude oil price with an EMD-based neural network ensemble learning paradigm. Energy Economics 30 (5), 2623-2635.

Mechanism of directed technological investment on energy productivity and energy structure: A unified theoretical framework

SUN Xiaojun[1,3,5], FAN Yee Van[2,3], LEI Yalin[1,4,5,*], PAN Ting[3], VARBANOV Petar Sabev[6]

1. *School of Economics and Management, China University of Geosciences, Beijing* 100083, *China*
2. *Environmental Change Institute (ECI), University of Oxford, Oxford OX*1 3*QY, United Kingdom*
3. *Sustainable Process Integration Laboratory-SPIL, NETME Centre, Faculty of Mechanical Engineering, Brno University of Technology-VUT Brno, Technická* 2896/2, 616 69 *Brno, Czech Republic*
4. *School of Economics and Management, Beijing University of Chemical Technology, Beijing* 100029, *China*
5. *Key Laboratory of Carrying Capacity Assessment for Resource and Environment, Ministry of Natural Resources of the People's Republic of China, Beijing* 100083, *China*
6. *Széchenyi István University, Györ* H-9026, *Hungary*

Abstract The mechanisms and effects of technological investment on energy productivity and energy structure in the petrochemical industry remain unclear due to the directional nature of technological progress. This study proposes a unified theoretical framework for the impact of directed technological investment on energy productivity and energy structure by incorporating energy factors into the theory of technological progress bias. The aim is to elucidate the impact of technological progress on energy productivity and energy structure, and to unravel the underlying effect mechanisms. A fixed effects model that includes moderating effects is also developed to support the assessment. The study found that the petrochemical industry's technological investment in China was initially biased towards enhancing labour-augmenting technological progress. The mechanism analysis revealed that technological investment, under the moderating effects of price and environmental governance, preferred a capital-energy bias, leading to insignificant improvements in energy productivity but a substantial increase in labour productivity. In addition, the technological investment, influenced by the moderating effect of environmental governance, led to some improvement in the energy structure during the sample period. This study integrates the mechanisms of directed technological investment on energy productivity and energy structure into a unified analytical framework, systematically investigating the reasons, effect mechanisms, and consequences of bias, while providing empirical evidence that supports low-carbon development in the petrochemical industry.

原文刊于 *Energy Economics*, 2024 年第 40 卷。原文发表时为彩色。

* 为通信作者。

Keywords Directed technological investment, Factor-augmenting technological progress, Factor-biased technological progress, Energy productivity, Energy structure

1 Introduction

Global climate warming has become a severe issue, exacerbated by carbon emissions from energy activities. Existing studies generally concur that technological investment is crucial in enhancing energy productivity and improving the energy structure (Guo et al., 2016; Lin and Teng, 2024; Yu et al., 2023). However, some scholars hold different views, arguing that current industrial technological investments are driven by profit motives to develop energy-consuming technologies rather than energy-saving ones, highlighting the directional characteristics of technological investment (Acemoglu et al., 2012; Lyu and Ke, 2022). These directional characteristics lead to mechanistic divergences in how technological investment impacts energy productivity and energy structure.

The existing literature on the impact of technological progress on energy productivity can be divided into two main categories: the first category explores the effect of neutral technological progress on energy productivity within the framework of neoclassical growth theory (Keen et al., 2019; Oryani et al., 2021; Sun et al., 2012; Yuan et al., 2009; Zhu et al., 2019), while the second category investigates the impact of biased technological progress (BTP) on energy productivity based on endogenous growth theory (Chen and Liu, 2021; Qiu et al., 2023; Zhang and Yang, 2024). However, neutral and exogenous technological progress in the neoclassical growth theory overlooks the fact that technological progress exhibits biases towards different factors (Hou et al., 2021) and the issue of factor substitution caused by BTP (Li and Lin, 2018). Therefore, neutral technological progress fails to explain whether technological progress in a country, region, or industry is shifting from extensive (energy-consuming) to energy-saving modes (Lyu and Ke, 2022; Zhang and Ke, 2022).

Some scholars have proposed new insights and perspectives in response to the economic phenomena that are difficult to explain using neutral technological progress within the neoclassical growth theory framework. These scholars argue that the neoclassical growth model, in order to derive steady-state growth, assumes a production function in Cobb-Douglas form and assumes technological progress to be neutral, which does not correspond to reality (Acemoglu, 2002; Acemoglu et al., 2018, 2012; León-Ledesma et al., 2010). In reality, technological progress is mostly not neutral. It will be affected by price effects and market scale effects and tend to increase the productivity of a certain factor, that is, technological progress is biased (Acemoglu, 2007; Zhang and Yang, 2024).

This assertion has been supported by extensive research. For example, studies by Sanstad et al. (2006) and Hassler et al. (2012) suggest that industrial technological progress in India,

South Korea, and the United States is biased towards energy conservation. In the case of factor complementarity, energy - enhancing technological progress is beneficial for improving productivity in energy-intensive industries. Hou et al. (2021), based on a growth model of directed technical change, found that in Portugal, most economic sectors undergo technological changes that are not biased towards energy conservation, but rather towards improving labour productivity. Li et al. (2024) studied the impact of BTP on energy productivity and labour productivity. The research found that current technological progress in China is biased towards labour and energy conservation, resulting in the fastest increase in labour productivity, followed by energy productivity, and the weakest in capital productivity.

From the above, it is apparent that recent literature, after incorporating energy factors into the framework of endogenous growth theory, has conducted more in-depth analyses on the mechanisms through which the bias of technological progress influences energy productivity. However, it still predominantly focuses on the effects of market scale and price effects on the bias of technological progress, while neglecting the environmental governance effects brought about by energy factors, which could also influence the bias of technological progress.

On the other hand, existing research on the impact of technological progress on energy structure improvement is relatively limited, but its importance should not be underestimated (Yang et al., 2020). Dong et al. (2021) pointed out that the development of new energy technologies can enhance the use of clean energy sources like photovoltaic and wind power, thereby improving the energy structure. Chen et al. (2021) argued that technological progress is one of the main driving forces for optimizing and adjusting the energy structure and developing a low-carbon economy. Zhan (2023), based on input-output data from 30 provinces and cities in China, used the Fixed Effects Least Squares Dummy Variables (FELSDV) methods to study and found that improvements in the level of technological progress can drive an increase in the proportion of renewable energy consumption. Zhao et al. (2024) conducted research based on national panel data from 77 countries between 2002 and 2020. They explored how technological progress can synergize with climate aid to promote energy structure transformation and industrial structure upgrading, thereby mitigating carbon lock-in.

The literature mentioned above shows that research and development (R&D) investment biased towards clean technologies can optimize energy consumption structure. This is achieved by influencing the marginal substitution between energy sources, prompting industry producers to increase inputs of clean energy types to replace inputs of non-clean energy. However, there is currently a lack of literature on research into the highly energy-intensive petrochemical industry (Rootzén et al., 2023). It remains unclear whether R&D investment in this sector tends towards technological progress in clean energy or non-clean energy, and the mechanism behind this bias in R&D investment (i.e., why technological progress tends towards this type of energy) remains unclear.

The petrochemical industry, as a pillar of national economic growth, not only drives sustained economic development but also provides essential raw material support to various sectors. However, as a typical energy-intensive industry, the directional characteristics of its technological investments have a particularly pronounced impact on energy-related activities. During the period 2000—2022, technological investments in the Chinese petrochemical industry (such as R&D investment, patent applications, and the number of effective invention patents) have shown a rapid upward trend. Nevertheless, the improvement in energy productivity has been minimal, and the gap between energy productivity of the petrochemical industry and that of the overall industrial sector has widened further since 2009. Concurrently, the proportion of coal consumption has exhibited an upward trend followed by a decline from 2000 to 2021, indicating a less pronounced improvement in the energy structure (see Fig. 1). This raises a significant question: what are the underlying mechanisms of the impact of technological investment on energy productivity and energy structure? How can these three factors be integrated into a unified theoretical framework for research purposes?

The marginal contributions of this paper compared to existing literature lie in three aspects:

(1) In terms of theoretical framework, existing scholars predominantly explore environmental issues within the technological progress bias theory framework, focusing more on the bias of technological progress between energy and non-energy (capital, labour) production factors. This paper aims to investigate whether technological progress in the petrochemical industry exhibits bias among different production factors (capital-labour-energy) and examine whether there is bias in the selection of inputs for different types of energy (clean energy or non-clean energy) within this theoretical framework. This integrates energy productivity and energy structure mechanisms into a unified theoretical framework.

(2) In terms of theoretical mechanisms, this paper incorporates the environmental governance effect within the framework of technological progress bias theory, thereby complementing and expanding upon the original mechanisms (market scale effect and price effect) within the theoretical framework.

(3) In terms of research focus, relatively little literature has explored the directional characteristics of technological investment in the Chinese petrochemical industry, particularly whether these characteristics have driven technological progress toward energy consumption, leading to a further widening gap between the petrochemical industry and overall industrial energy productivity. Additionally, there is a lack of investigation into whether there is a bias in allocating different energy inputs. Thus, this issue remains unresolved.

The structure of the remainder of this paper is as follows: In Section 2, we construct a theoretical framework based on the unresolved key issues identified and the existing literature gaps in Section 1. Section 3 outlines the design of econometric models and data preprocessing based on the theoretical framework. Section 4 presents the results and discussions. This is

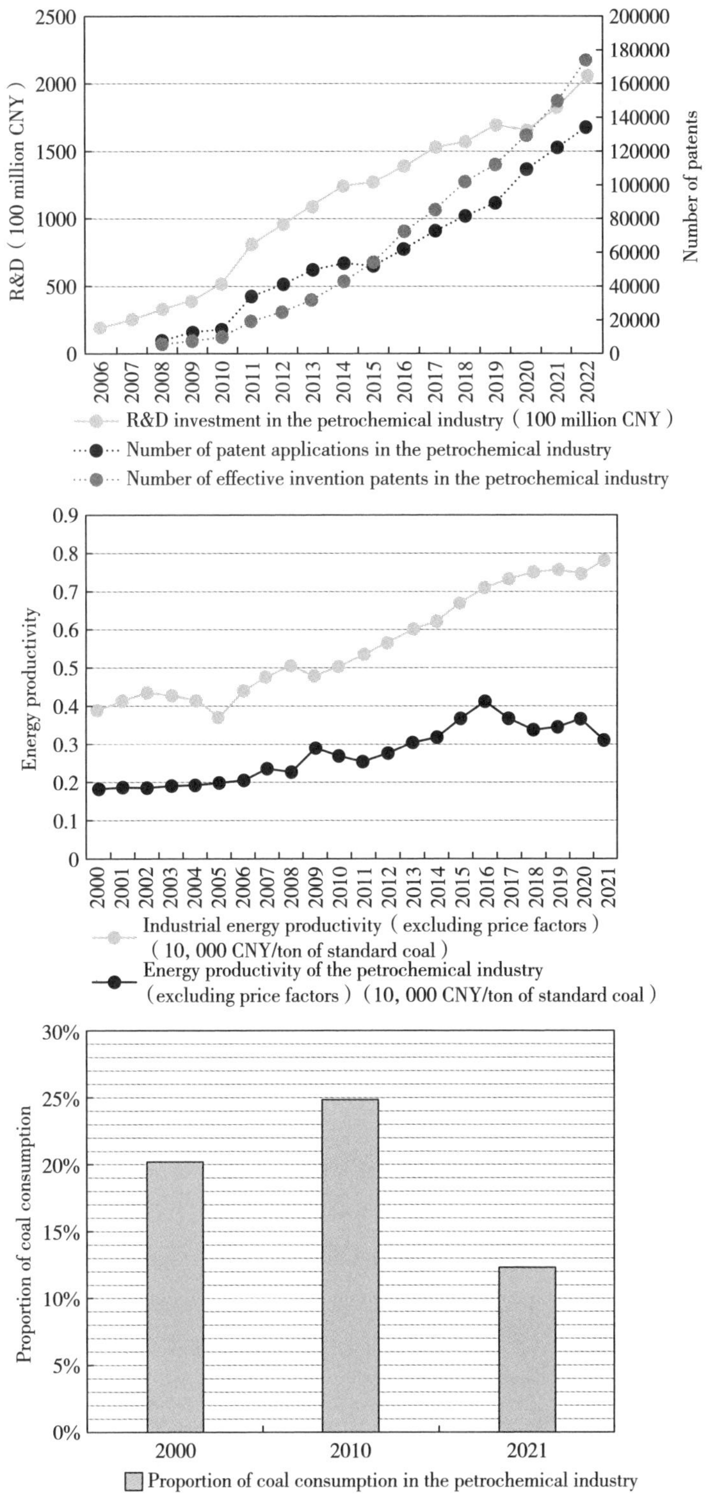

Fig. 1　Diagram of technological investment, energy productivity, and energy structure in the Chinese petrochemical industry

Data Source: National Bureau of Statistics of China (NBSC) and Carbon Emission Accounts and Datasets (CEADs)

followed by Section 5, which contains the conclusions and recommendations.

2 Theoretical framework

To investigate the mechanism and effects of directed technological investment on energy productivity and energy structure of the Chinese petrochemical industry, this study integrates the analytical framework of technological progress bias proposed by Acemoglu (2002), Acemoglu et al. (2016, 2012), Hémous (2016), and Zhou et al. (2024) with energy factors. It constructs a theoretical model of technological progress biased for labour, capital, and energy factors in the Chinese petrochemical industry, guided by the core principles of endogenous economic growth theory. In the original framework of biased technological progress, which primarily focused on labour and energy as the two key factors, the aim was to explore the effects of technological investment on unemployment (Adachi et al., 2019), income distribution (Perera-Tallo, 2017), and skill premium (Mallick and Sousa, 2017). By incorporating the energy factor, this paper not only introduces the mechanism of environmental governance effects but also investigates how directed technological investment influences the energy structure. This study integrates the impact mechanisms of energy productivity and energy structure into a unified theoretical framework, thus providing a constructive expansion of the original biased technological progress framework (see Fig. 2).

The specific mechanisms of influence are as follows: (1) The impact of technological investment (R&D investment) on energy productivity involves two transmission processes. Initially, technological investment enhances the different factor-augmenting technological progress (ATP) directionally under the moderation of three major effects, as described in equations (18) and (23). Subsequently, as detailed in equations (4) and (5), different factor-ATP transforms into different factor-BTP influenced by factor substitution elasticity, thus affecting energy productivity. (2) The improvement of energy structure is directly influenced by energy prices, and also indirectly affected by the indirect influence of technological investment under the moderation of environmental governance effects, as detailed in equation (26).

This paper, based on the CES production function with capital and labour as the two key factors constructed by Farajzadeh et al. (2023), incorporates an energy factor to establish a three-factor constant elasticity of substitution (CES) production function for the petrochemical industry as follows:

$$Y_t(L_t,\ K_t,\ E_t) = \{\gamma(A_{L_t}L_t)^{\frac{\sigma-1}{\sigma}} + (1-\gamma)[\tau(A_{K_t}K_t)^{\frac{\varphi-1}{\varphi}} + (1-\tau)(A_{E_t}E_t)^{\frac{\varphi-1}{\varphi}}]^{\frac{(\sigma-1)/\sigma}{(\varphi-1)/\varphi}}\}^{\frac{\sigma}{\sigma-1}} \tag{1}$$

Due to the addition of energy factors, the nesting pattern of factor combinations has changed to three types: L (KE)-type, E (KL)-type, and K (EL)-type. Sun et al. (2024b) conducted

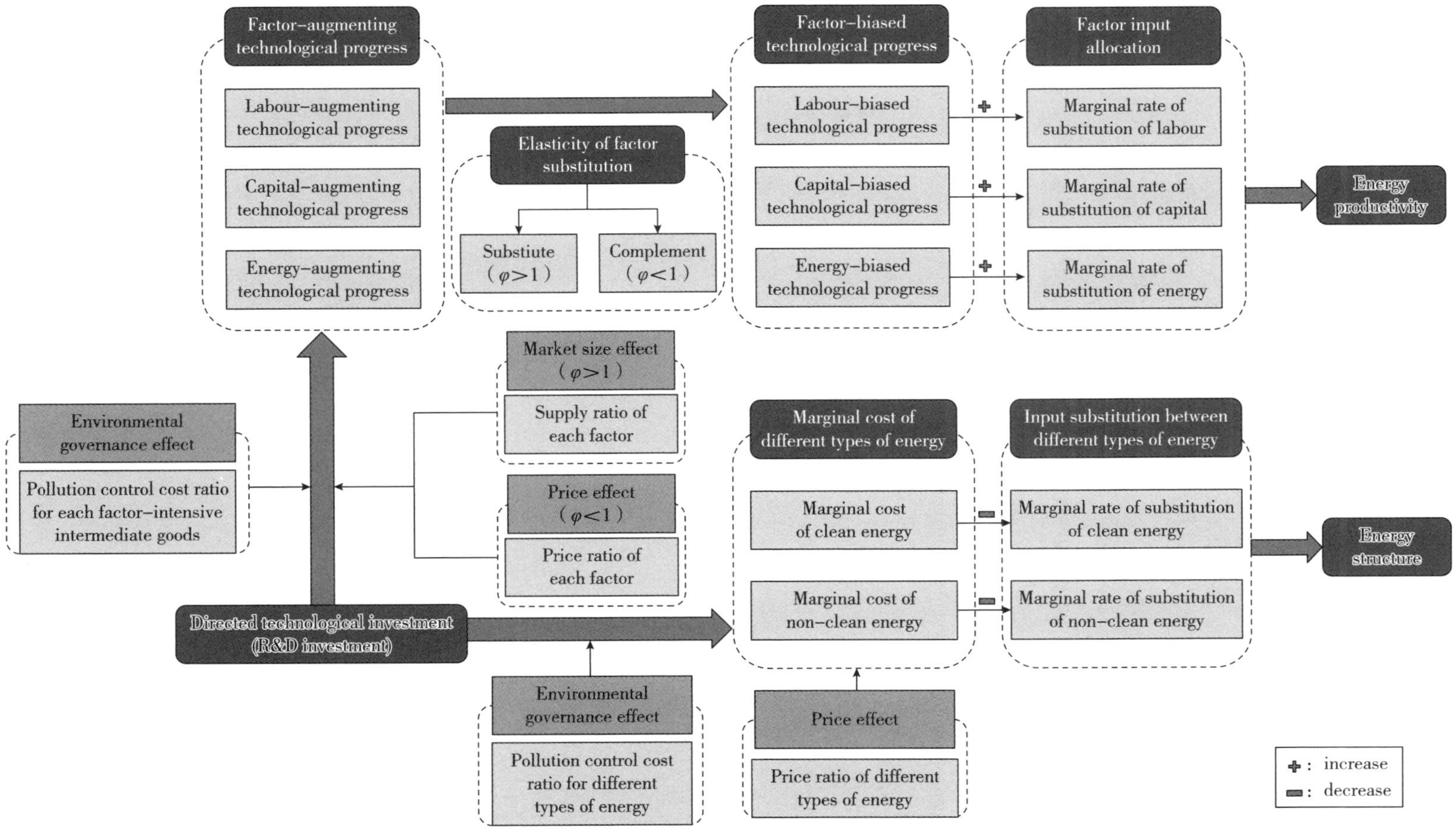

Fig. 2 Mechanism of directed technological investment on energy productivity and energy structure

parameter estimation for all three nesting patterns and found that the L (KE)-type is the most suitable production substitution model for the current Chinese petrochemical industry and its sub-industries. Therefore, this paper sets the CES production function for the Chinese petrochemical industry as L (KE)-type.

In the equation (1), Y_t represents the economic value added of the Chinese chemical industry at time t, while L_t, K_t, and E_t denote the inputs of labour, capital, and energy factors in production, respectively. A_{L_t}, A_{K_t}, and A_{E_t} represent labour-ATP, capital-ATP, and energy-ATP during the same period, respectively. Parameter σ denotes the elasticity of substitution between labour and non-labour factors, parameter φ denotes the elasticity of substitution between capital and energy factors, γ represents labour intensity, and τ represents capital intensity.

Taking the derivatives of the factor inputs in the above equation separately, we can derive that the relative marginal productivity of capital and energy factors (taking the relative marginal productivity of capital and energy factors as an example) is jointly determined by factor-ATP and factor substitution elasticity:

$$\frac{MP_{K_t}}{MP_{E_t}}=\left(\frac{\tau}{1-\tau}\right)\cdot\left(\frac{A_{K_t}}{A_{E_t}}\right)^{\frac{\varphi-1}{\varphi}}\cdot\left(\frac{K_t}{E_t}\right)^{-\frac{1}{\varphi}} \tag{2}$$

Subsequently, this paper integrates Acemoglu's (2002) definition of the direction of technological progress and introduces the technological progress biased index, denoted as TB_t:

$$\begin{aligned} TB_t\cdot\frac{MP_{K_t}}{MP_{E_t}}&=\left(\frac{MP_{K_{t+1}}}{MP_{E_{t+1}}}-\frac{MP_{K_t}}{MP_{E_t}}\right)\Bigg|_{\frac{K_t}{E_t}}=\frac{\partial(MP_{K_t}/MP_{E_t})}{\partial(A_{K_t}/A_{E_t})}\cdot\frac{\mathrm{d}(A_{K_t}/A_{E_t})}{\mathrm{d}t}\\ &=\frac{MP_{K_t}}{MP_{E_t}}\cdot\frac{\varphi-1}{\varphi}\cdot\left(\frac{A_{K_t}}{A_{E_t}}\right)^{-1}\cdot\frac{\mathrm{d}(A_{K_t}/A_{E_t})}{dt}\end{aligned} \tag{3}$$

$$\Rightarrow TB_t=\frac{\varphi-1}{\varphi}\cdot\left(\frac{A_{K_t}}{A_{E_t}}\right)^{-1}\cdot\frac{\mathrm{d}(A_{K_t}/A_{E_t})}{\mathrm{d}t} \tag{4}$$

The economic interpretation of TB_t is the rate of change in the ratio of the marginal productivity of capital to energy when the factor structure remains constant, driven by the difference in the advancement rates of capital-augmenting and energy-augmenting technologies. When the factor substitution elasticity $\varphi>1$, the advancement of capital-augmenting technology leads to an increase in the marginal productivity of capital relative to energy ($TB_t>0$). This indicates a technological bias towards the capital, favouring capital-intensive usage to conserve energy, thereby increasing energy productivity. Conversely, when the factor substitution elasticity $\varphi<1$, the advancement of capital-augmenting technology results in an increase in the marginal productivity of energy relative to capital ($TB_t<0$). In this case, the technological bias favors energy, promoting energy-intensive usage to save capital, consequently leading to a decrease in energy productivity. Finally, when the factor substitution elasticity $\varphi=1$, neutral technological progress in the industry causes proportional increases in the marginal productivity of labour, capital, and energy factors.

To visually demonstrate the cumulative effects of current factor-BTP, this paper introduces the Technological Biased Summation Index TBS_t.

$$TBS_{t+1} = (1 + TB_t)\ TBS_t \tag{5}$$

This paper sets the base period TBS equal to 1. The economic interpretation of TBS is that the cumulative impact of factor-BTP on the marginal productivity of each factor is TBS_{t+1} times that of the base period. If TBS increases, technological progress tends to favour capital; conversely, it favours energy. Consequently, this paper proposes the first hypothesis.

Hypothesis H1: When the Technological Biased Summation Index TBS increases, technological progress tends to favour the capital factor, resulting in the intensive use of capital to save energy and labour, leading to increased energy productivity and labour productivity; conversely, the opposite occurs.

In the subsequent section, this paper constructs a unified theoretical framework based on the theories of technological progress bias and endogenous growth. It investigates how the directional improvement of factor - ATP is affected by technological investment (R&D investment), factor substitution elasticity, as well as the impact of the three moderating effect mechanisms of price effect, market scale effect and environmental governance effect, which in turn affects energy productivity and energy structure.

2.1 Baseline model

The basic setup of the baseline model assumes that in an infinite horizon of economic activity, the petrochemical industry consists of labour-intensive, capital-intensive, and energy-intensive intermediate goods sectors, which produce final products. The production of final products adopts the CES production function. Additionally, the petrochemical industry has a dedicated R&D department that provides factor - augmenting new machines to the intermediate goods sectors. The number of machine types determines the level of technical enhancement for labour, capital, and energy factors. Each intermediate goods sector pays rent for factor - augmenting machines to the R&D department, which provides R&D expenditures. The relative profitability of different factor - augmenting technologies and factor substitution elasticity determine which factor the technological progress is biased towards.

2.1.1 Final products production

Assuming that the final product Y produced in the economy comes from labour - intensive intermediate good Y_L, capital - intensive intermediate good Y_K, and energy - intensive intermediate good Y_E, then the production function of the final product Y is given by:

$$Y \equiv \left\{\gamma Y_L^{\frac{\varepsilon-1}{\varepsilon}} + (1-\gamma)\left[\tau Y_K^{\frac{\delta-1}{\delta}} + (1-\tau) Y_E^{\frac{\delta-1}{\delta}}\right]^{\frac{(\varepsilon-1)/\varepsilon}{(\delta-1)/\delta}}\right\}^{\frac{\varepsilon}{\varepsilon-1}} \tag{6}$$

The final product Y adopts the CES production function form proposed by Solow (1956). ε

represents the substitution elasticity between labour-intensive intermediate goods and non-labour-intensive intermediate goods, while δ represents the substitution elasticity between capital-intensive intermediate goods and energy-intensive intermediate goods. γ represents the distribution parameter of the importance of labour-intensive intermediate goods in the final product Y, and τ represents the distribution parameter of the importance of capital-intensive intermediate goods in the non-labour-intensive intermediate goods, $0<\gamma$, $\tau<1$.

2.1.2 Intermediate goods production

Assuming that the production of each intermediate good requires the extensive use of specific production factors in conjunction with machines, and production exhibits constant returns to scale characteristics. Referring to Xue and Yip (2013), the production functions for labour-intensive intermediate good Y_L, capital-intensive intermediate good Y_K, and energy-intensive intermediate good Y_E can be represented as:

$$\begin{aligned} Y_L &= \frac{1}{1-\mu}\left(\int_0^{N_L} x_L(j)^{1-\mu}\,\mathrm{d}j\right) L^{\mu} \\ Y_K &= \frac{1}{1-\mu}\left(\int_0^{N_K} x_K(j)^{1-\mu}\,\mathrm{d}j\right) K^{\mu} \\ Y_E &= \frac{1}{1-\mu}\left(\int_0^{N_E} x_E(j)^{1-\mu}\,\mathrm{d}j\right) E^{\mu} \end{aligned} \tag{7}$$

Where, $\mu \in (0, 1)$, A, B, and C represent the inputs of labour, capital, and energy factors, respectively. $x_L(j)$, $x_K(j)$, and $x_E(j)$ represent the quantities of the j-th type of labour-augmenting, capital-augmenting, and energy-augmenting machines used in production, respectively. N_L, N_K, and N_E represent the number of types of machines used in the production of intermediate goods Y_L, Y_K, and Y_E, respectively. Since different types of machines are used in the production of intermediate products in the three sectors, this implies that technological progress is directional. The variations in N_L, N_K, and N_E represent changes in factor productivity and thus can represent labour-augmenting, capital-augmenting, and energy-ATP[1].

2.1.3 R&D department production

The R&D department, which invents new machines, holds exclusive patents and serves as the sole supplier of technology in the market. It is assumed that each intermediate goods sector uses machines supplied by the R&D department. The following technological production functions are set based on Tassey (2005):

$$\begin{aligned} N_L &= \beta_L RD_L \\ N_K &= \beta_K RD_K \\ N_E &= \beta_E RD_E \end{aligned} \tag{8}$$

[1] Aligned with Acemoglu (2002), in the intermediate goods production function and the new technology (machinery) production function, represent the number of machine types used, and they are expressed as factor-ATP.

Here, RD_L, RD_K, and RD_E represent the R&D expenditures on producing labour-intensive, capital-intensive, and energy-intensive intermediate goods machines, respectively. Parameters β_L, β_K, and β_E indicate that R&D expenditures on different factor-intensive intermediate goods result in the formation of different numbers of new machine types. Equation (8) implies that R&D expenditures used to develop directionally different factor-augmenting technologies (machines) can generate β number of new types of factor-augmenting machines.

2.2 Endogenous determination of factor-ATP.

In a self-sufficient economy with competitive markets, the profit function of the R&D department is a function of machine rent, R&D costs, and the quantity of machines demanded by the intermediate goods sector. The profit function of firms producing final products is a function of the quantity of machines, factor prices, intermediate goods prices, and pollution control costs. Both the R&D department and production firms maximize their respective profits, and market equilibrium is achieved. For ease of analysis, first, consider N_L, N_K, and N_E as given. Assuming that machines operate each year continuously and depreciate after use, with the marginal cost of machine production denoted by N_E, we further discuss the endogenous determination of N_L, N_K, and N_E.

2.2.1 Optimal decision of final product

In a competitive market, applying the first-order condition (FOC)[1] to equation (6) yields the relative price P_{KE} of capital-intensive intermediate good Y_K and energy-intensive intermediate good Y_E (taking capital and energy-intensive intermediate goods as an example):

$$P_{KE} \equiv \frac{P_K}{P_E} = \frac{\tau}{1-\tau}\left(\frac{Y_K}{Y_E}\right)^{-\frac{1}{\delta}} \tag{9}$$

Since the intermediate goods compete in the market, market clearance implies that their relative prices are satisfied: the more the supply of Y_K relative to Y_E, the lower the relative price P_{KE}, as shown in equation (9). The responsiveness of relative prices to relative supply depends on the substitution elasticity δ between capital-intensive intermediate goods and energy-intensive intermediate goods.

Using the price of the final product as a benchmark, it follows:

$$\tau P_K^{1-\delta} + (1-\tau) P_E^{1-\delta} = 1 \tag{10}$$

Assuming continuous operation of machinery in each department annually, the labour-intensive, capital-intensive, and energy-intensive departments each face maximization problems:

$$\max.\ \pi_{L,\{x_L(j)\}} = P_L Y_L - wL - \int_0^{N_L} \chi_L(j) \cdot x_L(j)\, dj - \zeta_L$$

$$\max.\ \pi_{K,\{x_K(j)\}} = P_K Y_K - rK - \int_0^{N_K} \chi_K(j) \cdot x_K(j)\, dj - \zeta_K$$

[1] Taking the first-order derivative of the equation and setting it equal to zero to find the optimal value.

$$\max.\ \pi_{E,\{x_E(j)\}} = P_E Y_E - sE - \int_0^{N_E} \chi_E(j) \cdot x_E(j)\, dj - \zeta_E$$

$$\zeta_i = \int_0^{N_i} \zeta_i(j) \cdot x_i(j)\, dj \tag{11}$$

In equation (11), P_L, P_K, and P_E respectively represent the intermediate product prices of the labour-intensive, capital-intensive, and energy-intensive departments. w, r, and s represent the prices of labour, capital, and energy factors. $\chi_L(j)$, $\chi_K(j)$, and $\chi_E(j)$ represent the rental prices of machinery in the labour-intensive, capital-intensive, and energy-intensive departments, respectively, determined by equation (7). ζ_L, ζ_K, and ζ_E respectively represent the pollution control costs of the labour-intensive, capital-intensive, and energy-intensive departments. These costs are determined by the total pollution treatment expenses incurred from the production of products by different types of machinery. $\zeta_i(j)$ represents the pollution treatment expenses required for the i-intensive department using the j-th type of machinery to produce products.

Substituting equation (7) into equation (11) and using the FOC regarding $x_L(j)$, $x_K(j)$, and $x_E(j)$ in the maximisation problem faced by the firms, it obtains the demand for machinery by the labour-intensive, capital-intensive, and energy-intensive departments.

$$x_L(j) = \left(\frac{P_L}{\chi_L(j)\ \zeta_L(j)}\right)^{1/\mu} L$$

$$x_K(j) = \left(\frac{P_K}{\chi_K(j)\ \zeta_K(j)}\right)^{1/\mu} K \tag{12}$$

$$x_E(j) = \left(\frac{P_E}{\chi_E(j)\ \zeta_E(j)}\right)^{1/\mu} E$$

Equation (12) indicates that the demand for machinery increases with higher levels of intermediate product prices and factor inputs, while it decreases with increases in machinery rental and pollution treatment costs.

Using equation (7) and taking the FOC with respect to L, K, and E, it can derive the price functions (unit factor reward functions) for labour, capital, and energy:

$$w = \frac{\mu}{1-\mu} P_L \left(\int_0^{N_L} x_L(j)^{1-\mu} \mathrm{d}j\right) L^{\mu-1}$$

$$r = \frac{\mu}{1-\mu} P_K \left(\int_0^{N_K} x_K(j)^{1-\mu} \mathrm{d}j\right) K^{\mu-1} \tag{13}$$

$$s = \frac{\mu}{1-\mu} P_E \left(\int_0^{N_E} x_E(j)^{1-\mu} \mathrm{d}j\right) E^{\mu-1}$$

2.2.2 Optimal decision of the R&D department

Assuming that the marginal cost of producing machines by the R&D department is equal to ψ, the profit functions for producing labour-augmenting, capital-augmenting, and energy-augmenting machines can be expressed as follows:

$$\pi_L = \int_0^{N_L} (\chi_L(j) - \psi) \cdot x_L(j)\, dj$$

$$\pi_K = \int_0^{N_K} (\chi_K(j) - \psi) \cdot x_K(j)\, dj \tag{14}$$

$$\pi_E = \int_0^{N_E} (\chi_E(j) - \psi) \cdot x_E(j)\, dj$$

In equation (14), taking the FOC of the profit functions with respect to each type of machine price respectively, yields the machine price functions.

$$\chi_L(j) = \chi_K(j) = \chi_E(j) = \frac{\psi}{1-\mu} \tag{15}$$

Standardizing the marginal cost to $\psi = 1-\mu$ implies that in equilibrium, all machine prices $\chi_L(j) = \chi_K(j) = \chi_E(j) = 1$.

Using the machine price functions from equation (15) along with the machine demand functions from equation (12), the profit function of the R&D department can be further rewritten:

$$\pi_L = \mu\left(\frac{P_L}{\zeta_L}\right)^{\frac{1}{\mu}} L$$

$$\pi_K = \mu\left(\frac{P_K}{\zeta_K}\right)^{\frac{1}{\mu}} K \tag{16}$$

$$\pi_E = \mu\left(\frac{P_E}{\zeta_E}\right)^{\frac{1}{\mu}} E$$

The net present value of the profit of the research and development department is represented by the standardized dynamic programming equation:

$$r'V_L - \hat{V}_L = \pi_L$$

$$r'V_K - \hat{V}_K = \pi_K \tag{17}$$

$$r'V_E - \hat{V}_E = \pi_E$$

In equation (17), r' represents the interest rate, which may vary over time. The equation relates the discounted value of future period profits V to current period profits π, where $\hat{V}$ represents the change in the discounted value of future period profits.

2.2.3 Equilibrium Equation

Under the assumption of unchanged current profits and interest rates, at equilibrium when $\hat{V}=0$, the profits of the R&D department from producing labour-augmenting, capital-augmenting, and energy-augmenting machines are as follows:

$$V_L = \frac{\mu P_L^{1/\mu} L}{r'\zeta_L^{1/u}}$$

$$V_K = \frac{\mu P_K{}^{1/\mu} K}{r' \zeta_K{}^{1/u}}$$

$$V_E = \frac{\mu P_E{}^{1/\mu} E}{r' \zeta_E{}^{1/u}} \tag{18}$$

From the above, it is evident that the larger V_i is, the greater the motivation for the R&D department to develop i-augmenting machines. Equation (18) reveals three forces determining the direction of technological progress:

(1) Price effect: The R&D department will have greater motivation to develop technologies that produce more expensive products, as evidenced by the fact that higher prices for intermediate goods P_L, P_K, and P_E lead to greater profits.

(2) Market scale effect: The R&D department will have greater motivation to innovate technologically with more abundant factors. This can be observed from the fact that profits V_L, V_K, and V_E increase as factor supplies L, K, and E increase.

(3) Environmental governance effect: The R&D department will have greater motivation to develop technologies that produce intermediate goods with lower pollution control costs. This is evidenced by the fact that profits increase when pollution control costs ζ_L, ζ_K, and ζ_E required for the production of intermediate goods Y_L, Y_K, and Y_E are reduced.

This discussion emphasizes the roles of the price effect, market scale effect, and environmental governance effect, but the factor prices are not explicitly included in equation (18). However, the theory of induced technological innovation suggests that R&D departments typically innovate on the productivity of "more expensive" factors (Fellner, 1961; Habakkuk, 1963; Hicks, 1963). Using equations (12) and (13), the equation (18) can be rewritten as follows:

$$V_L = \frac{(1-\mu)\, wL}{r' N_L \zeta_L}$$

$$V_K = \frac{(1-\mu)\, rK}{r' N_K \zeta_K} \tag{19}$$

$$V_E = \frac{(1-\mu)\, sE}{r' N_E \zeta_E}$$

Equation (19) indicates that R&D departments will have greater incentives to innovate technologies that enhance the productivity of expensive factors, as speculated by the theory of induced technological innovation.

Next, this paper will delve deeper into how the forces of price effect, market scale effect, and environmental governance effect play different roles influenced by factor substitution elasticity. Substituting equation (12) into equation (7) yields:

$$Y_L = \frac{1}{1-\mu}\left(\frac{P_L}{\zeta_L}\right)^{\frac{1-\mu}{\mu}} N_L L$$

$$Y_K = \frac{1}{1-\mu}\left(\frac{P_K}{\zeta_K}\right)^{\frac{1-\mu}{\mu}} N_K K$$

$$Y_E = \frac{1}{1-\mu}\left(\frac{P_E}{\zeta_E}\right)^{\frac{1-\mu}{\mu}} N_E E \tag{20}$$

Substituting equation (20) into equation (9), it obtains the relative product price function expressed in terms of the relative technological level N_K/N_E and the relative factor supply K/E (taking capital and energy-intensive intermediate goods as examples):

$$P_{KE} = \frac{P_K}{P_E} = \left(\frac{\tau}{1-\tau}\right)^{\frac{\mu\delta}{\varphi}} \left(\frac{\zeta_K}{\zeta_E}\right)^{\frac{1-\mu}{\varphi}} \left(\frac{N_K}{N_E} \cdot \frac{K}{E}\right)^{-\frac{\mu}{\varphi}} \tag{21}$$

In equation (21), δ represents the substitution elasticity between capital-intensive intermediate goods Y_K and energy-intensive intermediate goods Y_E, while φ denotes the substitution elasticity between the capital factor K and the energy factor E, satisfying: $\varphi = \delta - (\delta-1)(1-\mu)$. Obviously, it is evident that $\varphi>1$ only when $\delta>1$, that is, if capital-intensive intermediates and energy-intensive intermediates are substitutes, capital factors and energy factors also show a substitution relationship. The opposite is true.

On the balanced growth path, where intermediate goods' relative prices remain constant and factor-ATP N_K and N_E grow at the same rate, it implies that $\hat{V}$ in equation (17) equals zero and V_K/V_E remains unchanged. Moreover, V_K/V_E should be equal to and must be equal to β_K/β_E, indicating that the R&D department is willing to supply technology to both capital-intensive and energy-intensive sectors. The technological market clearing condition is:

$$\beta_K \pi_K = \beta_E \pi_E \tag{22}$$

Further combining equations (8), (16), and (21), the technological market clearance condition can be rewritten (taking capital and energy-augmenting machines as examples):

$$\frac{N_K}{N_E} = \left(\frac{\beta_K}{\beta_E}\right)^{\varphi} \left(\frac{\tau}{1-\tau}\right)^{\delta} \underbrace{\left(\frac{\zeta_K}{\zeta_E}\right)^{-\delta}}_{\text{Environmental governance effect}} \underbrace{\left(\frac{K}{E}\right)^{\varphi-1}}_{\text{Market scale effect}}, \ if\ \varphi > 1(\delta > 1)$$

$$= \left(\frac{\beta_K}{\beta_E}\right)^{\varphi} \left(\frac{\tau}{1-\tau}\right)^{\delta} \underbrace{\left(\frac{\zeta_K}{\zeta_E}\right)^{-\delta}}_{\text{Environmental governance effect}} \underbrace{\left(\frac{r}{s}\right)^{1-\varphi}}_{\text{Price effect}}, \ if\ \varphi < 1(\delta < 1) \tag{23}$$

$$\frac{N_K}{N_E} = \frac{\beta_K}{\beta_E} \cdot \frac{RD_K}{RD_E}, \ \text{where}\ RD = RD_K + RD_E + RD_L$$

Equation (23) implies that the bias of R&D investment towards enhancing which factor-ATP is endogenous, determined by the moderation of price effect, market scale effect, and environmental governance effect. The magnitude of factor substitution elasticity determines which one dominates between the price effect and the market scale effect, while the environmental governance effect operates independently of the price effect and the market scale effect.

Depending on the value of substitution elasticity, the following three scenarios are discussed:

Scenario (1): When the factor substitution elasticity $\varphi>1$ ($\delta>1$), the market scale effect dominates. Under the moderation of the market scale effect, the R&D department tends to develop factor - augmenting technology that factor is abundant. Additionally, the R&D department will lean towards developing factor-augmenting technologies with lower pollution control costs under the moderation of the environmental governance effect.

Scenario (2): When the factor substitution elasticity $\varphi<1$ ($\delta<1$), the price effect dominates. Under the moderation of the price effect, the R&D department tends to develop factor-augmenting technology that factor is expensive. Similarly, the R&D department will also favour developing factor-augmenting technologies with lower pollution control costs under the moderation of the environmental governance effect.

Scenario (3): When the factor substitution elasticity $\varphi=1$ ($\delta=1$), both the market scale effect and the price effect are simultaneously eliminated, but the environmental governance effect still exists.

After clarifying the mechanism of the directional improvement of factor-ATP influenced by factor substitution elasticity and the three major moderating effects, this paper proposes the following hypotheses:

Hypothesis H2: R&D investment is directional in enhancing different types of factor-augmenting progress.

Hypothesis H3: (Ⅰ) When the factor substitution elasticity $\varphi>1$ ($\delta>1$), if the relative supply of capital to energy factor K/E increases, R&D investment tends to enhance factor-augmenting progress in the abundant (capital) factor under the moderation of the market scale effect; (Ⅱ) When the factor substitution elasticity $\varphi<1$ ($\delta<1$), if the relative price of capital to energy factor r/s increases, R&D investment tends to enhance factor-augmenting progress in the expensive (capital) factor under the moderation of the price effect.

Hypothesis H4: R&D investment always tends to enhance factor-augmenting progress in the lower pollution cost factor under the moderation of the environmental governance effect. Specifically, when the pollution governance cost of energy-intensive intermediates relative to capital-intensive intermediates ζ_E/ζ_K increases, the R&D department tends to develop low-pollution (capital) factor-ATP under the influence of the environmental governance effect.

2.3 Endogenous determinants of substitution between energy sources

In a self-sufficient economy, producers are rational. The cost function of the energy sector is a function of the prices, consumption, and pollution control costs of clean and non-clean energy. The energy sector aims to minimise costs while meeting the demands of producers. Additionally, the energy sector will provide clean and non-clean energy to the intermediate goods sector based on different production technologies (machines) (Acemoglu et al., 2012).

The pollution control costs associated with different production technologies (machines) vary. Therefore, the substitution of clean energy for non-clean energy depends on energy prices, production technologies (machines), and pollution control costs.

2.3.1 Cost of the energy sector

To meet production needs, the energy sector provides energy to manufacturers. Assuming the machines operate continuously every year, following the approach of Acemoglu et al. (2012), the energy supplied to the intermediate goods sector is categorised into "clean" (c) and "non-clean" (n) based on the different types of energy used in the production process. Therefore, the cost function of the energy sector can be represented as:

$$\begin{aligned} C_E &= s_c E_c + \zeta_{Ec} E_c + s_n E_n + \zeta_{En} E_n \\ &= \sum_{j=0}^{N_{Ec}} [s_c(j) E_c(j) + \zeta_{Ec}(j) E_c(j)] + \sum_{j=0}^{N_{En}} [s_n(j) E_n(j) + \zeta_{En}(j) E_n(j)] \end{aligned} \tag{24}$$

Where, s_c and s_n represent the prices of clean and non-clean energy, respectively. E_c and E_n represent the consumption of clean and non-clean energy, ζ_{Ec} and ζ_{En} represent the pollution control costs incurred by consuming clean and non-clean energy, N_{Ec} and N_{En} represent the number of types of machines using clean and non-clean energy, respectively.

Referring to equations (8), (19), and (22), when the market is in equilibrium, the profits obtained by the energy sector from clean energy machines and non-clean energy machines are equal. It follows that the pollution control costs of clean (non-clean) energy are inversely proportional to the R&D investment in clean (non-clean) energy machines.

$$\begin{aligned} &\beta_{Ec} RD_{Ec} = \frac{(1-\mu) s_c E_c}{r' V_{Ec} \zeta_{Ec}},\ \beta_{En} RD_{En} = \frac{(1-\mu) s_n E_n}{r' V_{En} \zeta_{En}},\ \beta_{Ec} V_{Ec} = \beta_{En} V_{En} \\ &\Rightarrow RD_{Ec} = \frac{\kappa_{Ec}^2}{\zeta_{Ec}},\ RD_{En} = \frac{\kappa_{En}^2}{\zeta_{En}},\ \text{where } \frac{\kappa_{Ec}}{\kappa_{En}} = \frac{s_c E_c}{s_n E_n} \end{aligned} \tag{25}$$

2.3.2 Endogenous determination of marginal substitution between energy sources

This section examines the endogenous determination of the substitution of clean energy for non-clean energy in the petrochemical industry by comparing the marginal costs of clean and non-clean energy. If the marginal cost of clean energy is higher than the marginal cost of non-clean energy, producers are willing to invest in non-clean energy to replace clean energy, and vice versa. According to equations (24) and (25), the ratio of marginal costs of clean energy to non-clean energy in the petrochemical industry can be derived as:

$$\frac{MC_{Ec}}{MC_{En}} = \frac{s_c + \zeta_{Ec}}{s_n + \zeta_{En}} = \frac{s_c + \sqrt{\zeta_{Ec}} \cdot \dfrac{\kappa_{Ec}}{\sqrt{RD_{Ec}}}}{s_n + \sqrt{\zeta_{En}} \cdot \dfrac{\kappa_{En}}{\sqrt{RD_{En}}}}$$

$$\text{where}, \ \sqrt{\zeta_{Ec}} = \frac{\kappa_{Ec}}{\sqrt{RD_{Ec}}}, \ \sqrt{\zeta_{En}} = \frac{\kappa_{En}}{\sqrt{RD_{En}}} \tag{26}$$

Equation (26) indicates that the ratio of marginal costs between energy sources is influenced by energy prices, R&D investment, and pollution control costs. Clean energy's marginal cost relative to non-clean energy decreases as the ratio of energy prices and the ratio of pollution control costs decrease. It also decreases as the ratio of R&D investment in production technology (machinery) increases. Moreover, the impact of R&D investment on energy structure is moderated by pollution control costs. Therefore, hypothesis H5 is proposed.

Hypothesis H5: The improvement in energy structure is influenced directly by energy prices and indirectly by the moderating effect of technological investment on pollution control costs. Specifically, when the relative price of non-clean energy compared to clean energy s_n/s_c is higher, indicating a stronger marginal substitution of clean energy for non-clean energy, the energy structure improves. Similarly, when the relative pollution control costs of non-clean energy compared to clean energy ζ_{En}/ζ_{Ec} are higher, indicating a directional increase in the marginal substitution of clean energy for non-clean energy under the moderating effect of technology investment on pollution control, the energy structure improves.

3 Econometric model and data

3.1 Econometric model

Based on the theoretical framework of this paper, which analyses the mechanisms through which directed technological investment impacts energy productivity and energy structure, two-dimensional panel data at the petrochemical sub-sector and time level are used to construct two sets of fixed baseline models and three sets of fixed mechanism models. The two fixed baseline models test hypotheses H1 and H2 proposed in Section 2 of the study, while the three fixed mechanism models test hypotheses H3 to H5.

To test the hypothesis H2 that R&D investment is directional in enhancing different types of factor-augmenting progress, the following baseline model is constructed:

$$\begin{aligned} ALE_{it} &= \alpha_0 + \alpha_1 RD_{it} + \sum \lambda_1 X_{it} + u1_{it} + v1_{it} \\ AEK_{it} &= \beta_0 + \beta_1 RD_{it} + \sum \lambda_2 X_{it} + u2_{it} + v2_{it} \end{aligned} \tag{27}$$

Where ALE_{it} represents the ratio of labour-ATP to energy-ATP for sub-sector i at time t; AEK_{it} represents the ratio of energy-ATP to capital-ATP; RD_{it} denotes R&D investment; X_{it} denotes control variables, which include foreign direct investment FDI, operating revenue IR, energy structure ES, labour input L, capital input K, and energy input E; $u1_{it}$ and $u2_{it}$ represent industry-year fixed effects; $v1_{it}$ and $v2_{it}$ represent random error terms.

To test hypothesis H3 (Ⅱ) that R&D investment is biased towards enhancing factor-augmenting progress in the expensive factor under the moderation of the price effect ($\varphi<1$), and hypothesis H4 that R&D investment is biased towards enhancing technologies with lower pollution control costs under the environmental governance effect, the following mechanism model is constructed:

$$ALE_{it} = \alpha_0 + \alpha_1 RD_{it} + \alpha_2 sw_{it} + \alpha_3 PCLE_{it} + \alpha_4 RD_{it} \times sw_{it} + \alpha_5 RD_{it} \times PCLE_{it} + \sum \lambda_1 X_{it} + u1_{it} + v1_{it}$$

$$AEK_{it} = \beta_0 + \beta_1 RD_{it} + \beta_2 sr_{it} + \beta_3 PCKE_{it} + \beta_4 RD_{it} \times sr_{it} + \beta_5 RD_{it} \times PCKE_{it} + \sum \lambda_2 X_{it} + u1_{it} + v2_{it} \tag{28}$$

Where ALE_{it} represents the ratio of labour-ATP to energy-ATP for sub-sector i at time t; AEK_{it} represents the ratio of energy-ATP to capital-ATP; RD_{it} denotes R&D investment; sw_{it} represents the ratio of energy price to labour price; sr_{it} represents the ratio of energy price to capital rental; $PCLE_{it}$ represents the ratio of pollution control investment in labour-intensive intermediate goods to energy-intensive intermediate goods; $PCKE_{it}$ represents the ratio of pollution control investment in capital-intensive intermediate goods to energy-intensive intermediate goods; X_{it} denotes control variables, which include foreign direct investment FDI, operating revenue IR, energy structure ES, labour input L, capital input K, and energy input E; $u1_{it}$ and $u2_{it}$ represent industry-year fixed effects; $v1_{it}$ and $v2_{it}$ represent random error terms.

To test the hypothesis H3 (Ⅰ) that R&D investment tends to enhance factor-augmenting progress in the abundant factor under the moderation of the market scale effect ($\varphi>1$), and the hypothesis H4 that R&D investment is biased towards enhancing technologies with lower pollution control costs under the environmental governance effect, the following mechanism model is constructed:

$$ALK_{it} = \alpha_0 + \alpha_1 RD_{it} + \alpha_2 LK_{it} + \alpha_3 PCLK_{it} + \alpha_4 RD_{it} \times LK_{it} + \alpha_5 RD_{it} \times PCLK_{it} + \sum \lambda_1 X_{it} + u1_{it} + v1_{it}$$

$$AKE_{it} = \beta_0 + \beta_1 RD_{it} + \beta_2 KE_{it} + \beta_3 PCKE_{it} + \beta_4 RD_{it} \times KE_{it} + \beta_5 RD_{it} \times PCKE_{it} + \sum \lambda_2 X_{it} + u2_{it} + v2_{it} \tag{29}$$

Where ALE_{it} represents the ratio of labour-ATP to energy-ATP for sub-sector i at time t; AKE_{it} represents the ratio of capital-ATP to energy-ATP; RD_{it} denotes R&D investment; LK_{it} represents the ratio of labour input to capital input; KE_{it} represents the ratio of capital input to energy input; $PCLK_{it}$ represents the ratio of pollution control investment for labour-intensive intermediate goods relative to capital-intensive intermediate goods; $PCKE_{it}$ represents the ratio of pollution control investment for capital-intensive intermediate goods relative to energy-intensive intermediate goods; X_{it} denotes control variables, which include foreign direct investment FDI, operating revenue IR, energy structure ES, labour price w, capital rent r, and energy price s; $u1_{it}$ and $u2_{it}$ represent industry-year fixed effects; $v1_{it}$ and $v2_{it}$ represent random error terms.

Since the transformation of factor-ATP into factor-BTP influenced by factor substitution elasticity has been thoroughly discussed in Sun et al. (2024b), the basic conclusion is that when the factor substitution elasticity is greater than 1, factor-ATP transforms into BTP towards the same factor; otherwise, it transforms into BTP towards other factors. Building on this research, this paper further explores the impact of the factor-BTP on energy productivity and labour productivity to test hypothesis H1. The constructed baseline model is as follows:

$$\begin{aligned} EP_{it} &= \alpha_0 + \alpha_1 TBS_{it} + \sum \lambda_1 X_{it} + u1_{it} + v1_{it} \\ LP_{it} &= \alpha_0 + \alpha_1 TBS_{it} + \sum \lambda_2 X_{it} + u2_{it} + v2_{it} \end{aligned} \tag{30}$$

Where EP_{it} represents the energy production of sector i at time t; LP_{it} represents the labour productivity; TBS_{it} represents the cumulative index of BTP. X_{it} denotes control variables, which include foreign direct investment FDI, operating revenue IR, energy structure ES, labour input L, capital input K, energy input E, and energy price s; $u1_{it}$ and $u2_{it}$ represent industry-year fixed effects; $v1_{it}$ and $v2_{it}$ represent random error terms.

To test hypothesis H5 that the improvement in energy structure is influenced directly by energy prices and indirectly by the moderating effect of technological investment on pollution control costs, the following mechanism model is constructed:

$$\begin{aligned} ES_{it} = \alpha_0 + \alpha_1 snc_{it} + \alpha_2 RD_{it} + \alpha_3 PCENC_{it} + \alpha_4 RD_{it} \times PCENC_{it} + \\ \sum \lambda X_{it} + u1_{it} + v1_{it} \end{aligned} \tag{31}$$

Where, ES_{it} represents the energy structure of sub-sector i at time t; snc_{it} represents the ratio of non-clean energy prices to clean energy prices; RD_{it} represents R&D investment; and $PCENC_{it}$ represents the ratio of pollution abatement investment for non-clean energy-intensive intermediate goods to clean energy-intensive intermediate goods. X_{it} denotes control variables, which include foreign direct investment FDI, operating revenue IR, economic value added Y, labour input L, capital input K, energy input E, labour price w, capital rent r, and energy price s; $u1_{it}$ represents industry-year fixed effects; $v1_{it}$ represents random error terms.

3.2 Data sources and preprocessing

The data utilized in this study covers the period from 2000 to 2021 and includes various indicators pertaining to five sub-industries within the Chinese petrochemical industry❶. These indicators encompass total output, labour input, capital input, energy consumption, labour productivity, energy productivity, labour price, capital rent, energy price, energy structure, R&D investment, pollution control investment, pollution control investment for

❶ Following the research of Sun et al. (2024a), the five sub-sectors—oil and gas mining industry (B07), petroleum, coal, and other fuel processing (C25), chemical raw materials and chemical products manufacturing (C26), chemical fibre manufacturing (C28), and rubber and plastic products (C29) —are aggregated into the Chinese petrochemical industry.

labour - intensive intermediate goods, pollution control investment for capital - intensive intermediate goods, pollution control investment for energy - intensive intermediate goods, foreign direct investment, operating revenue, labour - ATP, capital - ATP, energy - ATP, technological progress bias index, and patent applications. Regarding the primary data sources and preprocessing: (1) The total output Y_t represents the economic value added by the petrochemical industry and its sub-industries. This data is sourced from the NBSC and the Big Data Information System of the China Petroleum and Chemical Industry Federation (http://www.cpcifdata.org.cn/). Additionally, the Producer Price Index (PPI) of the five sub-industries within the petrochemical industry is utilized to adjust for price changes, thereby converting it into constant-price value added for the year 2000. (2) Labour input L_t is the number of employees in petrochemical industry and its five sub-industries, derived from the China Labour Statistics Yearbook (CLSY). (3) Capital input K_t is the capital stock of the Chinese petrochemical industry and its five sub-industries as a proxy variable. The data is sourced from Sun et al. (2024a). (4) Energy consumption E_t is derived from China Energy Statistics Yearbook (CESY). (5) Labour productivity LP_{it} denotes the value added of the industry created per unit of labour input. (6) Energy productivity EP_{it} denotes the industry value added created per unit of energy input. (7) Labour price w_t is obtained by dividing labour remuneration w_tL_t by labour input, and the data on labour remuneration are obtained from China's Input-Output Table (CIOT) 2000—2021. (8) Energy price s_{it} is the price of coal, coke, crude oil, gasoline, diesel, fuel oil, natural gas, and electricity in China region from 2000—2021. The prices of coal, coke, gasoline, diesel, natural gas, and electricity in 2000—2019 are obtained from the China Price Yearbook, and the prices in 2020—2021 are obtained from the China Price Statistics Yearbook (CPSY). Crude oil and fuel oil prices are first collected for 2020—2021 through the CPSY, and then calculated for 2000—2019 through the industrial producer ex - factory price index for major products. (9) Ca - pital rent r_t is obtained by dividing capital remuneration r_tK_t by capital input. Capital remuneration r_tK_t is obtained by the following calculation. In the system of national income accounting, value added consists of labour compensation, net taxes on production, depreciation of fixed assets and operating surplus. Since the producer's remaining income after paying labour compensation is used to pay for capital and energy, the remuneration of capital is the net production tax, depreciation of fixed assets plus operating surplus minus the remuneration of energy. Data are from CIOT 2000—2021, with missing years filled in by interpolation. (10) Energy structure ES_{it} is the ratio of energy consumed by petrochemicals and its subsectors, other than coal and coke, to energy consumption. Specific energy types include coal, coke, crude oil, gasoline, diesel, fuel oil, natural gas, and electricity, and the data are derived from the CESY. When calculating the ratio, the above eight energy types are converted into standard coal, and the reference coefficients of energy conversion to standard coal used are from CESY. (11) R&D

input RD_{it} is from NBSC. Since the NBSC did not count R&D inputs before 2006, the data of R&D inputs from 2000 to 2005 are calculated by the method of year-on-year comparison with the value-added. (12) Pollution control investment PC_{it} is from the NBSC. Since the NBSC only counts the investment in industrial pollution control, the pollution control investment in petrochemical industry and its subsectors is obtained by multiplying investment in industrial pollution control by the ratio of the value added of the petrochemicals subsector to the value added of industry. (13) Energy-intensive intermediate goods pollution control investment PCE_{it} is calculated. The NBSC statistics on industrial pollution control investment include wastewater control investment, exhaust gas control investment, solid waste control investment, noise control investment and other project investment. Since the pollution generated by energy-intensive intermediate goods is mainly exhaust gas, the PCE_{it} is obtained by multiplying the proportion of exhaust gas control investment in industrial pollution control investment by the investment in petrochemical industry pollution control. (14) Labour-intensive intermediate goods pollution control investment PCL_{it} in accordance with the proportion of solid waste management investment in industrial pollution control investment multiplied by the petrochemical industry pollution control investment. (15) Capital-intensive intermediate goods pollution control investment PCK_{it} is obtained by multiplying the ratio of the remaining project treatment investment to the industrial pollution control investment by the petrochemical industry pollution control investment. (16) Clean energy pollution control investment $PCEC_{it}$ is obtained by multiplying the petrochemical industry energy pollution treatment investment according to the ratio of clean energy carbon emissions to total petrochemical industry carbon emissions. (17) Non-clean energy pollution control investment $PCEN_{it}$ in accordance with the proportion of non-clean energy carbon emissions in the total carbon emissions of the petrochemical industry multiplied by the petrochemical industry energy pollution control investment. (18) Foreign direct investment FDI is derived from the NBSC. Since the NBSC only counts FDI in the mining and manufacturing industries, the FDI in each subsector of petrochemical industry is obtained by multiplying the FDI in the mining and manufacturing industries by the ratio of the value added of the petrochemicals subsector to the value added of the mining and manufacturing industries. (19) Operating Revenue IR is from the NBSC. (20) Labour-ATP AL_{it}, capital-ATP AK_{it}, energy-ATP AE_{it}, and technological progress bias index TB_{it} are from Sun et al. (2024b). (21) Patent applications PA_{it} are from the NBSC. Since the NBSC did not count the number of patent applications before 2008, the data of patent applications from 2000 to 2007 are obtained by using the calculation method of year-on-year with the value added.

4 Results and discussion

4.1 Impact mechanism of technological investment on energy productivity

According to the theoretical model constructed in this paper, the impact of technological investment (R&D investment) on energy productivity involves two transmission processes. First, R&D investment influences different factor-ATP under the moderation of three major effects. Then, this factor-ATP transforms into different factor-BTP, influenced by factor substitution elasticity, thereby affecting energy productivity. The following sections empirically test these two transmission processes individually.

4.1.1 Baseline regression of technological investment on factor-ATP

Since equation (27) to be estimated belongs to simultaneous equations, this paper adopts the Three-Stage Least Squares (3SLS) method for estimation. The main reason is that the 3SLS method, compared to Ordinary Least Squares (OLS) and Two-Stage Least Squares (2SLS) methods, can effectively handle complex endogeneity issues (Singh, 2022). This method forecasts endogenous explanatory variables in the system of equations, transforming them into instrumental variables that are correlated with endogenous explanatory variables but uncorrelated with the error term, and then eliminates the endogeneity effects through a multi-stage estimation process (Dong et al., 2021; Zou et al., 2023). Table 1 presents the baseline regression results of technological investment (R&D investment) on various factor-ATP using the 3SLS method.

Table 1 Baseline regression results of the impact of R&D investment in the Chinese petrochemical industry on factor-ATP

VARIABLES	PI		SUB		COM	
	(1)	(2)	(3)	(4)	(5)	(6)
	lnALE	lnAEK	lnALK	lnAKE	lnALE	lnAEK
lnRD	2.395***	0.495*	2.251**	-1.093***	3.531**	1.327
	(3.223)	(1.659)	(2.427)	(-4.433)	(2.380)	(1.015)
Control	√	√	√	√	√	√
Chi2	58.14	132.23	63.43	97.97	119.49	192.89
P>chi2	0.0000	0.0000	0.0000	0.0000	0.0000	0.0000
RMSE	3.315	1.332	3.391	0.901	1.230	1.083
Observations	110	110	66	66	44	44
R-squared	0.346	0.546	0.490	0.597	0.731	0.814

z-statistics in parentheses: ***$p<0.01$, **$p<0.05$, * $p<0.1$

Note: PI refers to the petrochemical industry; SUB refers to the sub-sectors where labour and non-labour factors (capital, energy) exhibit substitution relationships, including B07, C28, C29; COM refers to the sub-sectors where labour and non-labour factors (capital, energy) exhibit complementary relationships, including C25, C26. This classification is based on Sun et al. (2024b).

(i) From the entire sample of the petrochemical industry, it is observed that R&D investment exerts varying enhancing effects on different types of factor-ATP, with the strength of influence ranked as AL > AE > AK. The coefficients of the key explanatory variable lnRD in columns (1) and (2) are significant at the 1% and 10% levels, respectively. This signifies that a 1% increase in R&D investment enhances labour-ATP by 2.395% relative to energy-ATP, and enhances energy-ATP relative to capital-ATP by 0.495%. In economic terms[1], a 1% increase in the standard deviation of R&D investment yields a marginal effect of a 3.051% enhancement of labour-ATP relative to energy-ATP, and a 0.631% enhancement of energy-ATP relative to capital-ATP. Overall, both statistically and economically, R&D investment exhibits a bias towards enhancing different types of factor-ATP, thus confirming hypothesis H2. This conclusion not only supports the widely held consensus that R&D investment can boost technological progress (da Silva, 2023; Dai et al., 2022; Zhou et al., 2024) but also further reveals the selective enhancement of specific types of factor-ATP by R&D investment.

(ii) When analysed by sub-industry, in the sub-sectors (SUB) where labour exhibits substitution relationships with capital and energy, a 1% increase in R&D investment enhances labour-ATP relative to capital-ATP by 2.251%, while capital-ATP decreases by 1.093% relative to energy-ATP. In economic terms, a 1% increase in the standard deviation of R&D investment results in a marginal effect of a 2.379% enhancement of labour-ATP relative to capital-ATP, and a 1.155% decrease in capital-ATP relative to energy-ATP.

(iii) In the sub-sectors (COM) where labour exhibits complementary relationships with capital and energy, a 1% increase in R&D investment enhances labour-ATP relative to energy-ATP by 3.531%, and enhances energy-ATP relative to capital-ATP by 1.327%. In economic terms, a 1% increase in the standard deviation of R&D investment results in a marginal effect of a 5.173% enhancement of labour-ATP relative to energy-ATP, and a 1.944% increase in energy-ATP relative to capital-ATP.

4.1.2 Mechanism regression of technological investment on factor-ATP

This section further examines the mechanism by which technological investment (R&D investment) affects different types of factor-ATP. According to the theoretical model constructed in this paper, when the elasticity of factor substitution is less than 1 (indicating a complementary relationship between factors), the impact of R&D investment on factor-ATP is influenced by the price and environmental governance effects. Therefore, this section introduces interaction terms between R&D investment and the factor price ratio, as well as between R&D investment and the pollution control investment ratio, to test this mechanism of influence. Table 2 reports the results of the mechanism regression using the 3SLS method to estimate the impact of R&D investment on factor-ATP in the petrochemical industry.

---

[1] Economic terms represent the impact of a 1% standard deviation increase in the explanatory variable on the explained variable (calculated as coefficient × standard deviation × 1%). The values of the explanatory variables are provided in the Supplementary Material, and the corresponding standard deviations can be calculated.

Table 2 Mechanism regression results of the impact of R&D investment on factor-ATP

VARIABLES	PI	
	(1)	(2)
	lnALE	lnAEK
lnRD×lnsw	2.010***	
	(4.718)	
lnRD×lnPCLE	0.438	
	(1.129)	
lnRD×lnsr		0.214***
		(3.243)
lnRD×lnPCKE		-0.774***
		(-4.468)
Control	√	√
Chi2	100.70	937.70
P>chi2	0.0000	0.0000
RMSE	2.960	0.638
Observations	110	110
R-squared	0.479	0.896

z-statistics in parentheses: *** $p<0.01$, ** $p<0.05$, * $p<0.1$

(i) From the regression results, the coefficients of the key explanatory variables lnRD× lnsw and lnRD×lnsr interaction terms are both significant at the 1% level. This indicates that the impact of R&D investment on factor-ATP is indeed moderated by the price effect. Since the energy-labour price ratio in the petrochemical industry showed a decreasing trend during the sample period (see Fig. 3a), a 1% increase in R&D investment enhances labour-ATP relative to energy-ATP by 2.010%. Conversely, as the energy-capital price ratio showed an increasing trend during the sample period (see Fig. 3b), a 1% increase in R&D investment enhances energy-ATP relative to capital-ATP by 0.214%. This is mainly due to the fact that current fossil fuel subsidies in China distort energy prices (Li et al., 2024), making companies more willing to invest in energy for production while slowing down the uptake of energy-enhancing technologies. This validates hypothesis H3 (Ⅱ). This finding also aligns with the conclusion of Dasgupta and Roy (2015) that rising energy prices promote technological progress in the context of Indian manufacturing while providing a more in-depth analysis.

(ii) Under the moderating effect of environmental governance, R&D departments tend to focus on developing factor-augmenting technologies with lower pollution control costs. According to the regression coefficients of the key explanatory variables lnRD×lnPCLE and lnRD×lnPCKE interaction terms, since the pollution control investment ratio of labour-intensive intermediates to energy-intensive intermediates in the petrochemical industry showed a decreasing trend during the sample period (see Fig. 3c), a 1% increase in R&D investment enhances labour-ATP relative to energy-ATP by 0.438%. Additionally, as the pollution control investment ratio of capital-intensive intermediates to energy-intensive intermediates also showed a decreasing trend during the sample period (see Fig. 3d), a 1% increase in R&D investment decreases energy-ATP relative to capital-ATP by 0.774%. This validates hypothesis H4. This finding suggests that under the pressure of government environmental regulations, industries are compelled to choose technological progress with lower pollution costs (i.e., green technological progress). This is consistent with the conclusion reached by Xie and Teo (2022).

Specifically, for the sub-sectors (SUB) within the petrochemical industry where the elasticity of substitution between labour and non-labour factors is greater than 1, this section examines whether the impact of R&D investment on factor-ATP is influenced by the market scale effect and the environmental governance effect as moderating mechanisms. Table 3 reports the mechanism regression results using the 3SLS method to estimate the impact of R&D investment on factor-ATP in these petrochemical sub-sectors (SUB).

Table 3 Mechanism regression results of the impact of R&D investment on factor-ATP in petrochemical sub-sectors (SUB) (σ>1)

VARIABLES	SUB	
	(1)	(2)
	lnALK	lnAKE
lnRD×lnLK	-0.967*	
	(-1.861)	
lnRD×lnPCLK	1.103*	
	(1.860)	
lnRD×lnKE		0.912***
		(7.046)
lnRD×lnPCKE		0.347***
		(2.590)
Control	√	√
Chi2	351.88	2426.11
P>chi2	0.0000	0.0000
RMSE	1.894	0.235
Observations	66	66
R-squared	0.841	0.973

z-statistics in parentheses: ***p<0.01, **p<0.05, * p<0.1

According to the regression coefficient results of the key explanatory variables lnRD×lnLK and lnRD×lnKE interaction terms, since the labour-capital input ratio showed a decreasing trend during the sample period (see Fig. 4a), a 1% increase in R&D investment decreases labour-ATP relative to capital-ATP by 0.967%. Conversely, as the capital-energy input ratio

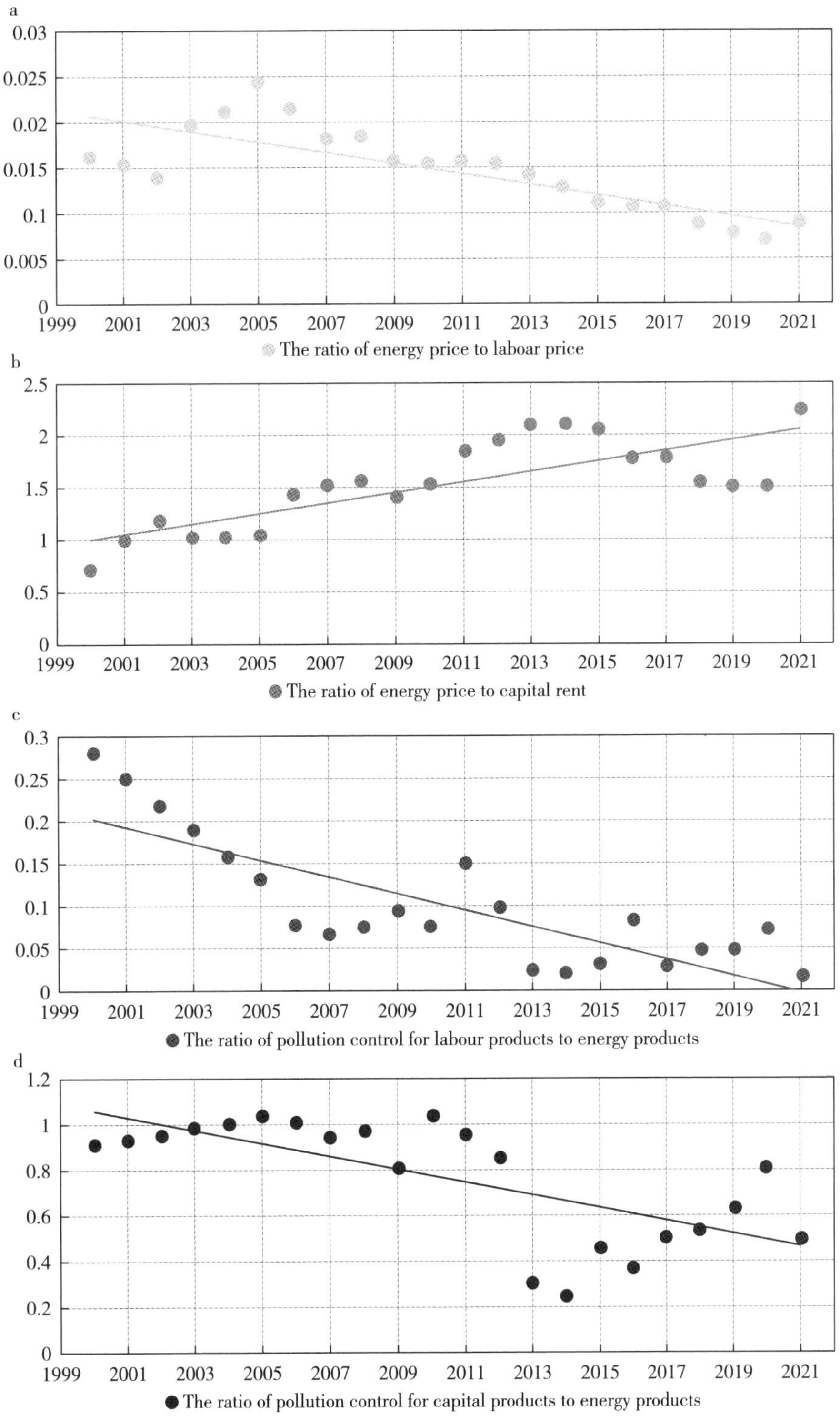

Fig. 3 Factor price ratios and pollution control investment ratios in the petrochemical industry

showed an increasing trend during the sample period (see Fig. 4b), a 1% increase in R&D investment enhances capital-ATP relative to energy-ATP by 0.912%. This validates hypothesis H3 (Ⅰ). This indicates that in the presence of substitutable production factors, the moderating effect of market scale predominates as firms tend to enhance technological progress with a larger factor market, as innovation, to some extent, is demand-driven (Xu et al., 2008).

According to the regression coefficient results of the key explanatory variables lnRD×lnPCLK and lnRD×lnPCKE interaction terms, since the pollution control investment ratio of labour-intensive intermediates to capital-intensive intermediates showed a decreasing trend during the sample period (see Fig. 4c), a 1% increase in R&D investment enhances labour-ATP relative to capital-ATP by 1.103%. Additionally, as the pollution control investment ratio of capital-intensive intermediates to energy-intensive intermediates also showed a decreasing trend during the sample period (see Fig. 4d), a 1% increase in R&D investment enhances capital-ATP relative to energy-ATP by 0.347%, which is consistent with hypothesis H4.

For petrochemical sub-sectors (COM) with an elasticity of substitution between labour and non-labour factors less than 1, this section examines whether the impact of R&D investment on factor-ATP is moderated by the effects of price and environmental governance. Table 4 reports the results of the mechanism regression using the 3SLS method to estimate the impact of R&D investment on factor-ATP in the petrochemical sub-sectors (COM).

Table 4 Mechanism regression results of the impact of R&D investment on factor-ATP in petrochemical sub-sectors (COM) (σ<1)

VARIABLES	COM	
	(1)	(2)
	lnALE	lnAEK
lnRD×lnws	0.415	
	(1.010)	
lnRD×lnPCLE	0.387 * *	
	(2.289)	
lnRD×lnsr		0.486 * * *
		(5.971)
lnRD×lnPCEK		-0.561 * * *
		(-4.147)
Control	√	√
Chi2	417.64	2729.14
P>chi2	0.0000	0.0000
RMSE	0.733	0.323
Observations	44	44
R-squared	0.904	0.984

z-statistics in parentheses: * * * p<0.01, * * p<0.05, * p<0.1

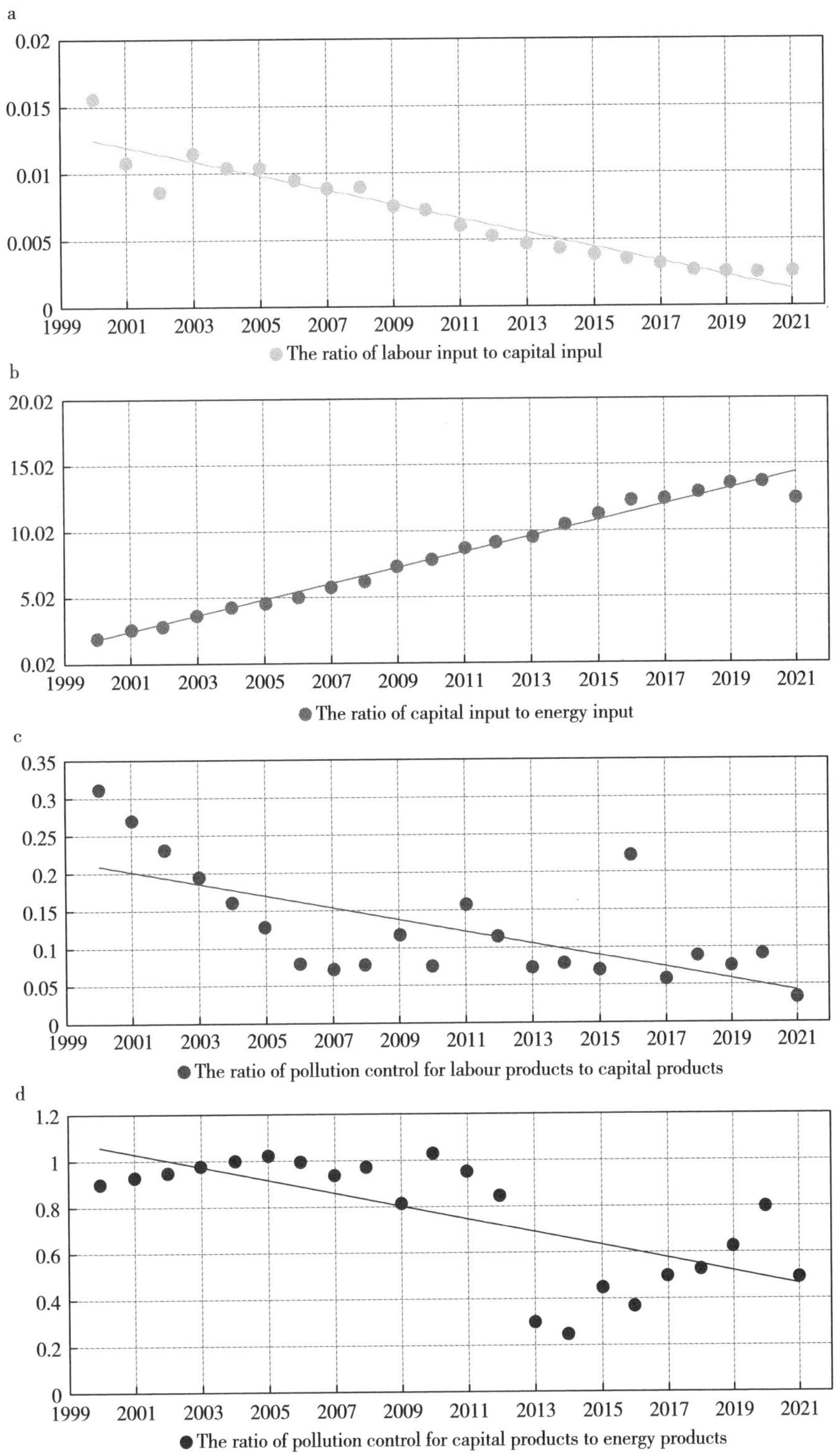

Fig. 4 Factor input ratios and pollution control investment ratios in the petrochemical sub-sector (SUB)

Based on the regression coefficient results of the interaction terms lnRD×lnws and lnRD×lnsr, it is observed that due to the upward trend in the labour-energy price ratio (see Fig. 5a), a 1% increase in R&D investment enhances labour - ATP relative to energy - ATP by 0.415%. Additionally, as the energy-capital price ratio also exhibits an upward trend (see Fig. 5b), a 1% increase in R&D investment enhances energy-ATP relative to capital-ATP by 0.486%, supporting hypothesis H3 (Ⅱ).

Furthermore, based on the regression coefficient results of the interaction terms lnRD × lnPCLE and lnRD × lnPCEK, it is observed that due to the declining trend in the pollution control investment ratio between labour - intensive intermediates and energy - intensive intermediates (see Fig. 5c), a 1% increase in R&D investment enhances labour-ATP relative to energy-ATP by 0.387%. Conversely, as the pollution control investment ratio between energy-intensive intermediates and capital - intensive intermediates shows an upward trend (see Fig. 5d), a 1% increase in R&D investment decreases energy-ATP relative to capital-ATP by 0.561%, supporting hypothesis H4.

4.1.3 Impact of factor-BTP on energy productivity

Due to the influence of factor substitution elasticity, the transformation of factor-ATP into factor-BTP has been extensively discussed by Sun et al. (2024b). The fundamental conclusion is that when the factor substitution elasticity is greater than 1, factor-ATP transforms into factor-BTP towards the same factor; conversely, it transforms into factor-BTP towards other factors. Building upon this research, this study further investigates the impact of factor-BTP on energy productivity and labour productivity to test hypothesis H1. Table 5 presents the comparative regression results using the 3SLS method to estimate the effect of factor-BTP on energy productivity and labour productivity in the petrochemical industry.

Table 5 Comparative regression results of the impact of factor-BTP on energy productivity and labour productivity

VARIABLES	PI		SUB		COM	
	(1)	(2)	(3)	(4)	(5)	(6)
	lnEP	lnLP	lnEP	lnLP	lnEP	lnLP
lnTBS	0.156***	0.240***	0.465***	0.527***	0.162***	0.231***
	(3.282)	(9.900)	(11.91)	(13.40)	(8.284)	(5.448)
Control	√	√	√	√	√	√
Chi2	885.91	582.44	666.97	711.93	1945.45	464.61
P>chi2	0.0000	0.0000	0.0000	0.0000	0.0000	0.0000
RMSE	0.322	0.266	0.204	0.228	0.077	0.154
Observations	110	110	66	66	44	44
R-squared	0.887	0.840	0.879	0.907	0.974	0.910

z-statistics in parentheses: ***p<0.01, **p<0.05, * p<0.1

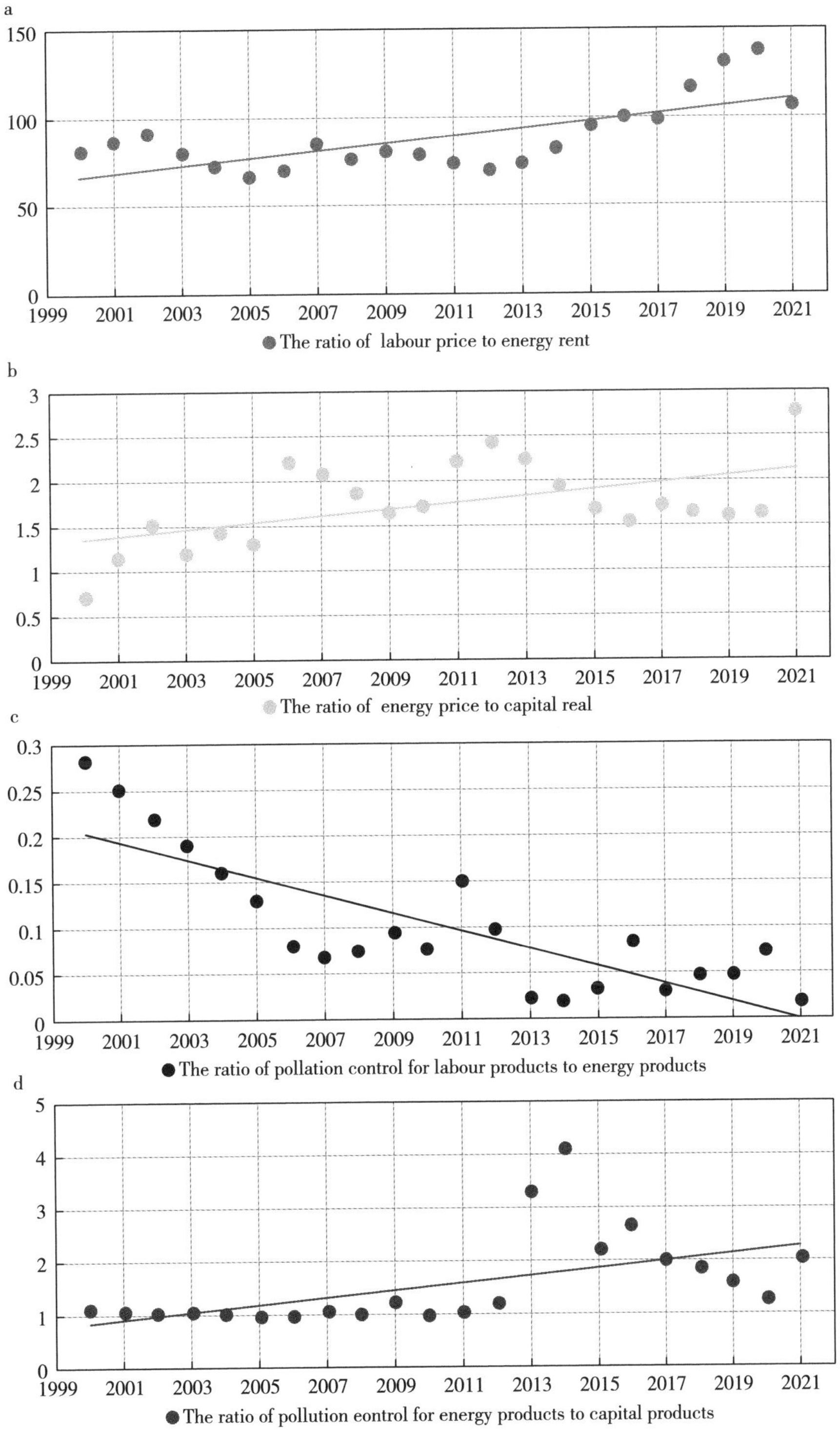

Fig 5 Factor price ratios and pollution control investment ratios in the petrochemical sub-sector (COM)

(i) From the entire sample of the petrochemical industry, a 1% increase in the technological biased summation index leads to a 0.156% increase in energy productivity and a 0.24% increase in labour productivity. In economic terms, a 1% increase in the standard deviation of technological biased summation index results in a marginal effect of a 0.2234% increase in energy productivity and a 0.3436% increase in labour productivity. This result confirms hypothesis H1 and also reveals that technological progress in the petrochemical industry currently favours improving labour productivity rather than energy productivity. This conclusion is supported by the research of Xiu et al. (2019) on the entire Chinese industry, which found that current technological progress is more biased towards energy use and carbon emissions relative to labour.

(ii) Examining the industry sectors separately, in the sub-industries (SUB) where labour substitutes for capital and energy, a 1% increase in the technological biased summation index results in a 0.465% increase in energy productivity and a 0.527% increase in labour productivity. In economic terms, a 1% increase in the standard deviation of technological biased summation index results in a marginal effect of a 0.5817% increase in energy productivity and a 0.6592% increase in labour productivity.

(iii) In the sub-industries (COM) where labour complements capital and energy, a 1% increase in the technological biased summation index leads to a 0.162% increase in energy productivity and a 0.231% increase in labour productivity. In economic terms, a 1% increase in the standard deviation of technological biased summation index results in a marginal effect of a 0.2561% increase in energy productivity and a 0.3651% increase in labour productivity.

4.2 Impact mechanism of technological investment on energy structure

This section introduces the ratio of non-clean and clean energy prices to test for the direct effect of price effects, and uses the ratio of pollution control investment between energy sources to form an interaction term with R&D inputs to test for the moderating mechanism. Table 6 reports the regression results using the 3SLS method to estimate the mechanism of the effect of R&D inputs on energy mix in the petrochemical industry.

Table 6 Regression results of the impact mechanism of R&D investment on energy structure in the Chinese petrochemical industry

VARIABLES	PI	SUB	COM
	(1)	(2)	(3)
	lnES	lnES	lnES
lnsnc	0.0811**	0.0112	0.00984
	(2.392)	(0.654)	(0.284)
lnRD×lnPCENC	0.0173**	0.0257***	0.0272**
	(2.369)	(4.890)	(2.147)

Continued

VARIABLES	PI	SUB	COM
	(1)	(2)	(3)
	lnES	lnES	lnES
Control	√	√	√
Chi2	678.64	2267.74	3375.65
P>chi2	0.0000	0.0000	0.0000
RMSE	0.054	0.018	0.019
Observations	110	66	44
R-squared	0.861	0.972	0.987

z-statistics in parentheses: ***p<0.01, **p<0.05, * p<0.1

(i) From the full sample of the petrochemical industry, an increase of 1% in the price ratio of non-clean to clean energy improves the energy structure by 0.0811%. This result indicates that when enterprises do not have a specific preference for energy types (i.e., different types of energy are strong substitutes), they tend to choose relatively cheaper energy (Liu and Li, 2011; Wu et al., 2024). Additionally, due to the increasing pollution control investment ratio of non-clean to clean energy (see Fig. 6a), an increase of 1% in R&D investment improves the energy structure by 0.0173%. This finding confirms that the environmental governance effect between different types of energy dominates enterprises' R&D investment directionally towards improving the marginal substitution of energy with lower pollution control costs (Chang et al., 2023). In economic terms, an increase of 1% standard deviation in the price ratio of non-clean to clean energy leads to a marginal improvement of 0.0247% in the energy structure; an increase of 1% standard deviation in R&D investment results in a marginal improvement of 0.022% in the energy structure. This result confirms hypothesis H4.

(ii) By industry segment, for sub-sectors (SUB) where labour and non-labour factors are substitutes, a 1% increase in the price ratio of non-clean to clean energy improves the energy structure by 0.0112%. Additionally, duc to the rising pollution control investment ratio of non-clean to clean energy during the sample period (see Fig. 6b), a 1% increase in R&D investment improves the energy structure by 0.0257%. In economic terms, a 1% standard deviation increase in the price ratio of non-clean to clean energy leads to a marginal improvement of 0.0034% in the energy structure; a 1% standard deviation increase in R&D investment results in a marginal improvement of 0.0272% in the energy structure.

(iii) For sub-sectors (COM) where labour and non-labour factors are complements, a 1% increase in the price ratio of non-clean to clean energy improves the energy structure by 0.0098%. Additionally, due to the rising pollution control investment ratio of non-clean to clean energy during the sample period (see Fig. 6c), a 1% increase in R&D investment improves the energy structure by 0.0272%. In economic terms, a 1% standard deviation increase in the price ratio of non-clean to clean energy leads to a marginal improvement of 0.003%; a 1% standard deviation increase in R&D investment results in a marginal improvement of 0.0398% in the energy structure.

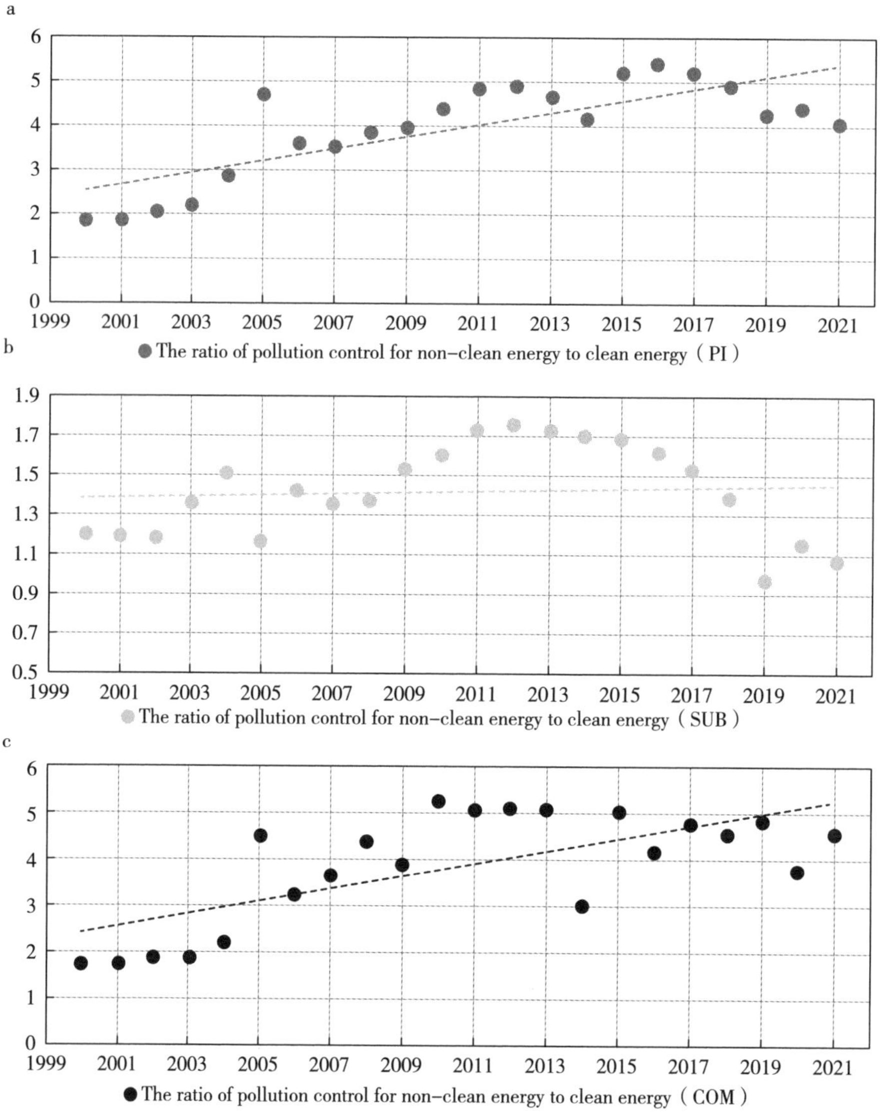

Fig. 6　Pollution control investment ratios for non-clean and clean energy in the petrochemical industry

Note: a represents the pollution control investment ratio for non-clean and clean energy in the petrochemical industry (PI); b represents the pollution control investment ratio for non-clean and clean energy in the petrochemical sub-sectors (SUB); c represents the pollution control investment ratio for non-clean and clean energy in the petrochemical sub-sectors (COM)

4.3 Robustness tests

To further verify the robustness of the results regarding the impact of technological investment on factor-ATP and energy structure, this section replaces the original proxy variable for technological investment, R&D investment, with the number of patent applications (PA) (Li et al., 2024). The simultaneous equations are re-estimated using the 3SLS method (see Table 7).

Table 7 Regression results of the number of patent applications (PA) in the Chinese petrochemical industry on factor-ATP and energy structure

VARIABLES	PI		
	(1)	(2)	(3)
	lnALK	lnAEK	lnES
lnPA	1.305**	0.421**	
	(2.514)	(2.067)	
lnsnc			0.0789**
			(2.330)
lnPA×lnPCENC			0.0140**
			(2.345)
Control	√	√	√
Chi2	52.46	135.50	727.18
P>chi2	0.0000	0.0000	0.0000
RMSE	3.373	1.324	0.051
Observations	110	110	110
R-squared	0.323	0.552	0.869

z-statistics in parentheses: $^{***}p<0.01$, $^{**}p<0.05$, $^{*}p<0.1$

The results from the table indicate that after replacing R&D investment with the number of PA, all statistical results remain significant at the 5% level. The regression results in columns (1) and (2) indicate that a 1% increase in PA leads to a 1.305% enhancement in labour-ATP relative to energy-ATP and a 0.421% enhancement in energy-ATP relative to capital-ATP. The order of impact intensity is AL>AE>AK, consistent with the results in Table 1.

The regression results in column (3) demonstrate that after replacing R&D investment with the number of PA, a 1% increase in the price ratio of non-clean to clean energy results in a 0.0789% improvement in energy structure, which is nearly identical to the 0.0811% improvement observed in Table 6. Additionally, due to the increase in the pollution control investment ratio of non-clean to clean energy (as depicted in Fig. 6a), a 1% increase in PA leads to a 0.0140% improvement in energy structure, which closely matches the 0.0173% improvement observed in Table 6. These results indicate that the empirical findings regarding the impact of technological investment on factor-ATP and energy structure are robust, and the insights gained are reliable.

5 Conclusions and policy suggestions

5.1 Conclusions

The mechanisms of technological investment on energy productivity and energy structure in the petrochemical industry remain unclear due to the directional nature of technological progress. Addressing this scientific issue, this paper proposes a unified theoretical framework for the impact of directed technological investment on energy productivity and energy structure. Utilising a two-dimensional panel dataset at the petrochemical sub-industry and time levels, this study empirically investigates the effects of directed technological investment on energy productivity and energy structure through baseline regression, mechanism regression, and robustness tests. The main findings are as follows:

(1) From the full sample of the Chinese petrochemical industry, technological investment has different enhancing effects on various factor-ATP, with the strength of the effects ordered as follows: labour-ATP > energy-ATP > capital-ATP. Specifically, a 1% increase in technological investment enhances labour-ATP by 2.395% relative to energy-ATP, and energy-ATP by 0.495% relative to capital-ATP. Additionally, influenced by factor substitution elasticity, the overall technological progress in the petrochemical industry shifts from labour-augmenting to capital-biased. With a 1% increase in the technological biased summation index, energy productivity improves by 0.156% and labour productivity by 0.24%. This finding explains that the current technological progress in the Chinese petrochemical industry has brought about an increase in energy productivity, but it is much lower than the increase in labour productivity, which makes producers choose to replace labour with more capital and energy while satisfying economic growth, resulting in irrational allocation of factor inputs.

(2) The role of overall technological investment in the petrochemical industry on factor-enhanced technological progress is moderated by the price effect (Reason: factor elasticity of substitution is less than 1). Specifically, because the energy-labour price ratio in the petrochemical industry has shown a downward trend during the sample period, a 1% increase in technological investment enhances labour-ATP by 2.010% relative to energy-ATP. Conversely, with the energy-capital price ratio showing an upward trend during the sample period, a 1% increase in technological investment enhances energy-ATP by 0.214% relative to capital-ATP.

(3) The impact of overall technological investment on factor-ATP in the petrochemical industry is also moderated by the environmental governance effect, consistently enhancing the technological progress of factors with lower pollution control costs. Specifically, as the pollution control investment ratio of labour-intensive intermediates to energy-intensive intermediates showed a decreasing trend during the sample period, a 1% increase in technological investment enhances labour-ATP by 0.438% relative to energy-ATP. Additionally, as the pollution control investment ratio of capital-intensive intermediates to energy-intensive intermediates also showed a decreasing trend during the sample period, a 1% increase in technological investment decreases energy-ATP by 0.774% relative to capital-ATP.

(4) The overall energy structure of the petrochemical industry is directly influenced by

energy prices and indirectly affected by technological investment under the moderation of the environmental governance effect. Specifically, a 1% increase in the price ratio of non-clean to clean energy improves the energy structure by 0.0811%. Furthermore, due to the rising trend of the pollution control investment ratio of non-clean energy relative to clean energy from 2000 to 2021, a 1% increase in technological investment improves the energy structure by 0.0173%.

5.2 Policy suggestions

Based on the research findings, the following policy recommendations are proposed.

Firstly, companies should further enhance energy-ATP given the current complementary relationship between production factors in the petrochemical industry. This would encourage companies to invest more labour to replace energy in the production process. Additionally, the current technological investment in labour-ATP has led to a continuous decline in sector employment rates. It is recommended that this portion of technological investment be redirected towards enhancing energy-augmenting technologies, which would not only stabilize employment but also reduce carbon emissions.

Secondly, between 2000 and 2021, the rapid increase in labour prices compared to energy prices in the petrochemical industry has incentivised companies to pursue labour-ATP. The government should expedite the process of removing energy subsidies, allowing energy prices to be primarily determined by supply and demand. Such a rebound in energy prices would motivate firms to shift their technological investments towards enhancing energy-augmenting technologies while reducing labour-augmenting technologies, thereby orienting overall technological progress towards labour and capital.

Thirdly, the government should strengthen environmental regulations by enforcing strict emission standards, compelling companies to allocate more funds for environmental management, thereby indirectly driving their willingness to improve energy-augmenting technologies and clean energy technologies. Moreover, the government should enhance its role in guiding firms' selection of energy-augmenting technologies by establishing a comprehensive system of constraints and incentives.

Finally, government subsidies for new energy have incentivized the petrochemical industry to adopt cleaner energy sources as production inputs, improving the industry's energy structure between 2010 and 2021. However, the government should exercise caution when intervening in market price controls. It is advisable to adopt a market-oriented approach to further improve environmental governance policies, thereby compelling firms to voluntarily direct their technological investments towards enhancing clean production technologies in order to reduce pollution control costs.

5.3 Future Work

Although this study has made some marginal contributions based on previous work, there is still room for further research. This paper reveals the complex channels through which technological investment impacts energy productivity and energy structure. However, how to influence technological investment through different policy scenarios to further enhance energy productivity and improve energy structure warrants deeper exploration.

Acknowledgements This work was supported by the National Key Research and Development Program of China (2022YFC2105401) and the China Scholarship Council (202306400059).

References

Acemoglu, D., 2007. Equilibrium bias of technology. Econometrica 75, 1371-1409.

Acemoglu, D., 2002. Directed technical change. Rev. Econ. Stud. 69, 781-809.

Acemoglu, D., Aghion, P., Bursztyn, L., Hemous, D., 2012. The environment and directed technical change. Am. Econ. Rev. 102, 131-166.

Acemoglu, D., Akcigit, U., Alp, H., Bloom, N., Kerr, W., 2018. Innovation, reallocation, and growth. Am. Econ. Rev. 108, 3450-3491.

Acemoglu, D., Akcigit, U., Hanley, D., Kerr, W., 2016. Transition to clean technology. J. Polit. Econ. 124, 52-104.

Adachi, H., Inagaki, K., Nakamura, T., Osumi, Y., 2019. Technological progress and unemployment. Technol. Prog. Income Distrib. Unempl. Theory Empirics 47-68.

Chang, J. J., Huang, C. Y., Wong, C. Y., Yang, Y., 2023. Environmental regulation stringency and allocation between R&D and physical capital: A two-engine growth model. J. Econ. Behav. Organ. 216, 733-753.

Chen, Y., Liu, Y., 2021. How biased technological progress sustainably improve the energy efficiency: An empirical research of manufacturing industry in China. Energy 230, 120823.

Chen, Y., Wong, C. W., Yang, R., Miao, X., 2021. Optimal structure adjustment strategy, emission reduction potential and utilization efficiency of fossil energies in China. Energy 237, 121623.

da Silva, M. A., 2023. Cobb-Douglas R&D production function, appropriability and opportunity: effects on R&D, technological progress and knowledge sharing. Econ. Innov. New Technol. 32, 1069-1086.

Dai, L., Zhang, J., Luo, S., 2022. Effective R&D capital and total factor productivity: Evidence using spatial panel data models. Technol. Forecast. Soc. Change 183, 121886.

Dasgupta, S., Roy, J., 2015. Understanding technological progress and input price as drivers of energy demand in manufacturing industries in India. Energy Policy 83, 1-13.

Dong, K., Dong, X., Jiang, Q., Zhao, J., 2021. Assessing energy resilience and its greenhouse effect: A global perspective. Energy Econ. 104, 105659.

Farajzadeh, Z., Ghorbanian, E., Tarazkar, M. H., 2023. The impact of climate change on economic growth: Evidence from a panel of Asian countries. Environ. Dev. 47, 100898.

Fellner, W., 1961. Two propositions in the theory of induced innovations. Econ. J. 71, 305-308.

Guo, P., Wang, T., Li, D., Zhou, X., 2016. How energy technology innovation affects transition of coal resource-based economy in China. Energy Policy 92, 1-6.

Habakkuk, H. J., 1963. American and British technology in the nineteenth century: The search for labour-saving inventions. J. Econ. Hist. 23, 104-106.

Hassler, J., Krusell, P., Olovsson, C., 2012. Energy-saving technical change. National Bureau of Economic Research, Working Paper No. w18456.

Hémous, D., 2016. The dynamic impact of unilateral environmental policies. J. Int. Econ. 103, 80-95.

Hicks, J., 1963. The theory of wages. Springer.

Hou, Z., Roseta-Palma, C., dos Santos Ramalho, J. J., 2021. Does directed technological change favor energy? Firm-level evidence from Portugal. Energy Econ. 98, 105248.

Keen, S., Ayres, R. U., Standish, R., 2019. A Note on the Role of Energy in Production. Ecol. Econ. 157, 40-46.

León-Ledesma, M. A., McAdam, P., Willman, A., 2010. Identifying the elasticity of substitution with biased technical change. Am. Econ. Rev. 100, 1330-1357.

Li, K., Lin, B., 2018. How to promote energy efficiency through technological progress in China? Energy 143, 812-821.

Li, L., Wang, Lie, Zhang, X., Wang, Lan, 2024. Technological Composition and Innovation Factors in Inventive Yangtze River Delta: Evidence from Patent Inventions. Appl. Sci. 14, 1842.

Li, Z., Zhao, Q., Guo, H., Huang, R., 2024. Impact of fossil fuel subsidies on energy-saving technological change in China. Energy 286, 129265.

Lin, B., Teng, Y., 2024. Industrial chain division and carbon emission intensity: The moderating effect of digitization. Energy 286, 129573.

Liu, W., Li, H., 2011. Improving energy consumption structure: A comprehensive assessment of fossil energy subsidies reform in China. Energy Policy 39, 4134-4143.

Lyu, X., Ke, H., 2022. Dynamic threshold effect of directed technical change suppress on urban carbon footprint in China. Int. J. Environ. Res. Public. Health 19, 5151.

Mallick, S. K., Sousa, R. M., 2017. The skill premium effect of technological change: New evidence from United States manufacturing. Int. Labour Rev. 156, 113-131.

Oryani, B., Koo, Y., Rezania, S., Shafiee, A., 2021. Investigating the asymmetric impact of energy consumption on reshaping future energy policy and economic growth in Iran using extended Cobb-Douglas production function. Energy 216, 119187.

Perera-Tallo, F., 2017. Growing income inequality due to biased technological change. J. Macroecon. 52, 23-38.

Qiu, Y., Han, W., Zeng, D., 2023. Impact of biased technological progress on the total factor productivity of China's manufacturing industry: The driver of sustainable economic growth. J. Clean. Prod. 409, 137269.

Rootzén, J., Nyberg, T., Karltorp, K., Åhman, M., 2023. Turning the tanker? Exploring the preconditions for change in the global petrochemical industry. Energy Res. Soc. Sci. 104, 103256.

Sanstad, A. H., Roy, J., Sathaye, J. A., 2006. Estimating energy-augmenting technological change in developing country industries. Energy Econ. 28, 720-729.

Singh, A., 2022. Oil Price uncertainty and labor investment efficiency. Energy Econ. 116, 106407.

Solow, R. M., 1956. A contribution to the theory of economic growth. Q. J. Econ. 70, 65-94.

Sun, G., Huang, Y., Tian, H., Wang, F., 2012. Total factor productivity, inputs substitution and regional energy efficiency disparity. Econ. Res. J. 9, 99-112.

Sun, X., Van Fan, Y., Lei, Y., Zhao, J., Chen, W., Cao, Z., 2024a. Towards decoupling in chemical industry: Input substitution impacted by technological progress. J. Clean. Prod. 452, 142040.

Sun, X., Lei, Y., Wang, X., Zhao J., Petar, S., 2024b. Directional Nature of Technological Progress in the Petrochemical Industry Prompting Energy Marginal Substitution. Energy 308, 132997.

Tassey, G., 2005. The disaggregated technology production function: a new model of university and corporate research. Res. Policy 34, 287-303.

Wu, L., Xu, C., Zhu, Q., Zhou, D., 2024. Multiple energy price distortions and improvement of potential energy consumption structure in the energy transition. Appl. Energy 362, 122992.

Xie, R., Teo, T. S., 2022. Green technology innovation, environmental externality, and the cleaner upgrading of industrial structure in China—Considering the moderating effect of environmental regulation. Technol. Forecast. Soc. Change 184, 122020.

Xiu, J., Zhang, G., Hu, Y., 2019. Which kind of directed technical change does China's economy have? From the perspective of energy-saving and low-carbon. J. Clean. Prod. 233, 160-168.

Xu, C., Wang, J., Wang, H., 2008. Demand-oriented innovation of firms in China: An empirical study. Front. Econ. China 3, 548-559.

Xue, J., Yip, C. K., 2013. Aggregate elasticity of substitution and economic growth: A synthesis. J. Macroecon. 38, 60-75.

Yang, G., Zha, D., Zhang, C., Chen, Q., 2020. Does environment-biased technological progress reduce CO_2 emissions in APEC economies? Evidence from fossil and clean energy consumption. Environ. Sci. Pollut. Res. 27, 20984-20999.

Yu, L., Wang, Y., Wei, X., Zeng, C., 2023. Towards low-carbon development: The role of industrial robots in decarbonization in Chinese cities. J. Environ. Manage. 330, 117216.

Yuan, C., Liu, S., Wu, J., 2009. Research on energy-saving effect of technological progress based on Cobb-Douglas production function. Energy Policy 37, 2842-2846.

Zhan, J., 2023. Regional differentiation in influencing factors of clean renewable energy consumption from the perspective of air pollution prevention and control. 3c Tecnol. Glosas Innov. Apl. Pyme 12, 331-345.

Zhang, C., Yang, S., 2024. The synergy effect of energy security and carbon-haze collaborative management: From the perspective of biased technological progress. Environ. Res. 252, 118741.

Zhang, H., Ke, H., 2022. Spatial spillover effects of directed technical change on urban carbon intensity, based on 283 cities in China from 2008 to 2019. Int. J. Environ. Res. Public. Health 19, 1679.

Zhao, C., Ma, X., Che, S., Liu, Z., Song, C., 2024. Does climate aid alleviate carbon lock-in? A global perspective. J. Clean. Prod. 449, 141782.

Zhou, G., Xu, Y., Zhang, F., 2024. Measurement of innovation efficiency in logistic enterprises: Evidence from China based on the three-stage DEA-Malmquist index model approach. Am. J. Econ. Sociol. 83, 331-381.

Zhou, Y., Tian, L., Yang, X., Wan, B., 2024. Robust green Schumpeterian endogenous growth model and spatial Kuznets curve. Energy Econ. 133, 107520.

Zhu, W., Zhang, Z., Li, X., Feng, W., Li, J., 2019. Assessing the effects of technological progress on energy efficiency in the construction industry: a case of China. J. Clean. Prod. 238, 117908.

Zou, R., Yang, J., Feng, C., 2023. Does informatization alleviate energy poverty? A global perspective. Energy Econ. 126, 106971.

矿产资源经济

Assessing nickel sector index volatility based on quantile regression for Garch and Egarch models: Evidence from the Chinese stock market 2018—2022

LU Linna[1,*], LEI Yalin[2], YANG Yang[3], ZHENG Haoqi[4], WANG Wen[5], MENG Yan[6], MENG Chunhong[7], ZHA Liqiang[8]

1. *School of Management*, *Hebei University*, *Baoding* 071002, *China*
2. *Principal's Office*, *China University of Geosciences* (*Beijing*), *Beijing* 100029, *China*
3. *Department of Commerce*, *Hubei Water Resources Technical College*, *Wuhan* 430070, *China*
4. *School of Mathematics and Information Science*, *Hebei University*, *Baoding* 071002, *China*
5. *Development and Research Center of China Geological Survey*, *Beijing* 100037, *China*
6. *Research Institute of Baoshan Iron and Steel Co.*, *Ltd.*, *Shanghai* 201999, *China*
7. *Air Satellite* (*Shandong*) *Technology Group Co.*, *Ltd.*, *Jinan* 250031, *China*
8. *School of Basic Medical Sciences*, *Hebei University*, *Baoding* 071002, *China*

Abstract Assessing nickel sector index volatility in China is essential to observe the price dynamics of China's nickel industry and to promote its sound development. This study proposed a volatility analysis method based on quantile regression for Garch and Egarch models to depict the fluctuation characteristics of the nickel sector index from June 12, 2018 to June 1, 2022. Garch, Egarch, Garch-Quantile Regression (QR) and Egarch-Quantile Regression (QR) models were established. The results indicated that the nickel sector market had been weak-form efficient and it could respond to market information effectively. An ARCH effect and volatility agglomeration characteristics exist while the nickel sector index is closely related to the global economic climate and geopolitics. Alienated conditional standard deviation shows the role of special events in fueling stock prices and the abnormal rise or fall of asset fluctuations in the current period. The sensitivity of the nickel sector index to negative information is not necessarily greater than that to positive information. By comparing the volatility fitted by different estimation methods, it can be observed that the EGARCH-QR estimation method has the best fitting effect on the volatility. Therefore, a more robust estimator can be obtained. EGARCH-QR model can more accurately reflect market fluctuations and improve the robustness of risk measurement estimates. Listed companies in the domestic nickel industry should grasp the dynamics of nickel supply and demand at home and abroad, and reasonably arrange production and sales.

Keywords Volatility, Nickel sector index, Garch model, Egarch model, Quantile regression (QR)

原文刊于 *Resources Policy*, 2023 年第 82 卷。原文发表时为彩色。

* 为通信作者。

1 Introduction

Nickel is the fifth most common element found on Earth. By far the most important use of new nickel is the production of stainless steels. This use accounts for over two thirds of first use nickel up from one-third in the past three decades. The price of nickel has shown considerable volatility over the last forty years. The rapid Chinese economic recovery during the second half of 2020 was a major driver behind the increase in nickel prices which reached 18568 US dollars in February 2021 (International Nickel Study Group, 2023). Assessing nickel sector index volatility in China is essential to observe the price dynamics of China's nickel industry and to promote its sound development.

Assessing volatility and investigating the stock price behavior is an essential step in many financial decision makings (McDonald and Fisher, 1972; Ball and Philip, 2002; Batmunkh et al., 2020; Le and Luong, 2022; Sharma et al., 2021). Engle constructed an autoregressive conditional heteroscedasticity model, namely the ARCH model, which effectively reflects the time-varying variance (Engle, 1982). Since then, academic circles have widely used it in the study of stock price volatility (Naik and Mohan, 2021). Diverse models have been applied to describe stock price volatility. The approximate factor model focuses on the synchronization of stock prices (Zhao et al., 2021). MCMC method, the TVP-SV-VAR model and Kinetic Ising Model are good at studying the dynamic impact effects of price-type and quantity-type monetary policies on stock prices (Sun and Zhu, 2021; Campajola et al., 2022). Agent-based artificial market simulation platform can be used to simulate the impact of heterogeneous herd behavior on stock price fluctuations (Ying and Yin, 2021). GARCH family of models has been extensively used in finance and economics, particularly for estimating volatility (Christofferson and Jacobs, 2004; Kim and Won, 2018; Nguyen and Nguyen, 2019; Zhao et al., 2019; Hu et al., 2020; Endri et al., 2020; Fakhfekh and Jeribi, 2020; Mohsin et al., 2020; Wang et al., 2020; Xing et al., 2021; Li, 2022; Endri et al., 2021; Liu et al., 2021; Naik and Mohan, 2021; Sen et al., 2021; Wang et al., 2022). GARCH models rarely consider time-varying volatility and heteroskedasticity (Tse, 2000; Fathian et al., 2019; Gong et al., 2019; Cai and Wu, 2022). Meanwhile, the quantile regression (QR) introduced by Koenker and Bassett (1978) has proven to be a very precise tool to model the impact of the conditional variables on the dependent variable (Koenk er and Hallock, 2001; Koenker, 2005). QR is suitable for the situation in the case of heteroscedasticity (Dieobold, 1986; Waldmann, 2018; Hu et al., 2022). Furthermore, the QR estimation of the Garch model can weaken the influence of extreme fluctuations and obtain a more accurate and effective estimation (Zhong and Li, 2022; Joo and Park, 2021; Lee and Noh, 2013). In view of this, this paper combined Garch models and QR estimation to study the nickel sector index fluctuations in China's stock market.

We aimed to: (1) establish the GARCH, GARCH-QR and EGARCH-QR models of nickel sector index in China's stock market from June 12, 2018 to June 1, 2022. (2) compare the regression accuracy of those models and describe the market characteristics of the volatility of the nickel sector index. The rest of the paper is organized as follows. Materials and methods introduce the model design of GARCH, GARCH-QR and EGARCH-QR models and data description. The regression results are presented and discussed in Results and Discussion, followed by the Conclusion.

2 Materials and methods

2.1 GARCH model

GARCH model further simulates the variance of random error term, which is suitable for time series volatility analysis and can effectively eliminate excessive peaks in data fluctuations (Lundbergh and Teräsvirta, 2002; Francq and Zakoian, 2019; Lin et al., 2020; Wang and Yan, 2022). Its mathematical expression is as follows:

$$\begin{aligned}\sigma_t^2 &= a_0 + \sum_{i=1}^{q} a_i\varepsilon_{t-i}^2 + \sum_{j=1}^{p} \theta_j h_{t-j} \\ &= a_0 + a(B)\varepsilon_t^2 + \theta_j(B)h_t \end{aligned} \tag{1}$$

Where, a_0 denotes the constant term, ε_t^2 is the random error term and σ_t^2 is the conditional variance, $p \geqslant 0$, $q > 0$, $a_0 > 0$, $a_i \geqslant 0$, $i = 1, \cdots, q$, $\theta_j \geqslant 0$, $j = 1, \cdots, p$. The Sequence follows GARCH (p, q) process. In order to ensure the wide and flat stability of GARCH (p, q) model, the parametric constraint is $a(B) + \theta(B) < 1$.

GARCH (1, 1) model is used to analyze nickel sector index fluctuation. GARCH (1, 1) is a distribution lag model about variance, and the standard form is shown in equation (2).

$$\begin{aligned} y_t &= x_t\gamma + u_t \\ \sigma_2^t &= a_0 + a_1\varepsilon_{t-i}^2 + \theta_1\sigma_{t-1}^2 \end{aligned} \tag{2}$$

Therefore, equation (2) can also be called conditional variance equation. Where, a_0 denotes the constant term, and the random error term of the previous period is ε_{t-1}^2 (i. e. ARCH term). The prediction variance of the previous period is σ_{t-1}^2 (i. e. GARCH item). a_1 and θ_1 are the coefficients of ε_{t-1}^2 and σ_{t-1}^2 respectively.

Applying the weighted average of invariant variance a_0, predicted value of variance in the previous period σ_{t-1}^2 and the residual square of the previous period ε_{t-1}^2 is of practical economic significance to predict the variance of the current period. When the current period's assets fluctuate abnormally, investors will deepen their psychological expectations of the variance in the next period, and the sharp fluctuations of assets may lead to further fluctuations.

Therefore, it is reasonable to use the GARCH model to analyze the fluctuation law of nickel sector index.

2.2 EGARCH model

In order to better explain the impact of positive and negative information on price fluctuations, the GARCH model is evolved into an exponential GARCH model, namely the EGARCH model, whose form is as follows:

$$\ln(\sigma_t^2) = \omega + \beta\ln(\sigma_{t-1}^2) + \alpha\left|\frac{\varepsilon_{t-1}}{\sigma_{t-1}}\right| + \gamma\left[\frac{\varepsilon_{t-1}}{\sigma_{t-1}}\right] \tag{3}$$

Compared with GARCH model, the advantage of EGARGH model is that it can distinguish the different effects of positive and negative information.

In equation (3), β reflects the degree of fluctuation and γ shows asymmetry. If $\gamma=0$, it means that the effect of positive information and negative information is symmetric. If $\gamma\neq0$, it means that the information effect is asymmetric. And only when $\gamma<0$, the leverage effect is significant. Besides, $\gamma<0$ means the impact of negative information is obvious.

2.3 GARCH-QR and EGARCH-QR models

Quantile regression (QR) estimation is to estimate the parameters of the model by minimizing the weighted sum of the absolute values of the residuals (Koenker and Bassett, 1978). Compared with traditional estimation methods, quantile regression is not easily affected by outliers, and does not require any error terms in the model (Meng et al., 2020; Korkmaz et al., 2021; Wang and Jv, 2021). It can reflect data from the selection of different quantiles and provide more effective information (He et al., 2019).

Assuming a random variable y, the sample composition time series is $\{y_t\}$, $t=1, 2, 3\cdots n$, for any $0<\tau<1$, the quantile regression estimation coefficient of the τ quantile of the sample series is:

$$\hat{\beta}^{QR} = \operatorname{argmin}\sum_{t=1}^{n}\rho_t(y_i - \beta) \tag{4}$$

Therefore, the GARCH model based on quantile regression estimation (GARCH-QR) is as follows:

$$(\hat{\omega}, \hat{\beta}, \hat{\alpha}) = \operatorname{argmin}\sum_{t=2}^{n}\rho_\tau(\sigma_t^2 - \omega - \beta\varepsilon_{t-1}^2 - \alpha\sigma_{t-1}^2) \tag{5}$$

The EGARCH model based on quantile regression estimation (EGARCH-QR) is as follows:

$$(\hat{\omega}, \hat{\beta}, \hat{\alpha}, \hat{\gamma}) = \operatorname{argmin}\sum_{t=2}^{n}\rho_\tau\left(\ln(\sigma_t^2) - \omega - \beta\ln(\sigma_{t-1}^2) - \alpha\left|\frac{\varepsilon_{t-1}}{\sigma_{t-1}}\right| - \gamma\frac{\varepsilon_{t-1}}{\sigma_{t-1}}\right) \tag{6}$$

2.4 Data description

Garch and Egarch models require large sample data. In order to ensure sufficient sample size, the time series used is the daily closingprice of the nonferrous nickel index from June 12, 2018 to June 1, 2022, with a total of 962 observations. The nonferrous nickel index comes fromthe statistical data provided by the wind data base(http://www.wind.com.cn/). Inorder to slow down the fluctuation of the series, the nickel sector index was treated with naturall ogarithm. In addition, in order to eliminate the inflation factor of mineral products, according to the World Bank(http://www.worldbank.org/), the nickel sector index was deflated by the inflation rate measured by the consumer price index based on June 2018.

3 Results

3.1 Stability test

Before constructing the ARCH model, it is necessary to use the unit root test to test the stationarity of the nickel sector index time series. Due to the autocorrelation of the nickel sector index time series, the ADF test was performed on it (Table 1). The results showed that the nickel sector index series rejected the unit root hypothesis and it was a stationary process.

Table 1 The result of the ADF test

Variable	T statistic	P value	1% statistical value	5% statistical value
np_t	-30.9252	0.0000	-3.9676	-3.4145

3.2 ARCH effect test

An ARIMA model was estimated for the nickel sector index series using the OLS method, and the result was shown in equation (7).

$$np_t = 6.6716 + 0.9972ar(1) + \varepsilon_t$$
$$(28.0418) \quad (403.4822) \tag{7}$$

$R^2 = 0.9941$, $\bar{R}^2 = 0.9941$, Log likelihood = 2359.819, AIC = -4.9070, SC = -4.8969

It can be seen from equation (7) that the statistics show that the model is very significant and the fitting validity is good. Whereas, the residuals of the model in Fig. 1 show the cluster characteristics of nickel sector index fluctuations, that is, the variance is relatively small in a certain period of time, but relatively large in another period, indicating that the error term has conditional heteroscedasticity.

Statistical characteristics of nickel sector index are shown in Fig. 2. The skewness of the time series of the nickel sector index is 1.0326, that is, the probability distribution of the time series is asymmetric, showing the characteristics of long left and thick tails and indicating that the negative deviation value is large. The kurtosis of the time series is 19.9687, showing an obvious distribution characteristic of sharp peaks and thick tails. The Jarque-Bera test also rejects the hypothesis that the time series is normally distributed at the 5% level of significance, indicating that the time series of nickel sector index fluctuates wildly.

Equation (6) is further subjected to the ARCH LM test for conditional heteroskedasticity with a lag of one order. The result is shown in Table 2.

The results of the ARCH LM test show that the corresponding P values of the F and LM statistics are both less than 0.05, rejecting the null hypothesis. Therefore, there is a first-

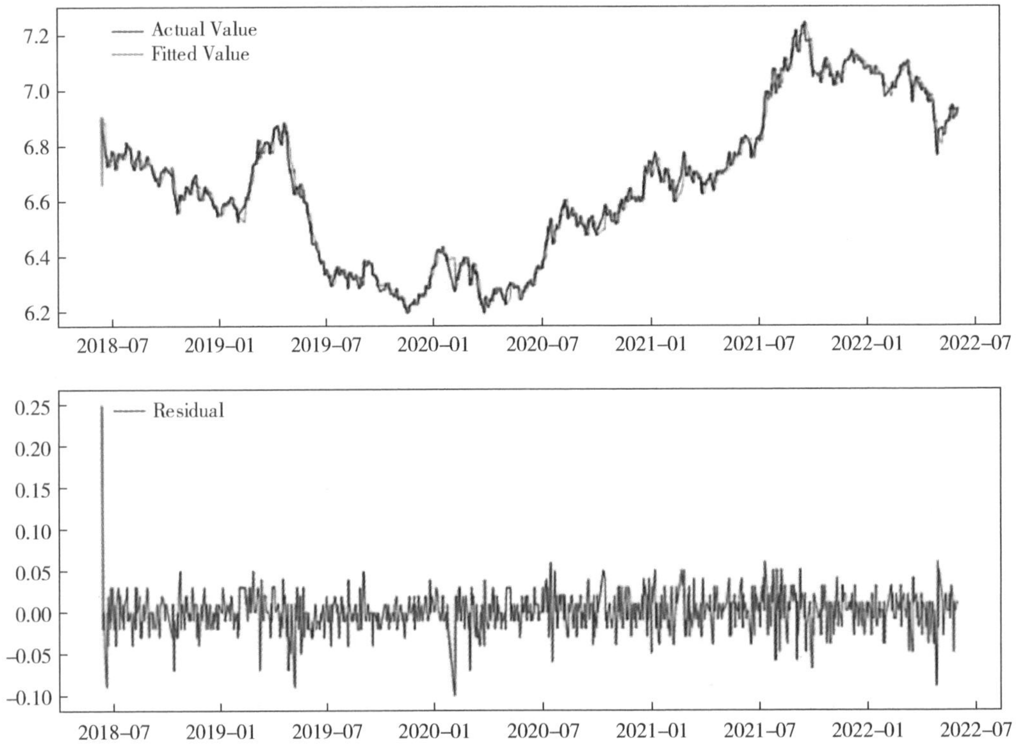

Fig. 1　Time series of nickel sector index

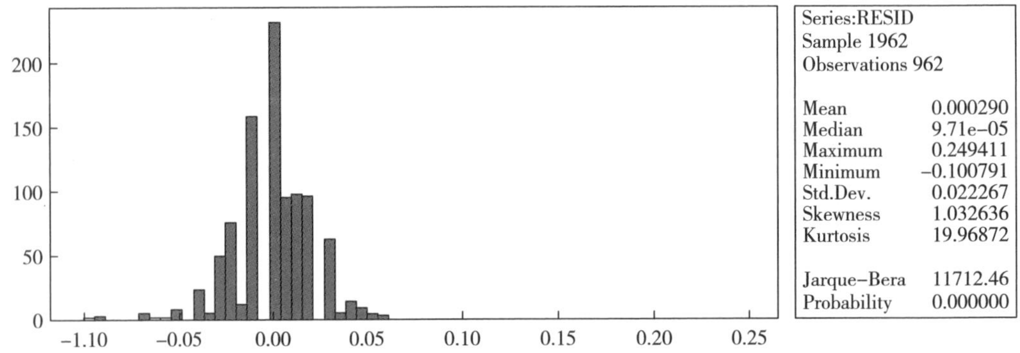

Fig. 2　Statistical characteristics of the nickel sector index

order autocorrelation in the squared sequence of the residuals, indicating that the residual sequence of the model has an ARCH effect.

Table 2　ARCH LM Test of the OLS fitting results

F-statistic	Scaled explained SS	F (1959)	Prob. Chi-Square (1)
6. 8378	12. 8192	0. 0091	0. 0091

3.3 GARCH model

The GARCH (1, 1) model is established as follows:

$$\sigma_t^2 = 0.0006 - 0.0265\varepsilon_{t-1}^2 - 0.4083\sigma_{t-1}^2$$
$$(11.9220)\ (-2.0927)\ (-3.6107) \quad (8)$$

$R^2=0.9941$, $\bar{R}^2=0.9941$, Log likelihood=2377.844, AIC=-4.9383, SC=-4.9130

The statistics corresponding to the coefficients of the ARCH term and the GARCH term in the conditional variance model are both significant, indicating that the conditional heteroscedasticity is obvious. The AIC and SC values of the GARCH (1, 1) model are both small, and it can be seen from the conditional variance estimates that the coefficients a_1 andθ_1 are both significantly different from 0, and $a_1+\theta_1=-0.4348<1$. Therefore, the GARCH (1, 1) model fits the nickel sector index sequence well. The GARCH term coefficient -0.4083 reflects the long memory of the system. Compared with θ_1, the fluctuation degree of α_1 is smaller, and the ARCH coefficient of -0.0265 reflects the impact of external shocks on nickel price fluctuations, indicating that the market fluctuations are more persistent. The Q statistics of the residual series and residual square series reflect that there is no autocorrelation and heteroscedasticity in the residual series. And the model fitting effect is good, showing the volatility aggregation characteristics of the nickel futures market, and the GARCH effect is relatively obvious.

After estimating the Garch model, standard deviation chart is as Fig. 3. The value fluctuates around -0.0188.

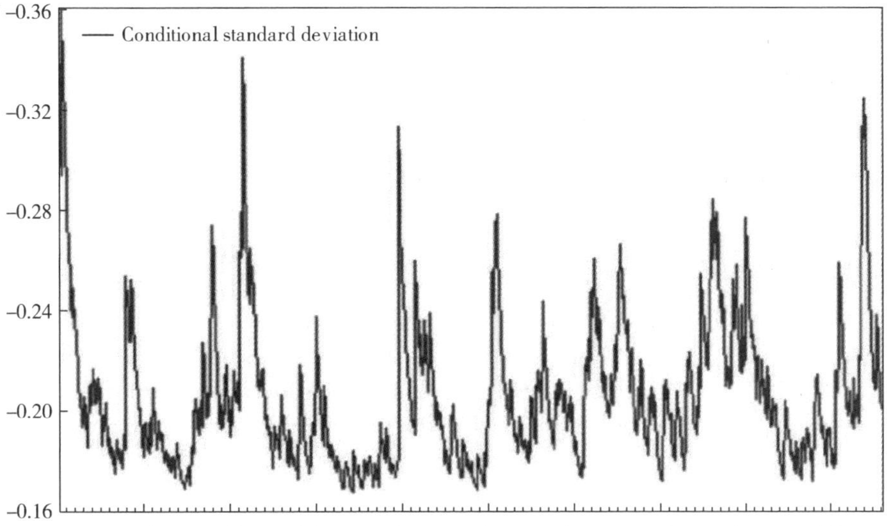

Fig. 3 Conditional standard deviation

3.4 EGARCH model

EGARCH model can be estimated as follows:

$$\ln(\sigma_t^2) = -0.8075 + 0.1721\left|\frac{\varepsilon_{t-1}}{\sigma_{t-1}}\right| + 0.0161\left[\frac{\varepsilon_{t-1}}{\sigma_{t-1}}\right] + 0.9137\ln(\sigma_{t-1}^2) \tag{9}$$

(-3.4889) (4.3602) (0.7848) (33.7107)

$R^2=0.9937$, $\bar{R}^2=0.9937$, Log likelihood=2154.766, AIC=-4.9832, SC=-4.9446

Compared the fitting results of ARIMA, GARCH and EGARCH models, their Log likelihoods were 2359.819, 2377.84, 2154.766, the AIC values were -4.9070, -4.9383, -4.9832, and the SC values were -4.8969, -4.9130, -4.9446, indicating that the overall fitting effect of EGARCH is better.

In addition, the significance of the coefficient term $\left[\frac{\varepsilon_{t-1}}{\sigma_{t-1}}\right]$ shows that it is reasonable to adopt the EGARCH model. The fitting result shows that $\gamma \neq 0$, indicating that the effect of information is asymmetric. Only when $\gamma < 0$, the leverage effect is significant, that is, the impact of negative information is greater. Therefore, if $\gamma > 0$ in this example, the leverage effect is very weak. As shown in the information impact curve in Fig. 4, although the impact of positive and negative information on the nickel sector index is asymmetrical, it is difficult to compare the severity. The nickel sector index does not necessarily have the phenomenon that the sensitivity to negative information is greater than the sensitivity to positive information.

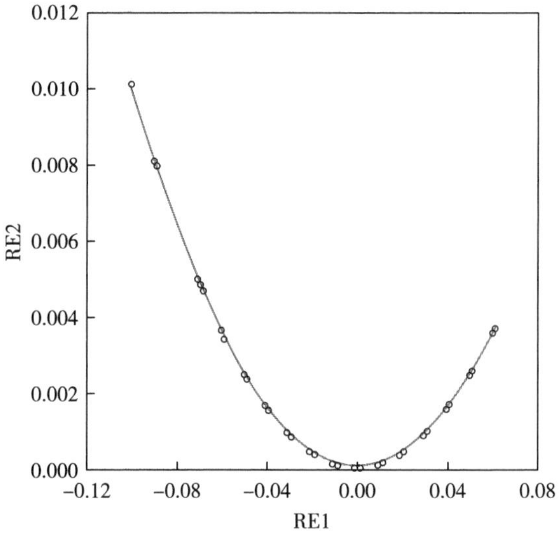

Fig. 4 The information impact curve

3.5 GARCH-QR model

The selected quantiles and estimated results are shown in Table 3.

Table 3 The selected quantiles and estimated results of GARCH-QR model

Quantiles	0.5	0.6	0.7	0.8	0.9
ω	1.806e-05	1.45e-05	1.349e-05	1.219e-05	5.536e-06
β	0.9575	0.9661	0.9689	0.9725	0.9887
α	7.477e-05	0.0001	0.0002	0.0003	0.0014

We estimated GARCH-QR model with different quantiles. When the quantile values selected are 0.5, 0.6, 0.7, 0.8, and 0.9, the coefficients ω, β, and α of the GARCH-QR model will also show varying degrees of change.

3.6 EGARCH-QR model

The selected quantiles and estimated results are shown in Table 4.

We estimated EGARCH - QR model with different quantiles. When the quantile values selected are 0.5, 0.6, 0.7, 0.8, and 0.9, the coefficients ω, β, α and γ of the EGARCH-QR model will also show varying degrees of change.

Table 4 The selected quantiles and estimated results of EGARCH-QR model

Quantiles	0.5	0.6	0.7	0.8	0.9
ω	-0.1613	-0.1511	-0.1412	-0.1113	-0.0538
β	0.9793	0.9806	0.9818	0.9856	0.9928
α	9.389e-06	8.502e-05	0.0001	0.0002	0.0009
γ	-0.0001	-0.0001	-0.0003	-0.0002	-0.0008

Fig. 5 shows how the parameters estimated by the EGARCH - QR model vary with the quantiles. With the quantile values of 0.5, 0.6, 0.7, 0.8, and 0.9, the coefficients of the EARGCH-QR model also show varying degrees of change. The value of α, β, ω increases with the increase of the quantile value, while the value of γ decreases with the increase of the quantile value. Comparing the variation range of α, β, ω, the order from large to small is α, ω, β.

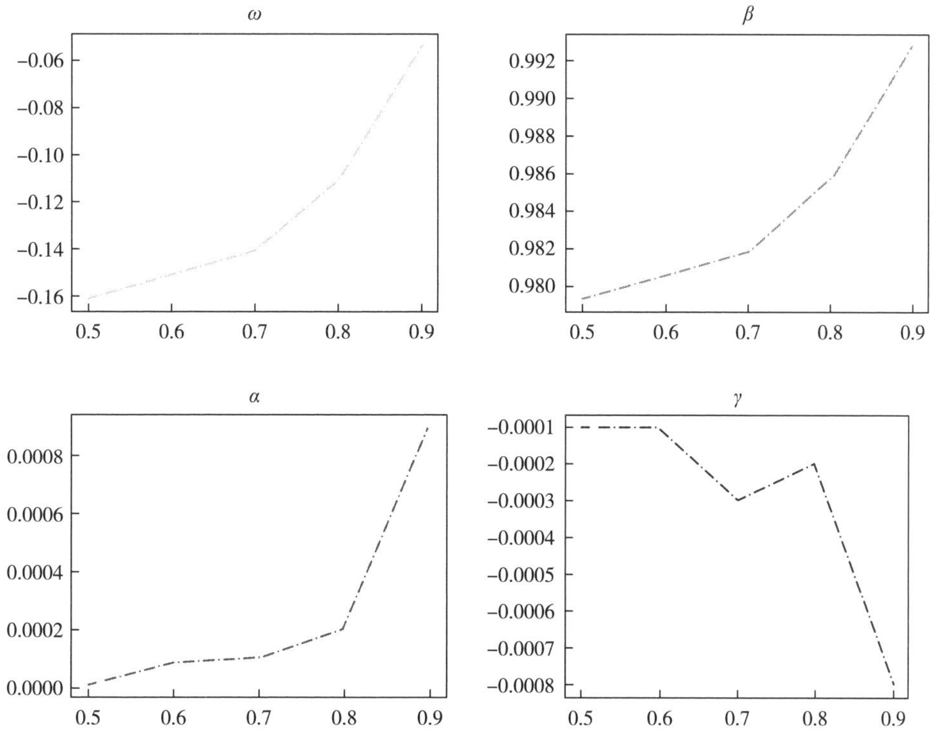

Fig. 5 The graph of the parameters estimated by the EGARCH-QR model changing with the quantile points

3.7 Result comparison

Considering that fluctuations often have outliers, the mean absolute value error (Mean Absolute Error, MAE) in the loss function is more robust; as a result, this paper uses the criterion of minimizing the mean absolute value error to test the fitting effect. The smaller the MAE value, the better the effect of the model. The specific results are shown in Table 5.

It can be seen from the results in Table 5 that the GARCH model estimated by the least square method has the largest MAE = 0.004067. In the GARCH-QR models, when the quantile point τ is 0.5, the MAE value, 1.4403564749192446e-05 reaches the minimum and the model is optimal at this time. In the EGARCH-QR models, when the quantile point τ is 0.9, the MAE value, 2.268232437708172e-06 is the smallest. The rest MAE values are around e-6, which is generally smaller than the MAE value of GARCH-QR models, indicating that EGARCH-QR model is more stable.

Table 5 Comparison of mean absolute error values of different estimation methods

Model	Quantiles	MAE value
GARCH-QR models	τ=0.5	1.4403564749192446e-05
	τ=0.6	1.4506421886380045e-05
	τ=0.7	1.4652821894144053e-05
	τ=0.8	1.4998266654269376e-05
	τ=0.9	1.6894037764141835e-05
EGARCH-QR models	τ=0.5	5.386113181874955e-06
	τ=0.6	4.95227230812616e-06
	τ=0.7	4.536865430212605e-06
	τ=0.8	3.58709291284951e-06
	τ=0.9	2.268232437708172e-06
GARCH (1, 1)		0.004067

Comparing the volatility fitted by different estimation methods in Fig. 6, it can be seen that although the volatility fitted by the least squares method and quantile regression can describe the general trend of yield fluctuations, in positions with large fluctuations the EGARCH-QR estimation method has the best fitting effect on the volatility, which is better than the GARCH-QR estimation method and the least squares method.

In summary, when the residual distribution is non-normal, the quantile regression estimation method does not require the residual item and is less susceptible to the influence of extreme fluctuation values. Therefore, a more robust estimator can be obtained. It can more accurately reflect market fluctuations and improve the robustness of risk measurement estimates.

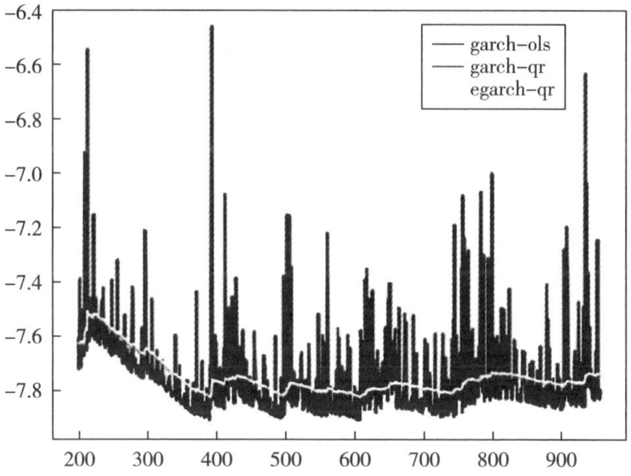

Fig. 6 Fitting volatility by different estimation methods

4 Discussion

We analyzed the statistical characteristics of the nickel sector index in China. The GARCH model is used to reflect the peak and heavy tail characteristics of the nickel sector index series while the EGARCH model is used to reflect the asymmetry of positive and negative information on the nickel sector index. GARCH-QR and EGARCH-QR can more accurately reflect market fluctuations and improve the robustness of risk measurement estimates.

4.1 The nickel sector market is weak-form efficient and it can respond to market information effectively

The time series chart of the nickel sector index (Fig. 1) shows that the nickel sector index in the past three years has obvious random walk characteristics and it has risen steadily while the fluctuation range has gradually increased. The measurement result indicates that there is an obvious autocorrelation relationship in the nickel sector index series. The nickel sector market is weak-form efficient, with a long and standardized operation time, and can respond to market information effectively. This further confirms the active role played by Shenzhen and Shanghai stock markets in the rational all ocation of financial resources and the pricing center.

4.2 An ARCH effect and volatility agglomeration characteristics exist while the nickel sector index is closely related to the global economic climate and geopolitics

There is an ARCH effect in the nickel sector index series, and the GARCH (1, 1) model can explain its volatility agglomeration characteristics, and the fluctuation of nickel prices shows that the impact of external shocks on the market is long-term. There are groups of price fluctuations. London nickel price was in a downturn between 2012 and 2016, hovering between 10000 and 20000 US dollars/ton. It started in 2017 and ushered in a rare market after years of sleep. Due to the tension between Russia and Ukraine and the influence of the speculative capital in 2022, the volatil ity of the international capital market has intensified, and the price fluctuation of the non-ferrous metal nickel is particularly violent. In March 2022, London nickel price took a "roller coaster". At the beginning of the month, it had been soaring and once rising to 100000 US dollars/ton. Correspondingly, the nickel sector index had also risen. Therefore, the nickel sector index had been closely related to the global economic climate and geopolitics. The nickel sector index reflects the prosperity and prosperity of the nickel industry. Listed companies in the nickel industry should rationally arrange investment in geological surveys as well as spot and futures trade of nickel in LME and Shanghai Futures Exchange, adjust output, stock pledge and repurchase in the financial market, etc., according to market needs. They should adjust nickel supply and equity in a timely manner to intervene in the stock price of listed companies in the nickel industry and the industry sector index, and maintain investor confidence in the nickel industry.

4.3 Alienated conditional standard deviation shows the role of special events in fueling stock prices and the abnormal rise or fall of asset fluctuations in the current period

It can be seen from Fig. 3 that the conditional standard deviation shows the characteristics of alienation and does not necessarily rise over time, but shows the rapid changes in the capital market, reflecting the role of special events in fueling stock prices and the abnormal rise or fall of asset fluctuations in the current period. The impact of psychological expectations on asset fluctuations in the next period is also reflected. The change in the value of the conditional standard deviation between 2019 and 2022 is the most dramatic, which also reflects the phenomenon of price fluctuations in the nickel sector caused by market climate change from another perspective, revealing that compared with other periods, there are greater risks and more opportunities for arbitrage in the market during these two periods. Investors should carry out stock transactions of listed companies in the nickel industry on the basis of sufficient technical analysis.

4.4 The EGARCH (1, 1) model has better fitting effect than ARIMA and GARCH models. The sensitivity of the nickel sector index to negative information is not necessarily greater than that to positive information

The overall fitting effect of the EGARCH (1, 1) model is better than the ARIMA model and the GARCH model. The information impact curve reflects that although the impact of positive and negative information on the nickel sector index is asymmetrical, it is difficult to compare its severity. The nickel sector index does not necessarily reveal a phenomenon that the sensitivity to negative information is greater than that to positive information. The investment decisions of listed companies in the nickel industry based on positive and negative information made by investors are not necessarily accurate, and more attention should be paid to technical analysis.

4.5 The EGARCH-QR estimation method has the best fitting effect on the volatility

Quantile regression is not easily affected by outliers, and does not impose any requirements on the error term in the model. It can reflect data from the selection of different quantile points and provide more effective information. Quantile regression can solve the problem when the residual distribution is non-normal. The EGARCH-QR estimation method has the best fitting effect on the volatility, which is better than the GARCH-QR estimation method and the least squares method.

4.6 Indonesian nickel products affect the total supply and the cost of nickel elements in the world

From the perspective of factors affecting the cost of nickel products, Indonesian nickel products play a pivotal role in the total supply of nickel elements in the world, and its cost changes are also crucial to the cost center of nickel products. Whether it is imposing tariffs or carbon taxes, it will increase the cost of Indonesian nickel products. With the gradual introduction of Indonesian policies, the price of related nickel products may rise accordingly, but the increase is limited. At the beginning of 2022, Indonesian official news stated that the carbon tax to be levied on April 1 this year was delayed due to issues such as the formulation of regulations and the drafting of government documents. Wait for a while. Therefore, for listed companies in the domestic nickel industry, it is necessary to always pay attention to changes in Indonesia's nickel product policy, calculate the marginal cost added value of nickeliron high matte and MHP products, and arrange production reasonably.

4.7 Listed companies in the domestic nickel industry should grasp the dynamics of nickel supply and demand at home and abroad, and reasonably arrange production and sales

From the perspective of nickel product supply and demand, according to the latest statistics from the International Nickel Study Group, global nickel consumption increased by 16.2% in 2021, and the nickel supply was short of about 168000 tons, the largest gap between supply and demand in at least 10 years. Nickel consumption was expected to grow a further 8.6% in 2022, surpassing the 3 million tons mark for the first time. Global nickel supply was expected to surge by 18.2%, driven by new production capacity in Indonesia. Nickel supply returned to a slight surplus in 2022, with a surplus of about 67000 tons. The two major downstream industries, stainless steel and new energy, both maintained positive growth. And the demand for nickel in new energy increased by more than 30% year-on-year. The alloy sector increased significantly during the year, and only the electroplating industry per-formed poorly during 2022. In 2023, primary nickel is expected to show a significant surplus which is still mainly concentrated on Nickel pig iron. And electrolytic nickel will also have a certain degree of surplus. In addition, news that the Indonesian government will increase export taxes on nickel products has repeatedly spread in the current market, which may also affect the pace of Nickel pig iron production and surplus expectations to a certain extent. Nevertheless, whether those will lead to a decline in nickel prices is still controversial in the market. Therefore, listed companies in the domestic nickel industry should carry out in-depth market research, fully grasp the dynamics of nickel supply and demand at home and abroad, and reasonably arrange production and sales.

4.8 In this round of wave-to-top market (2016—2022), supply disturbances can easily lead to a top-to-top market

Compared with the previous round of nickel cycle (2016—2022), changes in supply and demand have led to several small-band topping prices, and benefit from the increasing proportion of new energy demand and good growth expectations, nickel in each band in this round of cycle. The time for the price fall and callback is shorter while the center of gravity after each callback is higher than that of the previous round. With its outstanding resource endowment, Indonesia has rapidly increased reserves and expanded production, and its influence on nickel supply has gradually expanded. Changes in its trade and mining investment policies have caused several changes in nickel prices. In this round of wave-to-top market, we found that supply disturbances can easily lead to a top-to-top market.

5 Conclusion

Discussing the fluctuation rules of the nickel sector index in China will contribute to predict the market situation of the nickel industry sector more scientifically and reasonably, grasp the risks of the nickel market, and promote the healthy development of nickel and related industries and the capital market. This study created a combined GARCH modeling method applicable to general financial markets to make the modeling process more scientific. The feasibility of the method was verified in nickel sector index study in China's stock market. The main findings of the study are summarized as follows.

On the basis of the existence of stability and ARCH effect, this study established GARCH, EGARCH, GARCH-QR and EGARCH-QR models, and effectively assessed the fluctuation of the nickel sector index. Those revealed agglomerative features and asymmetrical characteristics of the nickel sector index sequence, and reflects the long-term impact of the external impact of fluctuations on the market. Nevertheless, the leverage effect of positive and negative news on the impact of nickel sector index is limited.

This study is based on the nickel sector index in China's stock market. In order to examine the problem more comprehensively, we have three main directions in further studies. (1) To measure the influence of the market news and government behavior on the nickel sector index can rationalize the investment strategy. (2) To discuss the mechanism of government institutional behavior in nickel listed companies specifically can ensure the validity of the nickel industrial policy. (3) To discuss the interaction of market news and nickel sector index can rationalize the stock pricing of the nickel listed companies.

Acknowledgements We are grateful to Xiong' an New Area Special Research Project of Hebei University (2023HXA025) for the financial support.

References

Ball, Philip, 2002. Stock market shock explained. Physicists model recent trading frenzy. Nature Science Update. https://www.nature.com/articles/news020923-18.

Batmunkh, M. U., Choijil, E., Vieito, J. P., etal., 2020. Doesherdingbehaviorexistinthe Mongolian stock market? Pac. Basin Finance J. 62, 101352.

Campajola, C., Gangi, D. D., Lillo, F., Tantari, D., 2022. Modelling time-varying interactions in complex systems: the score driven kinetic ising model. Sci. Rep. 12 (1), 19339.

Cai, Guanghui, Wu, Zhimin, 2022. Construction of functional EGARCH model and its volatil ity prediction. Stat. Res. 39 (5), 146-160.

Christoffersen, P., Jacobs, K., 2004. Which GARCH model for option valuation? Manag. Sci. 50 (9), 1204-1221.

Dieobold, F. X., 1986. Modeling the persistence of conditional variances: a comment. Econom. Rev. 5 (1), 51-56.

Endri, E., Abidin, Z., Simanjuntak, T. P., et al., 2020. Indonesian stock market volatility: GARCH model. Montenegrin J. Econ. 16 (2), 7-17.

Endri, E., Aipama, W., Septiano, R., 2021. Stock price volatility during the COVID-19 pandemic: the GARCH model. Invest. Manag. Financ. Innovat. 18 (4), 12.

Engle, R. F., 1982. Autoregressive conditional heteroscedasticity with estimates of the variance of United Kingdom inflation. Econometrica: J. Econom. Soc. 987-1007.

Fakhfekh, M., Jeribi, A., 2020. Volatility dynamics of crypto - currencies' returns: evidence from asymmetric and long memory GARCH models. Res. Int. Bus. Finance 51, 101075.

Fathian, F., Fard, A. F., Ouarda, T. B. M. J., et al., 2019. Modeling streamflow time series using nonlinear SETAR-GARCH models. J. Hydrol. 573, 82-97.

Francq, C., Zakoian, J. M., 2019. GARCH Models: Structure, Statistical Inference and Financial Applications. John Wiley & Sons.

Gong, X. L., Liu, X. H., Xiong, X., 2019. Measuring tail risk with GAS time varying copula, fat tailed GARCH model and hedging for crude oil futures. Pac. Basin Finance J. 55, 95-109.

He, Y., Qin, Y., Wang, S., et al., 2019. Electricity consumption probability density forecasting method based on LASSO-Quantile Regression Neural Network. Appl. Energy 233, 565-575.

Hu, J., Luo, Q., Tang, J., Heng, J., Deng, Y., 2022. Conformalized temporal convolutional quantile regression networks for wind power interval forecasting. Energy 248, 123497.

Hu, Y., Ni, J., Wen, L., 2020. A hybrid deep learning approach by integrating LSTM-ANN networks with GARCH model for copper price volatility prediction. Phys. Stat. Mech. Appl. 557, 124907.

International Nickel Study Group. Production, usageandpricereport. Available: https://www.insg.org/. (Accessed 7 February 2023).

Joo, Y. C., Park, S. Y., 2021. The impact of oil price volatil ity on stock markets: evidences from oil-importing countries. Energy Econ. 101, 105413.

Kim, H. Y., Won, C. H., 2018. Forecasting the volatility of stock price index: a hybrid model integrating LSTM with multiple GARCH-type models. Expert Syst. Appl. 103, 25-37.

Koenker, R., Bassett, G., 1978. Regression quantiles. Econometrica 46, 33-50.

Koenker, R., Hallock, K. F., 2001. Quantile regression. J. Econ. Perspect. 15 (4), 143–156.

Koenker, R., 2005. Quantile Regression. Econometric Society Monograph Series. Cambridge University Press, New York.

Korkmaz, M. Ç., Chesneau, C., Korkmaz, Z. S., 2021. Transmuted unit Rayleigh quantile regression model: alternative to beta and Kumaraswamy quantile regression models. Univ. Politeh. Buchar. Sci. Bull. Ser. Appl. Math. Phys. 83, 149–158.

Le, T. H., Luong, A. T., 2022. Dynamic spillovers between oil price, stock market, and investor sentiment: evidence from the United States and Vietnam. Resour. Pol. 78, 102931.

Lee, S., Noh, J., 2013. Quantile regression estimator for GARCH models. Scand. J. Stat. 40 (1), 2–20.

Li, H. F., Xing, D. Z., Huang, Q., Li, J. C., 2022. Roles of GARCH and ARCH effects on the stability in stock market crash. Europhys. Lett. 136 (4), 48003.

Lin, L., et al., 2020. Crude oil price forecasting based on a novel hybrid long memory GARCH-M and wavelet analysis model. Phys. Stat. Mech. Appl. 543, 123532.

Liu, J., Zhang, Z., Yan, L., et al., 2021. Forecasting the volatil ity of EUA futures with economic policy uncertainty using the GARCH-MIDAS model. Financial Innovat. 7, 1–19.

Lundbergh, S., Teräsvirta, T., 2002. Evaluating GARCH models. J. Econom. 110 (2), 417–435.

McDonald, J. G., Fisher, A. K., 1972. New-issue stock price behavior. J. Finance 27 (1), 97–102.

Meng, Q., Xiong, C., Mourshed, M., et al., 2020. Change-point multivariable quantile regression to explore effect of weather variables on building energy consumption and estimate base temperature range. Sustain. Cities Soc. 53, 101900.

Mohsin, M., Naiwen, L., Zia-Ur-Rehman, M., et al., 2020. The volatil ity of bank stock prices and macroeconomic fundamentals in the Pakistani context: an application of GARCH and EGARCH models. Oeconomia Copernicana 11 (4), 609–636.

Naik, N., Mohan, B. R., 2021. Stock price volatility estimation using regime switching technique–empirical study on the Indian stock market. Mathematics 9 (14), 1595.

Nguyen, C. T., Nguyen, M. H., 2019. Modeling stock price volatility: empirical evidence from the Ho Chi Minh City stock exchange in Vietnam. J. Asian Finance Econ. Bus. 6 (3), 19–26.

Sen, J., Mehtab, S., Dutta, A., 2021. Volatility modeling of stocks from selected sectors of the Indian economy using GARCH. In: 2021 Asian Conference on Innovation in Technology (ASIANCON). 1–9. IEEE.

Sharma, S., Aggarwal, V., Yadav, M. P., 2021. Comparison of linear and non-linear GARCH models for forecasting volatility of select emerging countries. J. Adv. Manage. Res. 18 (4), 526–547.

Sun, L., Zhu, Y. L., 2021. Measurement on dynamic impact effect of monetary policy on stock price volatility. Stat. Decis. 37 (18), 132–136.

Tse, Y. K., 2000. A test for constant correlations in a multivariate GARCH model. J. Econom. 98 (1), 107–127.

Waldmann, E., 2018. Quantile regression: a short story on how and why. Stat. Model. Int. J. 18 (3–4), 203–218.

Wang, L., Ma, F., Liu, J., Yang, L., 2020. Forecasting stock price volatility: new evidence from the GARCH-MIDAS model. Int. J. Forecast. 36 (2), 684–694.

Wang, X., Yan, L., 2022. Measuring the integrated risk of China's carbon financial market based on the copula model. Environ. Sci. Pollut. Control Ser. 29 (36), 54108–54121.

Wang, Y. L., et al., 2022. Volatility analysis based on GARCH-type models: evidence from the Chinese stock market. Economic Research-Ekonomska Istrazivanja 35 (1), 2530–2554.

Wang, Z. X., Jv, Y. Q., 2021. A novel grey prediction model based on quantile regression. Commun.

Nonlinear Sci. Numer. Simulat. 95, 105617.

Xing, D. Z., Li, H. F., Li, J. C., Long, C., 2021. Forecasting price of financial market crash via a new nonlinear potential GARCH model. Phys. Stat. Mech. Appl. 566, 125649.

Ying, J. L., Yin, W., 2021. Simulation of stock price volatility based on agent-based heterogeneous her d behavior. Comput. Simulat. 38 (5), 297-301, 333.

Zhao, G. M., Zheng, X., Liu, X. B., Zhang, X. Y., 2021. A new method to measure the synchronic ity of stock price fluctuation taking the financial industry as an example. Math. Pract. Theory 51 (19), 169-176.

Zhao, L. T., Liu, K., Duan, X. L., et al., 2019. Oil price risk evaluation using a novel hybrid model based on time-varying long memory. Energy Econ. 81, 70-78.

Zhong, Z. J., Li, T. T., 2022. On biweighted composite regression estimation of Garch model. J. Southwest China University (Natural Science Edition) 47 (5), 10-21.

Influencing factors analysis of China's iron import price: Based on quantile regression model

CHEN Wenhui[1,2], LEI Yalin[1,2,*], JIANG Yong[1,2]

1. *School of Humanities and Economic Management, China University of Geosciences, Beijing* 100083, *China*
2. *Key Laboratory of Carrying Capacity Assessment for Resource and Environment, Ministry of Land and Resources, Beijing* 100083, *China*

Abstract When encountering high import prices and price volatility, China does not have the power to affect prices, although China has ranked first in iron ore imports since 2003. The existing literature usually investigates the impact factors of iron ore prices using the averaging method. It is difficult to depict the detailed impact of various factors on prices accurately. To provide sounder basis for the Chinese government to enact policy, this paper develops a quantile regression model with the lagged variables to measure factors that affect the import prices of iron ore in China under high, medium and low price levels. The analysis uses monthly data through January 2003 to March 2015. The results indicate that the effect intensity of the factors on the prices are various under different quantiles. As prices rise, the degree of positive influence of previous period of crude steel production on iron ore prices is gradually decreasing; conversely, the strength of previous period of import volume's negative effect on prices is falling. Furthermore, it verifies that China has no voice in the international iron ore market. In low quantile, the strength of effect of prior period iron ore volume on prices is higher than that of China's production of iron ore on import prices because the grade of China's iron ore resources is low. Therefore, when the iron ore prices are at a low quantile, China should expand the import of iron ore appropriately and reduce the exploitation of low-grade iron ore resources. Additionally, China should optimize crude steel output and actively invest in overseas iron ore exploration and mining to reduce the effect of prices fluctuations by reducing the dependence on imported iron ore. China may also promote the development of an international iron ore futures market and innovate iron ore business models to hedge foreign exchange risks because of the US dollar index has greatest negative effect on the prices.

Keywords Iron ore prices, Impact factors, Quantile regression, Polynomial distributed lag model, Policy recommendations

1 Introduction

China represents one-third of the world's total iron ore consumption and has ranked first in

原文刊于 *Resources Policy*, 2016 年第 48 卷。原文发表时为彩色。
* 为通信作者。

the world in iron ore import since 2003 (Ma et al., 2013). In addition, China's iron-ore import volumes will increase in the future. According to the report released by China Iron and Steel Association (http://www.chinaisa.org.cn/gxportal/login.jsp), China's iron ore imports were 148 million tons in 2003, accounting for 24.46% of total global iron ore trade, and the external dependence is 56.97%. In 2014, China imported 933 million tons of iron ore accounted for 69.1% of world trade of iron ore, and the external dependence is high to 78.5%. Nevertheless, China does not have appropriate power in the global iron ore pricing system. As shown in Fig. 1, the average import prices of iron ore increased from $32.8/ton in 2003 to $131.1/ton in 2008.

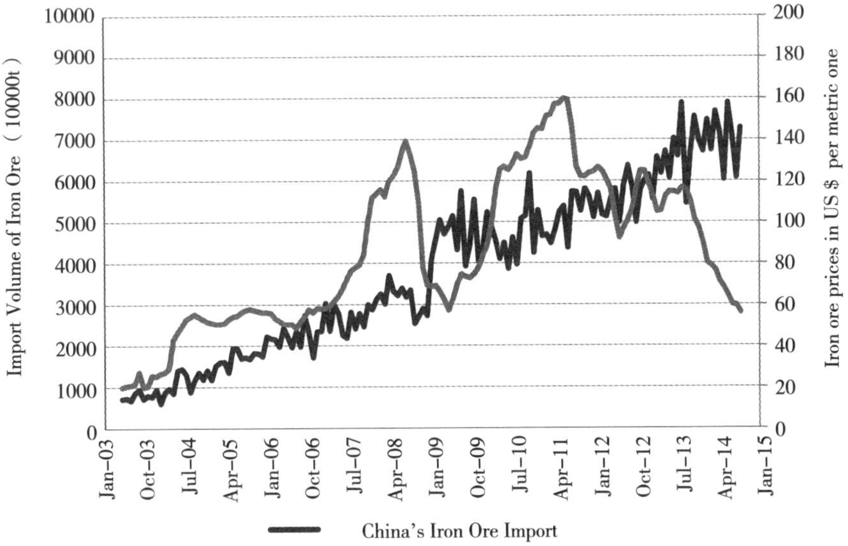

Fig. 1 Iron ore import and C. I. F. prices in US $ per metric one 2003—2015

Source: China Iron and Steel Association

The prices soared to $154.46/ton in August 2008 with a year-on-year growth rate of 72.08%. Then the prices followed sharp decline and increase. The average prices of iron ore imports dropped to $79.88/ton in 2009, whereas it rose to $163.64/ton in 2011. However, the average prices of China's iron ore imports fell by 21.6% in 2012. In 2014, iron ore prices remained in an overall downtrend; the annual average import prices were $100.42/ton, which was lower by $29.2/ton than last year. Although China's iron ore imports continue growing in recent years, the prices are in a state of constant fluctuation.

Scholars have conducted extensive research regarding the factors that affect the import prices of iron ore in China. Moreover, most presented methods use ordinary least squares (OLS) method to measure the overall data. In practice, OLS method may be inconsistent because it fails to address the information about the tails of a data distribution given that the OLS estimators concentrate on the mean as a measure of location and thus the information about the tails of a distribution is lost. Modeling the quantiles instead of the mean not only improves the robustness of the model but also reveals a fuller picture of the data in particular when it comes to the nonnormal distribution data.

This paper will perform a quantile regression analysis of the influencing factors on the iron ore prices. It will focus on two general issues. Firstly, why does this paper choose quantile re-gression to analyze the iron ore prices? Secondly, which factors have significant effect on the prices under the different quantile? This paper proceeds from the section that will describe the advantages of quantile regression model and select this model by testing the normal distribution of variables. The following section will focus on selecting the factors which have significantin fluence on iron ore prices, both using lagged variable and stepwise regression model. Thereafter a section of the paper calculates the various degree of the influencing factors on the prices under the different quantile. In the last section some interesting conclusions and policy suggestions are done.

2 Literature review

Scholars have conducted extensive research on the influencing factors of the imported iron ore price in China, primarily from the perspectives of the relation between supply and demand and market structure.

2.1 The view of supply and demand

It is widely accepted that iron ore demand is inelastic with respect to its price (Chang, 1994). Many scholars studied the prices of iron ore from the perspective of supply and demand. Sukagawa (2010) argued that when the demand of iron ore dramatically increased, it pushed iron ore prices to a high level. Wilson (2012) qualitatively analyzed the rapid development of China's steel industry, and he found that the expansion of the scale of China's steel industry increased the need for imports of iron ore and had increased prices. In addition, Kirk (2004) obtained the same conclusions. Some scholars used quantitative methods to discuss this issue. For example, Tcha and Wright (1999) used the ordinary least squares method and found that Australia's position as a major supplier of iron ore for China was contributed by the low price of Australia's iron ore. Ye and Li (2012) developed a dynamic regression model and used co-integration test and Granger test to analyze the import prices of iron ore. These researchers concluded that China's crude steel production had a positive influence on import prices of iron ore. US Dollar Index (USDX) was the Granger causality of import price. Xu (2014) got different results by using the same method. The conclusions were that there was a long-term co-integration relation among the China's iron ore import prices, global iron ore import and export volumes, degree of concentration of China's steel industry, the Baltic Dry Index (BDI) and USDX with the imported iron ore prices. Additionally, the effect of the USDX and the BDI on iron ore import prices were not obvious.

2.2 The view of market structure

Rajal (2002) held the opinion that Brazil and Australia can rely on market forces to trade iron. Zhu (2012) found a high level of industry consolidation with the BIG-3 players (Rio Tinto, BHP-B and Vale) represented close to 70% of the international iron ore trade. Pustov et al. (2013) indicated that the iron ore market is oligopolistic with a BIG-3 players that affected market prices and obtained a higher price premium compared with a market of perfect competition. Wårell and Lundmark (2008) believed the monopoly drives iron ore prices. Additionally, Wårell (2014) performed an Error-Correction Model (ECM) to analyze the short-run adjustments to the long-run equilibrium relationship. Scholars usually adopted co-integration test methods to study the relationship between import prices and the market concentration of China's iron and steel enterprise. For example, Xu and Xu (2014) used co-integration test methods for empirical analysis of the influencing factors of China's imported iron ore. The conclusion was that there was a long-term cointegration relationship between import prices of iron ore and China's steel industry concentration ratio.

2.3 Summary of literature review

Iron ore prices impact factors mainly include: China's crude steel production, degree of concentration of China's steel industry, the BDI, China's iron ore imports, global iron ore import and export volumes, production of domestic ore in China, USDX, and the degree of monopoly degree. Certain studies suggest that China's crude steel production, the BDI, and China's imports of iron ore have a positive correlation with iron ore import prices, and USDX has a negative effect on import prices. However, the strength of influencing factors on prices are various.

Qualitative and quantitative methods are used to analyze factors that influence the import prices of iron ore. However, qualitative method is certain subjectivity. In quantitative analysis, many scholars used ordinary least squares (OLS) method to study the relation among factors. There are two major advantages of using quantile regression instead of OLS. Firstly, rather than assuming the normal distribution of error terms conditional on the regressor, quantile regression has no assumption regarding the distribution of the error terms. This flexibility makes quantile regression estimates exhibit stronger model robustness compared to those obtained from OLS regression. Secondly, quantile regression is appropriate when the central location of conditional distribution and tails vary significantly with covariates. As import prices are volatility volatile, quantile regression allows inferences regarding the impact of explanatory variables conditional on the distribution of import prices. It is superior to select quantile regression model to analyze import prices.

3 Research methods

3.1 Model theory

3.1.1 The theory of quantile regression

Quantile regression, which was introduced by Koenker and Bassett (1978), estimates a range of marginal impacts for different quantiles of the distribution for the dependent or "response" variable. Compared with OLS, which assumes a constant marginal impact across the entire distribution of the dependent variable, quantile regression offers a systematic strategy for examining how covariates influence the location, scale, and shape of the entire response distribution. Consequently, quantile regression provides a significantly more complete picture of the true impact of the explanatory variable (Koenker and Hallock, 2001).

For a random variable Y with probability distribution function $F(y) = \text{Prob}(Y \leqslant y)$, the τ quantile of Y is defined as the inverse function $Q(\tau) = \inf\{y: F(y) \geqslant \tau\}$, $0<\tau<1$.

The sample median can be defined as the solution to the problem of minimizing a sum of absolute residuals as follows:

$$Q(\tau) = \arg\min_{\xi}\{\sum_i \rho_\tau(y_i - \xi)\} \tag{1}$$

where $\rho_\tau(.)$ is the tilted absolute value function that generates the τ-*th* sample quantile as its solution. Now suppose the conditional quantile of Y is represented by matrix X which is made up of k independent variables.

$$Q(\tau \mid x_i \beta(\tau)) = x_i{}'\beta(\tau) \tag{2}$$

Among (2), $x_i = (x_{1i}, x_{2i}, \cdots, x_{ki})'$ is the independent variables vector, and $\beta(\tau) = (\beta_1, \beta_2, \cdots, \beta_k)'$ is the coefficient vector under the quantile of τ. When τ changes on the (0, 1), we solve the following minimization problem to obtain a different quantile regression parameter estimation:

$$\arg\min\left[\sum_{y_i \geqslant x_i^T\beta} \tau \mid y_i - x_i^T\beta \mid + \sum_{y_i < x_i^T\beta} (1-\tau) \mid y_i - x_i^T\beta \mid\right]$$
$$\beta \in R \tag{3}$$

Generally, the quantile regression equation is:

$$Q_\tau(Y_i \mid X_i = x_i) = x'_i\beta(\tau) + \theta_i \tag{4}$$

where θ_i is a random disturbance.

3.1.2 The test of quantile regression method

Koenker and Machado (1999) proposed the concept of goodness-of-fit of quantile regression model. Quantile regression of goodness-of-fit is very similar to R^2 in general regression analysis.

Quantile regression is assumed to be $\hat{y}_{(\tau)} = X'\hat{\beta}_{(\tau)}$. The explanatory variables in the matrix and the parameter vector are divided into two parts, namely $X = (1, Z)'$ and $\hat{\beta}_{(\tau)} = (\hat{\beta}_{0(\tau)},$

$\hat{\beta}_{1(\tau)})'$. In this way, we get $\hat{y}_{(\tau)} = \beta_{0(\tau)} + Z'\beta_{1(\tau)}$.

Now we definition formula (5) and (6) as follow:

$$\hat{Q}_{(\tau)} = \min[-\sum_{t:\, y_i < X'\hat{\beta}_{(\tau)}}^{T} (1-\tau)(y_\tau - \hat{\beta}_{0(\tau)} - Z'\hat{\beta}_{1(\tau)}) + \sum_{t:\, y_i \geqslant X'\hat{\beta}_{(\tau)}}^{T} \tau(y_\tau - \hat{\beta}_{0(\tau)} - Z'\hat{\beta}_{1(\tau)})] \quad (5)$$

$$\widetilde{Q}_{(\tau)} = \min[-\sum_{t:\, y_i < X'\hat{\beta}_{(\tau)}}^{T} (1-\tau)(y_\tau - \hat{\beta}_{0(\tau)}) + \sum_{t:\, y_i \geqslant X'\hat{\beta}_{(\tau)}}^{T} \tau(y_\tau - \hat{\beta}_{0(\tau)})] \quad (6)$$

Formula (5) respective the minimal value of objective function of quantile regression with absence of restriction and formula (6) is the minimal value of objective function of quantile regression with constraint. The difference between formula (5) and formula (6) is that whether contains the constant term. According to Koenker and Machado, the goodness of fit criteria expression is formula (7):

$$R^*_{(\tau)} = 1 - \frac{\widehat{Q}_{(\tau)}}{\widetilde{Q}_{(\tau)}} \quad (7)$$

Obviously, the above statistics are very similar to the R^2 in the traditional regression model. Because $\hat{Q}_{(\tau)} \leqslant \widetilde{Q}_{(\tau)}$, the value of $R^*_{(\tau)}$ is between 0 and 1. When $\widehat{Q}_{(\tau)}$ is far less than $\widetilde{Q}_{(\tau)}$, the value of $R^*_{(\tau)}$ is closer to 1 and goodness-of-fit of the model is good. Therefore, $R^*_{(\tau)}$ can be used to examine the goodness-of-fit of the τ *-th* quantile regression is suitable or not.

3.2 Data and variables

3.2.1 Data sources

According to the analysis of the literature, this paper selects the imported prices of iron ore the dependent variable. This paper assumes that the prices are mainly affected by the crude steel production (*CSP*), production of domestic ore (*DOP*), *BDI*, China's iron ore import (*COI*), China Steel Industrial Concentration Ratio (*CR*10) and *USDX*. This paper selects the data sample interval of January 2003 to March 2015, which are based on monthly data. Prices are the C. I. F. average imported prices of iron ore (dollars/ton), obtained by measuring monthly customs statistics. *CSP* and *DOP* originate from the national bureau of statistics of the People's Republic of China (http://www.stats.gov.cn); the *BDI* data value is derived from the its official website (http://www.balticexchange.com/baltex/); The *USDX* data originates from the Federal Reserve's official website (http://www.federalreserve.gov); The *COI* is from General administration of Customs of the people's republic of China (http://www.customs.gov.cn/tabid/400/Default.aspx); The *CR*10 is from My steel website (www.mysteel.com) (Table 1).

Table 1 Description of the variables used in the analysis for the period Jan, 2003—Mar, 2015

Variable	Definition	Unit	Symbol
C. I. F. prices of iron ore imports	C. I. F. average price of iron ore imports	\$ /ton	*price*
China's iron ore import	The volumes of iron ore which China imports	104 ton	*COI*
China's iron ore production	Chinese domestic iron ore production	104 ton	*DOP*
China's crude steel production	Crude steel production in China's steel industry	104 ton	*CSP*
China Steel Industrial Concentration Ratio	The top ten of the largest enterprises of China's steel industry share of the market	%	*CR*10
Baltic Dry Index	BDI is the economic indicator of the shipping industry	—	*BDI*
US Dollar Index	Comprehensive reflection of the exchange rate of the dollar in the international exchange market	—	*USDX*

3. 2. 2 Tests of normal distribution

In order to eliminate the heteroskedasticity possibly existing in the model and facilitate hypothesis testing, all the factors take logarithmic form. Before modeling, we conduct a general statistical analysis of the time series of the above all variables. Table 2 presents the summary statistics of Ln*price*, Ln*CSP*, Ln*DOP*, Ln*BDI*, Ln*COI*, Ln*UDSX*, and Ln*CR*10 over the entire sample period. The results of statistical tests of the null hypothesis of normality (the Kolmogorov-Smirnova test and Shapiro-Wilk test) are also reported in Table 2. The skewness test measures the asymmetry of the data distribution about the mean, while kurtosis is less than three implies that it is rush. The second (skewness) and third (kurtosis) moments showed that the distributions of variables are negatively (positively) skewed and have rush tails. Both Kolmogorov-Smirnova test and Shapiro-Wilk test rejected the normality hypothesis, suggesting that variables do not follow the normal distribution.

Table 2 Tests of normal distribution

Statistics	Skewness	Kurtosis	df	Kolmogorov-Smirnov[a]		Shapiro-Wilk	
				Statistic	Sig.	Statistic	Sig.
Ln*price*	-0. 613	-0. 113	147	0. 118	0. 000	0. 930	0. 000
Ln*CSP*	-0. 724	-0. 539	147	0. 109	0. 000	0. 915	0. 000
Ln*DOP*	-0. 559	-0. 851	147	0. 102	0. 001	0. 922	0. 000
Ln*BDI*	0. 149	-0. 868	147	0. 081	0. 020	0. 971	0. 003
Ln*COI*	-0. 554	-0. 777	147	0. 133	0. 000	0. 933	0. 000
Ln*UDSX*	0. 412	-0. 054	147	0. 070	0. 073	0. 981	0. 040
Ln*CR*10	0. 681	-0. 827	147	0. 154	0. 000	0. 898	0. 000
Ln*PI*	-0. 737	-0. 33	147	0. 160	0. 000	0. 922	0. 000

a Lilliefors significance correction.

The kurtosis of normal distribution is 3, and the skewness is 0. The skewness value of the import prices of iron ore is - 0.613, which indicates that the sequence is a left skewed distribution and the distribution is not symmetrical. The kurtosis of iron ore import prices in China is -0.113, which is smaller than 3. The sequence of the tail of the distribution is thinner than the tail of the normal distribution. In addition, the probability p values of Kolmogorov-Smirnova test and Shapiro-Wilk test are both less than the test level (5%). This indicates the sequence does not form a normal distribution and has a left deviation phenomenon. With the exception of Ln*price*, the skewness values of the other variables sequence are not equal to zero, which indicates these sequences are nonsymmetrical. Moreover, the probability p values of Shapiro-Wilk test are both less than the test level (5%) that means the variables are not normal distribution. Thus, all of these variables are non-normal and non-symmetry.

The Fig. 2 is normal $Q-Q$ test chart of iron ore prices. The red line indicates the expectations of data that is normal distribution. It can be seen from the figure that the observation values depart from the red straight line. That means the variable (Ln*price*) does not obey the normal distribution and its partial degree has a tendency to increase in recent years. Furtherly the results illustrate the superiority of using a quantile regression model to analyze the characteristic of variable tail. Similarly, it can be seen from Figs. 2-7, the observation values of the variables are deviated from expected value of variable in normal distribution. Combined with the data analysis (Table 2), the results are that the variables of the model are non-normal distribution, and with the characteristics of skewness distribution. Then, the tail of the variables contains important information which cannot be researched comprehensively by the general regression method.

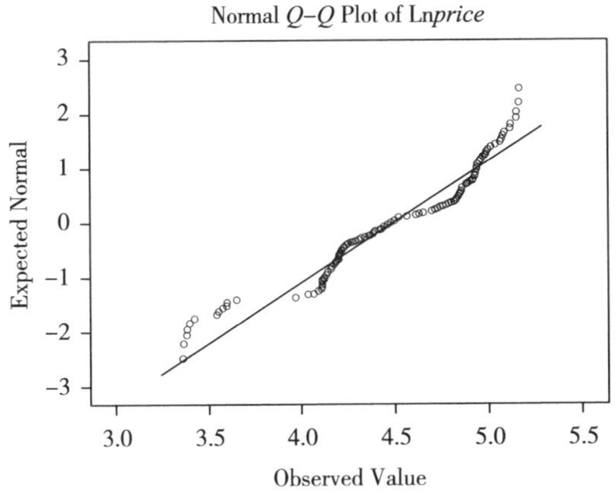

Fig. 2 The normal $Q-Q$ plot of Ln*price*
(For interpretation of the references to color in this figure legend, the reader is referred to the web version of this article)

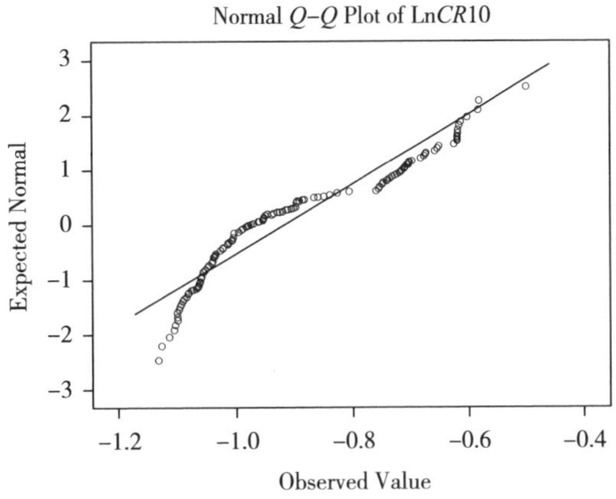

Fig. 3 The normal $Q-Q$ plot of LnCR10
(For interpretation of the references to color in this figure legend, the reader is referred to the web version of this article)

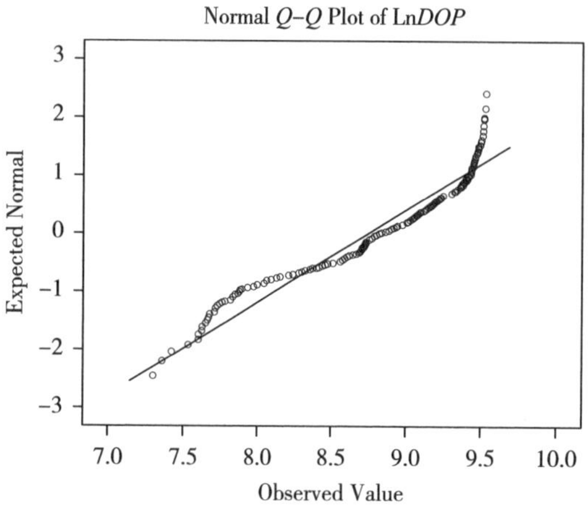

Fig. 4 The normal Q–Q plot of LnDOP
(For interpretation of the references to color in this figure legend, the reader is referred to the web version of this article)

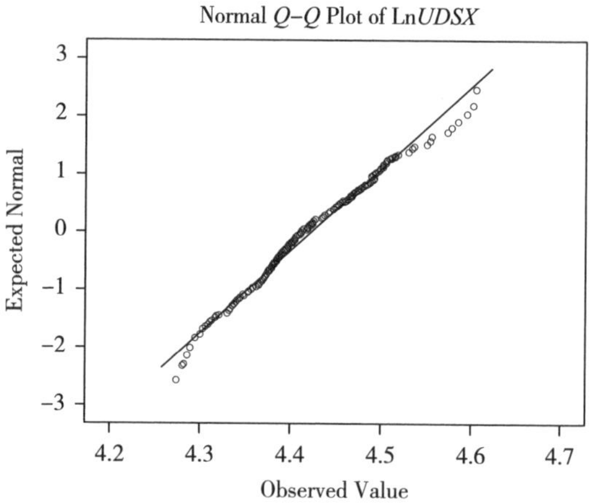

Fig. 5 The normal Q–Q plot of Ln$USDX$
(For interpretation of the references to color in this figure legend, the reader is referred to the web version of this article)

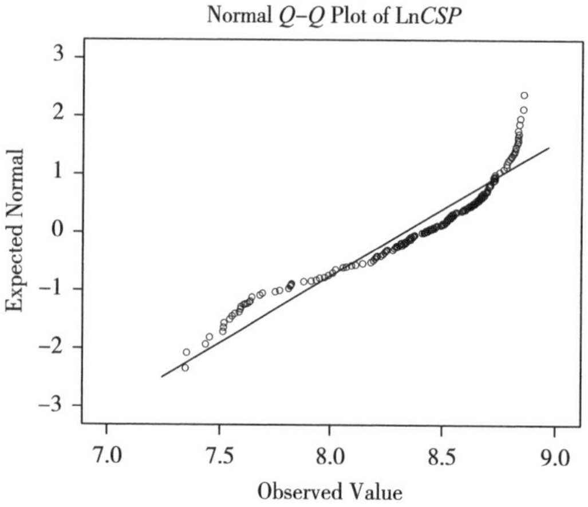

Fig. 6 The normal Q–Q plot of LnCSP
(For interpretation of the references to color in this figure legend, the reader is referred to the web version of this article)

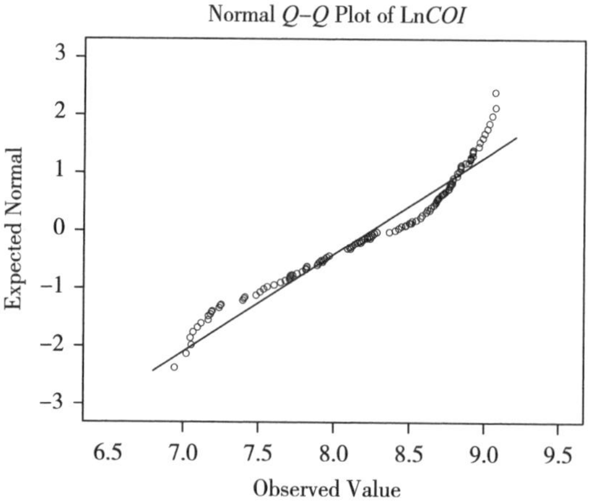

Fig. 7 The normal Q–Q plot of LnCOI
(For interpretation of the references to color in this figure legend, the reader is referred to the web version of this article)

3.3 Model

Since all the time series of economic variables are non-stationary, a differencing method is commonly used to eliminate the nonstationary trend in order to build a reasonable model. Then the stationary test of all the variables in first-order difference is conducted. Thus, it can be concluded that all the variables are first-order difference stationary, and can proceed with co-integration test. Both the eigenvalue and trace tests show that there are two cointegration equation among the variables. So these time series of economic variables are non-stationary, but they can be built a reasonable model. Then the quantile regression model equation is the formula (8).

$$Q_{\ln price_t}(\tau \mid X) = C_t(\tau) + \beta_1(\tau)\text{Ln}DOP + \beta_2(\tau)\text{Ln}CR10_t + \beta_3(\tau)\text{Ln}BDI_t + \beta_4(\tau)\text{Ln}USDX_t + \beta_5(\tau)\text{Ln}CSP_t + \beta_6(\tau)\text{Ln}COI_t + \varepsilon_t(\tau) \quad (8)$$

3.3.1 Stepwise regression model to select significant variables

For time series, economic policy transmission and behavior of mutual influence all require time in economic system. The value of time - series is determined by its own lag or other variable's. From the perspective of demand, we add the lagged variables included *COI* (-1), *DOP* (-1) and *CSP* (-1). Then the quantile regression model equation is the formula (9).

$$Q_{\ln price_t}(\tau \mid X) = C_t(\tau) + \beta_1(\tau)\text{Ln}DOP + \beta_2(\tau)\text{Ln}CR10_t + \beta_3(\tau)\text{Ln}BDI_t + \beta_4(\tau)\text{Ln}USDX_t + \beta_5(\tau)\text{Ln}CSP_t + \beta_6(\tau)\text{Ln}COI_t + \beta_7\text{Ln}DOP_{t-1} + \varepsilon_t(\tau)\beta_8\text{Ln}CSP_{t-1} + \beta_9\text{Ln}COI_{t-1} + \varepsilon_t \quad (9)$$

Then the explanatory variables which have significant effect on the dependent variable are selected by using the stepwise regression method. The explanatory variables should be eliminated in this article if its *p* value is greater than the significance level of the variables ($\alpha = 5\%$). After using the stepwise screening method, we obtain the results that fit the requirements of the model as shown in the Table 3.

Table 3 The results of stepwise regression.

Dependent variable: Ln*price*

Variable	coefficient	Std. error	*t*-Statistic	Prob.
Ln*COI* (-1)	-0.210314	0.098957	-2.125316	0.0353
Ln*CR*10	0.858192	0.104015	8.250648	0.0000
Ln*CSP* (-1)	0.944015	0.191940	4.918293	0.0000
Ln*USDX*	-2.432960	0.254772	-9.549572	0.0000
C	11.27861	1.430311	7.885423	0.0000
Ln*DOP*	-0.158129	0.085093	-1.858307	0.0452
R-squared	0.869143	*F*-statistic		185.9742
Adjusted *R*-squared	0.864470	Prob (*F*-statistic)		0.000000

The significance test results of the stepwise regression model are shown in the table below. From the Table 3, we determine the regression equation of *R* - squared is 0.869, and the Adjusted *R*-squared is 0.864. Both of these are very close to 1 that indicate the goodness-of-fit of regression model is very good. Additionally, the *F*-statistic of *F*-test is 185.9742 and its *p* value is 0.0000. This means that the model is fit.

3.3.2 Construct quantile regression model

According to the above factors which are selected by the stepwise regression model, the quantile regression model is established as follows:

$$Q_{\ln price_{et}}(\tau \mid X) = C_t(\tau) + \beta_2(\tau)\mathrm{Ln}CR10_t + \beta_7\mathrm{Ln}DOP_t + \beta_8\mathrm{Ln}CSP_{t-1} + \beta_9\mathrm{Ln}COI_{t-1} + \varepsilon_t(\tau) \quad (10)$$

$Q\tau$ (ln $price_i \mid X_i = x_i$) is the estimates of the dependent variable in different quantile regression models.

4 Results

4.1 Result of goodness-of-fit test of model

The goodness-of-fit results of the quantile regression model and the significance level of the regression coefficient of the variables under different quantile are run by the EViews6.0 software. The results of the goodness-of-fit test of the quantile regression model are shown in Table 4. According to goodness-of-fit method of Koenkerand Machado, we use the value of the adjusted R-squared to evaluate the model fitting degree.

Table 4 shows that $R^*_{(\tau)} > 0.8$, (τ =0.1, 0.2, ⋯, 0.9) and the value is close to 1, which indicates that the model fits well under τ (τ =0.1, 0.2, ⋯, 0.9) quantile.

Table 4 Results of goodness-of-fit test of quantile regression model

τ	0.1	0.2	0.3	0.4	0.5	0.6	0.7	0.8	0.9
Pseudo R-squared	0.898	0.890	0.893	0.869	0.902	0.899	0.887	0.870	0.856
Adjusted R-squared	0.893	0.886	0.888	0.864	0.897	0.895	0.883	0.865	0.850
Quasi-LR statistic	422.7	580.3	738.4	523.8	688.1	623.7	572.5	532.3	435.3
Prob (Quasi-LR stat)	0.000	0.000	0.000	0.000	0.000	0.000	0.000	0.000	0.000

4.2 Results of the regression coefficient

The results of the quantile regression model are shown in Table 5. We obtain the significant level of regression coefficient of different factors and the regression coefficient at different quantile regression levels. Under different quantiles, when the probability P value of the variable is less than the significance test level (5%), the variable has a significant effect on the dependent variable. The regression coefficient under different quantile determines the effect of the variable on prices is positive or negative. According to the absolute value of the regression coefficients, we get the influence intensity of the variable on the dependent variable.

The results indicate that the *CR*10 has a positive effect on the current imported iron ore prices in China under the quantile from 0.2 to 0.9, for which the coefficient is between 0.755 and 1.057. Accordingly, the probability P values of coefficients are less than the level of significance test (5%), which means that the regression coefficients of the variables are significant under those quantiles. In addition, the results show that the *USDX* has a negative effect on the current prices under the quantile from 0.1 to 0.9, and the coefficient is between −2.069 and −2.807. However, the previous period production of crude steel has a significant positive impact on the current imported prices iron ore under the quantile from 0. 1 to 0. 5, and 0. 8, and the coefficient is between 0. 595 and 1. 657. Moreover, China's iron ore production has a negative impact on the current prices of imported iron ore under the quantile from 0. 1 to 0. 3, while the coefficient is between −0. 245 and −0. 378. Besides the previous period iron ore import volume also exerts negative influence on iron ore prices and the intensity is from −0. 56 to −0. 36 in low (0. 1−0. 3) quantile.

Table 5 The results of the regression coefficient

Quantile	β_2		β_4		β_1		β_8		β_9	
	Coefficient	Prob.	Coefficient	Prob.	Coefficient	Prob.	Coefficient	Prob.	Coefficient	Prob.
0. 1	0. 771	0. 068	−2. 069	0. 000	−0. 378	0. 000	1. 533	0. 000	−0. 391	0. 0137
0. 2	0. 924	0. 000	−2. 369	0. 000	−0. 297	0. 003	1. 657	0. 000	−0. 561	0. 000
0. 3	1. 020	0. 000	−2. 735	0. 000	−0. 245	0. 028	1. 250	0. 001	−0. 360	0. 0517
0. 4	1. 057	0. 000	−2. 659	0. 000	−0. 198	0. 061	1. 046	0. 003	−0. 265	0. 1552
0. 5	0. 812	0. 000	−2. 807	0. 000	−0. 050	0. 576	0. 595	0. 048	−0. 043	0. 8071
0. 6	0. 761	0. 000	−2. 613	0. 000	−0. 027	0. 768	0. 587	0. 051	−0. 064	0. 7047
0. 7	0. 832	0. 000	−2. 192	0. 000	−0. 031	0. 781	0. 469	0. 060	0. 009	0. 94
0. 8	0. 894	0. 000	−2. 252	0. 000	−0. 141	0. 315	0. 535	0. 020	0. 043	0. 7126
0. 9	0. 755	0. 000	−2. 449	0. 000	−0. 033	0. 816	0. 326	0. 159	0. 088	0. 443
OLS	0. 858	0. 000	−2. 433	0. 000	−0. 158	0. 055	0. 944	0. 000	−0. 210	0. 035

From the Table 5, the result which is obtained by using OLS model is that the sequence of the size of effect on iron ore prices is *USDX*>*CSP* (−1)>*CR*10>*COI* (−1)>*DOP*. In low quantile, the ranking of these factors' influence is *USDX*>*CSP* (−1)>*COI* (−1)>*CR*10>*DOP*. *CR*10, *USDX* and *CSP* (−1) have significant ef−fect on prices and the sequence of the size of these factors' effect on iron ore prices is *USDX*>*CR*10>*CSP* (−1) under middle and high quantile.

As shown in Fig. 8, the degree of the influence of the factors on the prices are various under the different quantile. For example, the influence of the *USDX* on import prices is weaker in middle (0. 4−0. 6) quantile than the other. As the import prices decrease, the degree of previous period of crude steel production's positive influence on iron ore prices is gradually increasing. In addition, the results also show that the effect intension of previous period iron ore import volume on prices is gradually increasing with prices declining.

5 Discussion and conclusions

5. 1 Discussion

5. 1. 1 The demand

The previous period production of crude steel has a significant positive impact on the current

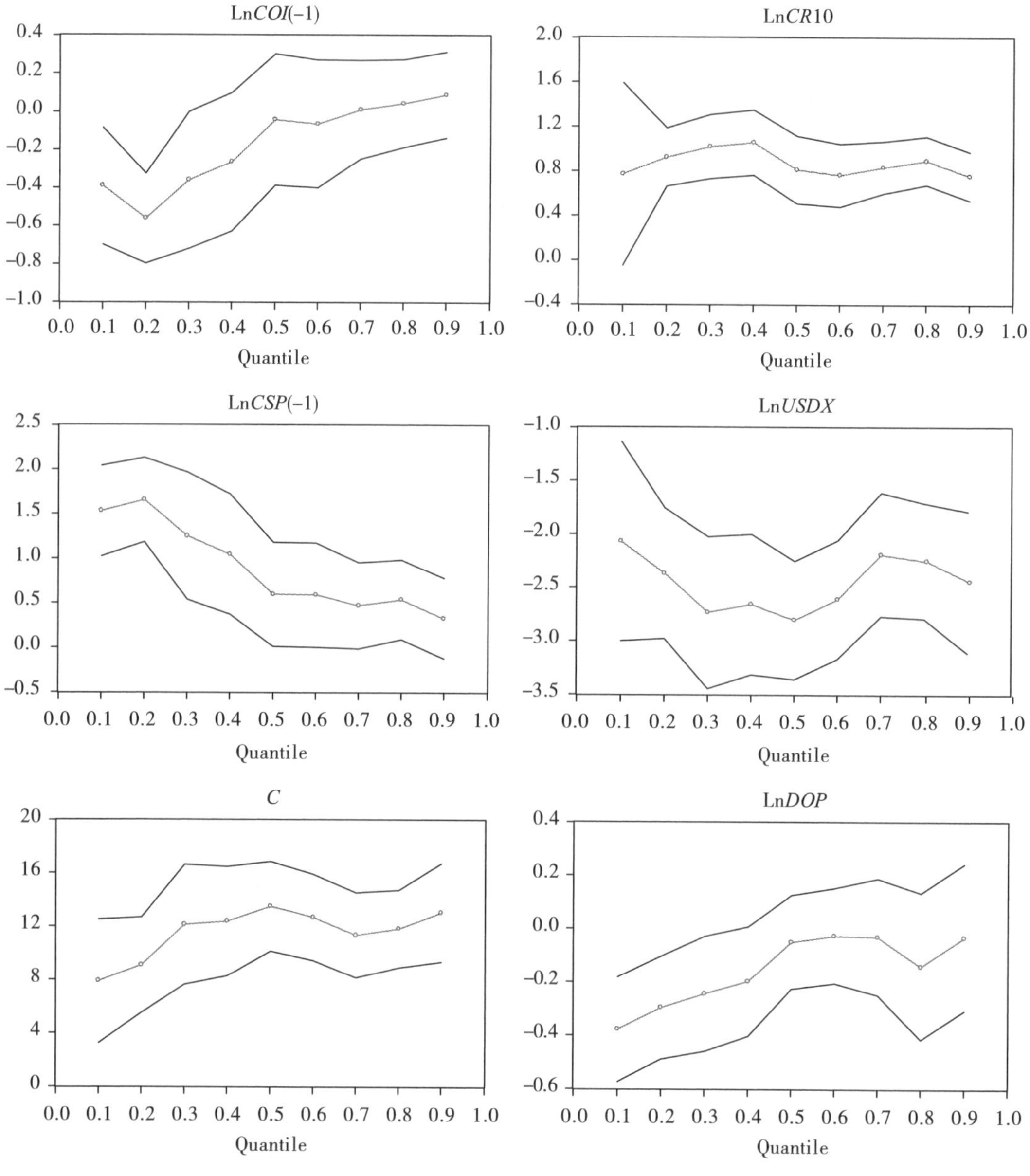

Fig. 8 Curves of coefficients of variables in different quantile

Note: the blue line is curve of coefficients of variables of quantile regression and the red line is the 5% confidence interval. (For interpretation of the references to color in this figure legend, the reader is referred to the web version of this article)

prices of imported iron ore in China, which is proved by the results of Marcus and Kirsis (2004), Kirk (2004). Some different results are that the influence strength which China's crude steel production on prices is in the opposite direction of prices change. As import prices fall, the positive influence degree of previous period of crude steel production on imported iron ore prices is gradually increasing. In the low prices, the average influence strength is 1.48 and higher than that of import volume on the prices which is -0.437. It indicates that the growth of crude steel production has led to the increase of iron ore imports, but does not exert the equal effect on reducing prices.

Although the imported iron ore prices were high and fluctuant, iron and steel enterprises largely imported iron ore to increase the steel production. However, this enterprises have been in the situation of overcapacity. As can be seen from Fig. 9, China's crude steel production accounted for the proportion of the world has been in an upward trend; it has risen from 22. 9% in 2003 to 49. 5% in 2014. China Steel Industry Association data showed that China's crude steel production over 822 million tons and the capacity has reached 1. 16 billion tons in 2014. China should reduce crude steel production reasonably to decrease the demand for iron ore, thereby inhibiting the rise in imports of iron ore prices.

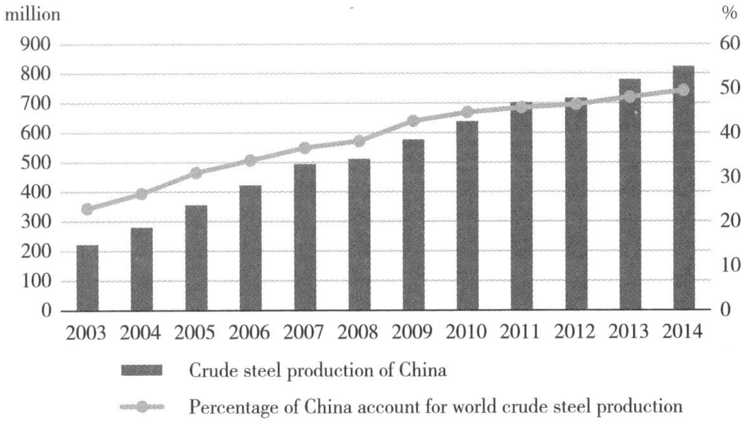

Fig. 9 Crude steel production of China and the percentage of China account forworld crude steel production

5. 1. 2 The supply

China's iron ore production and production of Chinese iron ore have negative impact on imported iron ore prices under the low quantile. The strength of effect of prior period iron ore volume on prices is higher than the production of Chinese iron ore because of the low-grade iron ore in China. Meanwhile, as import prices fall, the negative influence degree of the prior period iron ore volume prices is gradually increasing. In a result, China can expands the import of iron ore appropriately when the iron ore prices are at a low level. Moreover, China's iron ore resources is abundant but not rich. They are low-grade and the distribution is uneven. The grade of domestic iron ore is only 31. 3% on average and lesser than the average grade in the world which is 45. 79%, while the lean ore is 97. 2%. Thereby, China should reduce the exploitation of the domestic low-grade iron ore.

As can be seen from Fig. 10, the output of China's low-grade iron ore is much more than the number of imported high-grade iron ore. Although China has a high degree of dependence on foreign iron ore, China can adjust the source of iron ore resources properly in the premise of ensuring the safety. In general, when the iron ore prices are at a low level, China can expands the import of iron ore appropriately and reduce the exploitation of the domestic low-grade iron ore resources. Accordingly, when the international supply market of iron ore is shortage or the import prices are high, China can appropriately reduce the amount of import and cut the cost to ensure the development of iron and steel enterprises.

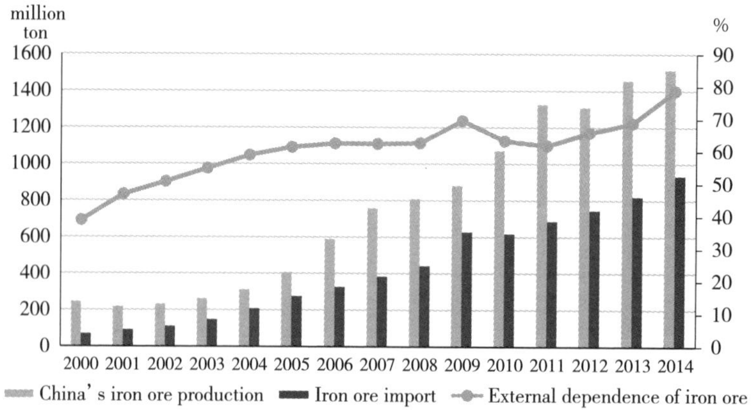

Fig. 10 The China's iron ore production,
iron ore import and the external dependence of iron ore

5.1.3 The concentration ratio of China's steel industry

Although China's steel industry concentration ratio has been adjusted to certain extent, the concentration ratio of China's steel. industry (*CR*10) is just between 30% and 50% and the concentration of *CR*4 is less than 30% (Fig. 11). In 2014, the *CR*4 of China's steel companies is only 17%, while that of Japan and United States is up to 78% and 67%. So, China's steel industry concentration is far behind that of other countries. In addition, there are 290 steel enterprises in China now. However, only 7 enterprises' annual steel production are more than 3

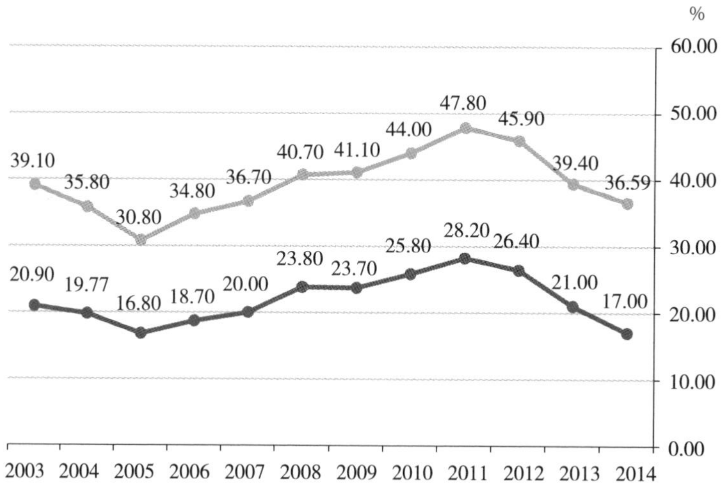

Fig. 11 The concentration ratio of China's steel industry

million tons. The market for China's steel industry is characterized by competition and that it displays the phenomenon of fragmented distribution and ill-or-ganized. As a result, the change of its concentration has not yet produced the expected effect on the imported iron ore prices.

China's iron and steel enterprises should integrate the resources and optimize the market structure. Expanding the scale of the production is not only way to improve the concentration degree of steel industry. Iron and steel enterprises should merge to reduce the number of iron and steel enterprises, and optimize the resources of them. Thus it can change the current market pattern with some characters of scattered, chaos and small. It is conducive to program Chinese steel output. Thereby, it could control the import volume of iron ore and port inventory, and then restrict the import prices of iron ore.

5.2 Conclusions

(1) The trend of China's iron ore demand is consistent with the trend of import prices. According to the aforementioned results, both the previous period iron ore import volume and China's iron ore production have negative impact on the current price of imported iron ore, while the previous period of crude steel production has a positive effect.

(2) The effect intensity of the factors on the prices are different under different quantiles. With the import prices declining, the degree of previous period of crude steel production's positive influence on iron ore prices is gradually increasing and the intensity of negative effect of previous period of import volume on prices is rising. Furthermore, it verifies that China has on voice on the international iron ore market.

(3) In low price quantile, the strength of effect of prior period iron ore volume on prices is higher than the production of iron ore in China because the grade of China's iron ore resources is low.

(4) *USDX* has a negative effect on the current prices of imported iron ore and its impact is relatively stable overall. It can be considered that the international iron ore market prices are not affected by the dollar index remarkably.

(5) Although *CR*10 has a positive influence on prices, it has not yet produced the desirable effect on the import prices of iron ore.

6 Policy recommendations

The paper provides the following policy recommendations:

(1) When the import iron ore prices are high, China should optimize crude steel output and reduce the consumption and import of iron ore. As the iron ore prices are at a low level, China should expand the import of iron ore appropriately; reduce the exploitation of the domestic low-grade iron ore resources and construct the mineral land reserve. Accordingly, when the iron ore in the international supply market is decreasing or the import prices are getting high,

China can appropriately reduce the amount of imports and cut the costs to develop iron and steel enterprises.

(2) To improve the status of China's iron ore trade, China should increase the overseas mining investment. By cooperating with foreign mining entities to ensure a stable supply of iron ore, China should break the monopoly. Therefore, China should actively invest in overseas iron ore exploration and mining interests to expand its iron ore supply so that China can reduce dependence on imported iron ore and the impact of imported iron ore prices fluctuations.

(3) The futures market plays an important role in the formulation of the prices and mitigating risks, so that many of the world's resources are priced by the futures market. Meanwhile, futures market to some extent can eliminate the influence of *USDX* which affects the cost of China's steel enterprises. In order to avoid price risk, government departments, steel companies, and investors should cooperate to establish and improve China's iron ore futures market. Building futures market which can represent the reality of the Chinese market and can be widely accepted by the market is conducive to take the initiative to safeguard the interests.

(4) As the China's steel industry concentration degree has not yet produced the desired effect on the prices of iron ore imports. Therefore, Chinese government could consider a variety of indicators such as the production of iron ore to measure the concentration of steel industry. On the other hand, iron and steel enterprises should be combined to reduce the number of iron and steel enterprises; optimize the resources of enterprises and change the market structure. At the same time, China's iron and steel enterprises should strengthen cooperation to unified planning the production of crude steel. Then restrict the import prices of iron ore though controlling the import volume of iron ore.

Acknowledgments The authors express sincere thanks for the support from Editor Prof. Campbell and the reviewers. The support from the following research funds is highly appreciated: Chinese Academy of Land & Resource Economics No. 12120113093200; the Development Research Center of China Geological Survey Bureau under Grant no. 12120114056601; the Fundamental Research Funds for the Central Universities under Grant no. 53200859534.

References

Chang, H., 1994. Estimating Japanese import shares of iron ore. J. Resour. Policy. 20 (2), 87–93.

Kirk, William S., 2004. China's emergence as the World's leading iron-ore-consuming country. J. Miner. Energy 19 (2), 16–27.

Koenker, R., Bassett, J. G., 1978. Regression quantiles. J. Econom. 46 (1), 33–50.

Koenker, R., Machado, Jose A. F., 1999. Goodness of fit and related inference processes for quantile regression. J. Am. Stat. Assoc. 94 (448), 1296–1310.

Koenker, R., Hallock, K., 2001. Quantile regression. J. Econ. Perspect. 15 (4), 143–156.

Ma, Weimin, Zhu, Xiaoxi, Wang, Miaomiao, 2013. Forecasting iron ore import and consumption of China using grey model optimized by particle swarm optimi zation algorithm. J. Resour. Policy 38, 613–620.

Marcus, P., Kirsis, K., 2004. Chinese Steel: Limits to Growth. World steel Dynamics, Englewood Cliffs.

Pustov, Alexander, Malanichev, Alexander, Khobotilov, Ilya, 2013. Long-term iron ore price modeling: marginal costs vs incentive price. J. Resour. Policy 38, 558-567.

Rajal, K., Deb, 2002. An Analysis of Generation Market Powering the Midwest International. J. Electricity, 15, 29-39.

Sukagawa, Paul, 2010. Is iron ore priced as a commodity? Past and current practice. J. Resour. Policy 35, 54-63.

Tcha, Moon Joong, Wright, Damione, 1999. Determinants of China's import demand for Australia's iron ore. J. Resour. Policy 25 (3), 143-149.

Wårell, Linda, 2014. The effect of a change in pricing regime on iron ore prices. J. Resour. Policy. 41, 16-22.

Wårell, Linda, Lundmark, Robert, 2008. Price effects of mergers in natural resources industries. J. Resour. Conserv. Recycl. 53, 57-69.

Wilson, Jeffrey D., 2012. Chinese resource security policies and the restructuring of the Asia-Pacific iron ore market. J. Resour. Policy 37, 331-339.

Xu, Bin, 2014a. Empirical study on the impact factors of the import price fluctuation of iron ore in China. J. Explor. Econ. Problems 07, 171-177, in Chinese.

Xu, Bin, Xu, Zhihong, 2014b. Analysis of the factors affecting the price fluctuations of China's imported iron ore. J. Price Monthly 446, 10-14, in Chinese.

Ye, Haiyan, Li, Jin, 2012. Analysis of the factors affecting the price of imported iron ore in China. J. Econ. Issues 10, 119-122, in Chinese.

Zhu, Zhirui, 2012. Identifying Supply and Demand Elasticities of Iron Ore. Thesis at Duke University, pp. 1-57.

China's provincial exhaustible resources rent and produced capital stock: Based on Hartwick's rule

YU Yun[1,2], LEI Yalin[1,3,*]

1. *Key Laboratory of Carrying Capacity Assessment for Resource and Environment, Ministry of Land and Resources, Beijing* 100083, *China*

2. *Chinese Academy of Land and Resource Economics, Beijing* 101149, *China*

3. *School of Humanities and Economic Management, China University of Geosciences, Beijing* 100083, *China*

Abstract Hartwick's rule, which concerns reinvesting resource rents in manmade capital, together with the concept of dynamic efficiency, dealt with by Hotelling's Rule, imply that locally constant consumption is sustainable. Based on Hartwick's rule, we have a hypothesis: if the 29 provinces in China (except Beijing and Shanghai) had invested their resource rents (from three energy resources: oil, natural gas, and coal; and nine mineral resources: iron, copper, lead, zinc, nickel, tin, antimony, molybdenum, and aluminum) in produced capital stock from 1995 to 2014, then the value of these assets should have grown commensurately over time. Using hypothetical estimates of capital stocks, the capital produced from a postulated series of historical investments, and a perpetual inventory model (PIM), the capital produced from exhaustible resources value reinvestment under Hartwick's rule is estimated. The empirical results prove that China's 29 provinces fare very well under Hartwick's rule from 1995 to 2014, and the results show that the provinces exhibit regional diffierences that are non-synchronous with local economic growth. Some provinces are striking: Xinjiang, Heilong, Shaanxi, Inner Mongolia, and Tianjin would have higher energy resources rent shares and lower levels of capital accumulation. Based on results, we provide managerial implications and suggestions for enhancing provincial performance under Hartwick's rule.

Keywords Exhaustible resources, Produced capital, Hartwick's rule, China

1 Introduction

In resource economics, there are two main intertemporal allocation rules: the Hotelling rule and the Hartwick rule (Asheim et al., 2002). This paper considers Hartwick's rule. Hartwick (1977) suggested that "investing all net returns from exhaustible resources in reproducible capital ... implies intergenerational equity." Scholars have subsequently called this Hartwick's rule for reinvesting resource rents in reproducible capital (Solow, 1993). Together with and

原文刊于 *Resources Policy*，2017 年第 52 卷。原文发表时为彩色。

* 为通信作者。

Hotelling's Rule covering dynamic efficiency, this implies locally constant consumption. Solow (1974) shows that exhaustible natural resource inputs can be substituted by the manmade capital in such a way that depleting these natural resources does not harm future generations. Thus, substitutability between natural and manmade capital, despite the exhaustibility of natural resources, may allow for equitable consumption for all generations. Hartwick (1977) seemed to have found the investment policy that would bring about sustainability in this way. Many scholars at that time, 40 years ago, established an extend proposition for Hartwick's rule. These are reviewed in the next section.

Based on Hartwick's rule, we have developed a hypothesis: if the 29 provinces in China (except Beijing and Shanghai) had invested their resource rents (from three energy resources: oil, natural gas, and coal; and nine mineral resources: iron, copper, lead, zinc, nickel, tin, antimony, molybdenum, and aluminum) in produced capital stock from 1995 to 2014, then the value of these assets should have grown commensurately over time. Using hypothetical estimates of capital stocks, the capital produced from a postulated series of historical investments, and a perpetual inventory model (PIM), the capital produced from exhaustible resources value reinvestment under Hartwick's Rule is estimated.

This paper is organized as follows: Section 2 reviews the literature and empirical cases; Section 3 gives the original propositions and the current hypothesis; Section 4 provides data and explains how these were obtained. Section 5 presents the empirical results and draws an initial conclusion regarding the implications for diffierent provinces in China. The final section presents and discusses conclusions and discusses the most important managerial implications and suggestions in relation to policy-making.

2 Literature review and empirical studies

Reviewing literature of Hartwick's rule on theory development for Hartwick's rule are three aspects. First: our paper draws on the literature extending the original rule. Hartwick (1977) proposed the original rule; Asheim (1986) and Hartwick (1995) expanded the Hartwick rule for open economies; and Cass and Mitra (1991) investigated the nature of a general production function that might support the constant consumption result; Hartwick (1996) analyzed and expanded the original proposition in "consumption is interest on capital"; Hartwick, Long and Tian (2003) showed a peaked time-path of consumption, wherein zero net investment occurred in advance of the peak; André and Cerdá (2005) extend the original rule in renewable resources. d'Autume and Schubert (2008) with the effects of technical progress; Martinet (2007) with the consumption and the price are not in the optimal path, and Amigues and Moreauxy (2008) consider a bisector model. Mitra et al. (2013) and Mitra (2015) under an additional assumption on the increasing of the initial resource stock, all non-trivial maximin paths are efficient. Second: our paper draws on the literature on established the converse to original proposed that under dynamic efficiency (Hotelling's Rule), constant consumption implies "investing exhaustible resource rents" (Dixit et al., 1980). Withagen and Asheim (2007) and Asheim (2013) offer a general proof of the converse of Hartwick's rule that includes positive discount factors; Cairns and Yang (2000) sought out this converse without "positive discount factors"; and Buchholz et al. (2005) show that, if a competitive path is equitable, it must satisfy Hartwick's rule. Third: the original rule implies the Instant consumption is zero, Hamilton (1995) offers the extended rule (invest resource rents) and dynamic efficiency, imply a positive constant (the Instant consumption is a positive); Sato and Kim (2002) prove Hamilton's extend proposed (by the Dasgupta and Heal) by the Cobb-Douglas model and consider relationships between Hartwick's rule and conservation laws; and Hamilton and Hartwick (2005) examined "maintaining capital intact" and locally unchanging consumption, and Bazhanov (2015) extends by inefficiency. Cairns (2006) considers relationships between Hartwick's rule and national-accounting prices. In China, the scholars just had the literature reviewed but not breakthrough in the rule in theory. Lihong (2007) thinks the profound influence of Hartwick's rule on sustainability, population growth, intergenerational equity, and the system of national accounts. Bo et al. (2013) think Hartwick's rule is the important one research of human well-being, the derived on math is very precise, this rule brings an important measuring sustainable method to measure the human well-being. This method could measure the total well-being by total income which came from various consumptions multiply shadow price.

There is a considerable body of literature presenting empirical studies of Hartwick's rule. Vincent et al. (1997) study the case of petroleum depletion in Indonesia and suggest that, to sustain their consumption levels, resource-rich countries will need to invest more than previously expected. Pearce and Atkinson (1993) measured 18 countries on the basis of a weak sustainability rule that allows for unconstrained elasticities of substitution between natural and man-made capital. Genuine saving (apositive constant) is the Hartwick rule's transform for empirical (Hamilton, 1994; Hamilton and Clemens, 1999; Hamilton and Hartwick, 2005). Hamilton et al. (2006) construct three alternative estimates of produced capital stock for the year 2000, using data covering 1970—2000. Genuine saving estimates were calculated by the World Bank (2005), based on Atkinson and Hamilton (2003). With negative genuine saving, a nation loses wealth, and social welfare falls (Dasgupta and MÄLer, 2000).

In China, the empirical studies focus on three facts. First, Chang and Yan (2011) evaluated the elasticity is two of the energy consumption and the manmade capital and would decline with technical progress. Second, Guoxia et al. (2009) evaluated the contribution of mineral resources to genuine savings in 31 provinces of China. Their results found that the value of genuine saving is negative in Shanxi, Guizhou, and Qinghai, which indicates that the development of the economy is not sustainable. Na and Yunyan (2008) compared the Jeffrey model and the Hamilton model to estimate the efficiency of China's genuine saving rate from the year 1960—2006, based on data provided by the World Bank, and found out that the main reason that is China's genuine saving by the World Bank does not conform to reality is diffierence in data formation method and being not suitable for China's data. Third, Some scholars focused on substitutions between exhaustible resources' capital and manmade capital (Angang and Yahua, 2005; Na, 2011; Qingyou and Xinfa, 2013).

3 Methodology and data

3.1 Generalizing the Hartwick rule

We assume that a simple economy with an exhaustible resource is essential for Hartwick's rule. The original "invest exhaustible resource rents" result, which was (a) investing resource rents: $\dot{K} = RF_R$, and (b) dynamic efficiency (Hotelling's Rule: $\dot{F}_R = F_K F_R$), implies constant consumption ($\dot{C}=0$) (Hartwick, 2003).

This can be deduced as follows: assume current output $Q = F(K, R, N)$, where K is produced capital from current stock; R is an explored exhaustible resources flow from current stock $S(t)$, i.e., $R(t) = -\dot{S}(t)$; and N is a constant for human resources. we set out a simple economy, with satisfied Dasgupta-Heal model (Dasgupta and Heal, 1980) and the Pezzey-Withagen model (Pezzey and Withagen, 1998), where

$$Q = F(K, R, N) = C + \dot{K} \tag{1}$$

Transform expression (1)

$$C = F(K, R)\dot{K}$$

Consider the time-derivative $\dot{C}$:

$$\dot{C} = F_K\dot{K} + F_R\dot{R} + F_N\dot{N} - \mathrm{d}\dot{K}/\mathrm{d}t$$

We should prove $\dot{C}=0$, that is

$$\dot{C} = F_K RF_R + F_R\dot{R} - \dot{R}F_R - R\dot{F}_R = [F_K F_R - \dot{F}_R]R = 0 \tag{2}$$

This produces a proposition:

Proposition (Hartwick's Rule). if human capital N is constant, in Hotelling Rule ($\dot{F}_R = F_K F_R$) and reinvesting exhaustible resource rents into other capitals implies constant consumption.

Hartwick (2003) notes that this original rule did not make use of constant returns to scale. Cass and Mitra (1991) support the Solow (1974) constant consumption result for "enough" substitutability between produced capital and exhaustible resources rents, and show that produced capital should be more productive than are exhaustible resources rents to maintain constant consumption.

3.2 The hypothetical methodology

3.2.1 A review of our hypothesis

Based on Hartwick's rule, we present a hypothesis: if the 29 provinces in China (except for Beijing and Shanghai) had invested their resource rents (from three energy resources: oil, natural gas, and coal; and nine mineral resources: iron, copper, lead, zinc, nickel, tin, antimony, molybdenum, and aluminum) in produced capital stock from 1995 to 2014, then the value of these assets should have grown commensurately over time.

Hartwick's rule concerns reinvesting exhaustible resources' rents into other assets, including production capital, human capital, and renewable resource capital. Hamilton et al. (2006) consider the scenario that most resource-rich countries were investing little resource value either in human or in foreign financial assets, and, therefore, just use produced capital. Majbouri (2015) calculate the counterfactual capital stocks and income in MENA based on the Hartwick's rule. Our research also uses only the produced capital, not only for the aforementioned reason but also because the Chinese economy has developed very rapidly over the past 20 years and has a huge account of produced assets.

3.2.2 Our hypothetical methodology

Using hypothetical estimates of capital stocks, the capital produced from a postulated series of historical investments, and a perpetual inventory model (PIM), the capital produced from exhaustible resources value reinvestment under Hartwick's Rule is estimated.

3.3 PIM model for (a) actual produced capital

We chose "investment stock" measuring the actual produced stock (Young, 2003; Zhang et al., 2004; Zhang, 2008) for our purpose. Through comparison with international results, we use historical investment data to calculate the capital produced by the Perpetual Inventory Method (PIM), which could be adopted by the OECD framework. In China, PIM is very widely used for empirical studies (Zhang et al., 2004; Haojie, 2008; Sun and Ren, 2014).

We use the PIM to estimate the actual produced assets stocks for (a). The expression (3) is based on Hamilton et al. (2006) K_t is the aggregate capital stock in period t, that is:

$$K_t = \sum_{i=0}^{T-1} I_{t-i}(1-\gamma)^i = I_t + (1-\gamma)K_{t-1} \tag{3}$$

where $t = 1995$ to 2014, a total of 20 years and I_t is the value of investment in constant prices in t. We assume, with the World Bank (2005, 2011) that the asset's service life is 20 years and the depreciation rate γ is 5% to be constant across provinces and over time. Thus, we get (a), the actually produced capital, from the expression (4):

$$K_{2014} = \sum_{i=0}^{19} I_{2014-i}(1-\gamma)^i \tag{4}$$

3.4 Following Hartwick's rule for (b) the produced capital

We follow the proposition for Hartwick's rule to calculate how many exhaustible resources value $\dot{K}=V_t=RF_R$ reinvestment into produced capital. So we establish expression (5):

$$K_{2014}^{*} = K_{1995} + \sum_{t=1995}^{2014} (V_t) \tag{5}$$

where the baseline year is 1995, we use the actual produced capital stock K_{1995} and assume that every province should reinvest its exhaustible resources' value V_t to K_{1995} until 2014. Then, following Hartwick's rule, we get K_{2014}^{*} as the aggregate capital stock.

We still discuss resources' value V_t, and we still use the World Bank's method (2005, 2011)

$$V_t = F_R R = \sum_{i=t}^{t+T-1} \frac{\pi_i q_i}{(1+r)^{(i-t)}} \tag{6}$$

where $\pi_i q_i$ is the aggregate rents in the moment i, π_i is the unit rent, q_i is the output, r is the social discount, and T is the lifetime of the resource that is an exhaustion time value. We assume, following the World Bank (2005, 2011) that the social discount is 4% and the lifetime, $T \leqslant 25$, are assumed to be constant across provinces and over time.

3.5 Data

3.5.1 The data of the expressions (4) and (5)

K_{1995} is the aggregate capital stock in 1995 in China. We use Zhang's database for 1995 (Zhang et al., 2004), taking the baseline as 1952 and avoiding some scholars' use of 1978 as the baseline to influence the investment price (Table 1). As Zhang's research lacks the data of Chongqing in 1995, we have used Zhang's method to estimate that Chongqing's produced capital stock is 42.5 Billion Yuan. The data of the value of investment I are taken from 29 provinces' Statistical Yearbooks.

Table 1 Estimate of China's provincial produced capital stock (unit: billion Yuan)

No.	Provincial name	Capital stock	No.	Provincial name	Capital stock	No.	Provincial name	Capital stock
1	Tianjin	81.2	11	Fujian	49.4	21	Sichuan	174.2
2	Hebei	134.5	12	Jiangxi	43.3	22	Guizhou	30.3
3	Shanxi	79.2	13	Shandong	264.6	23	Yunnan	7.1
4	Inner Mongolia	60	14	Henan	145.8	24	Xizang	7.1
5	Liaoning	53.2	15	Hubei	53.4	25	Shaanxi	82
6	Jilin	53.8	16	Hunan	72.3	26	Gansu	59.7
7	Heilongjiang	80.3	17	Guangdong	215.6	27	Qinghai	9.6
8	Jiangsu	261.8	18	Guangxi	46	28	Ningxia	11.2
9	Zhejiang	139.2	19	Hainan	20.1	29	Xinjiang	61.4
10	Anhui	36.2	20	Chongqing	42.5	30	China	2968

Source: Zhang et al. (2004), and the author.

3.5.2 The data of the expression (6)

We choose 29 provinces in China. We excluded Beijing and Shanghai, because these two cities did not have statistically significant data for energy and mineral development. Resource rents are calculated from three energy resources: oil, natural gas, and coal; and nine mineral resources: iron, copper, lead, zinc, nickel, tin, antimony, molybdenum, and aluminum.

For example; the oil value of Henan Province in 2014 is

$$V_{2014oilhenan} = \sum_{i=t}^{t+T-1} \frac{(p_i - c_i) * q_i}{(1+r)^{(i-t)}} \tag{7}$$

We calculate the resource value accumulation V_t of 12 resources and 29 provinces by expression (7).

Here we have difficult data for π_i in expression (7)

$$\pi_i = p_i - c_i \tag{8}$$

We use the same method of the World Bank that π_i is unit rent that unit price p_i minus marginal extraction cost c_i, we use the China's data. p_i Using the international price which should better reflect the social cost of the depletion of energy and mineral resources. Fuss et al. estimate the world price of units of specific commodities; the data are from the World Bank, the IMF, and other international organizations. In addition, we use some import prices for some minerals. The estimates of unit costs of extraction can vary, so we use average unit cost, rather than marginal cost, to determine the true changing cost (Fuss et al., 2016). We calculate the extraction cost between sales' incomes and unit profit (Guoxia et al., 2009); these data can highlight each province's characteristics. The data of the output sale income and profit of every resource, from a province, in every year, are from the China Nonferrous Metals Industry Association, the China National Bureau of Statistics, and the provincial National Bureau of Statistics.

4 Results and analysis

Table 2 shows two types of method used to estimate the produced capital, which is the actual produced assets stock on PIM and the Counterfactual assets stock on the Hartwick Rule. According to our hypothesis, 29 provinces ascribed their exhaustible resource value to their locally fixed assets under Hartwick's rule. Column four in Table 2 shows that a negative number is mean that 29 provinces are under the Hartwick's rule. But Table 2 also reveals some diffierent points.

Fig. 1 shows the exhaustible resources' value in 29 provinces in 1995—2014. The top five are Heilongjiang, Shandong, Shaanxi, Xinjiang, and Inner Mongolia; the aggregate value of these five provinces in China is 54%. The bottom five are Hubei, Chongqing, Hainan, Zhejiang, and Xizang; the aggregate value of these five provinces in China is 0.9%.

3.3.3 Data processing

The production curve of bastnasite mine experiences a significant fluctuation in 2006 and 2007. In 2006, the bastnasite mine in Sichuan Province of China was found of illegal and excessive exploitation, so that the production soared from 25709 t in 2005 to 37000 t in 2006. As a sanction, the central government imposed a production limitation on this mining area and the production in 2007 dived to 6800t. In order to diminish the impact of the dramatic production fluctuation in 2006 and 2007, their figures for those two years were set as missing value, and the linear interpolation method was employed to replace the original figures. The new figures for 2006 and 2007 are 21301 t and 22454 t, respectively (Fig. 1).

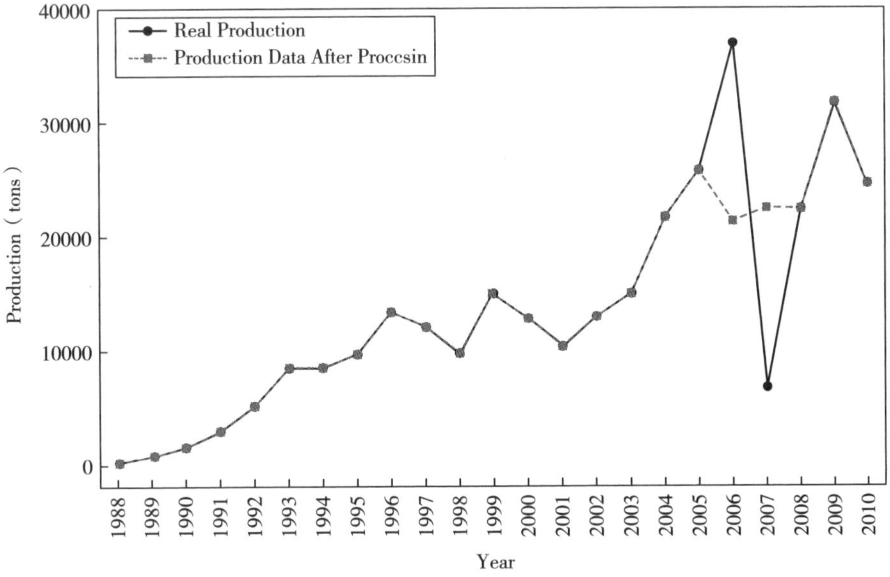

Fig. 1 Historical trend of China's bastnasite mine production

Data source: The National Development and Reform Commission (1998—2011)

4 Results and discussion

A systematic elaboration on the Generalized Weng model has been presented above. In this section, the Generalized Weng model is employed to predict the production trends and peaks of China's mixed rare earth, bastnasite and ion-absorbed rare earth respectively (Table 3 and Figs. 2-4).

Table 3 Generalized Weng model

Year	Mixed RE	Bastnasite	Ion-absorbed RE	Total
1988	1160.16	2.79	63.42	1226.37
1989	3542.75	34.35	335.20	3912.31

Continued

Year	Mixed RE	Bastnasite	Ion-absorbed RE	Total
1990	6633. 86	141. 80	863. 05	7638. 72
1991	10167. 93	374. 12	1655. 34	12197. 39
1992	13965. 29	772. 37	2701. 87	17439. 53
1993	17894. 87	1365. 43	3982. 16	23242. 46
1994	21857. 90	2168. 82	5470. 00	29496. 72
1995	25778. 95	3185. 29	7136. 24	36100. 48
1996	29600. 22	4406. 41	8950. 61	42957. 23
1997	33277. 68	5814. 60	10882. 98	49975. 26
1998	36778. 25	7385. 30	12904. 21	57067. 76
1999	40077. 63	9089. 10	14986. 66	64153. 39
2000	43158. 67	10893. 64	17104. 56	71156. 87
2001	46010. 02	12765. 30	19234. 26	78009. 58
2002	48625. 08	14670. 59	21354. 30	84649. 97
2003	51001. 10	16577. 27	23445. 48	91023. 85
2004	53138. 48	18455. 22	25490. 79	97084. 49
2005	55040. 16	20277. 06	27475. 41	102792. 63
2006	56711. 09	22018. 59	29386. 56	108116. 23
2007	58157. 84	23658. 98	31213. 44	113030. 25
2008	59388. 24	25180. 92	32947. 06	117516. 22
2009	60411. 09	26570. 56	34580. 16	121561. 80
2010	61235. 86	27817. 40	36107. 03	125160. 29
2011	61872. 55	28914. 10	37523. 39	128310. 05
2012	62331. 47	29856. 25	38826. 29	131014. 01
2013	62623. 08	30642. 08	40013. 92	133279. 08
2014	62757. 92	31272. 17	41085. 54	135115. 64
2015	62746. 46	31749. 18	42041. 34	136536. 98
2016	62599. 03	32077. 51	42882. 32	137558. 86
2017	62325. 76	32263. 04	43610. 22	138199. 02
2018	61936. 51	32312. 84	44227. 38	138476. 74
2019	61440. 85	32234. 94	44736. 69	138412. 48
2020	60848. 01	32038. 06	45141. 46	138027. 53
2021	60166. 85	31731. 41	45445. 39	137343. 65
2022	59405. 85	31324. 51	45652. 50	136382. 86
2023	58573. 10	30827. 02	45767. 01	135167. 12
2024	57676. 28	30248. 57	45793. 36	133718. 21
2025	56722. 70	29598. 67	45736. 09	132057. 45

Continued

Year	Mixed RE	Bastnasite	Ion-absorbed RE	Total
2026	55719.22	28886.59	45599.85	130205.65
2027	54672.33	28121.27	45389.32	128182.93
2028	53588.13	27311.27	45109.21	126008.61
2029	52472.31	26464.69	44764.18	123701.19
2030	51330.23	25589.15	44358.87	121278.25
2031	50166.85	24691.77	43897.81	118756.42
2032	48986.79	23779.12	43385.47	116151.38
2033	47794.33	22857.27	42826.19	113477.79
2034	46593.45	21931.72	42224.19	110749.36
2035	45387.79	21007.47	41583.56	107978.82
2036	44180.71	20089.00	40908.23	105177.94
2037	42975.29	19180.31	40201.97	102357.56
2038	41774.33	18284.89	39468.41	99527.63
2039	40580.40	17405.83	38711.00	96697.22
2040	39395.81	16545.75	37933.03	93874.59
2041	38222.65	15706.91	37137.63	91067.19
2042	37062.82	14891.18	36327.73	88281.73
2043	35917.98	14100.08	35506.14	85524.20
2044	34789.64	13334.84	34675.48	82799.95
2045	33679.12	12596.38	33838.19	80113.69
2046	32587.57	11885.39	32996.59	77469.55
2047	31516.00	11202.29	32152.81	74871.10
2048	30465.27	10547.31	31308.86	72321.45
2049	29436.12	9920.49	30466.59	69823.20
2050	28429.17	9321.69	29627.69	67378.55

4.1 Mixed rare earths

We chose the period of 1994 through 2010 from the data of China's mixed rare earths production which shows the best linear relations. By the linear goal seeking methods, the forecasting parameters were obtained: $a=1234.38$, $b=1.7$, $c=16.13$, and $R^2=0.9997$ for regression segment. When a, b and c were substituted into Eq. (1), the future productions of mixed rare earths could be predicted. The forecasts show that China's production of mixed rare earths will peak in 2014 at 62757 t, followed by a steady decline rate of 2% in the ensuing years (Fig. 2).

4.2 Bastnasite rare earths

We chose the period of 2000 through 2010 from the data of China's bastnasite rare earths production which shows the best linear relations. By the linear goal seeking methods, the forecasting parameters were obtained: $a = 3.15$, $b = 3.8$, $c = 8.13$ and $R^2 = 0.9994$ for

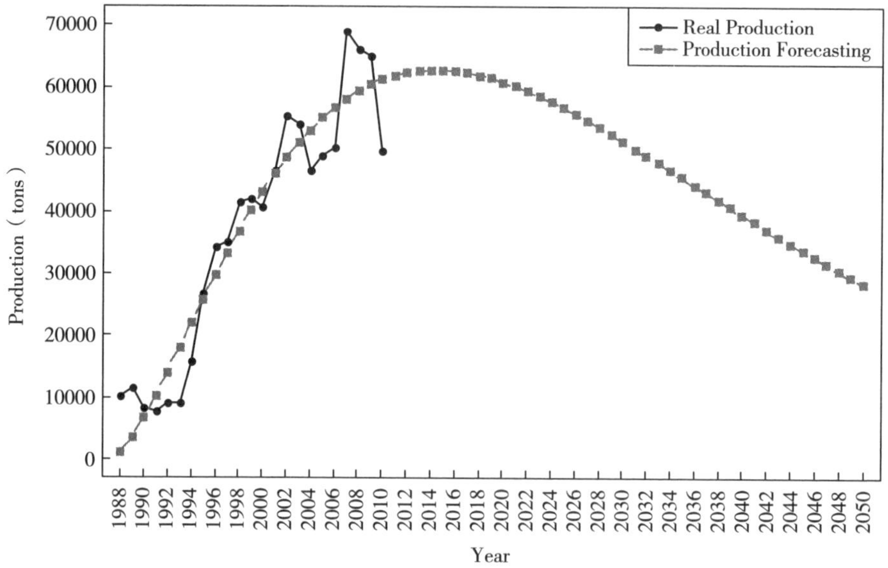

Fig. 2 China's mixed rare earths production forecasting

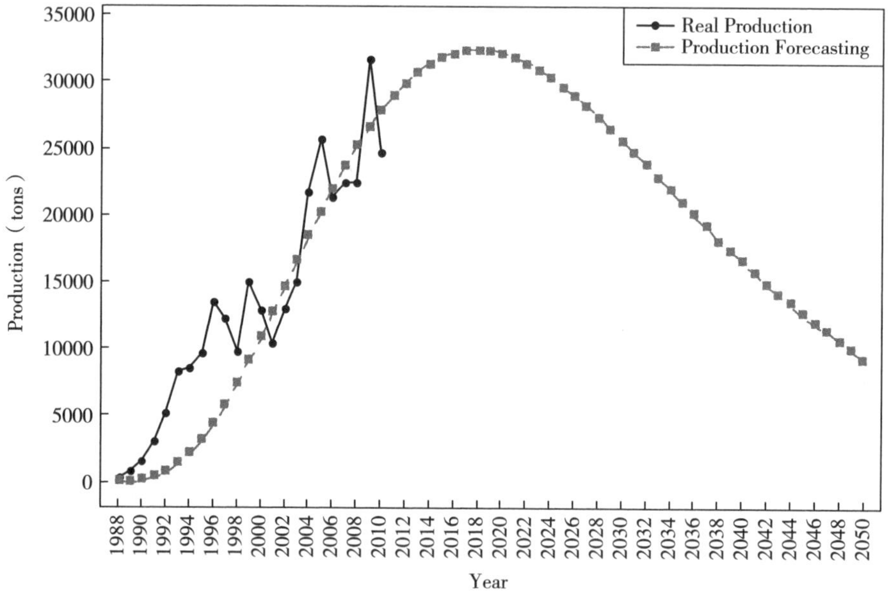

Fig. 3 China's bastna site production forecasting

regression segment. When a, b and c were substituted into Eq. (1), the future productions of bastnasite could be predicted. The forecasts show that China's production of bastnasite will peak in 2018 at 32312t, preceded by an annual increase of 1.67% and followed by a steady annual decline rate of 4% in the ensuing years (Fig. 3).

4.3 Ion-absorbed rare earths

We chose the period of 1994 through 2010 from the data of China's ionabsorbed rare earths production which shows the best linear relations. By the linear goal seeking methods, the forecasting parameters were obtained: $a = 67.88$, $b = 2.5$, $c = 14.72$ and $R^2 = 0.9962$ for regression segment. When a, b and c were substituted into Eq. (1), the future productions of ion-absorbed rare earths could be predicted. The forecasts show that China's production of ion-absorbed rare earths will peak in 2024 at 45793 t, preceded by an annual increase of 1.72% and followed by a steady annual decline of 4% in the ensuing years (Fig. 4).

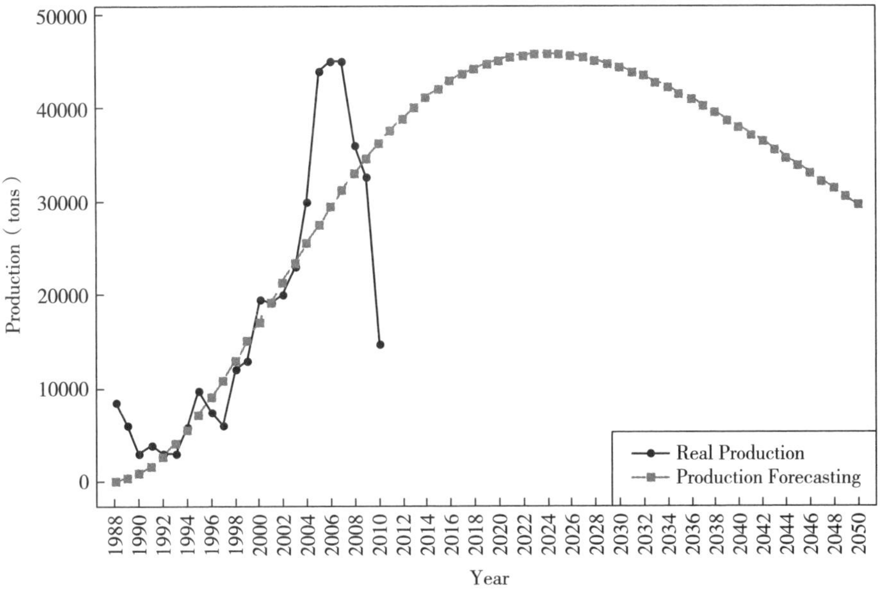

Fig. 4　China's ion-absorbed rare earths production forecasting

4.4　Discussion

First, from the comparison in Figs. 2-4, the predicted results of the three ores have a high degree of agreement with the actual productions from 1988 to 2010. The difference between the predicted and actual production for mixed rare earths, bastnasite and ion-absorbed rare earths averages to 1.4%, 1.1% and 2%, respectively. The forecasts are relatively reliable. The difference of total production between the predicted and the actual of the three types of rare earths may also be obtained which are averaged to me re 1.1% with its peak in 2018 at 138476 t (Fig. 5).

Second, the predicted results are quite close to the estimates of previous research. As stated in the review part, some research has been done and the productions of rare earths in the near future have been forecast by Kingsnorth (2010b), Chen (2011), Wübbeke (2013), Hurst (2010), He and Lei (2013) etc. The similar results proved the reliability of the prediction in this paper in some way since most of the previous results are based on practical experience or national or companies' plans though only for short term.

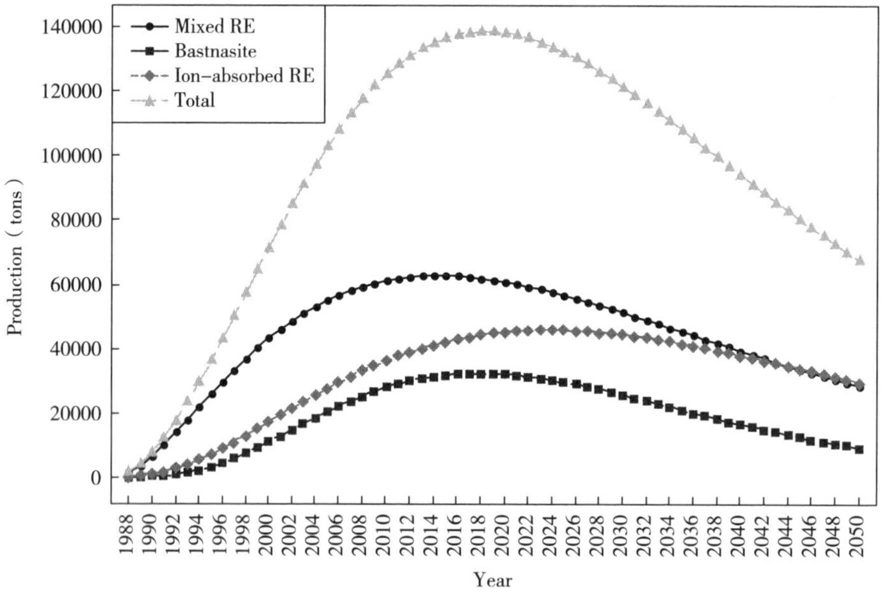

Fig. 5 Three types of rare earths production forecasting

5 Conclusions and recommendations

5.1 Conclusions

Based on the prediction and analysis of results in the Discussion, some conclusions are drawn.

(1) By 2020, China's supply of mixed rare earths and bastnasite will begin to decline steadily and China's dominant position will end. To be specific, the production of mixed rare earths in China will peak in 2014 at 62757 t, followed by an annual decline rate of 2%; that of bastnasite will peak in 2018 at 32312 t, followed by an annual decline rate of 4%.

(2) By 2024, China's supply of ion-absorbed rare earths will see a trend of continued growth at an annual rate of 1.72%, after which it will decrease on a yearly basis of 4%.

(3) By 2050, China's production of mixed rare earths, bastnasite and ion-absorbed rare earths will account for 45%, 28% and 64% of its peak value respectively, which indicates that ion-absorbed rare earths production will experience the smallest scale of decrease by 2050.

5.2 Policy recommendations

With the continuous development of rare earth terminal industries in and out of China, rare earths have been a target of world competition. Rare earths are distributed unequally in the world with 50% of the reserves located in China and about 30% in Russian and the United States. As for heavy rare earths (mainly refers to ion-absorbing type rare earths ore and xenotime), it is even incredibly endowed mostly in Longnan of China, Lehat of Malaysia and Strange Lake of Canada (Chen, 2011). Countries with or without rare earths should value the unrenewable resources and make suitable policies to ensure the supply for demand.

5.3 China

Chinese government should enforce environmental and resource exhaustible taxes soon and different domestic regulations for different rare earths according to the different production potential of different ores.

The monopolistic position of China in production and export of rare earths relies mostly on their low price in which the costs of environmental protection and resource remuneration have not been properly included. In fact, the rare earths for export are ridiculously cheaper than their own value, which results to that increases in export volume do not bring the same increases in export value (Ye and Wu, 2014). Low price coupled with overexploitation gives rise to a series of problems. He et al. (2014) mentioned that overproduction will accelerate the exhaustion of RE resources, and may well create a black market in RE products. So the system of environmental and resource taxes should be established and put into action soon.

As shown in the results, the three main rare earths in China will reach their peaks within only 10 years. However, more of other countries are endowed with light rare earths (including mixed rare earths, bastnasite) which implies that other countries will not depend on China's supply so keenly in the next few years as predicted by many previous research. It is a wise strategy for Chinese government to pay more attention to the research of substitute technologies of light rare earths and build longer industry chains to get more profit from rare earth resources. For heavy rare earths (mainly refers to ion-absorbed rare earth), China should first integrate the mining rights and mining companies to get the highest efficiency of exploitation and the best effect of environmental protection.

5.4 Other countries

Countries with RE resources need to start or restart their RE mines to gradually reduce dependence on China, such as the U. S., Russia, Australia, India, etc. (Table 4).

Table 4 Rare earth s supply situation oversea

Major supply mines	Total reserves (in 10000t)	Starting year of supply	Planned annual production(t)	Elements available
Mountain Pass, U. S.	180	From 2012	40000	La, Ce, Pr, Nd, Sm
Bear Lodge, U. S.	40	2015	10000	La, Ce, Pr, Nd, Sm
Bokan Mountain, U. S.	30	2014-2016	NA	La, Ce, Nd, Sm, Gd, Y, Dy
Tomtor, Russia	15000	After 2023(an estimate)	NA	Y, Nd oxides, Sc and Tb
Mt. Weld, Australia	120	2012	11000	Mainly light rare earths
Nolans, Australia	85	2012	20000	Rare-earth oxides
Dubbo, Australia	65	2013	2580	Light rare earths and yttrium
Total	15520	-	93580	-

Notes: The rare earth elements can be grouped as light rare earth elements-La, Ce, Pr, Nd, and middle and heavy rare earth elements-Sm, Eu, Gd, Tb, Dy, Ho, Er, Tm, Yb, Lu, Y. Sc is of dispersed elements, and Pm does not exist in nature (Xu, 1995).

Sources: Ye and Wu (2014)

Given the pattern of supply of REs in the world (China accounting for 97% of the world's supply of REs), a world-wide RE supply and demand gap will appear after the peak. Thus, countries with rare earths should consider starting or restarting their RE mines and relieving the world's RE supply pressure.

Countries without rare earth resources need to get prepared to face with the risk of price rising especially for ion-absorbed rare earths. It is welcome for these countries to find ways to recycle resources and consider alternative technologies in addition to keeping close political and economic ties with China and other countries with REs.

Acknowledgements We would like to thank the reviewers of this journal for helpful comments and suggestions. Thanks also goes to Ms Wenhui Chen for her help in data testing. This research is supported by the National Natural Science Foundation of China under Grant no. 71241027 and 71203203 and the support from the Development and Research Center of China Geological Survey under Grant no. 12120113093200.

References

Arps, J. J., 1945. Analysis of decline curves. Trans. Am. Inst. Min. Eng. 160, 228-247.

Bartlett, A. A., 2000. An analysis of US and world oil production patterns using Hubbert-style curves. Math. Geol. 32 (1), 1-17.

Brandt, A. R., 2010. Review of mathematical models of future oil supply: historical overview and synthesizing critique. Energy 35 (9), 3958-3974.

Brandt, A. R., 2007. Testing Hubbert. Energy Policy 35 (5), 3074-3088.

Castor, S. B., Hedrick, J. B., 2006. Rare Earth Elements. Industrial Minerals Volume, 7th ed. Society for Mining, Metallurgy, and Exploration, Littleton, CO, pp. 769-792.

Chen, Y. Q., Hu, J. G., 1996. Weng's model for review and the new derived. Chinas
Offshore Oil Gas 10 (05), 317-324 (in Chinese).

Chen, Z., 2011. Global rare earth resources and scenarios of future rare earth industry. J. Rare Earths 29 (1), 1-6.

Currie, A., 2012. Overview of Molycorp Californian Rare Earth Deposit. Retrived January 25, 2013, from http://rareearthinvestingnews.com/6796-overview-moyl corp-californian-rare-earth-deposit-mountain-pass-project-phoenix.htm).

Feng, L., Li, J., Pang, X., 2008. China's oil reserve forecast and analysis based on peak oil models. Energy Policy 36 (11), 4149-4153.

Feng, L., Tang, X., Zhao, L., 2007. Reasonable planning oil production in China based on peak oil model. Pet. Explor. Dev. 34 (4), 497-501 (in Chinese).

Feng, L., Zhao, L., Zhao, Q., et al., 2006. Models of peak oil theory and forecast of world oil peak. Acta Pet. Sin. 27 (5), 139-142 (in Chinese).

Feng, L., Wang, J., Zhao, L., 2010. Construction and application of a multicyclic model in the prediction of natural gas production. Nat. Gas Ind. 30 (7), 1-3 (in Chinese).

Gschneidner Jr., K. A., 2011. The Rare Earth Crisis— The Supply/ Demand Situation for 2010-2015, 6 (2), MaterialsMatters, 32-37. Retrieved July 19, 2014, from http://www.sigmaaldrich.com/.

He, C., Lei, Y., Ge, J., 2014. Assessment of trading partners for China's rare earth exports using a decision analytic approach. PLoS One 9 (7), e102870.

He, C., Lei, Y., 2013. Potential impact of US re-emerging rare earths industry on future global supply and demand trend. Int. Bus. Res. 6 (7), 44.

Höök, M., Aleklett, K., 2009. Historical trends in American coal production and a possible future outlook. Int. J. Coal Geol. 78 (3), 201-216.

Höök, M., Zittel, W., Schindler, J., Aleklett, K., 2010. Global coal production outlooks based on a logistic model. Fuel 89 (11), 3546-3558.

Hu, J., Chen., Y., Zhang, Z., 1995. A new model for Forecast oil and gas production and recoverable reserves. J. Oil 16 (1), 79-86 (in Chinese).

Hubbert, M. K., 1956. Nuclear energy and the fossil fuels, in Meeting of the Southern District, Division of Production, American Petroleum Institute, San Antonio, TX: Shell Development Company.

Hurst, C., 2010. China's Rare Earth Elements Industry: What Can the West Learn? Institute for the Analysis of Global Security, Washington DC.

Kingsnorth, D. J., 2010a. An Overview of the Rare Earths Market. IMCOA, Kingaroy Qld.

Kingsnorth, D. J., 2010b. Rare earths: facing new challenges in the new decade. In: Proceedings of the SME Annual Meeting.

Krautkraemer, J. A., 1998. Nonrenewable resource scarcity. J. Econ. Lit. 36 (4). 2065-2107.

Laherrere, J. H., 2000. Learn strengths, weaknesses to understand Hubbert curve. Oil Gas J. 98 (16), 63-76.

Lin, B., Liu, J., 2010. Estimating coal production peak and trends of coal imports in China. Energy Policy 38 (1), 512-519.

Lv, M. Y., Shi, H. F., Zheng, N., Zhou, G. Y., 2012. Coalbed methane production forecast method based on generalized Weng's model. China Coal Bed Methane6, 010 (in Chinese).

Massari, S., Ruberti, M., 2013. Rare earth elements as critical raw materials: focus on international markets and future strategies. Resour. Policy 38, 36-43.

McLellan, B. C., Corder, G. D., Ali, S. H., 2013. Sustainability of rare earths—an overview of the state of knowledge. Minerals 3 (3), 304-317.

Ministry of Commerce People's Republic of China. Principal of Department of Treaty and Law of MOFCOM, 2014. Gives a Speech on WTO's Panel Report on the Case of the U. S., EU and Japan's Action Against China's Export Management Measures of Relevant Rare Earth, Tungsten and Molybdenum Products. Retrived May 25, 2014, from http://english. mofcom. gov. cn/article/newsrelease/policyreleasing/201404/20140400538135. shtml), (accessed 31. 03. 14/02. 05. 14.).

Moore, C. L., 1966. Analysis and Projections of the Historic Patterns of U. S. Domestic Supply of Crude Oil, Natural Gas, and Natural Gas Liquids. United States Department of the Interior, Office of Oil and Gas, Washington, DC (82 pp.).

Moore, C. L., 1962. Method for Evaluating U. S. Crude Oil Resources and Projecting

Domestic Crude Oil Availability. United States Department of the Interior, Washington, D. C. (112 pp.).

Morrison, W. M., Tang, R., 2012. (Congressional Research Service Report). China's Rare Earth Industry and Export Regime: Economic and Trade Implications for the United States.

Qiang, S., Bing, D., Limei, M., 1995. Application of generalized Weng's model and Rayleigh model in recoverable reserves. J. Oil 16 (1), 79-86.

Tang, X., Feng, L., Zhao, L., 2009. Prediction and analysis of world oil supply pattern based on generalized Weng's model. Resour. Sci. 31 (2), 238-242.

The National Development and Reform Commission, Industry Coordination Department., 1998-2011. China's rare earths. Rare earths information (in Chinese).

USGS (U. S. Geological Survey)., 2009. Mineral commodity summaries, rare earths. Walls, M. A., 1992. Modeling and forecasting the supply of oil and gas: a survey of existing approaches. Resour. Energy 14 (3), 287-309.

Wang, B. G., 2010. Scarcity: Warren Buffett, "economic moat" another interpretation. Dyn. Anal. Stock Mark. 21, 005 (in Chinese).

Wang, J. L., Feng, L., Zhao, L., Snowden, S., Wang, X., 2011. A comparison of two typical multicyclic models used to forecast the world's conventional oil production. Energy Policy 39 (12), 7616-7621.

Weng, W. B., 1984. The Foundation of the Forecasting Theory. Petroleum Industry Press, Beijing, China (in Chinese).

Wübbeke, J., 2013. Rare earth elements in China: policies and narratives of reinventing an industry. Resour. Policy 38 (3), 384-394.

Xu, G. X., 1995. Rare Earths, 2nd ed. China Metallurgical Industry Press, Beijing (in Chinese).

Ye, R., Wu, Y., 2014. China's Rare Earth Strategy Development and Export Regulation Policy Research. Science Press, Beijing, China (in Chinese).

Zhao, L., Feng, L., Lu, X., et al., 2009. Comparison of two methods for generalized Weng model. Xinjiang Pet. Geol. 30 (5), 658-660 (in Chinese).

Zheng, M. G., Chen, Y. H., 2012. Supply and demand of global rare earth resources and China's rare earth industry policy. Nonferr. Met. Sci. Eng. 4, 70-74.

Zhu, J., Che, C. B., Liu, C. L., et al., 2008. Contrastive analysis of reserves increasing prediction models (Natural Science Edition). J. Xi' an Pet. Univ. 23 (5), 21-23 (in Chinese).

Zittel, W. and Schindler, J., 2007. Crude oil: the supply outlook. Energy Watch Group.

Research on financial early warning of mining listed companies based on BP neural network model

SUN Xiaojun[1,3], LEI Yalin [1,2,3,*]

1. *School of Economics and Management, China University of Geosciences, Beijing* 100083, *China*
2. *The College of Economics and Management, Beijing University of Chemical Technology, Beijing* 100029, *China*
3. *Key Laboratory of Carrying Capacity Assessment for Resource and Environment, Ministry of Natural Resources of the People's Republic of China*, *Bejing* 100083, *China*

Abstract Mining industry is the basic industry of the national economy. However, in recent years, listed mining companies have suffered serious financial risks due to special reasons such as poor spot market liquidity of their products, strong policy dependence, and long investment payback periods. In the previous studies, most of the financial crisis prediction focused on the whole industry and manufacturing industry. The research on the financial risk of mining enterprises focuses more on how to adjust R&D activities, environmental performance to improve the financial performance of enterprises. There is still a lot of room for indepth research on the systematic prevention and early warning of financial risks of listed mining companies. At the same time, in terms of research methods, many scholars used multivariate discriminant model, logistic regression model and support vector machine model. Compared with the Back-Propagation (BP) neural network model, these model methods have more or less defects. Therefore, we take mining listed companies as the research object, select the financial data of China's A-share mining listed companies in 2018, and construct the BP neura l network financial early warning model, trying to provide more practical means for the financial risk early warning of mining companies. The research conclusions of this paper are as follows: (1) The BP neural network financial early warning model constructed in this paper has high prediction accuracy, which can be well used in the practice of financial early warning of mining listed companies; (2) The financial situation of China's A-share mining listed companies in 2018 is generally in a good state. The companies with good financial status can effectively control the cost and have good debt paying ability while earning income; (3) For companies with financial status that require early warning, the root cause is mainly that they do not pay attention to the risk of bad debt losses, which makes current credit sales income and accounts receivable are at high levels, and they also do not have good profitability.

Keywords Financial early warning, Mining listed companies, Back-propagation (BP) neural network

---

原文刊于 *Resources Policy*，2021 年第 73 卷。原文发表时为彩色。

* 为通信作者。

1 Introduction

Mining industry is the foundation and pillar industry of national economy (Chand et al., 2020; Modak et al., 2017). However, in recent years, listed companies in the mining industry are facing many problems, such as the aggravation of overcapacity, the high level of important mineral inventory, and the serious decline of enterprise net profit (Fan et al., 2017; Tang et al., 2015; Himley, 2010). As a result, the financial crisis of Listed Companies in the mining industry has been increasing, which has greatly affected the sustainable and healthy development of the whole mining industry (Jonek Kowalska, 2019; Nel et al., 2018). It is very necessary for listed companies in mining industry to establish a scientific and reasonable financial early warning model with ideal prediction accuracy, so that managers of listed companies can find financial risks as soon as possible, and make judgments on the company's financial status and operation status (Shen et al., 2020; Geng et al., 2015; Zhou et al., 2015).

At present, most scholars use the way of building models for the study of financial early warning. The mature models mainly include multiple discriminant model, logistic regression model and support vector machine model. But these three models all have some defects. For example, multivariate discriminant model has very strict requirements on data quality, which requires normal distribution of independent variables and the same covariance (Kosmidou et al., 2006); Logistic regression model is not particularly ideal due to its more approximate processing in calculation (Jabeur et al., 2017; Irimiadieguez et al., 2015); Support vector machine model is difficult to operate and analyze large-scale training samples (Jiang et al., 2018; Zhou et al., 2016). TheBP neural network model can effectively overcome the shortcomings of the above-mentioned models, and has higher prediction accuracy, more stability, and more accurate results (Hoque and Rahman, 2020; Tavana et al., 2018; Sun and Xu, 2016; Yang et al., 2001; Cao et al., 2011; Koyuncaul and Ozgulbas, 2012), and has the advantages of no premise assumption, so it is a model with strong self-learning and adjustment ability. In the past research, few scholars used BP neural network model in financial early warning of mining listed companies. Therefore, we build a BP neural network financial early warning model to conduct forecast research on listed companies in the mining industry, hoping to help the sustainable and healthy development of mining companies.

The contribution of this paper is: (1) This paper establishes a scientific and reasonable BP neural network financial early warning model of mining listed companies, and systematically evaluates the financial security situation of China's A-share mining listed companies in 2018 through the model, and makes a detailed analysis of its internal causes, hoping to solve the practical problem of mining listed companies facing financial risks. (2) According to the evaluation results and the analysis of the internal reasons, the paper puts forward some targeted suggestions for the existing problems and future development of China's A-share mining listed companies. (3) It provides a detailed idea and a certain experience for the financial early warning of mining industry in the international market.

2 Literature review

2.1 Enterprise risk of mining industry

Mining production will be affected by many different risk factors (Jonek-Kowalska, 2019). For the overall mining risk assessment, the existing scholars have carried out relevant research from the national, regional and departmental perspectives. Ouyang and Lai (2021) constructed the attention LSTM neural network model to research the early warning of systemic risk in China. Scammacca et al. (2021) systematically assessed the mining risk of French Guiana at the regional level, and identified the main potential influencing factors of mining risk for each region. Jonek-Kowalska (2019) conducted an overall fuzzy assessment of the operational risk of Polish mining enterprises from the perspective of long-term and se ctoral research. Based on the research of environmental social governance (ESG)-risks, a few scholars have made a comprehensive assessment of the environmental, social and governance risks faced by phosphorus mining (Paat et al., 2021). On the other hand, scholars also pay great attention to the indepth study of a specific risk of mining enterprises (mining areas). Some scholars have made a systematic analysis of the environmental risks faced by mining companies (mining areas) by assessing the tailings da m break event of Vale (Saes and Muradian, 2021), arsenic pollution of tailings in Eastern Amazon gold mining area (de Souza Neto et al., 2020) and environmental pollution of qhorveh mining area in western Iran (Saedpanah and Amanollahi, 2019). Based on the theory of ecological risk assessment, some scholars have made a detailed quantitative assessm ent on the ecological risk of Pingshuo coal mining area and the mining areas in the south central and southeast of Ivory Coast (Wu et al., 2021; Kinimo et al., 2018). In addition, scholars have also made re levant research on the climate change risk (Sun et al., 2020), social risk (Bergeron, 2021) and foreign investment risk (Wang et al., 2020) faced by mining enterprises. As for the financial risks faced by mining enterprises, studies by relevant scholars have shown that mining enterprises are generally faced with financial risks such as poor cash liquidity, long payback period of investment, and serious decline of enterprise net profit (Fan et al., 2017; Tang et al., 2015; Himley, 2010).

2.2 Financial risk research status

Financial risk is the risk of bankruptcy caused by the insolvency of an enterprise (Beaver, 1966; Altman and Ratios, 1968; Deakin, 1972), which reflects the importance of financial risk compared with other risks. As for the research on financial risk faced by enterprises in different industries, scholars mainly focus on the influencing factors of financial risk, the construction of financial risk early warning model, and the prediction of financial distress. In terms of influencing factors of financial risk, scholars have studied the key factors to identify the financial risk of enterprises, and also studied the impact of a specific influencing factor on financial risk. Mahtani and Garg (2018) used the fuzzy analytic hierarchy process to analyze the key factors of the financial distress of Indian Airlines. Liu et al. (2021) used patent application data of Chinese listed companies from 2008 to 2017 to study how R&D innovation affects the financial distress risk of enterprises. Rafiq et al. (2016) and Abban (2021) respectively studied the impact of R&D activities and environmental performance on financial performance of mining enterprises. In the construction of financial risk early warning model, scholars have also done related research. Choi et al. (2018) used the integrated learning financial distress prediction model to make an early warning analysis of the construction industry in South Korea. Some scholars have also built PSO-SVM model (Zhou et al., 2020) and Z-score Model (Wang, 2012) to measure and warn the financial risk of China's real estate industry. In addition, scholars have also carried out research on the prediction of corporate financial distress from the perspective of the country and industry. Dinh et al. (2021) studied the financial distress of enterprises in six countries in the ASEAN economic community, and found that the main reason for the financial distress of enterprises is the difficulty in debt repayment. Crespí-Cladera et al. (2021) predicted the financial distress faced by Spanish hotel companies, and found that most of the hotel companies in financial distress are faced with solvency problems. Scalzer et al. (2019) studied the financial distress risk faced by Brazilian power companies, and found that the return on assets is outstanding in predicting the company's performance.

2.3 Financial early warning model

As for the research of financial early warning model, most scholars adopt the combination of model and qualitative analysis to research the company's financial early warning.

(1) Traditional single variable and multi variable early warning model

The earliest financial early warning model is the single variable early warning model proposed by Fitzpatrick (1932). Although this single variable early warning model is simple and easy to operate, its most prominent shortcomings are weak prediction ability and low accuracy. Therefore, in order to make up for this shortcoming, Professor Altman (1968) of New York University used the multivariate early warning model in the research of financial crisis early warning. He first selected 22 financial indicators as the early warning index system of primary selection, selected 66 companies as samples, analyzed the index discrimination level, and then selected five financial data indicators. Finally, through data weighting processing and linear regression, he got a comprehensive judgment value Z value of financial crisis, and created a complete Z-score Model.

(2) Logistic regression model

Although the prediction accuracy of multivariate early warning model is high, its application scope is limited due to the strict preconditions such as normal distribution of independent variables and the same covariance of sample data. Therefore, Martin (1977) introduced logistic regression model into the study of financial early warning model. The American scholar Zmijewski (1984) also used logistic regression method to select 76 bankrupt enterprises and 3880 enterprises in good financial condition from 1972 to 1978 to establish the early warning model of financial crisis.

(3) Artificial neural network model

With the rapid development of information technology, scholars in this field have also begun to use artificial intelligence technology to build financial crisis early warning model. The main research methods include neural network model, multi case reasoning system, support vector machine and other methods (Hosaka, 2019; Iturriaga and Sanz, 2015; Lee and Choi, 2013; Chen and Du, 2009; Sartori et al., 2016; Ahn and Kim, 2009; Chaudhuri and De, 2011; Yang et al., 2011). However, the most widely used and most widely used is neural network model. Odom and Shard a (1990) were the first to introduce neural network model to predict enterprise financial crisis. They selected the data of many enterprises and compared them with the traditional financial crisis discrimination method and neural network model, and found that the neural network model has higher prediction accuracy. In addition to the high prediction accuracy, the neural network model has the advantages of strong adaptability and robustness (TAM and Kiang, 1992). At present, most scholars will apply the artificial neural network model to the financial crisis prediction of enterprises or banks in the whole industry. For example, Barboza et al. (2017) selected the financial data of North American companies in the whole industry from 1985 to 2013, and used artificial intelligence model to predict the financial crisis of more than 10000 companies. Tseng and Hu (2010) used BP multilayer perceptrons and radial basis function networks to forecast UK companies across the industry. Iturriaga (2015) combined multi-layer perceptron and self-organizing map to construct a neural network model to evaluate the short-term, medium-term and long-term risks of Bank of America. Chauhan et al. (2009) used wavelet neural network model to predict the financial risks of Bank of America, Bank of Turkey and Bank of Spain. Although the prediction of financial crisis in the whole industry can give early warning to most enterprises in a country, some scholars have found that, due to the heterogeneity among industries, the effect of single industry prediction is often 6-12% better than that of the whole industry prediction (Lee and Choi, 2013). At present, most of the single industry forecasts using neural network model are concentrated in the manufacturing industry (Kim et al., 2016), and the research on mining listed companies needs to be further supplemented.

2.4 Research gaps

From the above analysis, it can be seen that the overall risk and a specific risk assessment of mining enterprises, the research of industry financial risk, and the research of financial early warning model have been widely concerned by the academia. However, the research on the financial risk of mining enterprises focuses more on how to adjust R&D activities (Rafiq et al., 2016; Liu et al., 2021), environmental performance (Abban and Hasan, 2021) to improve the financial performance of enterprises, and then reduce the financial risk of enterprises. There is still a lot of room for in-depth research on the systematic prevention and early warning of financial risks of listed mining companies. At the same time, in terms of research methods, many scholars used multivariate discriminant model, logistic regression model and support vector machine model. Compared with the BP neural network model, these models have more or less certain defects. Therefore, this paper uses BP neural network model which has high prediction accuracy, strong learning ability and efficiently deal with complex nonlinear problems to systematically study the financial early warning of mining listed companies, in order to contribute to the healthy development of mining enterprises.

3 Construction of financial early warning model

The financial early warning model constructed in this paper is based on the BP neural network main model with cluster analysis and close value method. The main reason for using these two methods is that the former scholars only considered whether the net profit continuously appears negative value in three years, which leads to the problems of inconsistent division of early warning interval and judgment of a company's financial status based on a single financial index. Cluster analysis combined with osculating value method can comprehensively reflect the real financial situation of enterprises from multiple angles, and make a detailed and coherent division of the warning interval of these sample companies, which can make the results of the main model more accurate and reliable.

3.1 BP neural network financial early warning model

BP neural network model is a multi-layer forward network model based on error back-propagation algorithm. It was first proposed by a team led by Rumelhart and McCelland in 1986 (Sun and Xu, 2016). It can simulate the information transmission mode of human brain neurons, and make nonlinear transformation and regression processing for complex information variables, and obtain the operation results with high fitting degree (Kaastra and Boyd, 1996). Just like the information transmission between neurons in the brain, after the input variables are input in the input layer, a linear combination of input variables will be obtained according to the initial set weight. When the weight is constantly modified to make the linear combination value exceed the threshold value, the information will be transferred to the output layer, as shown in Fig. 1.

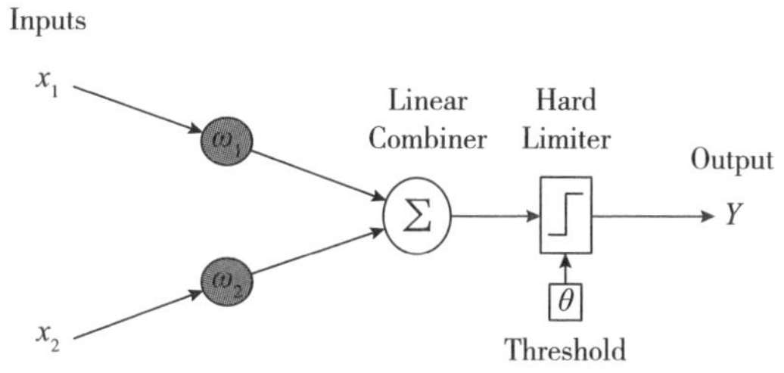

Fig. 1 BP neural network diagram

The structure of BP neural network includes input layer, hidden layer and output layer. A large number of tests show that the three-layer BP neural network is very suitable for regression analysis of the prediction model (Sun and Xu, 2016; Guo, 2020; Li and Cao, 2020; Du, 2021), so this paper selects the three-layer BP neural network to do the analysis of financial early warning. The following are the specific steps of model establishment:

Step 1. Prepare the sample of training network. From the factor analysis of the data preprocessing below, we can see that the company's financial crisis will be affected by 7 common factors, these 7 factors are network input variables, and the type of financial crisis is network output variables. Due to the limited number of samples, 40 groups of samples were randomly selected for training and testing, and the remaining 8 groups were used as simulation groups.

Step 2. Determines the initial parameters of the network. In this model, the following network training initial parameters are selected, as shown in Table 1. For the determination of the number of hidden layer nodes in the model, Section 5.2 of this paper makes a detailed description.

Step 3. Normalizes the input value of input layer. The common factors of 40 mining listed companies selected above are input into X matrix, and the data scale of the matrix is 7×40. The data were normalized and the result was $O_1 = X$.

Table 1 Initial parameters of network training

Initial parameter	Numerical value
Maximum training times	5000
Number of nodes in the hidden layer	10
Network learning speed	0.05
Tolerable error	3×10^{-4}

Step 4. Calculates the input and output values of the hidden layer. The threshold value of neurons in the hidden layer is assumed to be 1 (this value can be automatically adjusted by the subsequent error back-propagation algorithm). The input value of the hidden layer is equal to the sum of the weighted average of the output value of the input layer and the threshold value, i.e. $I_2 = w_{ij} \times X + B_{ij} \times ones$, where ones is a matrix of all 1 and I_2 is a matrix with data size of 10×40. Set the output function of the hidden layer as a logistic regression function, that is,

$f(x)=\frac{1}{1+e^{-x}}$, so the output value of the hidden layer is $O_2=\frac{1}{1+e_2^{-1}}$ The data scale of O_2 matrix is also 10×40.

Step 5. Calculates the input and output values of the output layer. The input of the output layer is very similar to that of the hidden layer, i. e. $I_3 = W_{jk} \times O_2 + B_{jk} \times \text{ones}$. I_3 is a 1×40 matrix. For the output of the output layer, it is usually assumed that the neuron connections in the output layer also transmit information through a linear function, so $O_3 = I_3$. The data scale of O_3 matrix is 1×40.

Step 6. Calculate the error function E. The error function E is calculated, that is, the sum of error squares between the output class value and the actual class value. When the error reaches the preset tolerance value, the training can be stopped. Assuming that the actual class value is y, from the definition of error function E, we can get the following results:

$$E = \sum (Y - O_3)^2 \tag{1}$$

Step 7. Calculates the weight and threshold adjustment between the hidden layer and the output layer. This step and the tenth step are the core of BP neural network. By using the principle of negative gradient descent, a chained partial differential is obtained.

$$\Delta w_{jk} = -\gamma \frac{\partial E}{\partial w_{jk}} = -\gamma \times (Y - O_3) \times O_2 \tag{2}$$

$$\Delta B_{jk} = -\gamma \frac{\partial E}{\partial B_{jk}} = -\gamma \times (Y - O_3) \times \text{ones} \tag{3}$$

Step 8. Calculates the weight and threshold adjustment between the input layer and the hidden layer. Firstly, the weight and threshold adjustment between the hidden layer and the output layer are calculated, and then the weight and threshold adjustment between the input layer and the hidden layer are calculated by using the error back-propagation algorithm. At the same time, we notice that the derivation of the logistic regression function $f(x)=\frac{1}{1+e^{-x}}$, we can get the following conclusion: $f'(x) = f(x)(1-f(x))$, so we can draw the following conclusion:

$$\Delta w_{jk} = -\gamma \frac{\partial E}{\partial w_{jk}} = -\gamma \times (Y - O_3) \times O_2 \times (1 - O_2) \times X \tag{4}$$

$$\Delta B_{jk} = -\gamma \frac{\partial E}{\partial B_{jk}} = -\gamma \times (Y - O_3) \times O_2 \times (1 - O_2) \times \text{ones} \tag{5}$$

Step 9. Calculates the adjusted threshold and weight. The threshold and weight of each layer at t time plus their respective adjustment amount are given to the threshold and weight of $t+1$ time.

$$w_{jk}(t + 1) = -\gamma \frac{\partial E}{\partial w_{jk}} + w_{jk}(t) = \Delta w_{jk} + w_{jk}(t) \tag{6}$$

$$B_{jk}(t + 1) = -\gamma \frac{\partial E}{\partial B_{jk}} + B_{jk}(t) = \Delta B_{jk} + B_{jk}(t) \tag{7}$$

$$w_{ij}(t+1) = -\gamma \frac{\partial E}{\partial w_{ij}} + w_{ij}(t) = \Delta w_{ij} + w_{ij}(t) \tag{8}$$

$$B_{ij}(t+1) = -\gamma \frac{\partial E}{\partial B_{ij}} + B_{ij}(t) = \Delta B_{ij} + B_{ij}(t) \tag{9}$$

Step 10. Restore the network output value of the output layer. Since the sample data are normalized in the previous neural network training process, it is necessary to reverse normalize O_3 to restore the real network output value.

3.2 Cluster analysis and osculating value method

Because we don't know the specific financial situation of the listed companies in mining industry, we can't get the output standard value of the BP neural network financial early warning model. Therefore, before that, it is necessary to divide the early warning range and financial status of listed companies. In this paper, K-means clustering analysis method is selected to divide the sample warning interval, because the K-means clustering analysis has a very good effect on the analysis and processing of cross-section data. After the division of the sample warning interval, the osculating value method is used to rank the financial status of the company. The calculation method is as follows: firstly, the data of 7 common factors of the selected sample companies are standardized, and the "best" and "worst points" of the sample indicators of various companies are determined. Then, the absolute distance between the sample of each company and the best and worst points, that is, the closeness degree, is calculated. Finally, the closeness value of each company sample is calculated, and the good and bad order is arranged according to the value.

The final evaluation method: the closeness value reflects the extent to which the sample set deviates from the best. The smaller the closeness value is, the closer the sample is to the best, and the farther away from the worst point, the better the financial condition of the sample.

4 Sample selection and index system construction

4.1 The selection of samples

The sample selected in this paper is China's A-share mining listed companies. The main reason is that the B-share and H-share listed companies follow international accounting standards in the formulation and audit of statements, which is quite different from China's accounting standards followed by A-share. Because the companies subject to different accounting standards cannot be mixed together for research and analysis, and the number and scale of A-share mining companies are enough to reflect the basic profile of China's mining listed companies, this paper selects Shanghai and Shenzhen A-share mining listed companies as research samples. By the end of 2018, there were 65 listed companies in the A-share mining industry sector of Shanghai and Shenzhen Stock Exchange, but the financial index data of 17 companies had not been provided and the financial statements had not been certified and audited. Therefore, considering the data integrity, authenticity, comparability and other principles, the invalid samples were deleted, and a total of 48 enterprises were selected as the research samples. The data came from CNINFO. ❶

After getting the data, we do the necessary preprocessing for the data. First, standardize the data. Secondly, Bartlett's sphere test and KMO test were carried out on the standardized data. Because there are many financial indicators to be selected in this paper, the calculation complexity of the model will be increased. Therefore, if we can reduce the number of variables without reducing the information content, the model will be more concise and the results will be more accurate. Factor analysis can do this. But before that, it is necessary to test the collected financial data to determine whether these data can be used for factor analysis. Bartlett's sphere test and KMO test are the methods to verify whether these financial data can be used for factor analysis. Finally, the correlation among variables is analyzed by factor analysis, and then the common factors are extracted from the whole variable group to replace the original variable.

4.2 Construction of financial early warning index system of mining listed companies

Based on the current research status of scholars on the selection of financial early warning indicators and the situation of China's listed mining companies, this paper determines to select targeted financial early warning indicators for China's mining listed companies from five aspects of enterprise profitability, operating capacity, solvency, development ability and cash flow (Kim et al., 2016; Sun and Xu, 2016; Ohlson, 1980; Deakin, 1972; Beaver, 1966; Fitzpatrick, 1932), as shown in Table 2. In order to understand the meaning of these five secondary indicators more clearly, we introduce them.

Table 2 Financial early warning index system of mining industry

Level indicators	Secondary indicators	Tertiary indicators
Financial early warning index system of mining industry	Profitability of listed mining companies	Operating profit margin (X_1)
		Return on total assets (X_2)
		Net profit margin (X_3)
		Cost profit margin (X_4)
	Operation capacity of listed companies in mining industry	Turnover rate of accounts receivable (X_5)
		Turnover rate of inventory (X_6)
		Turnover rate of total assets (X_7)

❶ CNINFO (www. cninfo. com. cn) isastatutoryinformationdisclosurewebsite andamulti - levelcapitalmarketinformation disclosureplatformapprovedbyChinaSecuritiesRegulatoryCommission.

Continued

Level indicators	Secondary indicators	Tertiary indicators
		Turnover rate of fixed assets (X_8)
		Turnover rate of current assets (X_9)
	Solvency of listed mining companies	Property right ratio (X_{10})
		Asset liability ratio (X_{11})
		Liquidity ratio (X_{12})
		Quick ratio (X_{13})
	Development ability of listed mining companies	Growth rate of operating revenue (X_{14})
		Growth rate of net assets (X_{15})
		Growth rate of total assets (X_{16})
	Cash flow of mining enterprises	Operating cash flow per share (X_{17})
		Net operating cash flow and debt ratio (X_{18})
		Net cash flow to net profit ratio of operating activities (X_{19})

(1) Profitability can reflect the ability of mining listed companies to obtain profits in a certain period of time. If the profitability is good, then the enterprise's capital flow will be more sufficient, there will be no capital chain fracture, and it is unlikely that there will be a financial crisis.

(2) Operating capacity reveals the capital operation turnover of an enterprise, and reflects the level of enterprise management and the efficiency of capital utilization. If the speed of capital operation turnover is faster, it means that the higher the liquidity of the enterprise capital, the stronger the solvency of the enterprise.

(3) Solvency refers to the ability of an enterprise to repay its debts by relying on the funds earned in daily operation. If the listed mining companies do not have enough solvency, they will not be able to repay the debts due and evolve into financial crisis.

(4) Development ability refers to the ability of enterprises to make long-term healthy and stable development in the process of operation, so as to ensure that enterprises will not be eliminated in the fierce external market environment.

(5) Cash flow refers to the amount of cash inflow and outflow of an enterprise, which reflects whether the enterprise is in good operating condition and has sufficient cash to repay debts.

5 Results and discussion

5.1 Data preprocessing results

After using SPSS software to perform Bartlett sphere test and KMO test on the data, the results are shown in Table 3.

Table 3 Bartlett spherical test and kmo test

Kaiser-Meyer-Olkin Measure of Sampling Adequacy.		0.532
Bartlett's Test of Sphericity	Approx. Chi-Square	899.563
	Df	171
	Sig.	0.000

It can be seen from the table that the Sig. Value of Bartlett's sphere test is 0.000, less than 0.05, indicating that the original hypothesis is rejected. There is a significant correlation between independent variables, which is suitable for factor analysis. It can also be seen from the first row of the table that KMO sampling suitability is 0.532, greater than 0.5, indicating that the correlation between independent variables is strong, and it is suitable for factor analysis. Through the above two kinds of tests, the financial indicators of the selected samples can be analyzed by factor analysis. Next, we use SPSS software to analyze the financial index data of 48 mining listed companies and extract common factors. It can be seen from the gravel map that there are seven characteristic roots greater than 1, and the total explanation degree of thcsc 7 components to the original variable is 84.85%, so the extracted 7 components are taken as common factors.

After seven common factors are obtained, the correlation test of these 7 common factors is conducted. It can be seen from Table 4 that the seven common factors are independent of each other and have no correlation, which perfectly solves the multicollinearity problem that may be encountered in the later model.

Table 4 Covariance matrix of common factors

Component	1	2	3	4	5	6	7
1	1.000	0.000	0.000	0.000	0.000	0.000	0.000
2	0.000	1.000	0.000	0.000	0.000	0.000	0.000
3	0.000	0.000	1.000	0.000	0.000	0.000	0.000
4	0.000	0.000	0.000	1.000	0.000	0.000	0.000
5	0.000	0.000	0.000	0.000	1.000	0.000	0.000
6	0.000	0.000	0.000	0.000	0.000	1.000	0.000
7	0.000	0.000	0.000	0.000	0.000	0.000	1.000

Then, using the principal component analysis of factor analysis in SPSS, the score coefficient matrix of the 7 common factors is obtained. According to the common factor score coefficient matrix, the functional relationship between the 7 common factors and the initial 19 financial indicators is as follows:

$$F_1 = 0.229X_1 + 0.230X_2 + 0.221X_3 + \cdots + 0.137X_{18} + 0.023X_{19}$$

$$F_2 = 0.019X_1 + 0.018X_2 + 0.030X_3 + \cdots + 0.047X_{18} + 0.060X_{19}$$

$$\vdots$$

$$\vdots \tag{10}$$

$$F_6 = -0.028X_1 + 0.035X_2 + 0.044X_3 + \cdots + 0.268X_{18} - 0.053X_{19}$$

$$F_7 = 0.091X_1 - 0.071X_2 + 0.098X_3 + \cdots - 0.001X_{18} - 0.097X_{19}$$

According to the above functional relationship, the scores of the 7 common factors of 48 sample companies are calculated, and the scores of these 7 common factors are the input variables to construct the financial early warning model of listed Companies in mining industry.

5.2 Model results

Using SPSS software and K-means cluster analysis method, the samples of mining listed companies are divided into five categories (health, good, general, mild warning, severe warning), as shown in Table 5. Then, the average values of the seven common factors corresponding to the above different types of companies are calculated.

Table 5 K-means cluster analysis results

Company category	Ⅰ	Ⅱ	Ⅲ	Ⅳ	Ⅴ
Number of Companies	4	1	12	29	2

Through osculating value method, the osculating values of five types of companies are 0.2377, 0.1860, 0.1576, 0.1136 and 0.1936, respectively, corresponding to category Ⅰ, category Ⅱ, category Ⅲ, category Ⅳ and category Ⅴ. Therefore, the least osculating value is category Ⅳ, followed by Category Ⅲ, followed by Category Ⅱ, then category Ⅴ, and the largest is category Ⅰ. According to the osculating value evaluation method of the company's financial status, it is concluded that category Ⅳ is in health financial condition, category Ⅲ is in good financial condition, category Ⅱ is in general financial condition, category Ⅴ is in mild warning and category Ⅰ is in severe warning.

In general, most of the listed companies in China's mining industry in 2018 are in a good or above state. However, there are several individual enterprises in financial crisis, including 4 enterprises with severe financial situation warning and 2 enterprises with mild financial situation warning, which need to be paid attention to. Fig. 2 represents the financial situation of 48 enterprises. Among them, the redder the z-axis corresponding to the grid is displayed, the more financial warning is needed for the enterprise; the bluer the z-axis corresponding to the grid is, the healthier the enterprise's finance is.

Then, according to the BP neural network algorithm, we compile the code through MATLAB software and run the results. For the determination of node number of hidden layer, considering the characteristics ofBP neural network model (the more nodes in hidden layer, the more unstable the model), we trained the number of nodes from 3 to 9 nodes once, and the results are shown in Fig. 3. From the results, the less the number of hidden layer nodes, the larger the MSE is, showing the lack of fitting ability; when the number of hidden layer nodes is greater than 7, it is not within the tolerance range initially set. Finally, according to the MSE minimum and model stability principle (Nguyen and Bui, 2019; Liu and Lei, 2018; Kanellopoulos and Wilkinson, 1997), hidden layer nodes-6 is selected as the number of hidden layer nodes in the model. The structure of the trained model is shown in Fig. 4.

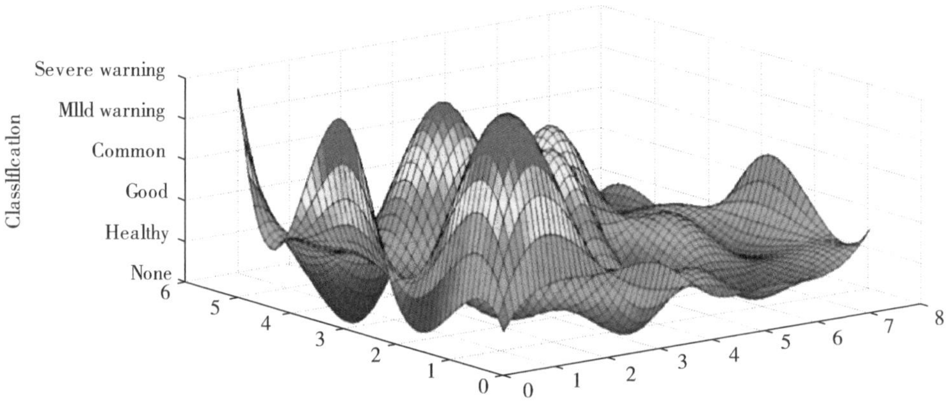

Fig. 2 Financial situation of the enterprise

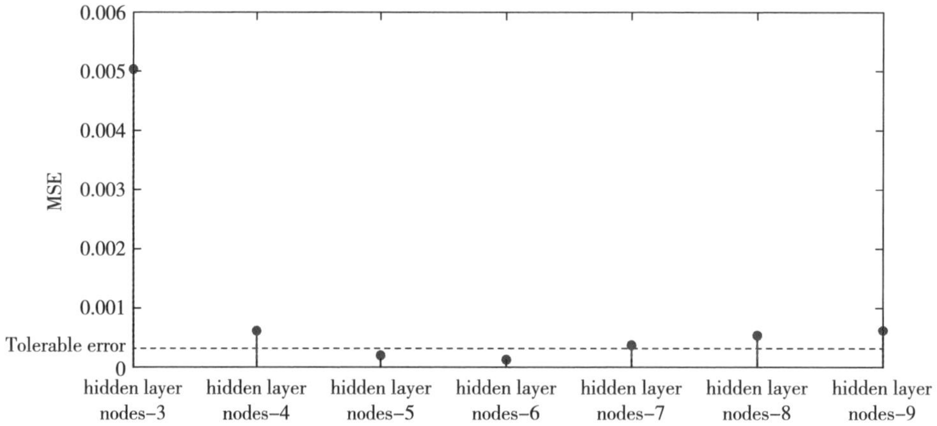

Fig. 3 Mean squared error from various hidden layer nodes

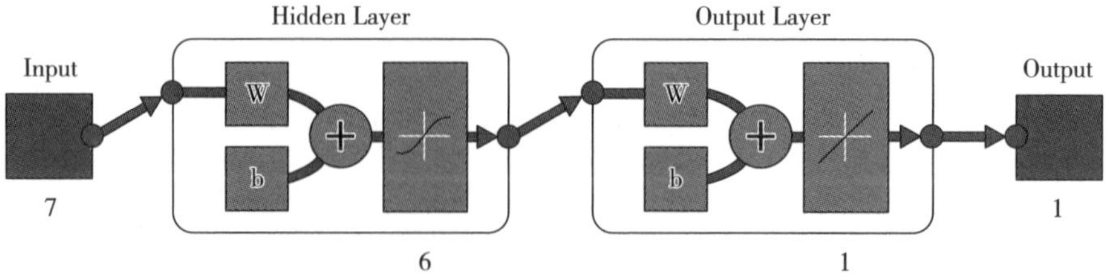

Fig. 4 Structure diagram of financial early warning BP neural network model

It can be seen from Fig. 5 that after 18 iterations, the training error value of the simulation training is 0.0001188, which is less than the target value of 0.0003, which meets the predetermined accuracy requirements, and the BP neural network model achieves convergence. In

Fig. 6, we can see that the R value is 0. 99975, which indicates that the model fits the observed values very well.

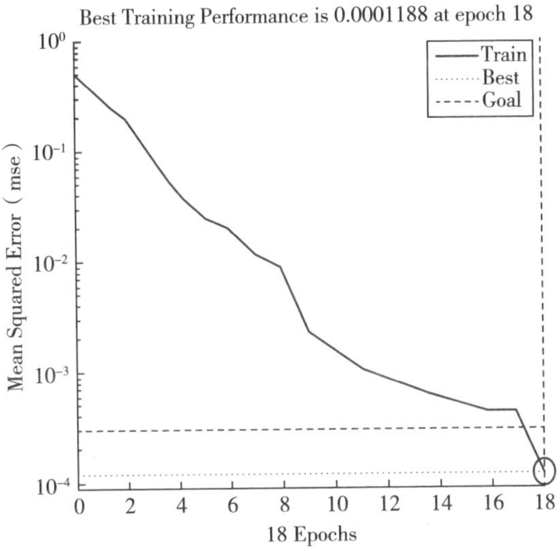

Fig. 5 Mean square error (MSE) diagram

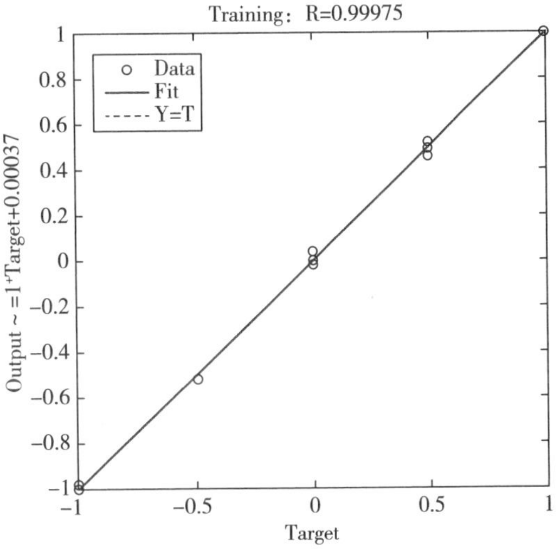

Fig. 6 Goodness of fit graph

As can be seen from Fig. 7, the actual class of the training set sample is exactly the same as the network output class, that is, the test accuracy is 100%. Therefore, the training effect of the model is very good. After the simulation test is completed, the remaining 8 groups of simulation samples after the previous training samples are extracted for prediction and simulation. The prediction results are shown in Table 6.

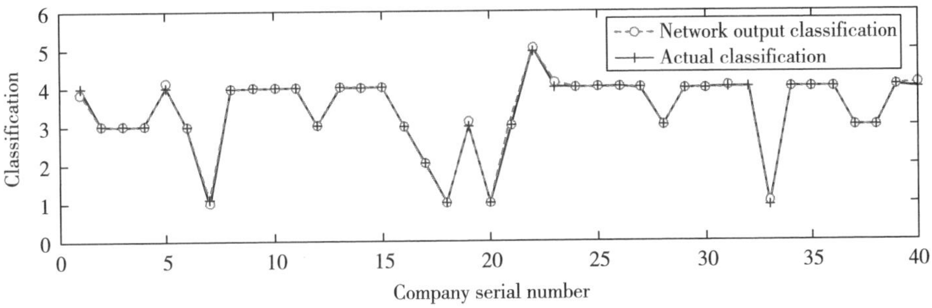

Fig. 7 Test results of training set

Table 6 Comparison of predicted results and actual categories of simulation group

Company serial number	5	7	9	12	18	20	23	33
Prediction result	4. 0081	3. 9839	4. 0050	4. 0032	4. 0031	4. 0077	3. 8571	3. 0463
Actual category value	Ⅳ	Ⅳ	Ⅳ	Ⅳ	Ⅳ	Ⅴ	Ⅳ	Ⅲ
Forecast category value	Ⅳ	Ⅳ	Ⅳ	Ⅳ	Ⅳ	Ⅳ	Ⅳ	Ⅲ
Discrimination	√	√	√	√	√	×	√	√

Through the observation and comparison of the training set, we determine that: when the prediction result is greater than 0.5 and less than 1.5, the financial crisis category is Ⅰ; when the prediction result is greater than 1.5 and less than 2.5, the financial crisis category is Ⅱ;...; when the forecast result is greater than 4.5 and less than 5.5, the financial crisis category is Ⅴ. Therefore, the predicted financial crisis categories of the above eight groups of samples are Ⅳ, Ⅳ, Ⅳ, Ⅳ, Ⅳ, Ⅳ, Ⅳ, Ⅳ, Ⅳ, Ⅲ. Compared with the actual financial crisis category, only one group of samples has deviation, and other samples are the same with the actual value, and the prediction accuracy rate is as high as 87.5%.

5.3 Results analysis and discussion

Further analysis, through the method level of reverse traceability, that is, from the judgment results to the common factors and then to the financial indicators, the main reason why most of the listed companies in China's A-share mining industry showed good financial status or above in 2018 was that the average cost profit margin of these enterprises reached 13.49%. This shows that the income of 1.135 units can be obtained by paying 1 unit of cost, which is 1.63 percentage points higher than the industry average level, and 5.8 times higher than that of enterprises with general financial status or below. In addition, the average current ratio and quick ratio (measure indicators of solvency) of these companies are also higher than the industry average, which shows that enterprises with good financial status pay attention to the ability to repay debts on time while earning high profits.

The main reason for the serious early warning of the 4m ining listed companies analyzed above is that the turnover rate of accounts receivable is as high as 40%, which indicates that the net income of credit sales of these 4 companies in 2018 accounts for 40% of all accounts receivable balance. In addition, the amount of accounts receivable is also increasing year by year, which makes the risk of bad debt loss of the company greatly increased. Among them, the turnover rate of accounts receivable of the top one is as high as 157%, and its accounts receivable increased from 120 million in 2017 to 570 million in 2018, which should be paid special attention to. In addition, the net profit margin of these four companies in 2018 is very low, and the highest one is only 2.234%, which is one to two percentage points lower than the average profit margin of A-share mining listed companies. Therefore, in the case of low net profit, its credit account is still so large that its overall financial situation presents a crisis situation, which requires severe early warning.

Through the analysis of the results, we can see that the results of the BP neural network financial early warning model constructed in this paper are consistent with the real financial situation of Listed Companies in mining industry, which shows that the model is suitable for the financial early warning of China's A-share mining listed companies, and the results obtained are accurate and reliable.

In addition, we also discuss how to reduce the risk of bad debt loss. For the whole mining industry, due to its own characteristics, generally speaking, the amount of a single transaction is relatively large, so few next enterprises can pay in full. Basically, they take the way of credit. Only when the sales products have capital return can they pay off the accounts receivable of mining companies. If the mining company adopts the way of cash and goods transaction on the spot, it will reduce the credit account and prevent the loss of bad debts, but at the same time, it will also lose many downstream customers, which makes the company's net profit decline seriously. Therefore, how to solve the contradiction between the two is also a problem worthy of deep discussion, such as equity pledge, real estate pledge or third-party guarantee. Although these can effectively avoid the occurrence of bad debts and ensure that the company must have funds to recover, whether the downstream companies are willing to and have sufficient assets mortgage has formed a new problem. This can be further discussed in the follow-up study.

6 Conclusions and recommendations

6.1 Conclusions

We take the listed companies in China's mining industry as the research object, and select 19 financial indicators from five aspects of profitability, operating capacity, solvency, development ability and cash flow to construct a financial early warning indicator system. We select the financial data of 48 Shanghai and Shenzhen A-share listed companies in mining industry in 2018 for analysis. Using the BP neural network model for simulation training, the results of 100% test accuracy and 87.5% simulation accuracy show that the BP neural network model for financial early warning of mining listed companies is feasible, and has a high prediction accuracy. The conclusions are as follows:

(1) The BP neural network financial early warning model constructed in this paper has high prediction accuracy, which can be well used in the practice of financial early warning of mining listed companies.

(2) The financial situation of China's A-share mining listed companies in 2018 is generally in a good state. Companies with good financial status and above often can control costs and expenses while earning income, so as to keep the cost profit margin at a high level; and the average current ratio and quick ratio of these companies are higher than the industry average, indicating that they have good solvency.

(3) For companies with financial status that require early warning, the root cause is mainly that they do not pay attention to the risk of bad debt losses, which makes current credit sales income and accounts receivable are at high levels; and the net profit rate of these companies is also lower than the average level of the industry, indicating that they do not have good profitability.

6.2 Policy implications

In view of the above three conclusions, we put forward some suggestions: (1) It is hoped that mining industry listed companies can reasonably and effectively use the BP neural network financial early warning model in the future. With the help of this model, they can get the results of financial risk early warning of enterprises, and accurately find the key financial indicators causing financial crisis. In this way, the enterprise management authorities can take timely governance measures according to the early warning results and key financial indicators, so that companies can avoid being hit by major financial risks; (2) For companies with healthy and good financial status, managers should continue to maintain good financial status, and at the same time, they should also be prepared for financial risk early warning; (3) For enter-prises that need early warning of financial status, the management authorities should pay special attention to the risk of bad debt loss. If the financial indicators of current credit sales income and accounts receivable increase sharply compared with the previous year or always maintain at a high level in the whole industry, enterprises should pay attention to it to avoid the huge financial risks caused by the break of cash flow in the future.

6.3 Future work

The BP neural network financial early warning model constructed in this paper has been applied to the research of China's A-share market and has achieved satisfactory results. In principle, it is also applicable to other international markets. However, if the overall characteristics of mining listed companies in other international markets are quite different from that of China's A-share market, further tests are needed for the index system. Therefore, whether to build an index system that can be applicable to multiple markets is a subject worthy of in-depth study in the future.

Acknowledgements The authors express their sincere thanks for the support from the National Science and Technology Major Project under Grant No. 2016ZX05016005-003 and the Beijing Propaganda Culture High-level Talent Training Subsidy Program under Grant No. 2017XCB031.

References

Abban, A. R., Hasan, M. Z., 2021. The causality direction between environmental performance and financial performance in Australian mining companies - A panel data analysis. J. Resources Policy 70, 101894.

Ahn, H., Kim, K., 2009. Bankruptcy prediction modeling with hybrid case-based reasoning and genetic algorithms approach. J. Applied Soft Computing 9 (2), 599-607.

Altman, E., Ratios, F., 1968. Discriminant analysis and the prediction of corporate bankruptcy. J. Journal of Finance 23 (4), 589-609.

Barboza, F., Kimura, H., Altman, E., 2017. Machine learning models and bankruptcy prediction. J. Expert Systems with Applications 83, 405-417.

Beaver, W. H., 1966. Financial ratios as predictors of failure. J. Journal of accounting research 71-111.

Bergeron, K. M., 2021. What can a hundred mining exploration projects in Canada tell us about social risk? Considering an area's trajectory to understand its sociogeological potential. J. The Extractive Industries and

Society 8 (1), 83-88.

Cao, Y., Chen, X., Wu, D. D., Miao, M., 2011. Early warning of enterprise decline in a life cycle using neural networks and rough set theory. J. Expert Systems with Applications 38 (6), 6424-6429.

Chand, Pushpendu, Jitesh, J., Thakkar, Kunal Kanti Ghosh, 2020. Analysis of supply chain performance metrics for Indian mining & earthmoving equipment manufacturing companies using hybrid MCDM model. J. Resources Policy.

Chaudhuri, A., De, K., 2011. Fuzzy support vector machine for bankruptcy prediction. J. Applied Soft Computing 11 (2), 2472-2486.

Chauhan, N., Ravi, V., Chandra, D. K., 2009. Differential evolution trained wavelet neural networks: application to bankruptcy prediction in banks. J. Expert Systems with Applications 36 (4), 7659-7665.

Chen, W. S., Du, Y. K., 2009. Using neural networks and data mining techniques for the financial distress prediction model. J. Expert systems with applications 36 (2), 4075-4086.

Choi, H., Son, H., Kim, C., 2018. Predicting financial distress of contractors in the construction industry using ensemble learning. J. Expert Systems with Applications 110, 1-10.

Crespí-Cladera, R., Martín-Oliver, A., Pascual-Fuster, B., 2021. Financial distress in the hospitality industry during the Covid-19 disaster. J. Tourism Management 85, 104301.

de Souza Neto, H. F., da Silveira Pereira, W. V., Dias, Y. N., et al., 2020. Environmental and human health risks of arsenic in gold mining areas in the eastern Amazon. J. Environ. Pollut. 265, 114969.

Deakin, E. B., 1972. A discriminant analysis of predictors of business failure. J. Journal of accounting research 167-179.

Dinh, D. V., Powell, R. J., Vo, D. H., 2021. Forecasting corporate financial distress in the Southeast Asian countries: a market-based approach. J. Journal of Asian Economics 74, 101293.

Du, G., Liu, Z., Lu, H., 2021. Application of innovative risk early warning mode under big data technology in Internet credit financial risk assessment. J. Journal of Computational and Applied Mathematic s 386, 113260.

Fan, S., Yan, J., Sha, J., 2017. Innovation and economic growth in the mining industry: evidence from China's listed companies. J. Resources Policy 54, 25-42.

Fitzpatrick, P. J., 1932. A Comparison of ratios of successful industrial enterprises with those of failed firms. J. Certified Public Accountant, October, November, December 598-605.

Geng, R., Bose, I., Chen, X., 2015. Prediction of financial distress: an empirical study of listed Chinese companies using data mining. J. European Journal of Operational Research 241 (1), 236-247.

Guo, Y., 2020. Credit risk assessm ent of P2P lending platform towards big data based on BP neural network. J. Vis. Commun. Image Represent. 71, 102730.

Himley, M., 2010. Global mining and the uneasy neoliberalization of sustainable development. J. Sustainability 2 (10), 3270-3290.

Hoque, M. M., Rahman, M. T., 2020. Landfill area estimation based on solid waste collection prediction using ANN model and final waste disposal options. J. Journal of Cleaner Production.

Hosaka, T., 2019. Bankruptcy prediction using image d financial ratios and convolutional neural networks. J. Expert Systems with Applications 117, 287-299.

Irimiadieguez, A. I., Blancooliver, A., Vazquezcueto, M. J., 2015. A comparison of classification/regression trees and logistic regression in failure Models☆. J. Procedia. Economics and finance 9-14.

Iturriaga, F. J. L., Sanz, I. P., 2015. Bankruptcy visualization and prediction using neural networks: a study of US commercial banks. J. Expert Systems with applications 42 (6), 2857-2869.

Jabeur, S. B., 2017. Bankruptcy prediction using partial least squares logistic regression. J. Journal of

Retailing and Consumer Services 197–202.

Jiang, H., Ching, W., Yiu, K. F., Qiu, Y., 2018. Stationary Mahalanobis kernel SVM for credit risk evaluation. J. Applied Soft Computing 407–417.

Jonek–Kowalska, I., 2019. Consolidation as a risk management method in the lifecycle of a mining company: a novel methodological approach and evidence from the coal industry in Poland. Resour. Pol. 60, 169–177.

Jonek–Kowalska, I., Nawrocki, T. L., 2019. Holistic fuzzy evaluation of operational risk in polish mining enterprises in a long–term and se ctoral research perspective. J. Resources Policy 63, 101464.

Kaastra, I., Boyd, M., 1996. Designing a neural network for forecasting financial and economic time series. J. Neurocomputing 10 (3), 215–236.

Kanellopoulos, I., Wilkinson, G. G., 1997. Strategies and best practice for neural network image classification. J. International Journal of Remote Sensing 18 (4), 711–725.

Kim, H. J., Jo, N. O., Shin, K. S., 2016. Optimization of cluster–based evolutionary undersampling for the artificial neural networks in corporate bankruptcy prediction. J. Expert Systems with Applications 59, 226–234.

Kinimo, K. C., Yao, K. M., Marcotte, S., et al., 2018. Distribution trends and ecological risks of arsenic and trace metals in wetland sediments around gold mining activities in central – southern and southeastern Côte d' Ivoire. J. Journal of Geochemical Exploration 190, 265–280.

Kosmidou, K., Pasiouras, F., Zopounidis, C., Doumpos, M., 2006. A multivariate analysis of the financial characteristics of foreign and domestic banks in the UK. J. Omega – international Journal of Management Science 34 (2), 189–195.

Koyuncugil, A. S., Ozgulbas, N., 2012. Financial early warning system model and data mining application for risk detection. J. Expert systems with Applications 39 (6), 6238–6253.

Lee, S., Choi, W. S., 2013. A multi–industry bankruptcy prediction model using back–propagation neural network and multivariate discriminant analysis. J. Expert Systems with Applications 40 (8), 2941–2946.

Liu, L., Lei, Y., 2018. An accurate ecological footprint analysis and prediction for Beijing based on SVM model. J. Ecological Informatics 33–42.

Li, Y. W., Cao, K., 2020. Establishment and application of intelligent city building information model based on BP neural network model. Comput. Commun. 153, 382–389.

Liu, B., Ju, T., Bai, M., et al., 2021. Imitative innovation and financial distress risk: the moderating role of executive foreign experience. J. International Review of Economics & Finance 71, 526–548.

Mahtani, U. S., Garg, C. P., 2018. An analysis of key factors of financial distress in airline companies in India using fuzzy AHP framework. J. Transportation Research Part A: Policy and Practice 117, 87–102.

Martin, D., 1977. Early warning of bank failure: a logit regression approach. J. Journal of banking & finance 1 (3), 249–276.

Modak, M., Pathak, K., Ghosh, K. K., 2017. Performance eva luation of outsourcing decision using a BSC and Fuzzy AHP approach: a case of the Indian coal mining organization. J. Resources Policy 181–191.

Nel, A. J., Vosloo, J. C., Mathews, M. J., 2018. Financial model for energy efficiency projects in the mining industry. J. Energy 546–554.

Nguyen, H., Bui, X., 2019. Predicting blast–induced air overpressure: a robust artificial intelligence system based on artificial neural network s and random forest. J. Natural resources research 28 (3), 893–907.

Odom, M. D., Sharda, R., 1990. A neural network model for bankruptcy prediction//1990 LJCNN International Joint Conference on neural networks. High Perform. Dev. IEEE Cornell Conf. 163–168.

Ohlson, J. A., 1980. Financial ratios and the probabilistic prediction of bankruptcy. J. Journal of accounting research 109–131.

Ouyang, Z., Lai, Y., 2021. Systemic financial risk early warning of financial market in China using

Attention-LSTM model. J. The North American Journal of Economics and Finance 56, 101383.

Paat, A., Roosalu, T., Karu, V., et al., 2021. Important environmental social governance risks in potential phosphorite mining in Estonia. J. The Extractive Industries and Society 100911.

Rafiq, S., Salim, R., Smyth, R., 2016. The moderating role of firm age in the relationship between R&D expenditure and financial performance: evidence from Chinese and US mining firms. J. Economic Modelling 56, 122-132.

Saedpanah, S., Amanollahi, J., 2019. Environmental pollution and geoecological risk assessment of the Qhorve h mining area in western Iran. J. Environ. Pollut. 253, 811-820.

Saes, B. M., Muradian, R., 2021. What misguides environmental risk perceptions in corporations? Explaining the failure of Vale to prevent the two largest mining disasters in Brazil. J. Resources Policy 72, 102022.

Sartori, F., Mazzucchelli, A., Di Gregorio, A., 2016. Bankruptcy forecasting using case-based reasoning: the CRePERIE approach. J. Expert Systems with Applications 64, 400-411.

Scalzer, R. S., Rodrigues, A., Macedo, M. á. S., et al., 2019. Financial distress in electricity distributors from the perspective of Brazilian regulation. J. Energy Policy 125, 250-259.

Scammacca, O., Gunzburger, Y., Mehdizadeh, R., 2021. Gold mining in French Guiana: a multi-criteria classification of mining projects for risk assessment at the territorial scale. J. The Extractive Industries and Society 8 (1), 32-43.

Shen, F., Liu, Y., Wang, R., Zhou, W., 2020. A dynamic financial distress forecast model with multiple forecast results under unbalanced data environment. J. Knowledge Based Systems.

Sun, W., Xu, Y., 2016. Financial security evaluation of the electric power industry in China based on a back propagation neural network optimized by genetic algorithm. J. Energy 101, 366-379.

Sun, Y., Yang, Y., Huang, N., et al., 2020. The impacts of climate change risks on financial performance of mining industry: evidence from listed companies in China. J. Resources Policy 69, 101828.

Tam, K. Y., Kiang, M. Y., 1992. Managerial applications of neural networks: the case of bank failure predictions. J. Management science 38 (7), 926-947.

Tang, J. R., Zhou, P., Shi, J. F., et al., 2015. Analysis and outlook of 2014—2015 global mining. China Min. Mag. 3, 42-47.

Tavana, M., Abtahi, A., Caprio, D. D., Poortarigh, M., 2018. An Artificial Neural Network and Bayesian Network model for liquidity risk assessment in banking. J. Neurocomputing 2525-2554.

Tseng, F. M., Hu, Y. C., 2010. Comparing four bankruptcy prediction models: logit, quadratic interval logit, neural and fuzzy neural networks. J. Expert Systems with Applications 37 (3), 1846-1853.

Wang, Y., 2012. Z-score model on financial crisis early-warning of listed real estate companies in China: a financial engineering perspective. J. Systems Engineering Procedia 3, 153-157.

Wang, D., Tong, X., Wang, Y., 2020. An early risk warning system for Outward Foreign Direct Investment in Mineral Resource-based enterprises using multi-classifiers fusion. J. Resources Policy 66, 101593.

Wu, J., Zhu, Q., Qiao, N., et al., 2021. Ecological risk assessment of coal mine area based on "source-sink" landscape theory-A case study of Pingshuo mining area. J. Journal of Cleaner Production 295, 126371.

Yang, B., Li, L. X., Ji, H., Xu, J., 2001. An early warning system for loan risk assessment using artificial neural networks. J. Knowledge-Based Systems 14 (5-6), 303-306.

Yang, Z., You, W., Ji, G., 2011. Using partial least squares and support vector machines for bankruptcy prediction. J. Expert Systems with Applications 38 (7), 8336-8342.

Zhou, L., Lu, D., Fujita, H., 2015. The performance of corporate financial distress prediction models with features selection guided by domain knowledge and data mining approaches. J. Knowledge Based Systems 52-61.

Zhou, Z., Xiao, T., Chen, X., Wang, C., 2016. A carbon risk prediction model for Chinese heavy-polluting industrial enterprises based on support vector machine. J. Chaos Solitons & Fractals 304-315.

Zhou, W., Chen, M., Yang, Z., et al., 2020. Real estate risk measurement and early warning based on PSO-svm. J. Socio-Economic Planning Sciences 101001.

Zmijewski, M. E., 1984. Essays ON corporate bankruptcy. J.

环境与健康

Public health effect and its economics loss of $PM_{2.5}$ pollution from coal consumption in China

CHEN Hong[1,2], LI Li[1,2,4,*], LEI Yalin[1,2],
WU Sanmang[1,2], YAN Dan[3], DONG Ziyu[1,2]

1. *School of Economics and Management*, *China University of Geosciences*, *Beijing* 100083, *China*
2. *Key Laboratory of Carrying Capacity Assessment for Resource and Environment*, *Ministry of Natural Resources of the People's Republic of China*, *Beijing* 100083, *China*
3. *Tsinghua-Berkeley Shenzhen Institute*, *Tsinghua University*, *Shenzhen* 518055, *China*
4. *State Key Laboratory of Water Resource Protection and Utilization in Coal Mining*, *Beijing* 100011, *China*

Abstract China's energy structure is based on coal resource and it accounts for main proportion in the primary energy consumption. Coal consumption produces $PM_{2.5}$ pollution, which seriously affects public health. Considering that there are few studies on the effect $PM_{2.5}$ pollution produced by coal consumption, this paper uses the Poisson Regression model to estimate the impacts on public health and the economic loss of $PM_{2.5}$ pollution produced by coal consumption using the data in 2015. Based on these results, the paper also predicts the impacts on public health effect and its economic loss caused by $PM_{2.5}$ pollution from coal consumption under the baseline scenario and total coal consumption control scenario in 2020 and 2030. Finally, based on the research conclusions, suggestions are proposed to reduce the public health economic loss from $PM_{2.5}$ pollution caused by coal consumption.

Keywords $PM_{2.5}$ pollution, Health effects, Economic loss, Coal consumption, Poisson regression model

1 Introduction

China has more coal and less oil. In the primary energy structure of China, coal has played a leading role. In 1957, coal consumption accounted for 92.3% of total energy consumption (China Statistics Press, 1999). Constrained by economic transformation and environmental protection, the growth rate of coal consumption has slowed down significantly in recent years. In 2005, the proportion of coal in energy consumption was 72.4% (China Statistics Press, 2006), and it had decreased to 59% in 2018 (China Statistics Press, 2019), but its proportion was still far higher than that of other energy consumption. It was predicted that China's energy consumption

原文刊于 *Science of the Total Environment*, 2020 年第 732 卷。原文发表时为彩色。
* 为通信作者。

demand would continue to grow steadily until 2030. The annual coal consumption would be 3.5 billion tons, accounting for about half of the total energy consumption (Lu et al., 2019).

Energy consumption structure with coal as its core is the key source of serious air pollution in China. Coal is basically composed of carbon, hydrogen, oxygen, sulfur and phosphorus. While most coal is used through direct combustion, once it is burned, it will produce a lot of harmful polluted substances including $PM_{2.5}$. A research displayed that coal burning and gasoline engine emissions were the two main sources of PAHs in $PM_{2.5}$ in Shanghai accounting for 30.5% and 20.9% respectively (Wang et al., 2015).

According to the Air quality guidelines 2005 issued by the World Health Organization, the risk of death increased by about 15% when the average annual $PM_{2.5}$ concentration reached 35 $\mu g/m^3$ compared with 10 $\mu g/m^3$. A statistical analysis showed that an increase of 10 $\mu g/m^3$ in $PM_{2.5}$ concentrations resulted in an increase of 0.40% in acute mortality rate, 1.43% in the death rate of respiratory disease and 0.53% in the death rate of cardiovascular disease, respectively (Xie et al., 2009). In 2015, 1.1 million people in China died from $PM_{2.5}$ exposure, making $PM_{2.5}$ the fifth leading risk factor for death globally (Fullman et al., 2017). A team from the energy policy institute at the Chicago University found that an increase in $PM_{2.5}$ concentration of 10 $\mu g/m^3$ resulted in a reduction of 0.98 years in life expectancy (Michael and Fan, 2018).

Based on this background, adopting the Poisson Regression model, this paper calculated the national residents' health effect and the economics loss of $PM_{2.5}$ pollution from coal consumption in 2015. It also predicted the impacts on public health effect and the economic loss in 2020 and 2030 in the baseline scenario and the total coal consumption control scenario, which was to reduce the health economic loss from coal consumption.

2 Literature review

In recent years, with the large amounts of problems of $PM_{2.5}$ pollution from energy consumption and people's increasing attention to their health, there are more and more research on the impacts of $PM_{2.5}$ pollution.

2.1 Literature on $PM_{2.5}$ pollution from energy production and consumption

Testing a newly formulated boiler briquette coal and a conventional raw coal when firing in China, Ge et al. (2001) analyzed the emission rates factors and collected the particulate matter emissions raw coal discharged. Han and Li (2006) analyzed energy consumption, energy structure and emissions of atmospheric pollutants in China. And the relationship between coal consumption and atmospheric pollutants was discussed as well. Chen et al. (2006) used nine kinds of coal with wide-ranged thermal maturities for measuring emission factors of residential burning coal-chunks. Wang et al. (2008) collected sixteen priority polycyclic aromatic compounds from $PM_{2.5}$ and PM_{10} samples at 20 locations in Beijing between December

2005 and January 2006. Through compositional analysis and principal component analysis, they concluded that they were the key sources for PAHs in particulate matter (PM). Khare and Baruah (2011) established the emission inventory of feed coal from Northeastern region of India by mass balance method. It proved the impact of coke oven on air quality. Lippmann (2014) investigated the effects from long-term $PM_{2.5}$ exposure on cardiovascular disease and found that they were from coal combustion for daily mortality. Sun et al. (2015) analyzed the toxic effects from $PM_{2.5}$ and proved its bad effects from coal combustion in Datong, Shanxi Province. Gautam et al. (2016) pointed out that India opencast coal mines were a major source of PM, and they also analyzed the traces of $PM_{2.5}$ the coal discharged. Finkelman and Tian (2018) stated that the use of coal in China caused severe health problems. Therefore, more measures were needed to minimize the health effects of coal use. Belis et al. (2019) used the tools to assess the geographical areas which had impacts on the $PM_{2.5}$ in the Danube and Western Balkans regions. They concluded that energy production and residential combustion had bigger impacts on the $PM_{2.5}$, accounting for 22% and 16% respectively. Samek et al. (2020) used energy dispersive X-ray fluorescence method and ion chromatography to test the elements. The results were that the contribution of combustion sometimes reached 55% with high $PM_{2.5}$ concentrations, while it was nearly 28% with low $PM_{2.5}$ concentrations. Lin et al. (2019) measured the concentration of carcinogenic elements in $PM_{2.5}$ samples which were collected in downtown Guiyang. They found that the annual mean concentration of As and Cr exceeded the standard limits. The carcinogenic elements might only come from coal burning and vehicle exhaust emissions.

2.2 Literature on the impacts on public health of $PM_{2.5}$ pollution

Bell and Davis (2001) developed new indicators of respiratory and other diseases mortality with deadly smog in the winter of 1952 in London. They finally estimated that about 12000 excess deaths happened because of the smog from December 1952 to February 1953. Greene and Morris (2006) concluded that people exposed to $PM_{2.5}$ in the long term were easier to be infected with disease in Washington. Zhang et al. (2008) analyzed the particulate matter in 111 cities in China in 2004, and they assessed its health effect and health risk of economic through statistical data and exposure-response. Bell et al. (2009) examined the differences in $PM_{2.5}$ chemical composition with the data of average concentrations of $PM_{2.5}$ from 2000 to 2005 to study the change of the risk of disease hospitalization it caused. The result was that higher $PM_{2.5}$ brought out higher risk of hospitalization. Boldo et al. (2011) used Community Multiscale Air Quality model to estimate the avoidable death toll which was related to the reduction of $PM_{2.5}$ concentrate in 2014 in Spain. Baxter et al. (2013) evaluated the $PM_{2.5}$ morality risk which could be explained by the variations in $PM_{2.5}$ composition and gaseous pollutant concentrations from 1997 to 2002 in 27 cities in the US. Cox et al. (2013) examined the empirical relationship between the variation of average $PM_{2.5}$ concentrate and the data from National Mortality and Morbidity Air Pollution Study in 100 cities of U. S. from 1999 to 2000. The result demonstrated that $PM_{2.5}$ was positively correlated with the mortality. Buonocore et al. (2014) analyzed $PM_{2.5}$-related mortality and used value-of-statistical-life to monetize health estimates using the Community Multiscale Air Quality model and concentration-response functions. Feng et al. (2016) discussed some adverse impacts from $PM_{2.5}$ on public health and the relation between $PM_{2.5}$ pollution and public health. Sun et al. (2015) analyzed the toxicological effect of $PM_{2.5}$ in Datong city of Shanxi Province and proved the effect of coal combustion on $PM_{2.5}$. Chen et al. (2017) used regression model to analyze the quantitative relationship between public health benefits and $PM_{2.5}$ reduction in Tianjin. The results showed that the reduction of $PM_{2.5}$ would reduce the health effect and health economic loss. Comparing the baseline scenario and intensified emission reduction scenario set in 2020, 2025 and 2030 respectively, the results from Li et al. (2018) showed that the intensified emission measure made great contributions. Dong et al. (2018) estimated that $PM_{2.5}$ pollution brought about 6105 premature deaths of residents, 10314 hospitalizations for respiratory diseases and 7972 hospitalizations for cardiovascular diseases, which resulted in the economic loss of 814 million yuan. Chen (2018) elaborated on the research progress about the health effect and its economic loss from air pollution in China, and put forward suggestions for more accurate measurement of health and economic loss. Shi (2018) elaborated on the components of $PM_{2.5}$ and the acute health risks caused by exposing to $PM_{2.5}$ in the short term at home and abroad. Chen et al. (2019) used computable general equilibrium and exposure-response model to quantify the health effects caused by $PM_{2.5}$. It was estimated that the premature deaths were 156588 and the health economic loss caused by $PM_{2.5}$ pollution was 3062.63 million yuan. Kim et al. (2019) used a random-effect model to derive the risk of all kinds of pollutants before 2019, and concluded that $PM_{2.5}$ had a significant harm on liver cancer population. Bian et al. (2019) compared the correlation between the respiratory system prevalence of residents in Langfang and Hohhot in 2016 and the average annual $PM_{2.5}$ concentration in those regions from 2013 to 2016. It was calculated that the respiratory system prevalence and average annual $PM_{2.5}$ level in Langfang were both higher than that in Hohhot.

Based on the above research, it can be seen the energy consumption, including coal consumption, produces a certain amount of $PM_{2.5}$. And $PM_{2.5}$ can bring seriously influence on the public health and even lead to premature death. It can also cause a huge health economic loss. There were fewer previous researches on the economic health loss of $PM_{2.5}$ from coal consumption. Also, China's coal consumption was larger, which had a great impact on $PM_{2.5}$ concentration and public health economic loss. And the current price of coal did not contain the economic loss of public health, the pollution and health economic loss caused by coal consumption are often overlooked. Therefore, analyzing the relationship between $PM_{2.5}$ from coal consumption and the public health economic loss is important for controlling coal consumption and reducing the public health effect loss.

3 Methods and data

3.1 Poisson regression model

The Poisson Regression model, widely used in epidemiological studies of air pollution, was adopted to assess the public health effect risk from $PM_{2.5}$ exposure (Huang and Zhang, 2013).

$$\Delta E = P \cdot \Delta R = P \cdot R \cdot \left\{1 - \frac{1}{\exp[\beta \cdot (c - c_0)]}\right\} \quad (1)$$

ΔE is the health effect caused by $PM_{2.5}$ pollution, that is, the number of people suffering from health damage. Different health terminals can cause different health effects. The selection of health terminals in this paper was based on: (1) epidemiological studies had proved that $PM_{2.5}$ pollution had effects on the human body functions; (2) the availability of the relevant health disease data; (3) the availability of the corresponding exposure-response relation coefficients. Thus the paper selected three types of deaths (lung cancer mortality, cardiovascular mortality, respiratory system disease mortality), two types of hospital (hospitalization rates of cardiovascular, hospitalization rates of respiratory system disease), two types of outpatient (outpatient rate of internal medicine, outpatient rate of pediatrics) and two kinds of diseases (incidence rate of acute upper respiratory tract infection, incidence rate of chronic bronchitis in old age).

P is the total exposure population. This paper selected the total population of China in 2015 as the exposure population (China Statistics Press, 2016).

C is the actual concentration of $PM_{2.5}$ in the air under normal circumstance. C_0 is the reference concentration of $PM_{2.5}$. Assuming without human behavior, there would also be a certain amount of $PM_{2.5}$ in the air, thus C_0 would not be zero. 35 $\mu g/m^3$ was selected as the reference concentration of $PM_{2.5}$ according to Ambient Air Quality Standards (GB 3095-2012) issued by China National Environmental Monitoring Centre in 2012 (Ministry of Ecology and Environment of the People's Republic of China, 2012).

R is the health risk of the exposure population under the actual concentration of $PM_{2.5}$. It means the mortality rate, the hospitalization rate, the outpatient rate and the morbidity of those diseases. The data were collected from China Health and Family Planning Statistical yearbook in 2016 (Peking Union Medical college Press, 2016).

β is the exposure-response relation coefficient. For different health terminals, the exposure-response relation coefficients are different. In order to determine the exposure-response relation coefficient, the paper took many relevant literatures for reference, which were shown in Table 1.

Table 1 The exposure-response relation coefficients of different health terminals

Health terminal		Values (95% confidence interval)
Mortality rate	Lung cancer	0.00556 (0.00302, 0.00811)
	Cardiovascular disease	0.00053 (0.00015, 0.0009)
	Respiratory disease	0.00143 (0.00085, 0.00201)
Hospitalization rate	Cardiovascular disease	0.00068 (0.00043, 0.00093)
	Respiratory disease	0.00109 (0, 0.00221)
Incidence rate	Acute upper respiratory infection	0.00494 (0.001924, 0.007946)
	Chronic bronchitis	0.01009 (0.00366, 0.01559)
Outpatient rate	Internal medicine	0.00049 (0.00027, 0.0007)
	Pediatric	0.000556 (0.0002, 0.0009)

Data source: Xie et al., 2009; Jing and Ren, 2000; Kan et al., 2004.

3.2 Evaluation method of health economic loss

Health economic loss refers to the sum of the additional medical expenses incurred by various physical diseases, which were caused by $PM_{2.5}$ pollution. And it also contains the production value delayed due to the treatment of diseases, including the opportunity cost (Fan et al., 2019). There are mainly three methods for the evaluation of health economic loss, which are the revised human capital method, the disease cost method and the willingness to pay method. The three methods have their own advantages and disadvantages. In order to estimate the health economic loss more accurately, different methods are selected for different evaluations according to the health terminals.

For the various disease mortality, this paper chose the revised human capital method to evaluate its economic loss. This method takes the value of the output into account. The output is created by people who produce less than their expectancy life after falling ill and dying prematurely due to $PM_{2.5}$ pollution.

3.2.1 The excess number of deaths

$$P_{\delta n} = P_n \cdot R_{\delta n} \cdot \left\{1 - \frac{1}{\exp[\beta_\delta \cdot (C - C_0)]}\right\} \tag{2}$$

$P_{\delta n}$ is the excess number of deaths at the n-th age for δ. According to the general age segmentation method, 5 years is divided as an age period. Thus, there are 20 age periods. $P_{\delta n}$ refers to the premature death toll in each age period compared with the reference concentration. P_n is the number of the n-th age group. $R_{\delta n}$ is the mortality rate of class diseases in the n-th age group. β_δ is the coefficient of exposure-response relation corresponding to the health terminal.

3.2.2 The discounted value of healthy economic loss

First, the paper estimated the expectancy life of each age group. It was used to calculate the number of years of premature death people of all ages. Using the excess deaths, per capita GDP and GDP growth rate, the total economic loss caused by premature death was discounted into the present value. The GDP growth rate was 6.1%, which was the mean value of the prediction rates of China from 2019 to 2021. The discount rate was taken by the long-term bank loan interest rate, which was 4.7% (World Bank, 2019).

For the health terminals of the hospitalization rate and outpatient rate, this paper chose the disease cost method to evaluate the health economic loss. This approach is similar to the revised human capital method, both taking into account the opportunity costs. Instead of the loss of the output value caused by premature death, it was the down-time in hospital, clinic and other corresponding delayed work time.

$$C_i = (C_{pi} + GDP_{pi}) \cdot E_i \tag{3}$$

C_i is the total health economic loss caused by $PM_{2.5}$ to health terminal. C_{pi} is the unit disease cost of each health terminal. It can be collected from China Health and Family Planning Statistical yearbook in 2016 (Peking Union Medical College Press, 2016). GDP_{pi} is the loss of total per capital GDP which can be calculated by the data of per capital delayed work time and daily per capital GDP. Considering some family members or friends accompanied, this paper assumed that each patient needed to be accompanied by an average of one person. E_i is the health effect caused by $PM_{2.5}$.

For the health terminals of the incidence of acute or chronic respiratory diseases, method of willingness to pay was chosen to evaluate the economic loss of health effect. Unlike other diseases, once people suffer these diseases, the physical and mental pains are difficult to be measured numerically. The benefits of the willingness to pay method can quantify the health effect and its economic loss caused by respiratory diseases indirectly by investigating and collecting the costs of the residents willing to pay. Considering the unavailable relevant statistics in China, this paper took the results of the residents' willingness to pay from Huang and Zhang (2013) for reference. As Huang and Zhang (2013) mainly focused on Beijing-Tianjin-Hebei Region, the disposable income and the ratio of consumer price index were used here to calculate the residents' willingness to pay in China based on the results of the Beijing-Tianjin-Hebei Region.

3.3 Scenarios setting in 2020 and 2030

Coal consumption control can reduce the public health effect caused by $PM_{2.5}$ pollution. China has fully implemented the coal consumption control policy since 2016 (National Development and Reform Commission, 2016). For predicting the public health effects, this paper adopted the set of the baseline scenario and total coal consumption control scenario in 2020 and 2030.

Under the baseline scenario, based on the plan and the data from World Bank, BP Energy Outlook, the outline of the national "13th Five-Year Plan" and the development goals of each province in 2020 and 2030, the data of GDP, population, industrial structure, energy consumption, elasticity coefficient of coal consumption, coal consumption per product, coal consumption per GDP and coal consumption per capita were set. According to the revolutionary strategic data on energy production and consumption in 2020 and 2030 formulated by the National Development and Reform Commission and the National Energy Administration, the total energy consumption in 2020 was controlled within 5 billion tons of standard coal. The proportion of coal consumption was further reduced, accounting for about 58%, which was about 4.1 billion tons. The efficiency of energy development and utilization had been greatly improved. The energy efficiency of major industrial products had reached or approached the international advanced level, and the energy consumption per unit of GDP would dropped by 15% compared with 2015. In 2030, the use of fossil energy would significantly be reduced. The total energy consumption was controlled within 6 billion tons of standard coal, and the proportion of coal consumption in total energy consumption was further reduced. The energy consumption per unit of GDP (current price) had reached the current world average level (National Development and Reform Commission, National Energy Administration, 2017).

Under the total coal consumption control scenario, the nation's coal consumption would be reduced by measures such as optimizing the industrial structure, eliminating excess production capacity, and reducing backward capacity in high coal-consuming industries such as cement. Based on the baseline scenario, China's total coal consumption would be controlled at around 4 billion tons in 2020 and within 3.7 billion tons in 2030 (Chen et al., 2015).

3.4 Data source

According to the results from Xue et al. (2016), it could be seen that $PM_{2.5}$ concentration from coal consumption accounted for 56% in China. Combining with $PM_{2.5}$ concentration in 2015 in China (China National Environmental Monitoring Centre, 2017), it could be calculated that the $PM_{2.5}$ concentration produced by coal consumption in 2015 was 30.8 $\mu g/m^3$. The result of $PM_{2.5}$ pollution from coal consumption in 2020 and 2030 was adopted from Chen et al. (2015) for reference, as shown in Fig. 1.

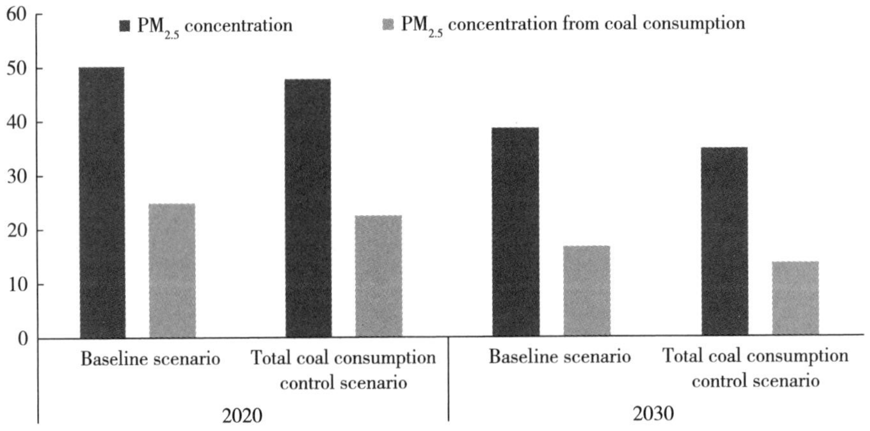

Fig. 1 Prediction of $PM_{2.5}$ concentration and $PM_{2.5}$ concentration from coal consumption in 2020 and 2030 ($\mu g/m^3$)

Source: Chen et al. (2015)

4 Results and discussion

4.1 Health effect and health economic loss caused by $PM_{2.5}$ from coal consumption in 2015

4.1.1 Health effect in four different health effect terminals

The health effect of $PM_{2.5}$ pollution from coal consumption on each health terminal in 2015 was shown in Table 2. The total value of health effect in China in 2015 was 33860 (12980, 53730) thousand people. The incidence of acute upper respiratory infection had the highest health effect of 25459 (10093, 40275) thousand people. The outpatient rate had the lowest health effect of 4 (2, 5) thousand people.

Table 2 The health effects value caused by $PM_{2.5}$ pollution caused by coal consumption in 2015

Health effect terminal		E value (thousand people)
Mortality rate	Lung cancer	441 (23, 59)
	Cardiovascular disease	22 (6, 38)
	Respiratory disease	16 (10, 22)
Hospitalization rate	Cardiovascular disease	2128 (1348, 2906)
	Respiratory disease	2219 (0, 4471)
Incidence rate	Acute upper respiratory infection	25459 (10093, 40275)
	Chronic bronchitis	3967 (1491, 5948)
Outpatient rate	Internal medicine	4 (2, 5)
	Pediatric	5 (2, 8)

Note: according to the 95% confidence interval of the coefficients (95% CI).

4.1.2 Health economic loss of mortality

First, the population of all age groups in 2015 was calculated based on the sample population of all age groups and 1.55% sample ratio in 2015 (China Statistics Press, 2016). The excess deaths of lung cancer disease, cardiovascular disease and respiratory disease at all ages were calculated. The national life table in 2015 could be made with the relevant data. Finally, the health and economic loss of lung cancer disease, cardiovascular disease and respiratory disease mortality from $PM_{2.5}$ pollution caused by coal consumption in each age group in 2015 was calculated. As shown in Fig. 2, it was obvious that the health economic loss of lung cancer mortality was the largest, which mainly concentrated on people aged 60-65. The health economic loss of cardiovascular disease mortality was similar to that of respiratory disease mortality, both of which increased with age.

4.1.3 Health economic loss of other three health effect terminals

Adopting the data from China Health and Family Planning Statistical yearbook in 2016 (Peking Union Medical College Press, 2016), the unit disease cost of hospitalization rate for cardiovascular disease, the hospitalization rate for respiratory disease, the outpatient rate for

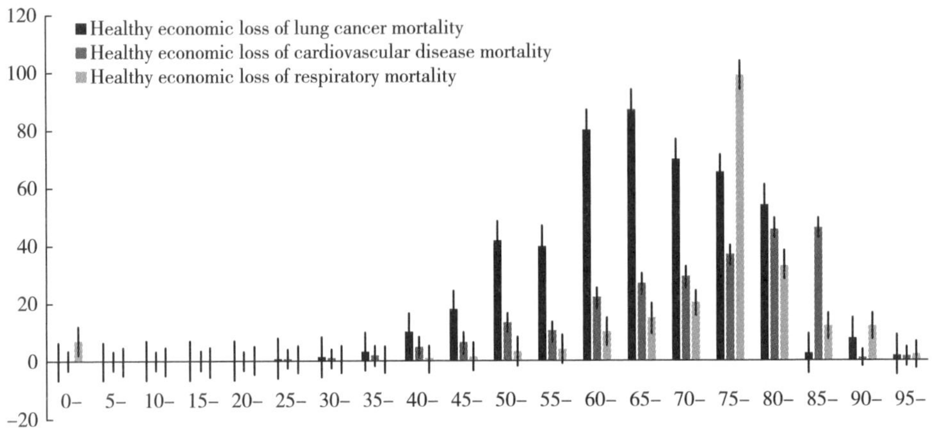

Fig. 2 Health economic loss of mortality in 2015 (Ten million yuan)

internal medicine and outpatient rate for pediatrics was calculated to be 222 yuan, 5667 yuan, 234 yuan and 234 yuan, respectively. The health economic loss value of hospitalization rate for cardiovascular disease and respiratory disease were the highest, which were 27640 (17500, 37740) million yuan and 17390 (0, 22780) million yuan, respectively. The incidence rate of chronic bronchitis was the lowest. It was 0.3 (0.2, 1) million yuan. The total health economic loss was 55210 (22640, 75750) million yuan in China in 2015 (Table 3).

Table 3 Health economic loss of other three health effect terminals in 2015 (million yuan)

Health effect terminal		Healthy economic loss
Hospitalization rate	Cardiovascular disease	27640 (17500, 37740)
	Respiratory disease	17390 (0, 22780)
Incidence rate	Acute upper respiratory infection	27 (24, 91)
	Chronic bronchitis	0.3 (0.2, 1)
Outpatient rate	Internal medicine	4 (2, 5)
	Pediatric	5 (2, 8)

Note: according to the 95% confidence interval of the coefficients (95% CI).

4.2 Economic loss change of health effects caused by $PM_{2.5}$ pollution from coal consumption in China between the two scenarios in 2020 and 2030

4.2.1 Health effect change in four different health effect terminals

The health effect declined under the total coal consumption control scenario. The reduction of health effect in 2020 and 2030 had been calculated (Table 4). The result illustrated that the total health effects were 7533 (3447, 11928) thousand people in 2020 and 10588 (3981, 17090) thousand people in 2030.

4.2.2 Prediction of health economic loss change of mortality in 2020 and 2030

After calculating the health economic loss of lung cancer mortality, cardiovascular disease mortality and respiratory disease mortality respectively in the baseline scenario and total coal consumption control scenario in 2020 and 2030, the change of health economic loss in each age group was obtained (Figs. 3, 4).

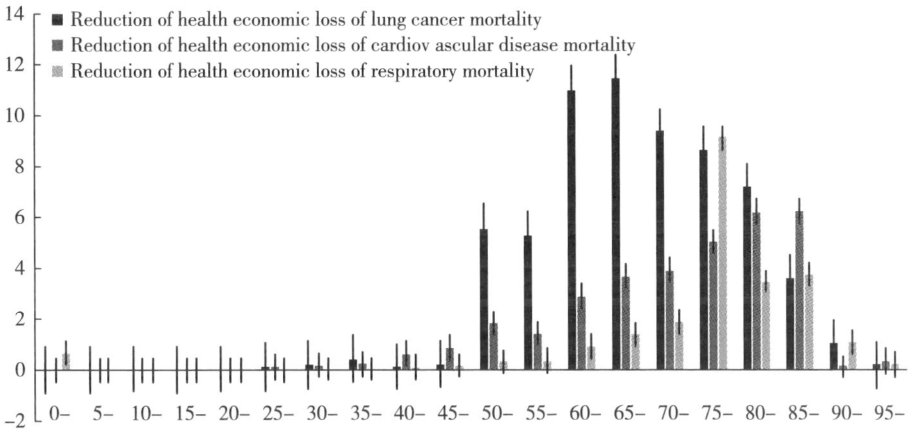

Fig. 3 Health economic loss reduction of mortality under the total coal consumption control scenario in 2020 (Ten million yuan)

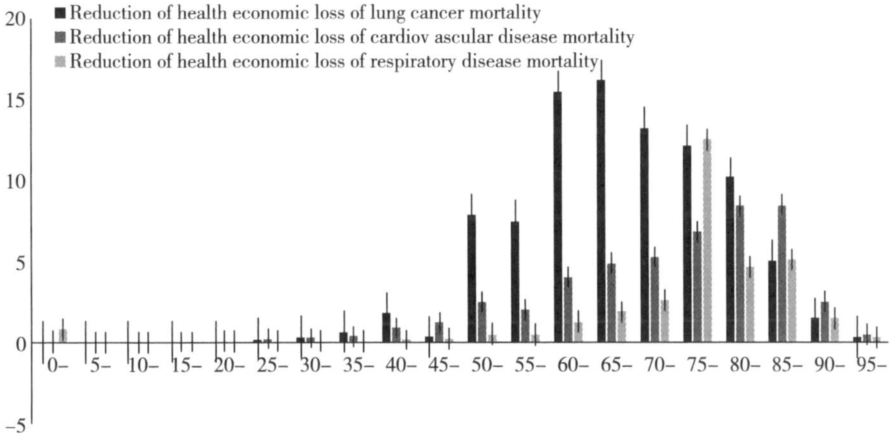

Fig. 4 Health economic loss reduction of mortality under the total coal consumption control scenario in 2030 (Ten million yuan)

From Fig. 3 and Fig. 4, the health economic loss variation of people aged 50-90 was much higher than that of people aged 0-40. The death rate of lung cancer caused the most health economic loss change. The health economic loss change of lung cancer mortality was nearly normally distributed, and the peak happened around age 65. Itspeak value was about 1 billion yuan in 2020 and 16 billion yuan in 2030. The health economic loss change of cardiovascular disease morality was positively correlated with age. The change of health economic loss of respiratory diseases mortality caused a peak aroundthe age of 75. Its peak value was about 9 billion yuan in 2020 and 13 billion yuan in 2030.

Table 4 Reduction of health effect between the baseline scenario and total coal consumption control scenario in 2020 and 2030 (thousand people)

Health effect terminal		Reduction of E value in 2020	Reduction of E value in 2030
Morality rate	Lung cancer	9 (5, 13)	13 (7, 19)
	Cardiovascular disease	5 (1, 8)	7 (2, 12)
	Respiratory disease	4 (2, 5)	5 (3, 7)
Hospitalization rate	Cardiovascular disease	475 (300, 648)	649 (410, 887)
	Respiratory disease	495 (0, 997)	679 (0, 1375)
Incidence rate	Acute upper respiratory infection	5664 (2248, 8941)	7957 (3096, 12810)
	Chronic bronchitis	879 (332, 1313)	1278 (463, 1977)
Outpatient rate	Internal medicine	0.8 (0.5, 1.2)	1.1 (0.6, 1.6)
	Pediatric	1.1 (0.4, 1.7)	1.5 (0.5, 2.4)

Note: according to the 95% confidence interval of the coefficients (95% CI).

4.2.3 Prediction of health economic loss change of other three health effect terminals

The paper predicted the health economic loss of the hospitalization rate of cardiovascular disease and respiratory disease, internal medicine rate and pediatric rate respectively under the baseline scenario and total coal consumption control scenario in 2020 and 2030. Thus, variation of health economic loss under the baseline scenario and the total coal consumption control scenario was obtained. The health economic loss changes were 7540 (4660, 10440) million yuan in 2020 and 10330 (6240, 14350) million yuan in 2030. The hospitalization rate of cardiovascular disease had the largest reduction of 6170 (3910, 8420) million yuan in 2020 and 8420 (5330, 11520) million yuan in 2030 (Table 5).

Table 5 Health economicloss reduction ofother three health effectterminalsbetween thebaseline scenario and total coal consumption control scenario in 2020 and 2030 (million yuan)

Health effect terminal		2020	2030
Hospitalization rate	Cardiovascular disease	6170 (3910, 8420)	8420 (5330, 11520)
	Respiratory disease	4950 (0, 9970)	4580 (0, 9290)
Outpatient rate	Internal medicine	8 (5, 10)	10 (4, 20)
	Pediatric	10 (6, 20)	10 (5, 20)
Incidence rate	Acute upper respiratory infection	120 (60, 190)	179 (80, 260)
	Chronic bronchitis	1.5 (0.6, 2.2)	2 (1, 3)

Note: according to the 95% confidence interval of the coefficients (95% CI).

Calculated by Poisson Regression model, the health effect produced by $PM_{2.5}$ emissions of national coal consumption in 2015 was 33860 (12980, 53730) thousand people. It accounted for 3% (0%, 4%) of the national population. The health economic loss was 55210 (22640, 75750) million yuan. It accounted for 0.08% (0.03%, 0.1%) of GDP in 2015.

Predicted that by 2020 and 2030, compared with the baseline scenario and total coal consumption control scenario, the proportion of $PM_{2.5}$ concentrations from coal consumption changed from 49.2% to 46.5% in 2020 and from 43.6% to 38.4% in 2030. The health effect would decrease by 7530 (3450, 11930) thousand people in 2020 and 10, 590 (3980, 17090) thousand people in 2030. They would separately account for 0.5% (0.2%, 0.8%) of the national population in 2020 and 0.8% (0.3%, 1.3%) of the national population in 2030. The health economic loss would decrease by 11840 (4210, 19520) million yuan in 2020 and 13980 (5740, 22360) million yuan in 2030. They accounted for 0.02% (0.01%, 0.03%) and 0.02% (0.001%, 0.03%) of GDP in 2015 respectively.

5 Conclusions and suggestions

5.1 Conclusions

According to the calculation results above, it was proved that $PM_{2.5}$ pollution produced by coal consumption had effects on public health and economic loss. It would cause great health effects and health economic loss.

5.1.1 The health effect of acute upper respiratory infection was the largest and the health effect of the internal medicine was the smallest in the health terminals

$PM_{2.5}$ emissions from coal consumption in 2015 increased the loss of health effect. It can be seen from the result that people in different ages would suffer different loss of health effect. People aged more than 40 years had more loss of health effect because they were easier to die of lung cancer, cardiovascular disease and respiratory diseases. The number of patients with acute upper respiratory infection was the largest, and its loss of health effect was the largest. It accounted for an average of 75.2% of the total health effect. While, the health effect of the other health terminals was relatively lowest. Especially the internal medicine had the smallest number of outpatient rates and the smallest loss of health effect. It accounted for an average of only 0.01% of the total health effect.

5.1.2 The economic loss from the hospitalization of cardiovascular disease and respiratory disease was relatively bigger

$PM_{2.5}$ emissions from coal consumption in 2015 increased the economic loss of health effect. The total cost of economic loss caused by the hospitalization of cardiovascular disease and respiratory disease was both bigger. Together they accounted for 81.6% of the total economic loss of health effect and they were the main contributors. While, the other health terminals had less economics loss of health effect. The incidence of chronic bronchitis caused the lowest economic loss of health effects which accounted for only 0.007% of the total economic loss of health effect.

5.1.3 Controlling coal consumption can reduce the loss of health effect and economic loss of health effect

Based on the baseline scenario and total coal consumption control scenario in 2020 and

2030, it can be seen that controlling the coal consumption can reduce the loss of public health effect and its economic loss. In the total coal consumption control scenario, the intensity of coal control in 2030 was greater than that in 2020. The results showed that the reduction of health effect loss and its economic loss in 2030 was greater than that in 2020. Therefore, controlling coal consumption to reduce $PM_{2.5}$ pollution can improve public health benefits. The greater the intensity of coal control, the higher the public health benefits.

5.2 Policy recommendations

5.2.1 Continuing to strengthen the policy of total coal consumption control

$PM_{2.5}$ pollution from coal consumption is an important cause of $PM_{2.5}$ emissions in China. From the result of baseline scenario and total coal consumption control scenario in 2020 and 2030, controlling total coal consumption would reduce the loss of public health effects and its economic loss. Besides, the reduction loss of health effect and its economic loss in 2030 would be more than that in 2020. Therefore, controlling total coal consumption is able to reduce $PM_{2.5}$ pollution and improve public health effects. The greater intensity of total coal consumption control, the stronger public health effect.

5.2.2 Promoting the technology of conserving energy to improve its efficiency

The development and innovation of technology for conserving energy and reducing emissions from coal should be promoted. Relevant studies have concluded that the performance of coal burning could be comprehensively improved through special processes, and the efficiency of coal burning could be increased to 95%. Mastering the new and high technology of improving the efficiency of coal burning can effectively control environmental pollution and reduce $PM_{2.5}$ emissions.

5.2.3 Promoting the use of clean energy

New and renewable energy sources should be actively utilized and optimized. Subsidies for clean energy should also be increased. Innovation of new energy technology can be accelerated and a clean and efficient energy system should be built. The development and utilization of wind power, solar energy, nuclear power and ocean energy should also be strengthened. These measures will be able to reduce $PM_{2.5}$ pollution from coal consumption and the loss of health effect.

Acknowledgements The authors express their sincere thanks for the support from the National Natural Science Foundation of China under Grant Nos. 71803182 and 71773118, Beijing Social Science Foundation Project under Grant No. 19YJC023, Open Fund of State Key Laboratory of Water Resource Protection and Utilization in Coal Mining under Grant No. SHJT-17-42.20.

References

Baxter, L. K., Duvall, R. M., Sacks, J., 2013. Examining the effects of air pollution composition on within region differences in $PM_{2.5}$ mortality risk estimates. J. Expo. Sci. Environ. Epidemiol. 23(5), 457-465.

Belis, C. A., Pisoni, E., Degraeuwe, B., et al., 2019. Urban pollution in the Danube and Western Balkans regions: the impact of major $PM_{2.5}$ sources. Environ. Int. 133, 105158.

Bell, M. L., Davis, D. L., 2001. Reassessment of the lethal London fog of 1952: novel indicators of acute and chronic consequences of acute exposure to air pollution. Environ. Health Perspect. 109 (Suppl. 3), 389-394.

Bell, M. L., Ebisu, K., Peng, R. D., et al., 2009. Hospital admissions and chemical composition of fine particle air pollution. Am. J. Respir. Crit. Care Med. 179 (12), 1115-1120.

Bian, Y. Y., Guo, W. Y., Xiu, M. W., Chen, G. Y., Chen, F., Li, J. Z., Liu, Y. C., 2019. Comparison of $PM_{2.5}$ concentration and prevalence of respiratory diseases between Langfang city and Hohhot city. Pract. Prev. Med. 26 (05), 513-517.

Boldo, E., Linares, C., Lumbreras, J., et al., 2011. Health impact assessment of a reduction in ambient $PM_{2.5}$ levels in Spain. Environ. Int. 37 (2), 342-348.

Buonocore, J. J., Dong, X., Spengler, J. D., et al., 2014. Using the Community Multiscale Air Quality (CMAQ) model to estimate public health impacts of $PM_{2.5}$ from individual power plants. Environ. Int. 68, 200-208.

Chen, J. Y., 2018. Research about the progress on the health and economic loss caused by air pollution in China. JNMU (Social Sciences Edition) 18 (04), 282-286.

Chen, Y., Zhi, G., Feng, Y., et al., 2006. Measurements of emission factors for primary carbonaceous particles from residential raw-coal combustion in China. Geophys. Res. Lett. 33 (20).

Chen, X. J., Jin, L., Lei, Y., Xue, W. B., Su, M., Yang, J. T., Wang, J. N., 2015. Research on total coal consumption control in China under the constraint of atmospheric environment. Chin. J. Environ. Manage. 7 (05), 42-49.

Chen, L., Shi, M., Li, S., et al., 2017. Combined use of land use regression and BenMAP for estimating public health benefits of reducing $PM_{2.5}$ in Tianjin, China. Atmos. Environ. 152, 16-23.

Chen, K., Wang, G., Wu, L., et al., 2019. $PM_{2.5}$ pollution: health and economic effect assessment based on a recursive dynamic computable general equilibrium model. Inter. J. Env. Res. Pub. Heal. 16 (24), 5102.

China National Environmental Monitoring Centre, 2017. Ambient air quality standard. www.cnemc.cn/jcgf/dqhj/201711/W020181008687883046768.pdf.

China Statistics Press, 1999. 1998 China Statistical Yearbook. www.stats.gov.cn/tjsj/ndsj/1999/indexch.htm.

China Statistics Press, 2006. 2005 China Statistical Yearbook. www.stats.gov.cn/tjsj/ndsj/2006/indexch.htm.

China Statistics Press, 2016. 2015 China Statistical Yearbook. www.stats.gov.cn/tjsj/ndsj/2016/indexch.htm.

China Statistics Press, 2019. 2018 China Statistical Yearbook. www.stats.gov.cn/tjsj/ndsj/2019/indexch.htm.

Cox, L. A., Popken, D. A., Ricci, P. F., 2013. Warmer is healthier: effects on mortality rates of changes in average fine particulate matter ($PM_{2.5}$) concentrations and temperatures in 100 U.S. cities. Regul. Toxicol. Pharmacol. 66 (3), 336-346.

Dong, Y., Xu, G. Z., Wang, A. H., Lu, P. P., 2018. Assessment of health hazard and economic loss of residents caused by $PM_{2.5}$ pollution in an urban area of a city from 2014 to 2016. Chuang. Hua. Yu. Fang. I. Hsueh. Tsa. Chih. 19 (8), 579-582.

Fan, F. Y., Wang, H. F., Fan, L. J., 2019. Assessment of health and economic loss of air pollution in Beijing-Tianjin-Hebei region. Ecol. Econ. 35 (9), 157-163.

Feng, S., Gao, D., Liao, F., et al., 2016. The health effects of ambient $PM_{2.5}$ and potential mechanisms. Ecotox. Environ. Safe. 128, 67-74.

Finkelman, R. B., Tian, L., 2018. The health impacts of coal use in China. Int. Geol. Rev. 60 (5-6), 579-589.

Fullman, N., Barber, R. M., Abajobir, A. A., et al., 2017. Measuring progress and projecting attainment on the basis of past trends of the health-related sustainable development goals in 188 countries: an analysis from the global burden of disease study 2016. Lancet 390 (10100), 1423-1459.

Gautam, S., Prasad, N., Patra, A. K., et al., 2016. Characterization of $PM_{2.5}$ generated from opencast coal mining operations: a case study of Sonepur Bazari opencast project of India. Enviro. Technol. Inno. 6, 1-10.

Ge, S., Bai, Z., Liu, W., et al., 2001. Boiler briquette coal versus raw coal: part I—stack gas emissions. J. Air Waste Manage. Assoc. 51 (4), 524-533.

Greene, N., Morris, V., 2006. Assessment of public health risks associated with atmospheric exposure to $PM_{2.5}$in Washington, DC, USA. Inter. J. Env. Res. Pub. Heal. 3 (1), 86-97.

Han, M. X., Li, H. M., 2006. Coal consumption and atmospheric pollutant emission in China. Coal. Eng. (03), 76-78.

Huang, D. S., Zhang, S. Q., 2013. Health benefit assessment of $PM_{2.5}$ pollution control in Beijing-Tianjin-Hebei region. Environ. Sci-China. 33 (01), 166-174.

Jing, L., Ren, C., 2000. Study on the relationship between air pollution and acute or chronic respiratory system disease in Benxi. Environ. Health 17 (5), 268-270.

Kan, H., Chen, B., Wang, H., 2004. The economic evaluation of the harm to public health for the urban atmospheric particulate matter in Shanghai, China. Health Econ. 23 (2), 8-11.

Khare, P., Baruah, B. P., 2011. Estimation of emissions of SO_2, $PM_{2.5}$, and metals released from coke ovens using high sulfur coals. Environ. Prog. Sustain. 30 (1), 123-129.

Kim, H. B., Shim, J. Y., Park, B., et al., 2019. Long-term exposure to air pollution and the risk of non-lung cancer: a meta-analysis of observational studies. Perspect. Public. Heal., 1-10 https://doi.org/10.1177/1757913919891751.

Li, L., Lei, Y. L., Wu, S. M., et al., 2018. Evaluation of future energy consumption on $PM_{2.5}$ emissions and public health economic loss in Beijing. J. Clean. Prod. 187, 1115-1128.

Lin, X. N., Dong, X., Chen, Z., 2019. Pollution characteristics and source analysis of carcinogenic and harmful elements in $PM_{2.5}$ in Guiyang city. Environ. Earth. 47 (03), 254-260.

Lippmann, M., 2014. Toxicological and epidemiological studies of cardiovascular effects of ambient air fine particulate matter ($PM_{2.5}$) and its chemical components: coherence and public health implications. Crit. Rev. Toxicol. 44 (4), 299-347.

Lu, Q. G., Li, S. Y., Huang, C. R., 2019. Current situation and development suggestions of coal clean and efficient combustion and utilization technology in industrial field. Bull. Chin. Acad. Sci. 34(4), 392-400.

Michael, G., Fan, Q., 2018. Introduction of AQLI: 12 Facts about Fine Particulate Air Pollution, Human Health, and Global Policy. R. EPIC.

National Development and Reform Commission, 2016. A notice on coal consumption reduction and replacement, requiring strict control of new production capacity of high - consumption coal projects. Chin. Equip Eng. 10, 1.

National Development and Reform Commission, National Energy Administration, 2017. Notice on printing and distributing the "revolutionary strategy for energy production and consumption (2016—2030)". www.gov.cn/xinwen/201704/25/5230568/files/286514af354e41578c57ca38d5c4935b.pdf.

Peking Union Medical College Press, 2016. China Health and Family Planning Statistical Yearbook. www.nhc.gov.cn/wjw/sjcx/sjcx.shtml.

Samek, L., Stegowski, Z., Styszko, K., et al., 2020. Seasonal variations of chemical composition of $PM_{2.5}$fraction in the urban area of Krakow, Poland: PMF source attribution. Air. Qual. Atmos. Heal. 13 (1), 89-96.

Shi, X. M., 2018. Progress of epidemiological research on the impact of atmospheric $PM_{2.5}$ and its components on acute health of population. J Shandong Univ Health Sci. 56 (11), 1-11.

Sun, L., Lin, Z., Liao, K., et al., 2015. Adverse effects of coal combustion related fine particulate matter ($PM_{2.5}$) on nematode Caenorhabditis elegans. Sci. Total Environ. 512, 251-260.

Wang, X., Cheng, H., Xu, X., et al., 2008. A wintertime study of polycyclic aromatic hydrocarbons in $PM_{2.5}$ and $PM_{2.5}$-PM_{10} in Beijing: assessment of energy structure conversion. J. Hazard. Mater. 157 (1), 47-56.

Wang, F., Lin, T., Feng, J., et al., 2015. Source apportionment of polycyclic aromatic hydrocarbons in $PM_{2.5}$ using positive matrix factorization modeling in Shanghai, China. Environ. Sci.: Processes & Impacts 17, 197-205.

World Bank, 2019. Global economic outlook. data. worldbank. org. cn/country/china? View =chart.

Xie, P., Liu, X. Y., Liu, Z. R., Li, T. T., Bai, Y. H., 2009. Study on the relationship between exposure and response of atmospheric particulate pollution in Chinese population. J. Environ. Sci-China. 29 (10), 1034-1040.

Xue, W. B., Wu, W. L., Fu, F., Ha, Y. M., Lei, Y., Chen, X. J., Wang, J. N., Xu, Y. L., 2016. Astudy on the impact of coal consumption on $PM_{2.5}$ pollution in China. Environ. Manag. 8(2), 94-98.

Zhang, M., Song, Y., Cai, X., et al., 2008. Economic assessment of the health effects related to particulate matter pollution in 111 Chinese cities by using economic burden of disease analysis. J. Environ. Manag. 88 (4), 947-954.

Synergistic effect of CO_2 and $PM_{2.5}$ emissions from coal consumption and the impacts on health effects

JIA Wanlin[1,2], LI Li[1,2,*], LEI Yalin[1,2], WU Sanmang[1,2]

1. *School of Economics and Management, China University of Geosciences, Beijing* 100083, *China*
2. *Key Laboratory of Carrying Capacity Assessment for Resource and Environment, Ministry of Natural Resources of the People's Republic of China, Beijing* 100083, *China*

Abstract China put forward that it would strive to achieve the carbon peaking in 2030 and carbon neutrality goals in 2060, which needed realizing the synergistic effect of pollution reduction and CO_2 reduction. CO_2 and $PM_{2.5}$ from coal consumption were "homogenous and synchronous" and they had significant synergistic effect. The synergistic reduction of $PM_{2.5}$ from coal consumption could have impacts on public health. The Logarithmic Mean Divisia Index (LMDI) was adopted to decompose the $PM_{2.5}$ from coal consumption in China from 2012 to 2017 based on data availability and quantified the synergy. $PM_{2.5}$ emissions from coal consumption were forecasted in 2030, 2035, 2050 and 2060, and its health effects of $PM_{2.5}$ were estimated. The main conclusions were: (1) the synergy of CO_2 and $PM_{2.5}$ was the largest contribution factor of $PM_{2.5}$ from coal consumption. (2) $PM_{2.5}$ from coal consumption in 2060 would achieve the WHO air quality guidelines. (3) The most health effects economic loss from $PM_{2.5}$ generated by coal consumption in China was in 2014. The health effects economic loss would have been decreased to negative by 2050. We should reinforce the synergistic effect of CO_2 and $PM_{2.5}$, improve energy efficiency and optimize energy structure to achieve $PM_{2.5}$ pollution control and reduce health effects economic loss.

Keywords Synergistic effect, Health effects, CO_2, $PM_{2.5}$, Coal consumption

1 Introduction

Climate change is a major problem in the world. CO_2 is the main greenhouse gas (GHG) causing climate change. CO_2 emissions from coal consumption (CC) were about 80% of the total CO_2 emissions (Guan et al., 2021). In 2020, China put forward the carbon peaking and carbon neutrality goals (Xinhua News Agency, 2020). Achieving the carbon peaking and carbon neutrality goals are a long-term and difficult process, which needs to be based on China's national circumstances. The guiding document on the country's work was unveiled to achieve the carbon peaking and carbon neutrality goals under the new development philosophy in 2021. It

原文刊于 *Journal of Environmental Management*，2023 年第 325 卷。原文发表时为彩色。

* 为通信作者。

clearly pointed out that CO_2 emissions per GDP would have decreased by more than 65% compared with the 2005 in 2030. The share of non-fossil energy consumption would be about 25% in 2030 and would be over 80% in 2060 (Central People's Government of the People's Republic of China, 2021). Future low-carbon targets will significantly reduce $PM_{2.5}$ concentrations, resulting in health benefits (Gschwind et al., 2015).

The control of $PM_{2.5}$ had achieved remarkable results in China, but it was still about five times higher than the WHO standard in 2020 (Ministry of Ecology and Environment of the People's Republic of China, 2020; WHO, 2021). $PM_{2.5}$ pollution still needed in-depth treatment and had not been fundamentally controlled. Primary $PM_{2.5}$ emissions were mainly generated by coal use in China, which was 63% of the total $PM_{2.5}$ emissions (China Coal Consumption Control Program and Policy Research Project Team, 2014). $PM_{2.5}$ emissions affected public health and labor mobility (Yue et al., 2021). Atmospheric $PM_{2.5}$-related premature deaths could be over 1 million in China each year, posing a serious threat to public health (Yin et al., 2020). $PM_{2.5}$-related premature deaths decreased by 177000 due to air quality policies in 2019 (Liu et al., 2022). $PM_{2.5}$ concentrations in prefecture-level and above cities by 2025 would be 10% lower than in 2020 (Central People's Government of the People's Republic of China, 2021). Coal continues to occupy an important position energy consumption in China. On the grounds of the National Energy Administration, the share of CC in total energy consumption was 57.7% in 2019 (National Bureau of Statistics of China, 2020). Although the proportion of CC decreased year by year, but it still occupied a dominant position. Coal use could seriously damage health and it was necessary to explore the necessity of $PM_{2.5}$ emissions from CC.

The Intergovernmental Panel on Climate Change (IPCC) clarified the definition of co-benefits at the beginning of the 21st century. It considered CO_2 and $PM_{2.5}$ were "homogenous". The synergistic effect could affect $PM_{2.5}$ emissions, which thus indirectly threatened human health. CO_2 control policies can generate considerable health effects (Cai et al., 2021; Ebi et al., 2017). The "Three-year action plan to fight against air pollution" released explicitly called for "synergistic reduction of GHG" (Central People's Government of the People's Republic of China, 2018). Carbon emissions reduction policies could affect $PM_{2.5}$ emissions from CC and the synergistic control of CO_2 and $PM_{2.5}$ could produce a synergistic effect of "1+1>2". This paper aimed to transform CO_2 emissions change into $PM_{2.5}$ emissions change based on the synergistic effect of CO_2 and $PM_{2.5}$ emissions and quantify the impacts of $PM_{2.5}$-related health effects.

Former researches on the effect of synergistic emissions reduction for CO_2 and $PM_{2.5}$ have been conducted. The focus of synergistic effect mainly was on a national level or city level as well as the industrial, transportation sectors and so on. There was a lack of research on the effect of synergistic emissions reduction for CO_2 and $PM_{2.5}$ from CC. In this paper, the LMDI method was innovatively adopted to analyze the influencing factors of $PM_{2.5}$ emissions from CC and the synergy of CO_2 and $PM_{2.5}$ emissions was quantified. $PM_{2.5}$ emissions from CC under the future carbon peaking and carbon neutrality goals were predicted according to the synergy of CO_2 and $PM_{2.5}$ emissions. Then figured out the health effects of $PM_{2.5}$ emissions reduction under the carbon peaking and carbon neutrality goals based on PC from CC. The generated health effects would reduce the abatement costs and provide a scientific basis for future climate change mitigation and atmospheric pollution prevention policy formulation.

2 Literature review

Previous papers on the synergistic effect of CO_2 and $PM_{2.5}$ emissions and the health effects have been conducted. Subsequently, different organizations defined synergistic effect in different ways. We argued that under the implementation of CO_2 control policies, the various policy processes produced additional effects beyond the control of CO_2 emissions.

2.1 The progress of the synergistic effect of CO_2 and $PM_{2.5}$ emissions

Many studies had confirmed that CO_2 and $PM_{2.5}$ have synergistic effects. $PM_{2.5}$ emissions reduction measures could significantly reduce CO_2 emissions. For example, Li and Song (2021) probed into the synergistic effect of CO_2 emissions and atmospheric pollutants from urban road mobile sources in Beijing under different policies using the Long-range Energy Alternatives Planning System (LEAP). The synergisticemissions reduction effect was found to be most significant for the integrated policy scenario. Gao et al. (2021) applied the emission factor method and the synergistic effect assessment index method to evaluate the measures of the Air Pollution Prevention Action Plan. The results indicated that all the emission reduction measures of pollutants had significant synergistic effect on CO_2. Xing et al. (2020) used the emission factor method and the co-control cross-elasticity analysis method to carry out the synergistic emissions reduction effect assessment of the measures of the Blue Sky Defense Action Plan in Tangshan. The results demonstrated that air pollution control measures could effectively reduce CO_2 emissions. Vennemo et al. (2009) adopted an epidemiological approach to assess the potential health effects of the Clean Development Mechanism (CDM) in China. The results were that 3000 to 40000 lives could be saved each year, and the estimated annual additional benefits could be more than 1 billion yuan.

Many researches had found that climate change mitigation policies and constraints on CO_2 reduction targets could accelerate the transformation of energy structure and mitigate $PM_{2.5}$ pollution (Dong et al., 2022a, 2022b). Bollen et al. (2009) applied Evaluating the Regional and Global Effects of GHG Reduction Policies (MERGE) model to simulate the combined local air pollutant and GHG reduction effects, which demonstrated that climate change mitigation policies could reduce atmospheric pollution. Plachinski et al. (2014) combined

MyPower model and Community Multiscale Air Quality (CMAQ) model, which found that low-carbon electricity policies in the U.S. could achieve synergistic effect in $PM_{2.5}$ emissions reduction. Yang et al. (2017) applied the China-MAPLE model to explore the association between CO_2 emissions and local air pollutants under a common control scenario of carbon peaking in China. The results displayed that the emissions of air pollutants compared to 2010 would be reduced by about 80% in 2030. It implied that the air quality goal and the carbon peaking target could be achieved. Lu et al. (2021) used the energy system model IMED | TEC to construct carbon-constrained scenarios with 1.5℃ and 2℃ targets. The synergistic effect of energy structure change and atmospheric pollution improvement from decarbonization of the transportation sector in China were probed. The results displayed that decarbonization could bring significant atmospheric pollution improvement benefits. Dong et al. (2019) probed into the synergistic reduction effect of CO_2 and $PM_{2.5}$ emissions by LMDI decomposition. The results indicated that CO_2 abatement activities had a significant contribution to $PM_{2.5}$ reduction. Jia et al. (2020) applied the emission factor method to screen and evaluated the CO_2 emissions reduction effects of 341 policy measures in 31 provinces in different sectors and different provinces. The results indicated that the CO_2 emissions reduction effects of two measures were obvious, which were industrial restructuring and layout and boiler renovation and treatment. The most fundamental causes of excessive CO_2 and $PM_{2.5}$ emissions were the imbalance of the fossil energy consumption, industrial structure and the lack of environmental treatment (Shao et al., 2016). According to the large factors of impact on $PM_{2.5}$ emissions, we should control $PM_{2.5}$ emissions in different provinces.

2.2 The progress of the health effects of synergistic emissions reduction

Unlike the long-term benefits achieved by CO_2 emissions reduction policies, synergistic emissions reduction of CO_2 and $PM_{2.5}$ could improve air quality and reduce mortality from respiratory and cardiovascular diseases in the short term, resulting in huge health benefits (Bell et al., 2008; Bollen et al., 2009a). Zhang et al. (2021) assessed the costs and atmospheric pollutant emission pathways of different carbon mitigation technologies in China using the Computable General Equilibrium (CGE) model and the air quality model. The results presented that the integration of climate and atmospheric pollution control policies could significantly improve air quality and extend life expectancy in China. Tang et al. (2022) integrated the Global Change Analysis Model (GCAM), Weather Research and Forecasting Model Coupled with Chemistry (WRF-Chem) and Global Exposure Mortality Model (GEMM) models to quantify ambient $PM_{2.5}$ concentrations and associated mortality under different carbon mitigation pathways. The cost of carbon mitigation in 2050 under the 2℃ target scenario could be offset by health synergistic benefits. Short-term health benefits could offset some costs of CO_2 emissions reduction. In China, some industries had launched CO_2 reduction actions. Wang et al. (2021) adopted the latest population and health data with a localized and calibrated atmospheric pollution exposure–response model. Then facility–level co–benefits were evaluated for each specific coal power resource location in China, using the Chinese coal power industry as an example. Peng et al. (2018) applied the WRF–Chem model and exposure–response model to assess the air quality improvements and health effects from an electrification strategy in China. The results demonstrated that switching to semi–decarbonized power sources for electrification in the transportation and residential sectors could reduce CO_2 emissions by 14%–16% and improve air quality and health level.

In the long run, the stricter the CO_2 emissions reduction policies were implemented, the greater the health benefits would obtained. Xie et al. (2018) used an integrated assessment framework to explore the co–benefits of air quality and health from climate mitigation in Asia. The results indicated a \$2.8 trillion reduction in health effects economic loss by 2050 under the 2℃ goal. He et al. (2010) combined multiple models on energy forecasting, emissions estimation, air quality and health effects evaluation. They evaluated comprehensively the CO_2 emissions reduction and air quality improvement synergistic effect of different energy policies in China. It displayed that 14.69 t of CO_2 emissions reduction, 12%–32% decline in atmospheric pollutant concentrations and \$100 billion health effects could be achieved by 2030. Liu et al. (2021) coupled multiple shared socioeconomic pathways, regional climate–air quality dynamic downscaling ensemble simulations and a health assessment model for atmospheric pollution and climate. The results demonstrated that clean air the synergistic benefits from the 1.5℃ abatement pathway was significantly higher than the direct climate (heatwave–related) benefits. Lara et al. (2022) integrated the assessment model World Induced Technical Change Hybrid. The model had two main types of measures: structural transformation of the energy system and end–of–pipe treatment. Health impacts were explored from the economy under different temperature control targets and atmospheric pollution control scenarios. The results indicated that the combination of climate policies and atmospheric pollution control policies could have more health effects. Both global and regional benefits were increased and there was no negative impact on global inequality. China's population aging would be worse in the future, which would greatly reduce the health risks $PM_{2.5}$–related in the elderly population.

2.3 Literature summary

The above researches confirmed the synergistic effect of CO_2 and $PM_{2.5}$ emissions and it could impact health. However, there were improvements in the existing researches: (1) Existing researches confirmed that CO_2 and $PM_{2.5}$ emissions had synergistic effect. Domestic re–searches were more concerned with one–way synergistic effect. In other words, the synergistic effect of a single policy was explored by combining some models such as the emission factor method, LEAP, CGE, co–control effects coordinate system, WRF–Chem and so on. These models needed a large number of economic, social and environmental basic data, so the models were relatively complex. LMDI method had advantages of a few and available data. (2) Many papers considered the synergistic emissions reduction in a region or an industry, but a few researches focused on the CC. As CC still dominated energy consumption in China, it was strategic importance to explore the synergistic effect of CO_2 and $PM_{2.5}$ emissions and the health effects of synergistic emissions reduction from CC in China. The LMDI was applied to calculate the driving forces of CO_2 emissions on $PM_{2.5}$ emissions change and quantified the PC. Based on the PC, the future $PM_{2.5}$ emissions were forecasted and the exposure-response model was adopted to account for the health effects of $PM_{2.5}$ emissions change under the carbon peaking and carbon neutrality goals.

3 Models and data sources

3.1 Models

LMDI was adopted to resolve the driving factors of $PM_{2.5}$ emissions generated by China's CC. The $PM_{2.5}$ emissions reduction effects of six factors were probed, which were *PC*, CO_2 emissions intensity from coal (*CI*), energy structure (*ES*), energy intensity (*EI*), GDP per capita (GDP*C*) and population (*P*). The contribution of six factors to the change in $PM_{2.5}$ emissions was explored. $PM_{2.5}$ emissions from CC in 2030, 2035, 2050, and 2060 were forecasted based on PC. The exposure-response model was utilized to calculate the health effects economic loss caused by $PM_{2.5}$ from 2012 to 2060.

3.1.1 LMDI

The LMDI was devised by Ang et al. (Ang, 2015) in 1998, it was based on the premise of satisfying Kaya identity (Pan and Dong, 2022). It can solve the zero value problem in the decomposition results and the decomposition results do not contain residuals. It has been widely applied in the researches related to energy and CO_2 emissions.

LMDI was adopted to decompose the factors influencing $PM_{2.5}$ emissions from CC during the period of 2012—2017. China put forward the low carbon trial pilot in 2011 and the air pollution prevention and control action plan in 2013. According to the policy effect lag and the data availability, the data from 2012 to 2017 were selected to analyze the change of $PM_{2.5}$ emissions. According to the mathematics of Kaya identity, the influencing factors were decomposed into six factors: *PC*, *CI*, *ES*, *EI*, GDP*C* and *P*. The $PM_{2.5}$ emissions from CC in China could be expressed in the following equations (1) - (7):

$$PM = \sum_{i=1}^{n} PM_i = \sum_{i=1}^{n}\left(\frac{PM_i}{C_i} \times \frac{C_i}{Coal_i} \times \frac{Coal_i}{EN_i} \times \frac{EN_i}{\mathrm{GDP}_i} \times \frac{\mathrm{GDP}_i}{P_i} \times P_i\right)$$

$$= \sum_{i=1}^{n} PC_i \times CI_i \times ES_i \times EI_i \times \mathrm{GDP}C_i \times P_i \qquad (1)$$

$$\Delta PC = \sum_{i=1}^{n} \frac{PM_i^t - PM_i^{t-1}}{\ln PM_i^t - \ln PM^{t-1}} \times \ln \frac{PC_i^t}{PC_i^{t-1}} \tag{2}$$

$$\Delta CI = \sum_{i=1}^{n} \frac{PM_i^t - PM_i^{t-1}}{\ln PM_i^t - \ln PM^{t-1}} \times \ln \frac{CI_i^t}{CI_i^{t-1}} \tag{3}$$

$$\Delta ES = \sum_{i=1}^{n} \frac{PM_i^t - PM_i^{t-1}}{\ln PM_i^t - \ln PM^{t-1}} \times \ln \frac{ES_i^t}{ES_i^{t-1}} \tag{4}$$

$$\Delta EI = \sum_{i=1}^{n} \frac{PM_i^t - PM_i^{t-1}}{\ln PM_i^t - \ln PM^{t-1}} \times \ln \frac{EI_i^t}{EI_i^{t-1}} \tag{5}$$

$$\Delta \mathrm{GDP}C = \sum_{i=1}^{n} \frac{PM_i^t - PM_i^{t-1}}{\ln PM_i^t - \ln PM^{t-1}} \times \ln \frac{\mathrm{GDP}C_i^t}{\mathrm{GDP}C_i^{t-1}} \tag{6}$$

$$\Delta P = \sum_{i=1}^{n} \frac{PM_i^t - PM_i^{t-1}}{\ln PM_i^t - \ln PM^{t-1}} \times \ln \frac{P_i^t}{P_i^{t-1}} \tag{7}$$

Where, i was the i-th province, n was 30, and t was the year. PM_i was $PM_{2.5}$ emissions of the i-th province. C_i was CO_2 emissions of the i-th province. $Coal_i$, EN_i, GDP_i and P_i were coal consumption, energy consumption, GDP and population of the i-th province, respectively. PC was the synergy of CO_2 and $PM_{2.5}$ emissions. CI, ES, EI, GDPC denoted CO_2 emissions intensity from coal, energy structure, energy intensity and GDP per capita. ΔPC was the change in the synergistic effect, which reflected the contribution of synergy to the change in $PM_{2.5}$ emissions from CC. The remaining five factors ΔCI, ΔES, ΔEI, ΔGDPC and ΔP were calculated in the same way as ΔPC.

3.1.2 Linear extrapolation method

The Dynamic Prediction model for Emissions in China (DPEC) is a database of future emission dynamics of major atmospheric constituents in China under different policy contexts developed by Tsinghua University (Tong et al., 2020; Cheng et al., 2021). The forecasted CO_2 emissions from DPEC, a linear extrapolation method was utilized to forecast $PM_{2.5}$ emissions in 2030, 2035, 2050, and 2060 based on PC in 2017. Previous researches found that $PM_{2.5}$ emissions from CC accounted for 63% of the total $PM_{2.5}$ emissions (China Coal Consumption Control Program and Policy Research Project Team, 2014). The $PM_{2.5}$ emissions from CC prediction were expressed in the following equation (8):

$$PM_i = 63\% \times PC_i \times C_i \tag{8}$$

Where, i is the i-th province. PM_i is $PM_{2.5}$ emissions of the i-th province. PC_i is the synergy of CO_2 and $PM_{2.5}$ emissions of the i-th province in 2017. C_i is CO_2 emissions of the i-th province. Based on the results by Dong et al. (2019) and Men et al. (2020), it was assumed that the height of $PM_{2.5}$ concentrations monitoring was the same with each province. $PM_{2.5}$ concentrations were related to the area of the different provinces. The $PM_{2.5}$ concentrations data were calculated from dividing the $PM_{2.5}$ emissions by the area in each province.

3.1.3 Exposure-response model

The exposure-response coefficient was the change of the health effect ends due to the unit change in atmospheric pollutant concentrations and it was important for quantitative evaluation of atmospheric pollution on the health effects. The epidemiological exposure - response model (Dong et al., 2021) was applied to quantitatively assess the health effects of $PM_{2.5}$ emissions reduction in 2012—2017 and 2030, 2035, 2050, and 2060 under the carbon neutrality target. It was expressed in the following equation (9) - (11):

$$R = R_0 \cdot e^{[\beta \times (C - C_0)]} \tag{9}$$

$$\Delta R = R - R_0 = R\{1 - \frac{1}{e^{[\beta \times (C - C_0)]}}\} \tag{10}$$

$$E = P_e \times \Delta R = PR\{1 - \frac{1}{e^{[\beta \times (C - C_0)]}}\} \tag{11}$$

Where, E was the change in the health effect of $PM_{2.5}$-related health effect ends. Pe was the number of the exposed population. R was the health effect at actual $PM_{2.5}$ concentrations. R_0 was the health risk under the reference concentration of $PM_{2.5}$. β was the exposure-response coefficient. C and C_0 were the actual and the threshold concentrations of $PM_{2.5}$, respectively. It was calculated that the number of illnesses, hospitalizations and deaths at $PM_{2.5}$ exposure concentrations. The baseline AQG of 5 $\mu g/m^3$ in WHO (WHO, 2021) was selected as the baseline value for evaluating the health damage value of $PM_{2.5}$. According to ICD - 10 International Classification of Diseases coding and domestic reported studies on the $PM_{2.5}$-related health effect, health effect ends were selected (Xie et al., 2009). The $PM_{2.5}$ exposure health effect ends were determined respiratory disease mortality, cardiovascular disease mortality, lung cancer mortality, respiratory disease hospitalization rates, cardiovascular disease hospitalization rates, chronic bronchitis morbidity and asthma morbidity.

3.1.4 Health effects economic loss

The human capital approach, the value of statistical life (VSL), the cost of illness approach and willingness to pay method were the methods for assessing the economic loss of $PM_{2.5}$-related health effects (Li et al., 2018). VSL (Huang and Zhang, 2013) was utilized to define the health effects economic loss from premature death for an individual. It was expressed in the following equation (12):

$$V_{ti} = VSL_{ti} \times E_{ti} \tag{12}$$

Where, V_{ti} was the health effects economic loss of the i-th province in year t. VSL_{ti} was the VSL value of the i-th province in year t and E_{ti} was the change in health effects of the i-th province in year t.

The cost of illness approach was chosen to calculate the cost of hospitalization for respiratory disease and cardiovascular disease. The cost of chronic bronchitis was taken to be 40% of VSL (Viscusi et al., 1991). The cost of asthma chose the results by Huang et al. (Huang and

Zhang, 2013). The cost of chronic bronchitis and asthma in different provinces were discounted according to the GDP per capita of each province.

3.2 Data sources

CO_2 emissions from CC in 30 Chinese provinces (excepted for Hong Kong, Macao, Taiwan and Xizang) during the period of 2012—2017 came from the Carbon Emission Accounts and Datasets for emerging economies (CEADs) (Guan et al., 2021; Shan et al., 2018, 2020). CO_2 emissions in 30 Chinese provinces in 2030, 2035, 2050, and 2060 under the carbon neutrality target were derived from DPEC. $PM_{2.5}$ emissions in 30 provinces during the period of 2012—2017 came from Multi-resolution Emission Inventory for China (MEIC) (Li et al., 2017; Zheng et al., 2018). According to the ratio of as 63% (Xue et al., 2016), $PM_{2.5}$ emissions from CC were calculated. The data on CC, total energy consumption, GDP, and population for 30 Chinese provinces came from the Energy Statistical Yearbook and the Statistical Yearbook in China and each province. The VSL estimate of 5.54 million yuan given by Jin et al. (2020) was chosen to calculate the health effects economic loss from death for individuals. The baseline incidence of health effect ends was obtained from the China Health Statistical Yearbook (National Health Commission of the People's Republic of China, 2012—2017). White provinces without data were not included in the analysis. The data on health effect ends and exposure-response coefficient were displayed in Table 1.

Table 1 The health effect ends and $PM_{2.5}$ exposure-response coefficient

Health effect ends		Group	Exposure-response coefficient (%)		Baseline incidence (‰)	Reference
			β	95% *CI*		
Mortality	Respiratory disease	all	1.43	(0.85, 2.01)	0.65	Xie et al. (2009)
	Cardiovascular disease	all	0.53	(0.15, 0.90)	1.54	Xie et al. (2009)
	Lung cancer	all	9.55	(2.67, 24.86)	0.49	Xie et al. (2019)
Hospitalization rate	Respiratory disease	all	1.09	(0, 2.21)	21.90	Liu et al. (2010)
	Cardiovascular disease	all	0.68	(0.43, 0.93)	8.55	Liu et al. (2010)
Prevalence	Chronic bronchitis	all	10.09	(3.66, 15.59)	6.94	Wang et al. (2017)
	Asthma	all	2.10	(1.45, 2.74)	11.90	Xie et al. (2009)

4 Results

4.1 LMDI decomposition results

4.1.1 The decomposition of $PM_{2.5}$ emissions from CC

The LMDI was used to explore the influencing factors of $PM_{2.5}$ emissions from CC in China during the period of 2012—2017, and the results were demonstrated in Fig. 1. $PM_{2.5}$ emissions from CC in China decreased year by year. From the contribution of six influencing factors to the change of $PM_{2.5}$ emissions, *PC*, *ES*, and *EI* were positive effects on $PM_{2.5}$ emissions reduction. *EI* from 2012 to 2013 had the largest contribution (67.8%) to promote $PM_{2.5}$ emissions reduction. *PC* reduction contributed about 27.1%. *ES* emissions reduction contributed 5.1%. The factor contribution was the largest that *PC* promoted $PM_{2.5}$ emissions reduction from 2013 to 2017, which was 41.3%-64.7%. The following factor contribution was *EI*, which was 23.1%-37.7%. The contribution of *ES* emissions reduction was the smallest, which was 11.9%-19.4%. It could be concluded that the main influencing factor to promote $PM_{2.5}$ emissions reduction was *PC*, followed by *EI*. The main drivers in increasing $PM_{2.5}$ emissions were GDPC and *P*. The contribution of GDPC increased in $PM_{2.5}$ emissions from 2012 to 2017, which was 58.4%-95.0%. The contribution of *P* increased in $PM_{2.5}$ emissions, which was 2.4%-7.2%. Economic growth could significantly increase the burden of atmospheric pollution. The contribution of *CI* from 2012 to 2017 was positive and negative.

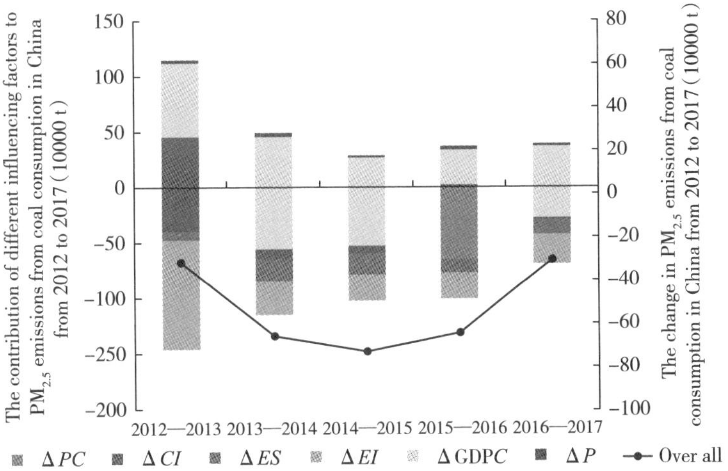

Fig. 1 The decomposition of factors influencing $PM_{2.5}$ emissions from CC in China during the period of 2012—2017

In summary, *PC*, *EI* and GDP*C* had significant effects on $PM_{2.5}$ emissions from CC. The effects of *CI*, *ES* and *P* on $PM_{2.5}$ emissions from CC were relatively weak. China's low-carbon policies and the clean air action plans were the main reasons for the significant reduction in $PM_{2.5}$ emissions during the period of 2012—2017.

4.1.2 The decomposition of $PM_{2.5}$ emissions from CC in 30 provinces

$PM_{2.5}$ emissions were highly localized. The difference of $PM_{2.5}$ emissions from CC varied widely among different provinces. ArcGIS was applied to visualize the change amount of $PM_{2.5}$ emissions from CC in 30 provinces from 2012 to 2017. And the contribution of $PM_{2.5}$ emissions from CC in each province was reclassified using Natural Breaks (Jenks) method. The decomposition results were demonstrated in Fig. 2❶, which reflectedthe decrease in the change in $PM_{2.5}$ emissions from CC in 30 provinces from 2012 to 2017. The darker bule provinces indicated the greater change in $PM_{2.5}$ emissions from CC in China. More economically developed provinces and resource-based provinces had the largest change in $PM_{2.5}$ emissions reduction. Hebei and Shandong were the two of largest changes in $PM_{2.5}$ emissions reductions, up to 268400 t and 241300 t. They were followed by Henan, Shanxi, Sichuan, Hubei, Jiangsu, Inner Mongolia and Liaoning, which had contributed positively to $PM_{2.5}$ emissions reduction from CC. Beijing-Tianjin-Hebei region and its surrounding areas were the key provinces in reducing $PM_{2.5}$ emissions. The measures such as the renovation of coal-fired boilers and power plants had been taken to reduce $PM_{2.5}$ emissions.

Fig. 3 expressed the decomposition results of the changes in $PM_{2.5}$ emissions from CC in 30 provinces from 2012 to 2017. The changes in $PM_{2.5}$ emissions from CC in the provincial spatial perspective varied greatly. The main influencing factors contributing to $PM_{2.5}$ emissions reduction from CC were *PC*, *ES* and *EI*. The contribution of each emissions reduction factor differed significantly among different provinces. The largest contribution of emissions reduction was *PC* in fourteen provinces, which were Shanxi, Hebei, Henan and so on. This indicated that CO_2 emissions reduction policies effectually promoted $PM_{2.5}$ reduction. The largest contribution of emissions reduction was the *EI* in seven provinces, which were Guizhou, Hubei, Sichuan and so on. It expressed that lower *EI* in each province could reduce $PM_{2.5}$ emissions, because each

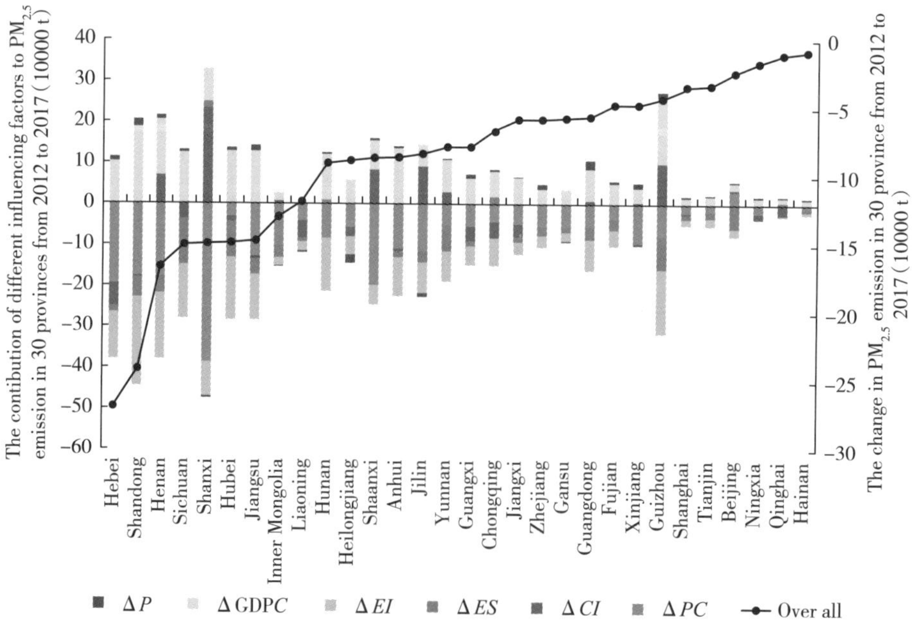

Fig. 3 The decomposition of factors influencing $PM_{2.5}$ emissions from CC in 30 provinces during the period of 2012—2017

❶ 图 2 见原文，刊于 *Journal of Environmental Management*，2023 年第 325 卷。原图发表时为彩色。

province had different CO_2 emissions reduction policies and energy consumption. The major influencing factors causing the increase in $PM_{2.5}$ emissions from CC were GDP*C* and *P*. GDP*C* had more contribution to the increase in emissions. The contribution of *CI* to the change in $PM_{2.5}$ emissions was positive and negative. It had positive effects on the increase of $PM_{2.5}$ emissions in six provinces, including Shanghai, Guizhou, Jilin, Shaanxi and so on. It demonstrated a weak contribution to $PM_{2.5}$ emissions reduction. It was mainly due to the regional complexity of $PM_{2.5}$ emissions.

4.2 The forecast of $PM_{2.5}$ emissions

The $PM_{2.5}$ emissions from CC in 30 provinces in 2030, 2035, 2050 and 2060 were predicted according to equation (8). In the DPEC data-base, the Ambitious-Pollution-neutral-goal scenario was a combination of energy economy transition and the best available end-of-pipe pollution control technologies (BHE). Under this scenario, the comprehensive policy would significantly reduce $PM_{2.5}$ emissions. In 2030 and 2035, the predicted value of the $PM_{2.5}$ emissions from CC in this paper were higher than the DPEC. Because China would be still in the early stage of energy structure transformation and the potential for synergistic effect would be not released yet. In 2050 and 2060 the predicted value of the $PM_{2.5}$ emissions from CC in this paper were smaller than DPEC. Because renewable energy would be dominated in China. The CO_2 recovery measures such as carbon capture technology and carbon sinks would also reduce $PM_{2.5}$ emissions.

Then the $PM_{2.5}$ concentrations from CC in 30 provinces were calculated in Fig. 4❶. The a-b plots displayed the distribution of $PM_{2.5}$ concentrations in 2030, 2035, 2050, and 2060. It could be seen that the $PM_{2.5}$ concentrations in 30 provinces would decrease year by year from 2030 to 2060. The dark blue areas indicated that the $PM_{2.5}$ concentrations were lower than the WHO AQG of 5 $\mu g/m^3$. The darker blue color demonstrated the lower $PM_{2.5}$ concentrations.

Fig. 4 expressed that $PM_{2.5}$ concentrations would be higher in central and eastern China in 2030, such as Beijing, Shanghai, Tianjin and Shandong. Fig. 5❶ expressed that $PM_{2.5}$ concentrations would remain higher in central and eastern China in 2035. $PM_{2.5}$ concentrations in southern China would improve significantly compared to 2030 and $PM_{2.5}$ concentrations in Jiangxi would decrease to below 5 $\mu g/m^3$. Fig. 6❶ demonstrated significant reduction in $PM_{2.5}$ emissions in central and eastern China in 2050. The $PM_{2.5}$ concentrations in Beijing, Shanghai and Tianjin would be 24.7 $\mu g/m^3$, 11.7 $\mu g/m^3$ and 10.6 $\mu g/m^3$, respectively. In addition, $PM_{2.5}$ concentrations from CC in 30 provinces would reduce to below 5 $\mu g/m^3$. As shown in Fig. 7❶, $PM_{2.5}$ concentrations from CC in 30 provinces would reduce to below 5 $\mu g/m^3$ by 2060. It was huge potential for the synergistic reduction of $PM_{2.5}$ emissions from CC under the carbon peaking and carbon neutrality goals.

❶ 图 4 至图 7 见原文，刊于 *Journal of Environmental Management*，2023 年第 325 卷。原文发表时为彩色。

4.3 Health effects economic loss of $PM_{2.5}$ emissions reduction

The number of deaths, hospitalizations and illnesses increased and then decreased from 2012 to 2060. The greatest number of deaths in 2013 was 207000 (95% CI: 73000-402000). The greatest number of hospitalizations in 2013 was 1057000 (95% CI: 132000-1975000). The greatest number of illnesses in 2013 was 3051000 (95% CI: 1479000-4241000).

Fig. 8 indicated the health effects economic loss of $PM_{2.5}$ emissions from CC in China from 2012 to 2060. The health effects economic loss of $PM_{2.5}$ emissions would be negative from 2050 to 2060. It did not mean that the "economic loss" caused by $PM_{2.5}$ pollution could turn into "economic benefit". It meant that $PM_{2.5}$ concentrations were below the standard (5 μg/m^3).

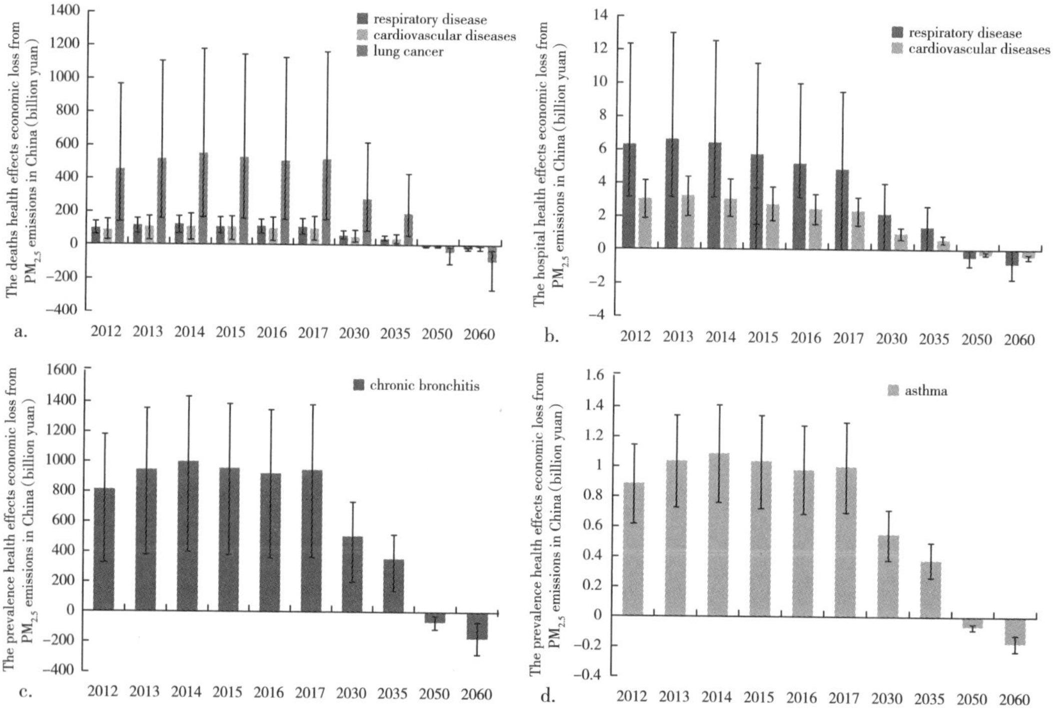

Fig. 8 The different health effects economic loss of $PM_{2.5}$ emissions from CC in China from 2012 to 2060

Fig. 8a demonstrated the health effects economic loss of deaths from respiratory diseases, cardiovascular diseases, and lung cancer. Among them, the $PM_{2.5}$-related economic loss of lung cancer deaths was the largest, up to 550.5 billion yuan (95% CI: 169.4-11750 billion yuan) in 2014. Fig. 8b indicated the $PM_{2.5}$-related health effects economic loss from hospitalizations of respiratory disease and cardiovascular disease. The larger hospitalization health effects economic loss was respiratory disease, up to 6.7 billion yuan (95% CI: 0-13 billion yuan) in 2013. Fig. 8c and Figured expressed the $PM_{2.5}$-related health effects economic loss of chronic bronchitis and asthma. Among the all $PM_{2.5}$-related health effect ends, the health effects economic loss of chronic bronchitis was 998.2 billion yuan (95% CI: 395.9-

143. 31 billion yuan) in 2014, which was the biggest. The health effects economic loss of asthma was the smallest economic loss in all health effect ends, which was 1. 1 billion yuan (95% CI: 8–1. 4 billion yuan) in 2014.

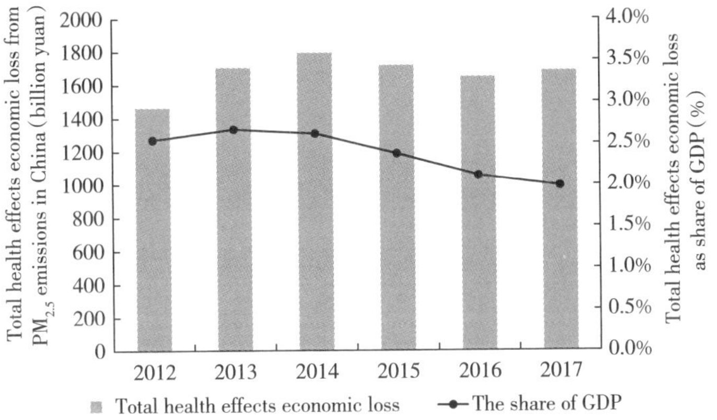

Fig. 9 Total health effects economic loss of $PM_{2.5}$ emissions from CC and the share of GDP in China from 2012 to 2017

Fig. 9 indicated total health effects economic loss of $PM_{2.5}$ emissions from CC in China from 2012 to 2017. The most total health effects economic loss was 179. 19 billion yuan in 2014, equivalent to 2. 6% of GDP in the year. Total health effects economic loss was 170. 2 billion yuan in 2013, which was 2. 7% of the maximum share of GDP. Health effects economic loss as a percentage of GDP had been decreasing since 2013. The proportion of health effects economic loss was 2. 0% in 2017. Huang et al. (2013) estimated that the health effects economic loss of $PM_{2.5}$ emissions in the Beijing–Tianjin–Hebei region accounted for 4. 7% of the regional GDP in 2009. Li et al. (2018) estimated that the health effects economic loss of $PM_{2.5}$ emissions accounted for 0. 52%–2. 35% of GDP in China's cities in 2015. Maybe a series of policies were enacted after 2013 to reduce emissions of $PM_{2.5}$ and CO_2. The improvement in energy efficiency and the change in energy structure could reduce $PM_{2.5}$ emissions.

5 Conclusion and recommendations

5. 1 Conclusion

China is facing the dual pressure of GHG reduction and atmospheric pollution prevention and control. The synergistic emissions reduction of CO_2 and $PM_{2.5}$ is of strategic importance. LMDI was used to resolve the influencing factors of $PM_{2.5}$ emissions from CC in China, which were *PC*, *CI*, *ES*, *EI*, GDP*C* and *P*. The $PM_{2.5}$ emissions from CC in 2030, 2035, 2050 and 2060 were predicted based on *PC*. The exposure–response model was applied to calculate the health effects of $PM_{2.5}$ emissions from 2012 to 2060. VSL was utilized to calculate the health effects economic loss.

The main findings were as follows: (1) $PM_{2.5}$ emissions from CC in China decreased year by year from 2012 to 2017. The largest contribution of $PM_{2.5}$ emissions reduction was *PC*. The largest contribution of increasing $PM_{2.5}$ emissions was GDP*C*. (2) $PM_{2.5}$ concentrations from CC would be higher in central and eastern China compared to other regions in 2030 and 2035. $PM_{2.5}$ concentrations from CC would be significantly lower by 2050. $PM_{2.5}$ concentrations from CC in 30 provinces by 2060 would be below the WHO AQG. It was huge potential for $PM_{2.5}$ emissions reduction from the synergistic effect under the carbon peaking and carbon neutrality goals. (3) The $PM_{2.5}$-related health effects were greatest in 2013. The total health effects economic loss in 2014 was the most. The health effects economic loss was negative by 2050. It indicated that $PM_{2.5}$ concentrations from CC were below 5 $\mu g/m^3$ in China. The future would continuously reduce health effects economic loss of $PM_{2.5}$ emissions. Chronic bronchitis prevalence had the most economic loss in all health effect ends. Asthma had the least economic loss in all health effect ends.

There were gaps in the contribution rates of CC to $PM_{2.5}$ concentrations among different provinces. Taking the same value for each province would bring some uncertainties. The change of energy structure in the future would change the contribution rates of CC to $PM_{2.5}$, which would have an impact on the synergistic effect of CO_2 and $PM_{2.5}$ emissions. In the future, it was necessary to strengthen the analysis of the contribution rates of CC to $PM_{2.5}$ concentrations among different provinces and accurately predict the impact of future energy structure on $PM_{2.5}$ emissions.

5.2 Recommendations

According to the above conclusions, we could get the following recommendations: (1) The decoupling of economic growth from atmospheric pollution should be further accelerated and a sound system of synergistic management of CO_2 and $PM_{2.5}$ emissions should be established. It was necessary that CO_2 emissions reduction measures needed to combine with $PM_{2.5}$ pollution management policies for comprehensive and systematic management. We should find the balance of economic growth and air pollution control. (2) According to the 4. 1. 1, energy efficiency would be improved and energy structure comprehensive would be optimized at the national level. We should develop clean coal technology and improve the energy conversion rate. Promoting the electrification and replacing coal power with renewable energy generation would speed up the adjustment of the energy structure. For example, industrial boilers and residential heating were converted from coal to electricity and from coal to gas. It released the potential for $PM_{2.5}$ emissions reduction from CC. (3) Research on the health effects of $PM_{2.5}$ emissions should be increased to supplement the basic epidemiological data about $PM_{2.5}$ in China. Effective $PM_{2.5}$ emissions control and emergency health protection measures should betaken to reduce health effects economic loss. We should improve the medical insurance system to reduce health risks for the elderly, which could actively respond to the population aging.

Acknowled9gements Thanks to the support from Beijing Social Science Fund Project (No. 19YJC023) and the Fundamental Research Funds for the Central Universities (No. 2-9-2019-084).

References

Ang, B. W., 2015. LMDI decomposition approach: a guide for implementation. Energy Pol. 86, 233-238.

Bell, M. L., Davis, D. L., Cifuentes, L. A., et al., 2008. Ancillary human health benefits of improved air quality resulting from climate change mitigation. Environ. Health 7, 1-18.

Bollen, J., van der Zwaan, B., Brink, C., et al., 2009. Local air pollution and global climate change: a combined cost-benefit analysis. Resour. Energy Econ. 31 (3), 161-181.

Bollen, J., Guay, B., Jamet, S., et al., 2009a. Co-benefits of Climate Change Mitigation Policies: Literature Review and New Results.

Cai, W. J., Zhang, C., Zhang, S. H., et al., 2021. The 2021 China report of the Lancet Countdown on health and climate change: seizing the window of opportunity. Lancet 6 (12), e932-e947.

Central People's Government of the People's Republic of China, 2018. Notice on the Issuance of the Three-Year Action Plan to Win the Blue Sky Defense [N/OL. -. http://www. gov. cn/zhengce/content/2018-07/03/content_ 5303158. htm.

Central People's Government of the People's Republic of China, 2021a. Working guidance for carbon dioxide peaking and carbon neutrality in full and faithful implementation of the new development philosophy. http://www. gov. cn/zhengce/2021-10/24/content_ 5644613. htm.

Central People's Government of the People's Republic of China, 2021b. A circular on further promoting the nationwide battle to prevent and control pollution. http://www. gov. cn/zhengce/2021-11/07/content_ 5649656. htm.

Cheng, J., Tong, D., Zhang, Q., et al., 2021. Pathways of China's $PM_{2.5}$ air quality 2015—2060 in the context of carbon neutrality. Natl. Sci. Rev. 8 (12) nwab078.

China Coal Consumption Control Program and Policy Research Project Team, 2014. Contribution of Coal Use to Air Pollution in China.

Dong, F., Yu, B., Pan, Y., 2019. Examining the synergistic effect of CO_2 emissions on $PM_{2.5}$ emissions reduction: evidence from China. J. Clean. Prod. 223, 759-771.

Dong, Z., Li, L., Lei, Y. L., et al., 2021. The economic loss of public health from $PM_{2.5}$ pollution in the Fenwei Plain. Environ. Sci. Pollut. Control Ser. 28 (2), 2415-2425.

Dong, F., Zhu, J., Li, Y., et al., 2022a. How green technology innovation affects carbon emission efficiency: evidence from developed countries proposing carbon neutrality targets. Environ. Sci. Pollut. Control Ser. 29 (24), 35780-35799.

Dong, F., Li, Y., Gao, Y., et al., 2022b. Energy transition and carbon neutrality: exploring the non-linear impact of renewable energy development on carbon emission efficiency in developed countries. Resources, Conservation and Recycling 177, 106002.

Ebi, K. L., Ogden, N. H., Semenza, J. C., et al., 2017. Detecting and attributing health burdens to climate change. Environ. Health Perspect. 125 (8), 085004.

Gao, Q. X., Gao, W. O., Ma, Z. Y., et al., 2021. The synergy effect assessment method and its application for air pollutants and greenhouse gases reduction. Climate Change Research 17 (3), 268-278.

Gschwind, B., Lefevre, M., Blanc, I., et al., 2015. Including the temporal change in $PM_{2.5}$ concentration in the assessment of human health impact: illustration with renewable energy scenarios to 2050. Environ. Impact Assess. Rev. 52 (2), 62-68.

Guan, Y. R., Shan, Y. L., Huang, Q., et al., 2021. Assessment to China's recent emission pattern shifts. Earth's Future 9, e2021EF002241.

He, K., Lei, Y., Pan, X., et al., 2010. Co-benefits from energy policies in China. Energy 35, 4265-4272.

Huang, D. S., Zhang, S. Q., 2013. Health benefit evaluation for $PM_{2.5}$ pollution control in Beijing-Tianjin-Hebei region of China. China Environ. Sci. 33(1), 166-174.

Jia, L. Y., Wang, Y. H., Wang, K., et al., 2020. Evaluation of carbon dioxide coordination emission reduction based on national air pollution control plan. Environ. Prot. Sci. 46 (6), 19-26+43.

Jin, Y., Andersson, H., Zhang, S., 2020. Do preferences to reduce health risks related to air pollution depend on illness type? Evidence from a choice experiment in Beijing, China. J. Environ. Econ. Manag. 103, 102355.

Lara, A. R., Laurent, D., Prof, M. T., 2022. Internalising health-economic impacts of air pollution into climate policy: a global modelling study. Lancet Planet. Health 6 (1), e40-e48.

Li, M., Liu, H., Geng, G. N., et al., 2017. Anthropogenic emission inventories in China: a review. Natl. Sci. Rev. 4, 834-866.

Li, Y. Y., Song, Y. D., 2021. Study on the synergetic emission reduction effect of CO_2 and air pollutants from the mobile source of urban roads in Beijing under the target of carbon neutralization [J]. Chinese Journal of Environmental Management 13 (3), 113-120.

Li, H. J., Zhou, D. Q., Wei, Y. J., 2018. An assessment of $PM_{2.5}$-related health risks and associated economic losses in Chinese cities. Environ. Sci. J. Integr. Environ. Res. 39 (8), 3467-3475.

Liu, X. Y., Xie, P., Liu, Z. R., et al., 2010. Economic assessment of acute health impact due to inhalable particulate air pollution in the Pearl River Delta. Acta Sci. Nauralium Univ. Pekin. 46(5), 829-834.

Liu, S., Xing, J., Wang, S., et al., 2021. Health benefits of emission reduction under 1.5℃ pathways far outweigh climate-related variations in China. Environ. Sci. Technol. 55, 10957-10966.

Liu, H., Liu, J., Li, M., et al., 2022. Assessing the evolution of $PM_{2.5}$ and related health impacts resulting from air quality policies in China. Environ. Impact Assess. Rev. 93, 106727.

Lu, P. T., Han, Y. L., Dai, H. C., 2021. Co-Benefits of decarbonizing China's transport sector in energy saving and emission reduction under 1.5-and 2-degree targets in 2050. Acta Sci. Nauralium Univ. Pekin. 57 (3), 517-528.

Men, D., Huang, X., Yi, X., et al., 2020. Study on driving effects and spatial characteristics of haze pollution in the Yangtze River Economic Zone. Environ. Sci. Technol. 43 (3), 10-20.

Ministry of Ecology and Environment of the People's Republic of China, 2020. China ecological environment status bulletin. https://www.mee.gov.cn/hjzl/sthjzk/zgh jzkgb/202105/P020210526572756184785.pdf.

National Bureau of Statistics of China, 2020. China Energy Statistics Yearbook. China Statistics Press, Beijing.

National Health Commission of the People's Republic of China, 2012—2017. China Health Care Statistics Yearbook. China Union Medical University Press, Beijing.

Pan, Y., Dong, F., 2022. Dynamic evolution and driving factors of new energy development: fresh evidence from China. Technol. Forecast. Soc. Change 176, 121475.

Peng, W., Yang, J., Lu, X., et al., 2018. Potential co-benefits of electrification for air quality, health, and CO_2 mitigation in 2030 China. Appl. Energy 218, 511-519.

Plachinski, S. D., Holloway, T., Meier, P. J., et al., 2014. Quantifying the emissions and air quality co-benefits of lower-carbon electricity production. Atmos. Environ. 94, 180-191.

Shan, Y. L., Guan, D. B., Zheng, H. R., et al., 2018. China CO_2 emission accounts 1997-2015. Sci. Data 5, 170201.

Shan, Y. L., Huang, Q., Guan, D. B., et al., 2020. China CO_2 emission accounts 2016—2017. Sci.

Data 7, 54.

Shao, S., Li, X., Cao, J., et al., 2016. China's economic policy choices for governing smog pollution based on spatial spillover effects. Econ. Res. J. 51 (9), 73–88.

Tang, R., Zhao, J., Liu, Y., et al., 2022. Air quality and health co-benefits of China's carbon dioxide emissions peaking before 2030. Nat. Commun. 13, 1–9.

Tong, D., Cheng, J., Liu, Y., et al., 2020. Dynamic projection of anthropogenic emissions in China: methodology and 2015—2050 emission pathways under a range of socioeconomic, climate policy, and pollution control scenarios. Atmos. Chem. Phys. 20 (9), 5729–5757.

Vennemo, H., Aunan, K., He, J., et al., 2009. Benefits and costs to China of three different climate treaties. Resour. Energy Econ. 31 (3), 139–160.

Viscusi, W. K., Magat, W. A., Huber, J., 1991. Pricing environmental health risks: survey assessments of risk-risk and riskdollar trade-offs for chronic bronchitis. J. Environ. Econ. Manag. 21(1), 32–51.

Wang, G. Z., Wu, L. Y., Chen, J. B., et al., 2017. A CGE-based analysis on $PM_{2.5}$-induced health-related economic effect in Beijing. China Environ. Sci. 37(7), 2779–2785.

Wang, P., Lin, C. K., Wang, Y., et al., 2021. Location-specific co-benefits of carbon emissions reduction from coal-fired power plants in China. Nat. Commun. 12, 1–11.

WHO, 2021. Global Air Quality Guidelines 2021. World Health Organization, Geneva.

Xie, P., Liu, X. Y., Liu, Z. R., et al., 2009. Exposure-response functions for health effects of ambient particulate matter pollution applicable for China. China Environ. Sci. 10, 1034–1040.

Xie, Y., Dai, H. C., Xu, X. H., et al., 2018. Co-benefits of climate mitigation on air quality and human health in Asian countries. Environ. Int. 119, 309–318.

Xie, Z. X., Qin, Y. C., Zheng, Z. C., et al., 2019. Assessing the death effect of $PM_{2.5}$pollution in cities of atmospheric pollution transmission channel in the Beijing-Tianjin-Hebei district. Acta Sci. Circumstantiae (3), 843–852.

Xing, Y. K., Mao, X. Q., Feng, X. Z., et al., 2020. An effectiveness evaluation of co-controlling local air pollutants and GHGs by implementing blue sky defense action at city level —a case study of Tangshan city. Chinese Journal of Environmental Management 12 (4), 20–28.

Xinhua News Agency, 2020. Statement by Xi Jinping President of the People's Republic of China at the General Debate of the 75th Session of the United Nations General Assembly. http://www.xinhuanet.com//mrdx/2020-09/23/c_139389875.htm.

Xue, W. B., Wu, W. L., Fu, F., et al., 2016. A Study on the Impact of Coal Consumption on $PM_{2.5}$ Pollution in China Environmental Management.

Yang, X., Teng, F., Wang, X., et al., 2017. System optimization and co-benefit analysis of China's deep decarbonization effort towards its INDC target. Energy Proc. 105, 3314–3319.

Yin, P., Brauer, M., Cohen, A. J., et al., 2020. The effect of air pollution on deaths, disease burden, and life expectancy across China and its provinces, 1990–2017: an analysis for the Global Burden of Disease Study 2017. Lancet Planet. Health 4 (9), e386–e398.

Yue, Q., Song, Y., Zhu, J., et al., 2021. Exploring the effect of air pollution on settlement intentions from migrants: evidence from China. Environ. Impact Assess. Rev. 91, 106671.

Zhang, S., An, K., Li, J., et al., 2021. Incorporating health co-benefits into technology pathways to achieve China's 2060 carbon neutrality goal: a modelling study. Lancet Planet. Health 5(11), e808–e817.

Zheng, B., Tong, D., Li, M., et al., 2018. Trends in China's anthropogenic emissions since 2010 as the consequence of clean air actions. Atmos. Chem. Phys. Discuss. 14095–14111.

Synergistic emission reductions and health effects of energy transitions under carbon neutrality target

WANG Zengchuan[1,2], LI Li[1,2,*], LEI Yalin[1,2], WU Sanmang[1,2]
CUI Yanfang[1,2], DONG Ziyu[1,2], JIA Wanlin[1,2], REN Chenxu[1,2]

1. *School of Economics and Management, China University of Geosciences, Beijing* 100083, *China*
2. *Key Laboratory of Carrying Capacity Assessment for Resource and Environment, Ministry of Natural Resources of the People's Republic of China, Beijing* 100083, *China*

Abstract China's climate and environmental problems remained prominent. Energy transition was the key to improve climate and pollution, and it was at a critical period. Previous studies mostly focused on the impact of climate policies on the reduction of air pollutants and the environmental health impacts of individual sector or pathway policies. Thus, focusing on the synergistic abatement of CO_2 and air pollutants and health effects of energy transition in carbon neutrality was significant for building beautiful and healthy China. LEAP-HEA model was constructed to project energy demand, CO_2 and air pollutants, and assess energy transition's environmental health effects from 2020 to 2060. LMDI model was adopted to decompose the abatement contribution. Synergistic assessment index was used to estimate the coordinated abatement effects. Cost-effectiveness of energy transition and regional heterogeneity of $PM_{2.5}$ toxicity were not considered. The results indicated: (1) energy demand would decrease and structure would be optimized significantly under deep energy transition scenario. (2) Deep energy transition could promote earlier realization of carbon neutrality. It would contribute significantly to CO_2 and air pollutants abatement with strong synergistic effects. (3) Energy transition would reduce health risks with large economic benefits. Recommendations were proposed based on the conclusions.

Keywords Synergistic emission reduction, Health effects, Energy transition, Carbon neutrality

1 Introduction

At present, global warming had aroused widespread concern among countries. As a world power and a major carbon emitter, China had actively responded to the issue of climate change, and proposed a "dual-carbon" goal in 2020. Since then, China had introduced a series of carbon emission reduction programs for key sectors, and completed the construction of "1+N" policy system (NDRC, 2023b). "1+N" policy system included a top-level design for

---

原文刊于 *Journal of Cleaner Production*，2024 年第 456 卷。原文发表时为彩色。
* 为通信作者。

achieving dual-carbon goals and an implementation plan for key areas or sectors. It contained both top-level design and foundation programs to promote dual-carbon goals. Energy was an important driving force for economic and social development, and it also was a major field to foster carbon neutrality. China's energy transition stood at an important window of time. It was urgent to promote energy transformation to meet the needs of the ecological environment and economic development (Central People's Government of the PRC, 2022b). Ministry of Ecology and Environment (MEE) published the "Implementation Program for Pollution Reduction, Carbon Reduction and Co-Efficiency" in 2022. It pointed out the characteristic that CO_2 and atmospheric pollutants had the same root and origin, and it was necessary to promote the reduction of pollutants and carbon emissions. The ability of synergistic management would be improved significantly in 2030, so as to achieve a win-win situation for environment, health and economy (MEE, 2022). State Council Office of the Central People's Government of the PRC (2022) published the "14th Five-Year Plan for National Health". It stated that government was actively promoting the construction of Healthy China. The "Healthy China 2030" plan was being implemented. Therefore, taking carbon neutrality as background was more in line with practical needs to research the environmental and health impacts of energy transition, which could supplement the basic theory of energy transition policymaking under carbon neutrality.

Since the 18th National Congress, China had accelerated its energy transition with remarkable results. Energy consumption per unit of GDP had dropped by 26.4% in 2022, but it still faced a severe situation (Central People's Government of the PRC, 2022a). China's energy consumption was dominated by non-fossil energy. Chinese energy demand reached 5.41 billion tce in 2022, with fossil energy accounting for 82.5% (NDRC, 2023a). According to the statistics, the average concentration of $PM_{2.5}$ in cities at prefecture level and above was 29 $\mu g/m^3$ in 2022. And the average concentration of SO_2 and NO_2 had been stabilized for three consecutive years. However, the task of ecological environmental protection was still arduous, and the construction of ecological civilization was still at a critical stage. Ecological health development required further consolidation of the foundations (MEE, 2023). In 1995, the IPCC proposed a synergistic reduction of atmospheric pollutants along with the reduction of greenhouse gas (IPCC, 1995). Former research had also fully confirmed that CO_2 and atmospheric pollutants had the same root and origin and they could be reduced in a synergistic manner. While energy fueled economic development, it also brought air quality problems in terms of emissions of air pollutants such as $PM_{2.5}$, SO_2 and NO_x. It aggravated health risks such as respiratory and cardiovascular diseases (Giorgini et al., 2016). China's $PM_{2.5}$ pollution caused 281000 premature deaths, 216000 COPD hospitalizations, 122000 cases of IHD hospitalization in 2017. The health economic loss amounted to 8.639 billion yuan, accounting for 1.48% of the 2017 GDP (Zeng and Ruan, 2020). Therefore, assessing the health effects and economic benefits of energy transition was of practical significance for building healthy China and making energy policy.

Previous studies mostly focused on the impact of climate policies on the reduction of air pollutants, but fewer researches analyzed the synergistic emission reduction contribution of the energy transition in the carbon neutral process. Researches on the environmental health effects of policies focused more on individual sector or pathway, but fewer researches comprehensively assessed the health effects of energy transition. Based on the previous studies, LEAP-HEA model that met China's situation was constructed by clarifying the internal logic of energy-environment-health. And the research gap on the contribution of energy transition to the synergistic emission reduction and the health effects in carbon neutrality processes was filled. It could provide theoretical basis for carbon neutral pathway and energy transition policy's formulation, which had practical significance for beautiful and healthy China.

2 Literature review

2.1 Research progress on synergistic emission reduction of CO_2 and air pollutants

The IPCC's Fifth Assessment Report suggested that carbon reduction pathways had positive synergies, and in 2018, it noted that synergies were the positive impacts of one policy on others (IPCC, 2015; IPCC, 2018). More scholars' studies proved the existence of synergistic abatement effects between CO_2 and atmospheric pollutants. These scholars had mainly explored the impact of climate policies on synergistic emission reductions. Rypdal et al. (2007) found that climate policies such as carbon taxes indirectly emit air pollutants by using CGE. Mittal et al. (2015) projected future transportation demand in India and analyzed the impact of low carbon policies using AIM model, suggesting that the combined use of global and local low carbon policies had a strong synergistic effect on air pollutant control. Dong et al. (2019) analyzed the influence of CO_2 abatement activities on $PM_{2.5}$ abatement and found that synergistic abatement had the greatest effect on $PM_{2.5}$ abatement. Mir et al. (2022) coupled GAINS and EnerNEO models to explore the impact of climate policies on air quality in Pakistan. The results showed that the implementation of sustainable climate policies resulted in both a half reduction in CO_2 and $PM_{2.5}$ emission reductions in 2050. Limaye et al. (2023) discovered that city-level climate policies promoted synergistic control of air quality by integrating data modeling in India. Yawale et al. (2023) analyzed the effects of the policy in various states of India by setting up multiple scenarios. It was found that power transition not only helped in achieving carbon neutrality but also led to significant reduction in SO_2, $PM_{2.5}$ and NO_x emissions.

Meanwhile, some scholars also focused on the impact of air quality policies or synergistic effect assessment. Song and Fu (2012) discovered that CO_2 abatement could be synergized with SO_2 abatement in major countries around the world. Dong et al. (2015) applied AIM and GAINS models in investigating the synergistic effects of CO_2 and atmospheric pollutants in the future of China, which revealed strong synergistic benefits and significant inter-provincial differences. Bhanarkar et al. (2018) conducted a comprehensive assessment of air quality and climate policies in Indian cities, and concluded that the two policies had synergistic effect on CO_2 and atmospheric pollutant reductions. Li and Song (2021) compared the synergistic abatement effects of CO_2 and atmospheric pollutants from mobile sources on urban roads in Beijing under different policy scenarios. They concluded that integrated policies had the strongest synergistic mitigation effects. Yu et al. (2023) estimated the synergistic control effects of CO_2 and atmospheric pollutants in Beijing. They suggested that the synergistic control results of cleaner production and control of production capacity in traditional industries were the best. Yang et al. (2023) quantified the synergistic mitigation potential of transportation in Tangshan based on the LEAP model. It was found that oil-to-gas conversion measures in transportation sector significantly improved the synergies. Xing et al. (2024) deemed that alternative power generation, such as nuclear and wind, had higher synergistic abatement effects than coal power plant optimization. It was concluded that former studies mainly focused on the synergistic abatement effect of climate policies on $PM_{2.5}$, but fewer researches focused on the synergistic abatement effect in the field of energy transition.

2.2 Research progress on health effects of air pollutants

There had been many epidemiological studies confirming the health risks associated with $PM_{2.5}$ exposure. Ostro et al. (2007) explored the relationship of $PM_{2.5}$ and mortality by using random effects model. The results showed that there were strong associations between $PM_{2.5}$ con-centrations and all-cause, cardiovascular, respiratory mortality. Giorgini et al. (2016) summarized previous studies on $PM_{2.5}$ and hypertension. It was found that prolonged exposure to $PM_{2.5}$ would raise the risk of emergency room visits and hospitalizations for hypertensive disorders. Han et al. (2020) concluded that the implementation of air pollution control programs was able to reduce $PM_{2.5}$ concentrations, and it also effectively reduced the risk of death. Wang et al. (2023) revealed the causal relationship of $PM_{2.5}$ exposure and all-cause mortality from a study of population-based big data in South China. Thc rcsults demonstrated that prolonged exposure to $PM_{2.5}$ resulted in higher risk of mortality. Wagner et al. (2023) identified a super-linear relationship between $PM_{2.5}$ and hospitalization and mortality rates through a comparative study. It also discovered a stage-specific effect of $PM_{2.5}$ concentration on premature mortality. Cheng et al. (2023) thought air pollution policies and carbon neutral policies together could reduce premature deaths by 2.62 million per year. Gatto and Chepeliev (2024) argued that reduction of global food waste could reduce air pollution and lower the risk of premature death.

Due to rapid economic development and the increasing prominence of climate issues, environmental health had become a major concern of society. Scholars had conducted a series of studies on the assessment of environmental health effects around different industries. Markandya et al. (2009) analyzed and found that the health benefits of electricity production transition were greater in India than in China and EU, with the largest amount of premature mortality averted. Wang et al. (2017) estimated the public health effects and economic losses of $PM_{2.5}$ in Beijing using CGE model. It was found $PM_{2.5}$ caused 22000 premature deaths, more than 1 million illnesses, and a total economic loss of 9.0 billion yuan in 2013. Lanzi et al. (2018) concluded that the economic cost of global outdoor air pollution in 2060 would be 1% of GDP. Yang et al. (2023) adopted REACH model to assess the environmental health effects of energy transition in China's BTHS regions. This study identified energy transition could reduce $PM_{2.5}$ concentrations by 6.3 $\mu g/m^3$ by 2035 and generate environmental health benefits which could offset implementation costs. Salimifard et al. (2023) employed CoBE tool to estimate the environmental health co-benefits of building energy consumption. They realized that for every unit of electricity saved, a health benefit of \$0.02-\$0.08 was generated. Hao et al. (2023) evaluated the health effects and economic losses of motor vehicles in China's BTH regions by coupling exposure-response model, cost-of-illness method and willingness-to-pay method. The study showed that $PM_{2.5}$ pollution from motor vehicles caused about 34000 premature deaths, 46000 hospitalizations and 439000 illnesses in 2020, with an economic loss of 129.97 billion yuan. It was revealed that researched on the environmental health effects of policies had focused more on individual sector or pathway, but fewer researches comprehensively assessed the health effects of the energy transition.

2.3 Literature summary

By analyzing literature, it was found that many scholars had conducted studies around the synergistic abatement effects of CO_2 and atmospheric pollutants and environmental health effects, but there were still inadequacies. (1) A number of researches had confirmed the synergistic effect of CO_2 and atmospheric pollutants emission reduction. Most of the previous studies had focused on the influence of climate policies on the reduction of air pollutants. Some scholars had also evaluated the synergistic effects of abatement using models such as CGE, GAINS and LEAP. However, the synergistic emission reduction contribution of energy transition under carbon neutrality had not been addressed. There was also a lack of synergistic abatement effects estimation research on the domestic energy transition as a whole. (2) Based on abundant epidemiological studies, it had been proved that $PM_{2.5}$ was hazardous to human health and caused a variety of health risks related to the respiratory system and cardiovascular system. Yet, there was still insufficient research data on the health of domestic populations. Many studies had been conducted to analyze the environmental health effects of various policies by constructing assessment models. But most researches focused on individual sector or individual pathway. Fewer researches had yet revealed the impact of energy transitions on health risks under carbon neutrality.

The scenarios were set based on China's economic, energy and other development trends and carbon reduction measures. LEAP-HEA model was constructed to predict future energy demand, CO_2 and air pollutants emissions. LMDI model was combined to decompose the emission reduction contribution of energy transition and assessed the synergistic abatement and health effects of energy transition. This study aimed to clarify the intrinsic connection between energy transition – environment – health. For China, a country at critical stage of energy transition, the conclusions could provide a reference for the formulation of policies related to energy transition, pollution and carbon reduction, and environmental health.

3 Models and data

3.1 LEAP-HEA model

3.1.1 Modeling framework

LEAP model was a software tool for bottom-up modeling and widely used for energy policy or climate change assessment. The disadvantage of LEAP model was that multiple industry details and data were involved. But it had a high degree of flexibility in its construction, which was suitable for scenario prediction with incomplete data. The combination of scenario analysis could show the results of different paths to achieve carbon neutrality. Moreover, its framework was stable and its layers could be adjusted according to research content. LEAP-HEA model was constructed sorting out the internal logic of energy – environment – health and combining the development characteristics of China. It integrated LEAP model, exposure-response, cost-of-illness method and VSL shown in Fig. 1. LEAP model was used to simulate and predict the energy demand, CO_2 and air pollutants emissions under future changes in technology mix and policy impacts. Finally, the HEA panel was used to assess the health effects due to $PM_{2.5}$ emission reductions from energy transition.

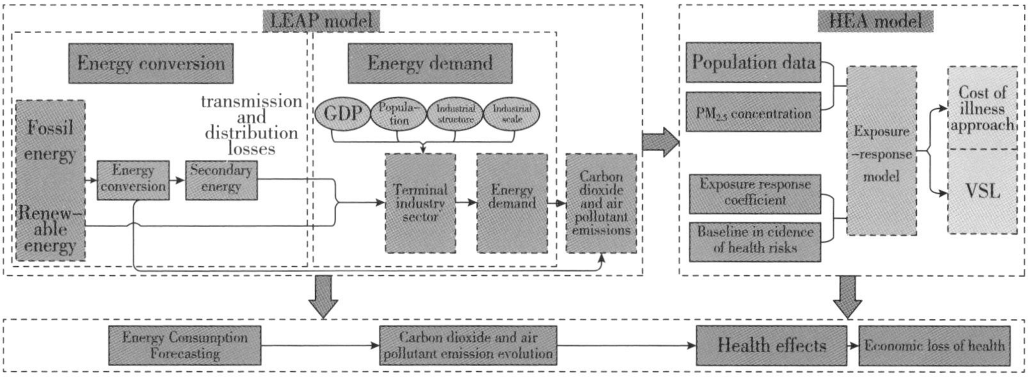

Fig. 1 The framework of LEAP-HEA model

3.1.2 Model formula

3.1.2.1 *LEAP model.* LEAP model could be applied to calculate energy consumption for different energy activities in different sectors. Its basic idea was to calculate the energy consumption at different levels of production and living through energy consumption per unit of production and living. Calculation method as shown in (1) and (2). CO_2 and pollutants emissions could be calculated by adding carbon and pollutants emission factors for different energy activities. The calculation was shown in (3).

$$ED_m = \sum_i (A_i \times EI_{i,m}) \tag{1}$$

$$E_m = ED_m + EL_m + ET_m \tag{2}$$

$$E_c = \sum_{m=1}^{n} (ED_m \times \delta_{mc}) \tag{3}$$

Where, ED_m denoted the activity consumption of energy categorym. A_i was the level of activity in category i. EI_i represented the energy intensity per unit of activity level. EL_m indicated the energy transmission and distribution losses. ET_m showed the processing and conversion consumption. E_m described total energy consumption. E_c expressed CO_2 or air pollutants emissions. δ_{mc} was the emission factor for the energy activity.

3.1.2.2 *HEA model.* HEA modeling was built on the outputs of the LEAP modeling. It tracked and assessed the environmental health effects of policy implementation and calculated health benefits. It consisted of exposure-response model and cost-of-illness methodology, and the value of a statistical life (VSL).

Exposure-response model was a way of measuring the relationship between the level of exposure to atmospheric pollutants and the health risks they caused (National Health Commission of the PRC, 2020). It was utilized to assess the health effects of $PM_{2.5}$ emission reductions due to energy transition under carbon neutral. The formula was shown in (4). Based on basic epidemiological studies and the availability of local exposure response factors in China. Seven major health risks due to $PM_{2.5}$ were investigated. These included premature death due to cardiovascular diseases (CVD) and respiratory disease, hospitalization for CVD and respiratory disease, and prevalence of acute and chronic bronchitis and asthma.

$$HE = P\Delta R = PR[1 - e^{-\alpha(V-V_0)}] \tag{4}$$

Where, P was the number of exposed population, R was the baseline incidence rate. α was the exposure-response coefficient. α and R referred to the researches of (Xie et al., 2009, 2019), Huang and Zhang (2013), and Liu et al. (2010), as shown in Fig. 2. V and V_0 were the actual and baseline concentrations of $PM_{2.5}$, respectively. The air mass criterion value of 5.0 $\mu g/m^3$ issued by WHO (WHO, 2021) was adopted for V_0.

The main methods for assessing the health economic losses of air pollution were modified human capital approach, cost-of-illness approach, illness and disability method, and VSL (Li et al., 2020). The cost-of-illness method was adopted to assess the $PM_{2.5}$ induced illness and hospitalization economic losses based on the advantage of its estimation structure being closer to the reality, as shown in (5). Considering its disadvantage of not being able to fully reflect

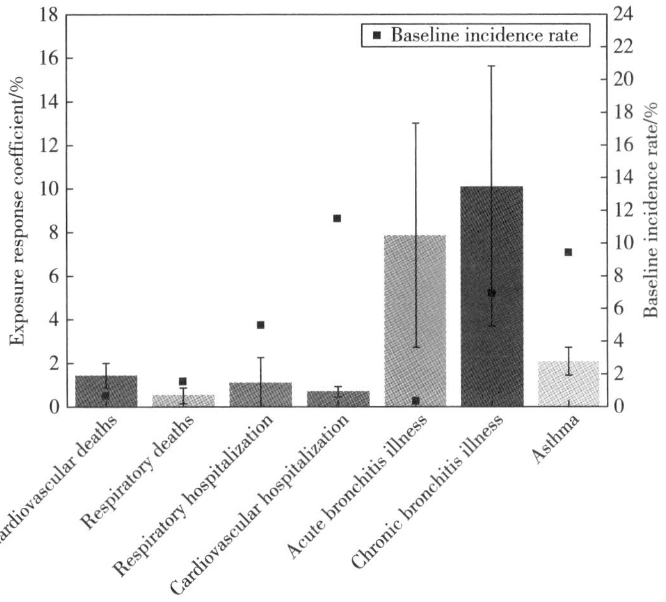

Fig. 2 Exposure response coefficients (95% confidence level) and baseline incidence rates for health risks

the economic losses of chronic diseases, VSL was supplemented to estimate the economic losses from premature mortality caused by $PM_{2.5}$, and the approach was shown in (6).

$$EL_i = (EL_{pi} + \mathrm{GDP}_p \times T_i) \times HE_i \tag{5}$$

$$EL_i = VSL_{ti} \times HE_{ti} \tag{6}$$

Where, EL_i meant the economic loss of health. EL_{pi} denoted unit cost of disease. GDP_p was GDP per capita. T_i indicated the loss of working days due to health risk i. VSL_{ti} was the VSL value for year t of i disease. VSL value and unit economic loss of each health terminal in this paper were based on the studies of Li et al. (2018) and Huang and Zhang (2013). The valuation level was adjusted by using the income elasticity effect method to make it consistent with the future economic development trend in China.

3.1.3 Parameter and scenario settings

The path to carbon neutrality was uncertain, depending on China's future economic development, energy transition, and deployment of carbon capture and storage (CCS). This paper focused on energy transition, supplemented by CCS and other technologies, to explore the energy transition as a possible pathway for China to achieve carbon neutrality.

3.1.3.1 *GDP growth rate setting.* China's future economic growth was uncertain in a optimistic trend, which owing to the complex global political and economic environment. World Economic Outlook and Medium and Long-term Goals, Strategies and Paths for China's Economic and Social Development projected future economic growth by considering the impact of multiple factors (International Monetary Fund, 2023; State Information Center, 2021). Referring to them, the future GDP growth rate was set, as shown in Table 1.

Table 1 GDP growth rate setup for China 2022—2060

Year	2022—2025	2026—2030	2031—2035	2036—2040	2041—2045	2046—2050	2051—2055	2056—2060
GDP growth rate (%)	5.5	5.0	5.0	4.0	3.5	3.5	3.0	2.5

Date source: refer to "World Economic Outlook" and "Medium and Long-Term Goals, Strategies and Paths for China's Economic and Social Development".

3.1.3.2 *Logistic growth model.* Environmental conditions and natural resources had a constraining effect on population growth. There existed an upper limit for population growth that was compatible with the environmental carrying capacity. Logistic growth model was constructed for population forecasting, then there was (7).

$$\begin{cases} \dfrac{dy}{dt} = r(y)y(t) \\ r(y) = r - ay(t) \end{cases} \tag{7}$$

Introducing the maximum population capacity y_m, there was (8).

$$\begin{cases} \dfrac{dy}{dt} = \left(r - \dfrac{ry(t)}{y_m}\right) y(t) \\ y(t_0) = y_0 \end{cases} \tag{8}$$

Separating the variables to solve for the generalization yielded (9).

$$y(t) = \frac{y_m}{1 + \left(\dfrac{y_m}{y_0} - 1\right) e^{-rt}} \tag{9}$$

Where, y was the population size; r denotes the intrinsic population growth rate, which was a constant. α represented a linear parameter of the growth rate. y_m and r were estimated by least squares fitting optimization combined with historical data.

3.1.3.3 *Scenario setting.* With 2020 as the base year and 2021—2060 as the projection period, BAU, Carbon neutral conventional energy transition (CN-ET1) and Carbon neutral deep energy transition (CN-ET2) were set up by integrating the energy transition intensity, carbon capture, utilization and sequestration (CCS). The variation between scenarios was mainly in the deployment or non-deployment of each mitigation measure and the intensity of implementation as shown in appendix Table 1. Chen et al. (2023), Wu et al. (2023) and Yang et al. (2023) were referenced for the scenario setting of carbon emission reduction pathways. BAU scenario continued current relevant policies and inertia development trend, and didn't take any other low-carbon measures. CN scenario adopted advanced technologies, energy transformation and other measures to reduce emissions, and proceeds in accordance with carbon neutral plan.

3.2 Synergistic emission reduction assessment index

In order to quantify the synergistic abatement effects of energy transition on CO_2 and atmospheric pollutants, synergistic emission reduction index was constructed by referring to the research approach of Gao et al. (2021). It was calculated as shown in (10).

$$I = \frac{\Delta Q_p}{\Delta Q_c} = \frac{Q'_p - Q_p}{Q'_c - Q_c} \tag{10}$$

Where, I was the synergistic emission reduction index. Q'_p and Q'_c were the atmospheric pollutants and CO_2 emissions under energy transition, respectively. Q_p and Q_c were the atmospheric pollutants and CO_2 emissions before energy transition, respectively. When $\Delta Q_p > 0$, it meant that the energy transition would result in an increment of pollutants emissions, and $\Delta Q_c >$ 0 meant that the energy transition would cause an increment of CO_2 emissions, and vice versa.

3.3 LMDI model

CO_2 and air pollutants emissions were affected by the implementation on multiple abatement initiatives. These abatement initiatives were categorized as emission intensity reduction, energy transition, energy efficiency improvement, energy demand reduction and CCS (Huang et al., 2022). The research on the initiatives contribution of each category of measures was conducted on this basis. LMDI model could decompose objectives into a collection of factors by qualitative method. Then it could analyze the contribution amount of each factor by quantitative analysis. It was suitable for target decomposition with a few influencing factors. Innovating on the basis of Kaya's constant equation, the emission reduction contribution of energy transition to CO_2 and air pollutants was estimated using the LMDI model. The expression was (11):

$$Q = \sum_{i=1}^{n} Q_i = \sum_{i=1}^{n} \frac{Q_i}{E_i} \times \frac{E_i}{E} \times \frac{E}{\mathrm{GDP}} \times \mathrm{GDP} \tag{11}$$

Where, Q was CO_2 or air pollutants emissions. Q_i represented CO_2 and air pollutants emissions from the ith energy. E_i denoted the consumption of the ith energy. E meant total energy consumption.

Let $EQ = \frac{Q_i}{E_i}$, $ES = \frac{E_i}{E}$, $EI = \frac{E}{\mathrm{GDP}}$, $G = \mathrm{GDP}$. Then the expression was (12)

$$Q = EQ_i \times ES_i \times EI \times G \tag{12}$$

Where, EQ denoted emission intensity. ES was energy structure. EI represented energy intensity. G meant gross product. The amount of change in the above four types of indicators could represent the remaining four types of emission reduction measures except CCS, respectively (Ma and Chen, 2017). Let the emissions be Q_0 in BAU scenario and Q_1 in CN-ET2 scenario, the contribution value of each emission reduction measure was calculated as shown in (13)-(18):

$$\Delta C_{EQ} = \sum_{i=1}^{n} A_i \ln(EQ_{i1} / EQ_{i0}) \tag{13}$$

$$\Delta C_{ES} = \sum_{i=1}^{n} A_i \ln(ES_{i1} / ES_{i0}) \tag{14}$$

$$\Delta C_{EI} = \sum_{i=1}^{n} A_i \ln(EI_{i1} / EI_{i0}) \tag{15}$$

$$\Delta C_G = \sum_{i=1}^{n} A_i \ln(G_{i1} / G_{i0}) \tag{16}$$

$$A_i = \frac{Q_{i1} - Q_{i0}}{\ln Q_{i1} / Q_{i0}} \tag{17}$$

$$\Delta Q = Q_1 - Q_0 = \Delta C_{EQ} + \Delta C_{ES} + \Delta C_{EI} + \Delta C_G \tag{18}$$

3.4 Data sources

The energy data were obtained through collating China Energy Statistics Yearbook published by National Bureau of Statistics. Carbon emission factors were from IPCC Guidelines for National Greenhouse Gas Inventories published by IPCC. Pollutants emission factors were from the LEAP modeling database. Air pollutants data were mainly from Multi-resolution Emission Inventory for China published by MEIC (MEIC, 2023; Li et al., 2017; Zheng et al., 2018). They were used to compare with model results, which in order to ensure the credibility of model. Scenario setting reference data for various parameter settings were mainly derived through collating various industrial development reports issued by Chinese Government, China Carbon Neutral Research Report before 2060 released by Global Energy Internet Development Cooperation Organization and government planning documents. CCS data were primarily obtained from Annual Report on Carbon Dioxide Capture co-authored by China Agenda 21 Management Center, Global Carbon Capture and Storage Institute and Tsinghua University. Exposure-response coefficients and health risk data were mainly from WHO and literature. The toxicity of $PM_{2.5}$ was different in different regions. In order to be on par with international standards, the air mass criterion value was adopted from the WHO standard of 5.0 $\mu g/m^3$.

4 Results and discussion

4.1 Energy consumption projections

Policies such as industry development plan, energy development plan and "dual-carbon" goal were combined. LEAP model was constructed to predict China's future primary energy consumption demand and structure under three scenarios. In terms of total demand, energy demand under the BAU scenario would continue to grow during 2020—2035 as shown in Fig. 3. It was expected to peak at about 6.30 billion tce in 2035 and decline to about 5.45 billion tce in 2060. Energy demand in CN-ET2 scenario would peak in 2030 at a lower level than in BAU scenario, and would decline to about 4.80 billion tce in 2060, owing to policy enhancements and technological advances. This meant future economic growth would achieve low energy consumption, which would be mainly driven by technological innovations and the greening of the economy. This could create a greater likelihood of achieving the "carbon neutral" goal. Because of the differences in forecasting methods, model design and scenario settings, there were some differences between the demand forecasting results in this paper and other scholars' studies. However, there was an overall trend of "rising to the peak and then slowing down" (CNPC Economics & Technology Research Institute, 2020; Huang et al., 2022).

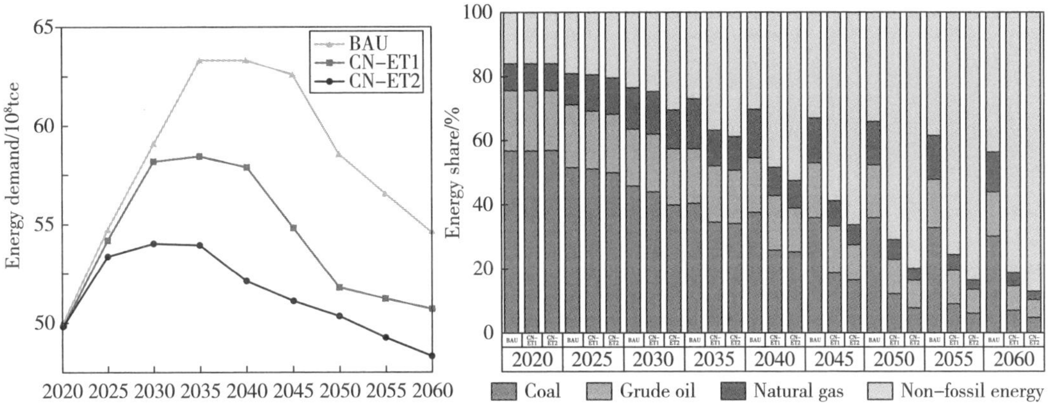

Fig. 3 Total primary energy consumption demand and structure under the three scenarios

Regarding energy consumption structure, the proportion of non-fossil energy under all three scenarios would show a gradual upward trend. In BAU scenario, coal would continue to decline in its consumption share from 2020 to 2060 as the main source of energy. However, it would account for 30.2% in 2060, and the share of non-fossil energy would rise from about 16% to 43.7%, with the energy mix shift not fully realized. Under two carbon-neutral scenarios, the share of coal and oil consumption would decline between 2020 and 2040. However, they would still account for about 25% and 15% respectively, ensuring the basic consumption of energy. With the deepening of the energy transition, the share of coal and oil consumption would decline significantly, non-fossil energy gradually would become the main consumption energy. The share of non-fossil energy under the CN-ET2 scenario was expected to reach 87.3% in 2060. It illustrated that China would face the challenge of rapid growth in electrical demand. China need to strengthen its power grid and pipeline optimization. The proportion of natural gas consumption would show an "upward and then downward" trend, assuming the role of a bridge in the energy transition process. It demonstrated that electrification was an effective way to develop a green economy in medium and long-term. With the driving force of the implementation of the accelerated green and low carbon innovation program, it would be gradually realized.

Under the deep influence of industrial transformation and energy transition, a pattern of energy flows would be formed in which primary energy would be dominated by non-fossil energy, terminal energy power, and terminal consumption industry and transportation as shown in Fig. 4 (not showing the destination of a small amount of energy consumption for the production of raw materials). Driven by carbon neutrality and deep energy transition, China's total end-use energy consumption would decline significantly, structure would be effectively transformed towards electrification and cleaner energy. It was consistent with China's current policy-driven aim to promote energy electrification. Electrification was a central measure to promote a low-carbon energy transition that could effectively improve climate issues. In CN-ET2 scenario, total end-use energy consumption would be about 4.6 billion tce in 2060. Among them, industry would remain the main sector, accounting for about 40%. With China's modernization and the promotion of technology in the industry, the transportation would develop rapidly. Its energy consumption would continue to rise. The structure of end-use energy consumption would undergo significant changes, with the overall proportion of electricity consumption reaching 84%. Electrification levels in all industries would increase significantly. Electrification levels would exceed 80% in industry and construction, and reach about 54% in transportation. The energy flowing in 2060 showed that all industries would successfully transform and upgrade under energy technology advancement and policy impetus. It could create favorable conditions for the realization of carbon neutral goal.

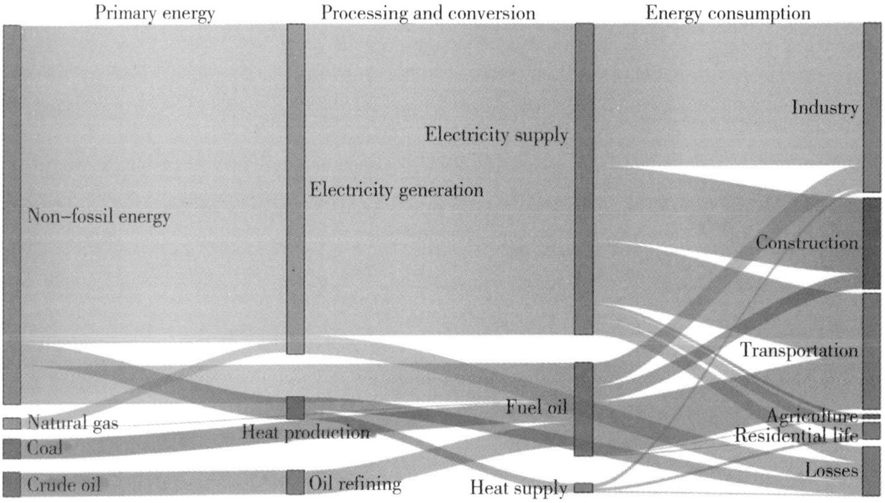

Fig. 4 Energy flowing map for 2060 under the CN-ET2 scenario

4.2 Synergistic emission reduction effects of CO_2 and air pollutants of energy transition

4.2.1 CO_2 and air pollutants emissions projections

In terms of CO_2 emissions, CO_2 emissions under BAU scenario would keep growing until 2035, peak at about 10.7 billion tons in 2035, and then begin to gradually decline shown in Fig. 5. Carbon neutrality might still not be achieved in 2060, and the gap would be large. The carbon emission trends under the two carbon-neutral scenarios could be summarized into peaking period and neutral period. CO_2 would peak on time in 2030 and ahead of schedule in 2025, at about 10.5 billion tons and 10.2 billion tons, respectively. Meanwhile, the two scenarios would achieve carbon neutrality around 2060 and 2055, respectively. Because of different scenario settings and different judgment, the values in this paper were different from those of Zhang and Chen (2022) and Zhang et al. (2022). But these results had a similar trend. This illustrated energy transition would have a positive effect in the achievement of the "dual-carbon" goal. The implementation of the deep energy transition policy could effectively promote carbon emission reduction in China. It could not only reduce the peak level, but also advance the time for realizing the "dual-carbon" goal.

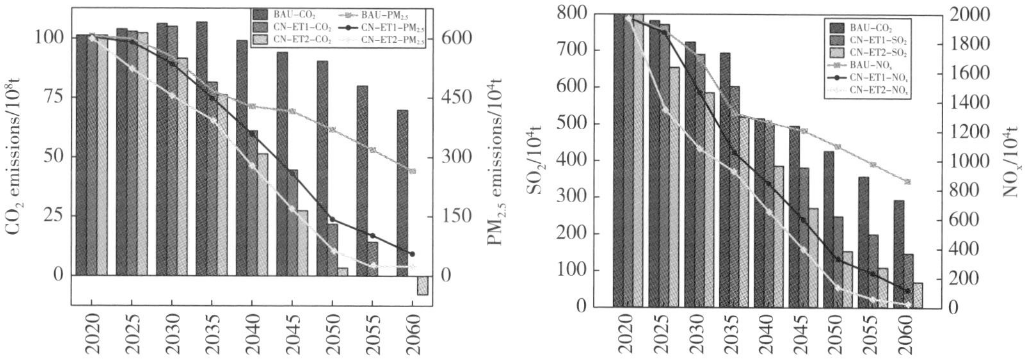

Fig. 5 The emissions of CO_2, $PM_{2.5}$, SO_2 and NO_x under three scenarios

The emissions of air pollutants in three scenarios would decline with increasing years, only the speed and magnitude of the decline differed. The emissions of various air pollutants under BAU scenario would show a rapid drop by 2035, followed by a slow decrease. $PM_{2.5}$ would fall to about 2.65 million tons, SO_2 would reduce to about 2.95 million tons, and NO_x would decrease to about 8.62 million tons in 2060. Driven by industrial upgrading and emission reduction policies, air pollutants would succeed in achieving substantial emission reductions. $PM_{2.5}$ would fall to about 0.25 million tons, SO_2 to about 0.70 million tons, and NO_x to about 0.24 million tons under CN-ET2 scenario in 2060. This illustrated energy transition would promote synergistic emission reductions of air pollutants when it promoted carbon emission reductions. It would promote the construction of beautiful China. This was coincided with China's policy objective of Synergistic emission reduction and efficiency increase.

4.2.2 Contribution of energy transition to CO_2 and air pollutants emission reductions

Using the projection results of the BAU and CN-ET2 scenarios, this paper further explored the emission reduction contributions to CO_2 and air pollutants by various emission reduction measures. Regarding CO_2 emission reduction, the emission reduction contribution of energy transition would rise and then fall during 2020—2060 shown in Fig. 6. Energy transition would become the primary CO_2 emission reduction contribution of the measures in 2040. The emission reduction contribution would maximize at 60.2×10^8 tons in 2050, and the total contribution of it and CCS would account for more than 90%. It could be concluded energy transition would be of great significance in realizing carbon neutrality. China should focus on the breakthroughs in energy transition technology and the implementation of policies to realize its important contribution to carbon emission reduction.

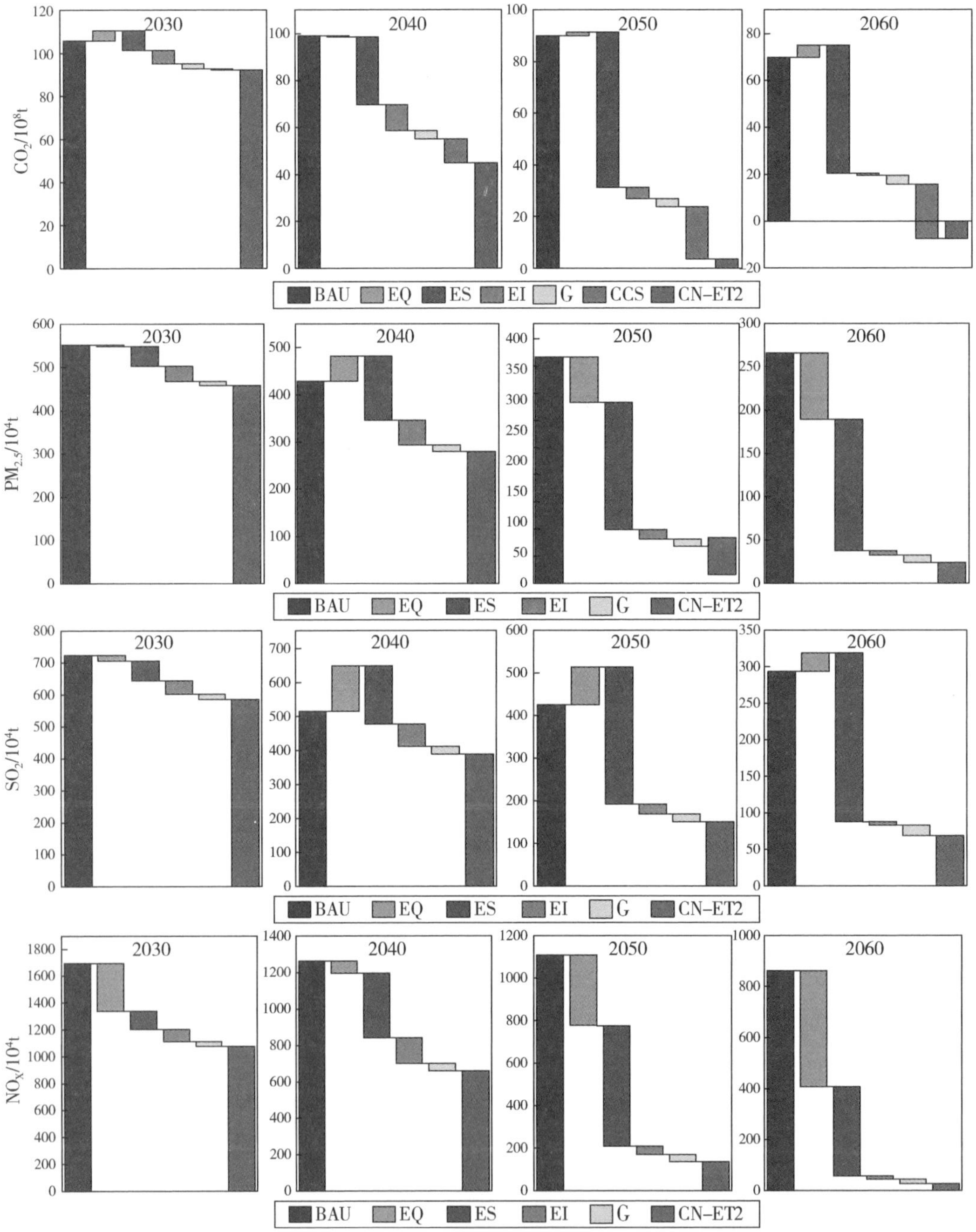

Fig. 6 Contribution of emission reduction measures to CO_2, $PM_{2.5}$, SO_2 and NO_x emission reductions

For air pollutants abatement, the temporal trend of the contribution of energy transition to the emission reduction of each air pollutant would be comparable with that of CO_2 emission reduction. The emission reduction contribution would reach the maximum in 2050, with the maximum contribution to $PM_{2.5}$, SO_2, and NO_x of 209.1×10^8t, 322.9×10^8t, and 562.9×10^8t, respectively. Further evidence of the important role of energy transition in promoting synergistic emission reductions of CO_2 and air pollutants. In addition, the optimal emission reduction measures for each air pollutant were different, except for energy transition. The measure that contributed most to $PM_{2.5}$ and NO_x emission reduction would be reducing emission intensity, while for SO_2 it would be reducing energy intensity. Therefore, air pollutants control in China needed to grasp the direction of focus.

4.2.3 Synergistic emission reductions of CO_2 and air pollutants

The synergistic emission reduction amount and index were calculated based on the emission reduction contribution of energy transition to CO_2 and air pollutants, revealing the synergistic emission reduction effect of energy transition on CO_2 and air pollutants. The points consisting of SO_2, $PM_{2.5}$, NO_x and CO_2 emission reductions were all located in the third quadrant in Fig. 7, indicating energy transition would have a significant impact on achieving synergistic emission reductions of CO_2 and atmospheric pollutants. At a same energy transition level, the syn-ergistic abatement indexof CO_2 and NO_x would be the highest, while the index of CO_2 and $PM_{2.5}$ would be the lowest. However, overall the synergistic emission reduction effect would be significant. The synergistic abatement index of NO_x would be up to 15.6×10^4, while the mean was 11.1×10^4. The synergistic abatement index of SO_2 would reach up to 7.0×10^4, with a mean of 5.8×10^4. The synergistic abatement index of $PM_{2.5}$ would max out at 5.4×10^4, and the average was 4.2×10^4. In time, the synergistic abatement index of air pollutants would gradually drop as energy transition progressed. It was mainly because the difficulty of CO_2 and atmospheric abatement reductions became more inconsistent as the abatement process advanced. The synergistic abatement in deep energy transition would be limited. It was necessary to choose the appropriate intensity in energy transition policy making to better utilize the pollutants and carbon reduction effects of energy transition (Wan et al., 2022).

4.3 Health effects of energy transition

The changes in health risks from energy transition were estimated based on exposure-response modeling. The results indicated energy transition would have significant health effects, reducing the risks of premature deaths, hospitalizations and illnesses. The reduction in premature deaths, hospitalizations, and illnesses from energy transition all would rise and then fall from 2021 to 2060. The maximum of premature deaths avoided in 2050 would be 255.0 (95% CI: 118.0-378.0) thousand, and the maximum of hospitalizations and illnesses would be 1964.0 (95% CI: 762.0-3010.0) thousand and 7362.0 (95% CI: 4930.0-7903.0) thousand, respectively. Judging from avoiding health risks, energy transition could effectively control $PM_{2.5}$ concentration in future. It would make a significant contribution to the improvement of the Chinese population's

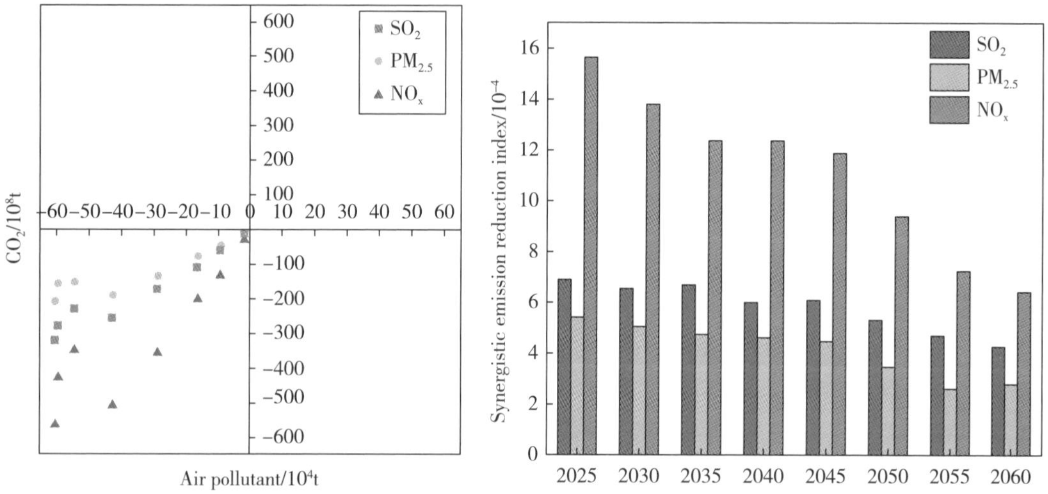

Fig. 7 Synergistic emission reductions and indexes of CO_2 and air pollutants from energy transition

health level. It could also promote the realization of healthy China.

Economic benefit was an important indicator of carbon-neutral energy transition (Cheng et al., 2023). The health economic effects of energy transition were estimated using the cost-of-illness method and VSL. Energy transition would produce large economic benefits at each health endpoint shown in Fig. 8. In terms of avoiding premature deaths and hospitalizations, the economic benefits of deaths and hospitalizations from respiratory and CVD would rise and then fall over the study period, and both were highest in 2050. They would reach the highest in 2050, with economic benefits of avoiding respiratory and CVD amounting to 993.2 (95% CI: 623.5-1322.1) billion yuan and 949.5 (95% CI: 278.6-1556.8) billion yuan, respectively, and reducing hospitalizations from respiratory and CVD would amount to 52.3 (95% CI: 0-95.4) billion yuan and 178.1 (95% CI: 115.3-237.9) billion yuan, respectively. For disease reduction, the economic benefits of energy transition across acute bronchitis, chronic bronchitis, and asthma would continue to grow over study period. The highest in 2060 would be 2.9 (95% CI: 1.3-3.8) billion yuan, 19488.5 (95% CI: 9738.8-23313.1) billion yuan, and 19.1 (95% CI: 13.7-24.1) billion yuan, respectively, which differed from the trend of premature deaths and hospitalizations because of the change in unit costs. Chronic bronchitis prevalence and acute bronchitis prevalence would be the largest and smallest health endpoints, respectively.

Energy transition would produce favorable health economic effects in long term, and the total health economic benefits of energy transition would continue to rise shown in Fig. 9. It would reach 21659.2 billion yuan in 2060, an increase of about 310 times over 2025. Meanwhile, the health economic benefits of energy transition as a percentage of GDP would fluctuate in 2050, but would maintain a continuous upward tendency, rising from 0.05% in 2025 to 4.44% in 2060. Therefore, energy transition under carbon neutral would not only promote the synergy of pollutants and carbon reduction, but also would have tremendous health economic effects.

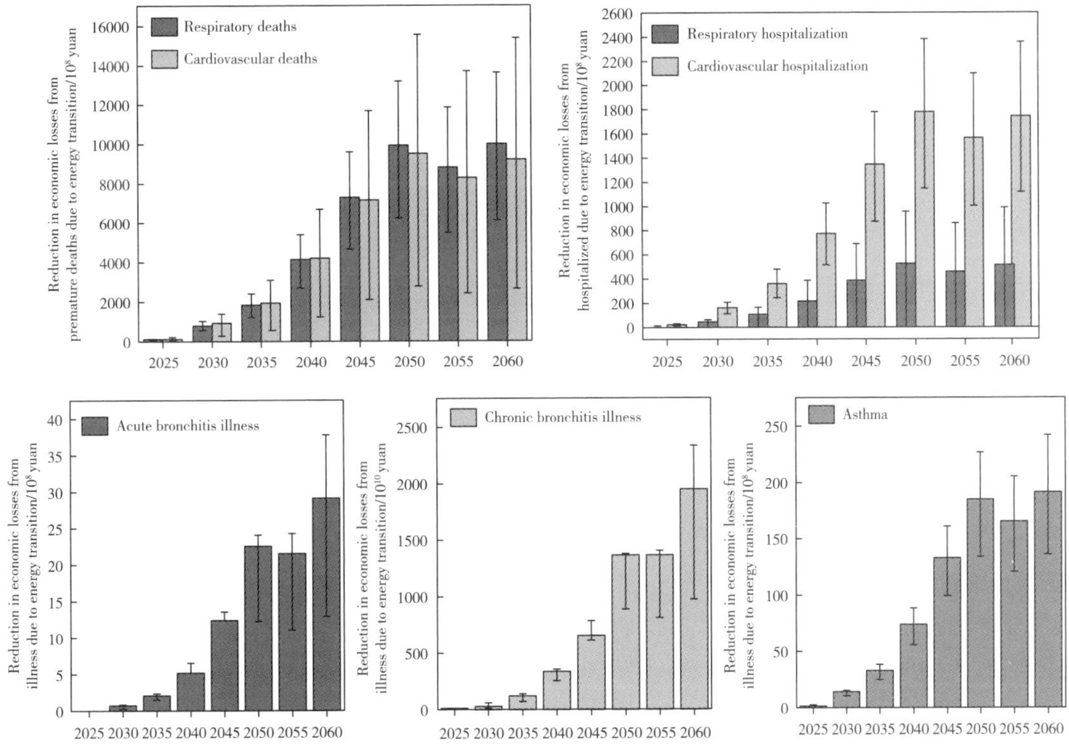

Fig. 8 Health economic effects of energy transition in reducing premature deaths, hospitalizations and illnesses

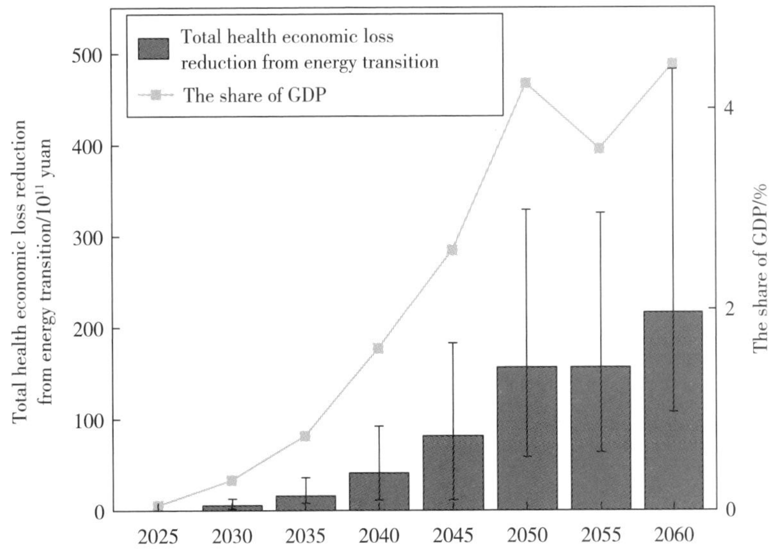

Fig. 9 Total health economic benefits and GDP share of the energy transition

4.4 The limitations and the prospects

The synergistic abatement and environmental health impacts from the energy transition were investigated with significant importance. There were two main limitations in model and data. First, due to the modeling and data limitations, a comprehensive assessment of cost-effectiveness was not performed. It was a direction for future research. Second, the LEAP model had a high degree of flexibility, and it was framed in a hierarchical distribution. The more subdivided the hierarchy, the more accurate the predictions. The lack of detailed data on energy activities might lead to a difference between the results and the reality. The above two limitations mainly originated from the model and data, which might make the results have a difference from the reality. Meanwhile, it also provided a direction for future research.

5 Conclusions and recommendations

5.1 Conclusions

China's future energy consumption, CO_2, $PM_{2.5}$, SO_2 and NO_x emissions were predicted by constructing LEAP-HEA model. And LMDI model was used to decompose the emission reduction contribution of energy transition. The synergistic emission reduction effect and health effect of energy transition were then explored. The following conclusions were obtained.

(1) In BAU scenario, energy demand would continue to grow to a high level. It would start to decline slowly after 2035, and it would still be dominated by fossil energy consumption. However, energy demand would peak early and decline significantly in CN-ET2 scenario, with 650 million tce lower in 2060 than the BAU scenario. Meanwhile, energy structure would be electrified and cleaned up. It would form a pattern of energy flow in which primary energy would be dominated by non-fossil energy sources, terminal energy would be dominated by electricity, and the sectors of use would be dominated by the industrial and transportation sectors.

(2) In BAU scenario, carbon neutrality could not achieve in 2060, while carbon neutrality could reach five years ahead of schedule in CN-ET2. Energy transition and CCS would be sustainable CO_2 emission reduction measures. Energy transition would contribute the most to abatement of air pollutants in 2050, with a 60% share of abatement for three air pollutants. Energy transition would be most beneficial for synergistic reduction of CO_2 and NO_x, but it would be least beneficial for synergistic reduction of CO_2 and $PM_{2.5}$.

(3) Energy transition was effective in reducing health risks and generating significant economic benefits. It would effectively reduce the risk of premature death, hospitalization and disease. The number of premature deaths, hospitalizations and illnesses reduced by energy transition would rise and then fall, while maintaining a high health effect. Energy transition also would generate large economic benefits at each health endpoint. The ratio of health economic benefits to GDP would maintain a continuous upward tendency.

5.2 Recommendations

Three recommendations were made according to the conclusions.

(1) It was necessary to optimize the construction of power grids and pipelines, improve the efficiency of energy transmission, and promote the electrification of energy structure. Because energy demand could not be effectively controlled and energy structure could not be effectively optimized in BAU scenario. Accelerating electrification was a major trend for high-quality development. Improving monitoring systems and implementing a low-carbon transformation of energy-consuming industries could effectively reduce energy demand. Non-fossil energy technologies should be actively developed to improve the applicability of cleaner technologies and to establish a non-fossil energy-driven energy consumption pattern.

(2) Energy transition paths should be optimized, and carbon neutrality should be steadily promoted with energy transition as the key. Energy transition could advance the realization ofcarbon neutrality, and it would have synergistic abatement effects on CO_2 and air pollutants. Synergistic emission reduction and efficiency increase was an important policy proposed by China to help build a beautiful China. Thus, synergistic assessments should be conducted regularly, and treatment programs should be dynamically adjusted based on the assessments. It could control the synergistic treatment of CO_2 and air pollutants at an optimal level.

(3) Energy low-emission technology development should be encouraged to reduce the intensity of emissions. Because energy transition could contribute to improving air quality and have significant health benefits. Supervision of high-emission enterprises should be strengthened. It could effectively promote the green development of industries. Pilot projects on environmental sanitation management should be carried out to clarify the path of synergistic management from practice. A win-win situation for environment, health and economy should be realized to help build a healthy China.

Acknowledgements Thanks to the support of Beijing Natural Science Foundation (No. 9232016), the support of Specialized Teaching Reform for Graduate Education in China University of Geosciences (Beijing) in 2024 (JG2024008), the support of Research on the Cultivation Model of Scientific Research Ability for Students in the Energy Economy Direction of Geoscience Universities (No. JG202306), and Graduate Innovation Funding Program in China University of Geosciences (Beijing) in 2024 (ZD2024YC046).

Appendix

Table 1 The scenario setup

Emission reduction measures	Typical approach or technology	BAU	CN-ET1	CN-ET2
Improvement of energy efficiency	Technology reforms on energy efficiency of industry, transportation. Development of energy efficiency technologies for household appliances	Continuation	Enhancement	Deep enhancement

Continued

Emission reduction measures	Typical approach or technology	BAU	CN-ET1	CN-ET2
Improvement of energy transmission efficiency	Improvement of energy transmission technology. Reduction in the rate of energy transmission and distribution losses	Continuation	Enhancement	Deep enhancement
Improvement of travel way	Increase in public transportation trips	Continuation	Enhancement	Deep enhancement
Electrification	Electrification of residential living. Increase in the proportion of new energy vehicles. Electrification of industrial production	Continuation	Enhancement	Deep enhancement
Improvement of the energy structure	Reduction in fossil energy use in industry and transportation. Increase in clean energy generation	Continuation	Enhancement	Deep enhancement
Optimization of industrial structure	Elimination of energy-intensive industries and development of green industries	Continuation	Enhancement	Deep enhancement
CCS	Use of CCS in steel mills and power plants	×	Deployment begins in 2030	Deployment begins in 2030

References

Bhanarkar, A. D. , Purohit, P. , Rafaj, P. , et al. , 2018. Managing future air quality in megacities: co-benefit assessment for Delhi. Atmos. Environ. 186, 158-177. https:// doi. org/10. 1016/j. atmosenv. 2018. 05. 026.

Central People's Government of the PRC, 2022a. Advancing the Energy Revolution in Depth, Accelerating the Construction of a Strong Energy Country, and Taking New Steps towards High-Quality Energy Development [EB/OL]. https: //www. gov. cn/ xinwen/2022-11/11/content_ 5726085. htm.

Central People's Government of the PRC, 2022b. Modern Energy System Planning for the 14th Five-Year Plan [EB/OL]. https://www. gov. cn/zhengce/zhengceku/2022-03/23/5680759/files/ccc7dffca8f24880a80af12755558f4a. pdf.

Chen, S. , Mai, X. Y. , Liu, Y. Y. , et al. , 2023. Synergistic analysis of carbon dioxide emission reduction and environmental health benefits under carbon neutral pathway in Gansu Province [J/OL]. J. Saf. Environ. 1-15. https: //doi. org/10. 3389/fenvs. 2022. 1042 344.

Cheng, J. , Tong, D. , Liu, Y. , et al. , 2023. A synergistic approach to air pollution control and carbon neutrality in China can avoid millions of premature deaths annually by 2060. One Earth 6 (8), 978-989. https: //doi. org/10. 1016/j. oneear. 2023. 07. 007.

CNPC Economics & Technology Research Institute, 2020. World and China's energy outlook in 2050 (2020 version) [R/OL]. https: //finance. sina. com. cn/tech/2020-12-29/doc-iiznezxs9471433. shtml.

Dong, F. , Yu, B. , Pan, Y. , 2019. Examining the synergistic effect of CO_2 emissions on $PM_{2.5}$ emissions reduction: evidence from China. J. Clean. Prod. 223, 759-771. https://doi. org/10. 1016/j. jclepro. 2019. 03. 152.

Dong, H. , Dai, H. , Dong, L. , et al. , 2015. Pursuing air pollutant co-benefits of CO_2 mitigation in China: a provincial leveled analysis. Appl. Energy 144, 165-174. https://doi. org/10. 1016/j. apenergy. 2015. 02. 020.

Gao, Q. X. , Gao, W. O. , Ma, Z. Y. , et al. , 2021. Methodology and application of synergistic effect assessment of atmospheric pollutants and greenhouse gas emission reduction. Progress in Climate Change Research 17 (3), 268. https: //doi. org/10. 12006/j. issn. 1673-1719. 2020. 206.

Gatto, A. , Chepeliev, M. , 2024. Reducing global food loss and waste could improve air quality and lower the risk of premature mortality. Environ. Res. Lett. 19(1),014080 https://doi.org/10.1088/1748-9326/ad19ee.

Giorgini, P., Di Giosia, P., Grassi, D., et al., 2016. Air pollution exposure and blood pressure: an updated review of the literature. Curr. Pharmaceut. Des. 22(1), 28-51. https://doi.org/10.2174/1381612822666151109111712.

Han, L., Sun, Z., Gong, T., et al., 2020. Assessment of the short-term mortality effect of the national action plan on air pollution in Beijing, China. Environ. Res. Lett. 15 (3), 034052 https://doi.org/10.1088/1748-9326/ab6f13.

Hao, Y. P., Song, X. W., Zhu, X. D., et al., 2023. Health impact analysis of fine particulate matter pollution from motor vehicles in Beijing-Tianjin-Hebei region. Environmental Science 44 (12), 6610-6620. https://doi.org/10.13227/j.hjkx.202302096.

Huang, D. S., Zhang, S. Q., 2013. Assessment of health benefits of controlling $PM_{2.5}$ pollution in Beijing-Tianjin-Hebei region. China Environ. Sci. 33 (1), 166-174. https://doi.org/10.3969/j.issn.1000-6923.2013.01.024.

Huang, Z., Xie, X. M., Zhang, T. T., 2022. Study on the medium and long-term energy demand forecast and transformation path in China under the background of "double carbon". China Engineering Science 24 (6), 8-18. https://doi.org/10.15302/J-SSCAE-2022.06.002.

International Monetary Fund, 2023. World economic outlook [R/OL]. https://meetings.imf.org/zh/IMF/Home/Publications/WEO/Issues/2023/10/10/world-economic-outlook-october-2023.

IPCC, 1995. IPCC Second Assessment Report: Climate Change 1995 (SAR) [R/OL]. https://www.ipcc.ch/languages-2/chinese/publications-chinese/.

IPCC, 2015. IPCC fifth assessment report: climate change 2014 [R/OL]. https://www.ipcc.ch/site/assets/uploads/2018/03/WG3AR5_ SPM_ brochure_ zh-1.pdf.

IPCC, 2018. Special peport: global warming of 1.5 ℃ [R/OL]. https://www.ipcc.ch /sr15/.

Lanzi, E., Dellink, R., Chateau, J., 2018. The sectoral and regional economic consequences of outdoor air pollution to 2060. Energy Econ. 71, 89-113. https:// doi.org/10.1016/j.eneco.2018.01.014.

Li, H. J., Zhou, D. Q., Wei, Y. J., 2018. Evaluation of health risk and economic loss of $PM_{2.5}$ pollution in urban China. Environmental Science 39 (8), 3467-3475. https://doi.org/10.13227/j.hjkx.201711237.

Li, H. J., Zhou, D. Q., Wei, Y. J., 2020. Progress of health economic loss assessment of air pollution. Research on Environmental Science 33 (10), 2421-2429. https://doi.org/ 10.13198/j.issn.1001-6929.2020.04.25.

Li, M., Liu, H., Geng, G., et al., 2017. Anthropogenic emission inventories in China: a review. Natl. Sci. Rev. 4 (6), 834-866. https://doi.org/10.1093/nsr/nwx150.

Li, Y. Y., Song, Y. D., 2021. Synergistic emission reduction of CO_2 and air pollutants from mobile sources on urban roads in Beijing under the carbon neutralization target. China Environmental Management 13 (3), 113-120. https://doi.org/10.16868/j.cnki.1674-6252.2021.03.113.

Limaye, V. S., Magal, A., Joshi, J., et al., 2023. Air quality and health co-benefits of climate change mitigation and adaptation actions by 2030: an interdisciplinary modeling study in Ahmedabad, India. Environ. Res.: Health 1 (2), 021003. https:// doi.org/10.1088/2752-5309/aca7d8.

Liu, X. Y., Xie, P., Liu, Z. R., et al., 2010. Evaluation of economic losses from acute health effects of respirable particulate matter pollution in the Pearl River Delta. J. Peking Univ. (Heal. Sci.): Natural Science Edition (5), 829-834. https://doi.org/10.13209/ j.0479-8023.2010.118.

Ma, D., Chen, W. Y., 2017. Peak carbon emission pathways based on China's TIMES model. Journal of Tsinghua University (Natural Science Edition) 57 (10), 1070-1075. https://doi.org/10.16511/j.cnki.qhdxxb.2017.25.047.

Markandya, A., Armstrong, B. G., Hales, S., et al., 2009. Public health benefits of strategies to reduce greenhouse-gas emissions: low-carbon electricity generation. Lancet 374 (9706), 2006—2015. https://doi.org/10.1016/s0140-6736(09)61715-3.

MEE, 2023. Minister of Ecology and Environment Huang Runqiu Delivers the Report of the State Council on the State of the Environment and the Achievement of Environmental Protection Targets in 2022 [EB/OL]. https://www.mee.gov.cn/ ywdt/hjywnews/202305/t20230506_ 1029130.shtml.

MEE, 2022. Ministry of Ecology and environment and 7 other departments jointly issued the implementation program for pollution reduction. Carbon Reduction and Co-Efficiency [EB/OL]. https://www.mee.gov.cn/ywgz/ydqhbh/wsqtkz/202206/t20 220617_ 985943.shtml.

MEIC, 2023. Multi-resolution emission inventory model for climate and air pollution research [DB/OL]. http://meicmodel.org.cn.

Mir, K. A., Purohit, P., Cail, S., et al., 2022. Co-benefits of air pollution control and climate change mitigation strategies in Pakistan. Environ. Sci. Pol. 133, 31-43. https://doi.org/10.1016/j.envsci.2022.03.008.

Mittal, S., Hanaoka, T., Shukla, P. R., et al., 2015. Air pollution co-benefits of low carbon policies in road transport: a sub-national assessment for India. Environ. Res. Lett. 10 (8), 085006 https://doi.org/10.1088/1748-9326/10/8/085006.

National Health Commission of the PRC, 2020. Technical specifications for health risk assessment of ambient air pollution [R/OL]. http://www.nhc.gov.cn/wjw/pgw/202003/fb4c325cb7a0491db7723fa91f6b9387/files/fd4bd51002584fa0b33185e3733b27ec.pdf.

NDRC, 2023a. Data related to energy production and consumption in China in 2022 [EB/OL]. https://www.ndrc.gov.cn/fggz/hjyzy/jnhnx/202303/t20230302_ 1350587.html.

NDRC, 2023b. NDRC Releases important results on the third Anniversary of the major declaration on carbon peak and carbon neutrality [EB/OL]. https://www.ndrc.gov.cn/fggz/hjyzy/stwmjs/202308/t20230817_ 1359890.html.

Ostro, B., Feng, W. Y., Broadwin, R., et al., 2007. The effects of components of fine particulate air pollution on mortality in California: results from CALFINE. Environ. Health Perspect. 115 (1), 13-19. https://doi.org/10.1289/ehp.9281.

Rypdal, K., Rive, N., Åström, S., et al., 2007. Nordic air quality co-benefits from European post-2012 climate policies. Energy Pol. 35 (12), 6309-6322. https://doi.org/10.1016/j.enpol.2007.07.022.

Salimifard, P., Rainbolt, M. V., Buonocore, J. J., et al., 2023. A novel method for calculating the projected health and climate co-benefits of energy savings through 2050. Build. Environ. 244, 110618 https://doi.org/10.1016/j.buildenv.2023.110618.

Song, F., Fu, J. F., 2012. The synergistic emission reduction of GHG and sulfur dioxide in the world's major countries and its revelation. Resour. Sci. 34 (8), 1439-1444. https://xueshu.baidu.com/usercenter/paper/show?paperid=13f5c1c478fedd22e6c4d5a48ca1f41c&site=xueshuse.

State Council Office of the PRC, 2022. Circular of the state council Office on the issuance of the 14th five-year national health plan[EB/OL]. https://www.gov.cn/zhengce/content/2022-05/20/content 5691424.htm.

State Information Center, 2021. Medium and long-term goals, strategies and paths for China's economic and social development [R/OL]. https://www.efchina.org/Reports-zh/report-lceg-20210207-4-zh.

Wagner, V., Pascal, M., Corso, M., et al., 2023. On the supra-linearity of the relationship between air pollution, mortality and hospital admission in 18 French cities. Int.

Arch. Occup. Environ. Health 96 (4), 551-563. https://doi.org/10.1007/s00420-022-01948-3.

Wan, Y. F., Cui, Y. Y., Wu, X. F., et al., 2022. Characterization of carbon dioxide emissions in the Beijing-Tianjin-Hebei region and its potential for synergistic emission reduction with air pollutants. J. Cap. Normal Univ. 43 (4), 46-52+74. https://doi.org/10.19789/j.1004-9398.2022.04.008.

Wang, G. Z., Wu, L. Y., Chen, J. B., et al., 2017. CGE analysis of health economic effects of $PM_{2.5}$ pollution in Beijing. China Environ. Sci. 37 (7), 2779-2785. http://www.zghjk x.com.cn/CN/Y2017/V37/I7/2779.

Wang, Y., Wei, J., Zhang, Y., et al., 2023. Estimating causal links of long-term exposure to particulate matters with all-cause mortality in South China. Environ. Int. 171, 107726 https://doi.org/10.1016/j.envint.2022.107726.

WHO, 2021. WHO global air quality guidelines [R/OL]. https://iris.who.int/bitstream/h andle/10665/345329/9789240034228-eng.pdf?isAllowed=y&sequence=1.

Wu, Q., Ma, H., Ren, H. B., et al., 2023. Medium and long-term carbon emission projections and emission reduction potential analysis of the Lingang special area based on the LEAP model [J/OL]. Environmental Science 1-17. https://doi.org/10.13227/j.hjkx.202301093.

Xie, P., Liu, X. Y., Liu, Z. R., et al., 2009. Exposure-response relationship of airborne particulate matter in China. China Environ. Sci. 29 (10), 1034-1040. https://doi.org/10.3321/j.issn:1000-6923.2009.10.006.

Xie, Z. X., Qin, Y. C., Zheng, Z. C., et al., 2019. Assessment of mortality effects of $PM_{2.5}$ pollution in cities along the Beijing-Tianjin-Hebei air pollution transmission corridor. Journal of Environmental Science 39 (3), 843-852. https://doi.org/10.13671/j.hjkxxb.2018.0420.

Xing, X. W., Huang, L., Hu, J. L., 2024. Benefit analysis of synergistic emission reduction of carbon dioxide and air pollutants from different low-carbon development paths in the power sector in Jiangsu Province [J/OL]. Environmental Science 1-19. https://doi.org/10.13227/j.hjkx.202311231[2024-04-07].

Yang, F., Zhang, G. Q., Sun, J., et al., 2023a. Peak carbon scenario of a city in Yangtze River Delta based on LEAP model [J/OL]. Environmental Science 1-19. https://doi.org/10.13227/j.hjkx.202301129.

Yang, R. P., Feng, X. Z., Wang, M., et al., 2023b. Analysis of synergic reduction of greenhouse gases and air pollutants emission in the urban transportation sector: taking Tangshan City as an example. J. Environ. Eng. Technol. 13 (6), 2033-2042. https://doi.org/10.12153/j.issn.1674-991X.20230055.

Yang, X., Sun, Y. S., Chang, S. Y., et al., 2023c. Assessing the environmental and health Co-benefits of accelerated energy transition and industrial restructuring: a assessing the environmental and health Co-benefits of accelerated energy transition and industrial restructuring: a case study of the BTHS region. Huanjing Kexue 44 (7), 3627-3636. https://doi.org/10.13227/j.hjkx.202207164.

Yawale, S. K., Hanaoka, T., Kapshe, M., et al., 2023. Direct and indirect air pollutant reductions as co-benefits of the energy transition toward carbon-neutrality in India's residential sector. Energy Strategy Rev. 49, 101143 https://doi.org/10.1016/j.esr.2023.101143.

Yu, S., Mou, J., Zhang, S., 2023. Benefit analysis of synergistic control of air pollutants and CO_2 in Beijing. Environ. Sci. Technol. 46 (7) https://doi.org/10.19672/j.cnki.1003-6504.2067.22.338.

Zhang, S., Chen, W., 2022. China's energy transition pathway in a carbon neutral vision. Engineering 14, 64-76. https://doi.org/10.1016/j.eng.2021.09.004.

Zhang, X. L., Huang, X. D., Zhang, D., et al., 2022. Research on the transition path and policy of energy economy under the goal of carbon neutrality. Manag. World 38 (1), 35-66. https://doi.org/10.19744/j.cnki.11-1235/f.2022.0005.

Zheng, B., Tong, D., Li, M., et al., 2018. Trends in China's anthropogenic emissions since 2010 as the consequence of clean air actions. Atmos. Chem. Phys. 18 (19), 14095-14111. https://doi.org/10.5194/acp-18-14095-2018.

Zeng, X. G., Ruan, F. F., 2020. National economic impact analysis of the health effects of $PM_{2.5}$ pollution in China. China Environ. Sci. 40 (7), 3228-3238. https://doi.org/10.19674/j.cnki.issn1000-6923.2020.0361.

Evolution of the spatiotemporal pattern of $PM_{2.5}$ concentrations in China: A case study from the Beijing–Tianjin–Hebei region

YAN Dan[1,2,3], LEI Yalin[1,2,*], SHI Yukun[4],
ZHU Qing[5], LI Li[1,2], ZHANG Zhien[6]

1. *School of Humanities and Economic Management*, *China University of Geosciences*, *Beijing* 100083, *China*
2. *Key Laboratory of Carrying Capacity Assessment for Resource and Environment*, *Ministry of Land and Resources*, *Beijing* 100083, *China*
3. *Research Center for Transition and Development of Resource–exhausted cities*, *Hubei Normal University*, *Huangshi* 435002, *China*
4. *Adam Smith Business School*, *University of Glasgow*, *Glasgow* G12 8QQ, *UK*
5. *Chinese Academy of Land & Resource Economics*, *Beijing* 101149, *China*
6. *School of Chemistry and Chemical Engineering*, *Chongqing University of Technology Chongqing* 400054, *China*

Abstract Atmospheric haze pollution has become a global concern because of its severe effects on human health and the environment. The Beijing–Tianjin–Hebei urban agglomeration is located in northern China, and its haze is the most serious in China. The high concentration of $PM_{2.5}$ is the main cause of haze pollution, and thus investigating the temporal and spatial characteristics of $PM_{2.5}$ is important for understanding the mechanisms underlying $PM_{2.5}$ pollution and for preventing haze. In this study, the $PM_{2.5}$ concentration status in 13 cities from the Beijing – Tianjin – Hebei region was statistically analyzed from January 2016 to November 2016, and the spatial variation of $PM_{2.5}$ was explored via spatial autocorrelation analysis. The research yielded three overall results. (1) The distribution of $PM_{2.5}$ concentrations in this area varied greatly during the study period. The concentrations increased from late autumn to early winter, and the spatial range expanded from southeast to northwest. In contrast, the $PM_{2.5}$ concentration decreased rapidly from late winter to early spring, and the spatial range narrowed from northwest to southeast. (2) The spatial dependence degree, by season from high to low, was in the order winter, autumn, spring, summer. Winter (from December to February of the subsequent year) and summer (from June to August) were, respectively, the highest and lowest seasons with regard to the spatial homogeneity of $PM_{2.5}$ concentrations. (3) The $PM_{2.5}$ concentration in the Beijing – Tianjin – Hebei region has significant spatial spillovers.

原文刊于 *Atmospheric Environment*, 2018 年第 183 卷。原文发表时为彩色。
* 为通信作者。

Overall, cities far from Bohai Bay, such as Shijiazhuang and Hengshui, demonstrated a high-high concentration of $PM_{2.5}$ pollution, while coastal cities, such as Chengde and Qinhuangdao, showed a low-low concentration.

Keywords Beijing-Tianjin-Hebei region, Air pollution, $PM_{2.5}$ concentration, spatial autocorrelation analysis

1 Introduction

China's serious haze problem has aroused widespread concern among the public. The increase in the concentrations of particulates (including PM_{10} and $PM_{2.5}$) in the atmosphere is the main cause of haze production (Xu and Lin, 2016; Hsu et al., 2017; Liao et al., 2017; Liu et al., 2015a). Studies have shown that high levels of $PM_{2.5}$ and PM_{10} are closely linked to high concentrations of fungi and bacteria in the air, which can cause serious harm to humans (Liu et al., 2017a; Liu et al., 2017b; Liu et al., 2015b). In 2012, China issued newly revised ambient air quality standards; $PM_{2.5}$, regarded as a routine indicator of atmospheric pollution, was included in the new standard and has become the key focus of atmospheric pollution research. $PM_{2.5}$ refers to fine particles with a dynamic diameter of less than 2.5 μm. These particles are composed of a wide variety of complex chemical substances emitted from various natural and anthropogenic sources (Alves et al., 2012; Wang et al., 2012; Zhao et al., 2014). A high concentration of $PM_{2.5}$ not only significantly reduces atmospheric visibility but also leads to increased incidences of respiratory and cardiovascular diseases (Dockery et al., 1993; Li et al., 2016; Chalbot et al., 2014; Hu et al., 2014).

At the beginning of 2013, China suffered from the most serious fog and haze pollution since the start of observational records, and this rare continuous high-intensity air pollution swept the middle and eastern parts of China. The most serious pollution occurred in the Beijing-Tianjin-Hebei (BTH) region, where the daily $PM_{2.5}$ concentration reached 500 $\mu g/m^3$ (Wang et al., 2014). The BTH region is China's political and cultural center and is an important core area of North China's economy.

Since the 1980s, the economy, society and culture of the BTH region have developed remarkably. The gross domestic product (GDP) increased approximately seven-fold from 593.342 billion yuan in 1980 to 4715.233 billion yuan in 2013 (Beijing Statistical Yearbook, 2014a; Tianjin Statistical Yearbook, 2014b; Hebei Economic Yearbook, 2014c). However, at the sametime, serious air pollution exists in the 13 cities within the BTH region. According to the Ministry of Environmental Protection, the worst ten cities in terms of air quality in 2015 were Baoding, Xingtai, Hengshui, Tangshan, Zhengzhou, Jinan, Handan, Shijiazhuang, Langfang and Shenyang. Seven of these cities are within the BTH region. The haze pollution in the BTH region cannot be overlooked, and haze governance has become the top priority for the BTH region (Cai et al., 2017). Therefore, studying the temporal and spatial characteristics of

$PM_{2.5}$ in this region has great practical value. Understanding the spatial variation of the $PM_{2.5}$ concentration will not only add to our knowledge of the mechanism of air pollution but also provide a scientific reference for implementing targeted control measures.

The consumption of fossil energy, such as coal and oil, generates a large amount of waste gas, which affects the climate and endangers human health. Therefore, energy-saving emission reduction is crucial (Ma et al., 2017a; Yan et al., 2017). Many scholars have focused on carbon emissions reduction in their research (Shuai et al., 2017; Ma et al., 2017b; Ma et al., 2017c), and now society as a whole is beginning to pay attention to the haze issue in China.

Existing research on the haze problem mainly focuses on two aspects. Initially, most scholars analyzed the composition of $PM_{2.5}$ from the physical and chemical perspectives (Bates and Sizto, 1987; Hussain et al., 2013; Jansen et al., 2014) and concluded that $PM_{2.5}$ was mainly composed of industrial waste gas, automobile and machine exhaust, cooking oil smoke and dust. These pollutants are closely related to socioeconomic factors such as GDP, population, energy consumption and industrial infrastructure. In light of these findings, a large number of studies have focused on the economic and social drivers of the $PM_{2.5}$ concentration. For example, based on the data on the $PM_{2.5}$ concentration and the air quality index (AQI) in 73 Chinese cities in 2013, Hao and Liu (2016) analyzed the influencing factors ofthe $PM_{2.5}$ concentration in China's cities and discussed how economic and social development could affect air quality. Their results show that the relationship between the $PM_{2.5}$ concentration and GDP per capita exhibits an inverted U shape and that car ownership and secondary industry have significant effects on the $PM_{2.5}$ concentration. Luo et al. (2017) explored the driving factors of the $PM_{2.5}$ concentration in China using a geographical regression weighting model, which confirmed the existence of the inverted U-shaped environmental Kuznets curve (EKC) for air quality and that the potential influencing factors of each significant area were different. By using the input-output framework and structural decomposition analysis, Guan et al. (2014) studied the socioeconomic drivers of China'sprimary $PM_{2.5}$ emissions and found that export is the only final demand category that led to emission growth between 1997 and 2010. The embodied $PM_{2.5}$ emissions from Chinese exports are mainly driven by consumption in OECD countries. The second studied aspect is the spillover effects of interregional air pollution. SO_2, NO_x and soot are often used as proxy variables for air pollution (Civan et al., 2015; Moroń et al., 2015; Zhao et al., 2017). Spatial variability analysis in geography is an important basis for simulating the spatial distribution of variables and revealing the spatial effects of variables (Liu et al., 2016; Zuo et al., 2015). Furthermore, spatial autocorrelation analysis is an important technical method that has recently been applied to the field of environmental pollution, particularly air pollution. For example, Zhang et al. (2016) examined the spatial clustering types of CO_2 emission efficiency in 30 provinces in China and confirmed that there is indeed a spatial spillover effect among Chinese provinces. To gain insight into the characteristics of air pollution, Wu et al. (2017) explored the characteristics and determinants of $PM_{2.5}$ pollution in China using spatial econometrics. Yan et al. (2017) used Moran's index to examine the spatial effects of the power industry in various regions of China. The empirical results showed that there is indeed a significant spatial agglomeration effect between carbon emission efficiencies, mainly for high-high and low-low agglomeration types.

The concentration of air pollutants is affected by the intensity of emission sources and by terrain and meteorological conditions; furthermore, it has remarkable temporal and spatial variability. In China, $PM_{2.5}$ started to attract attention in 2012.

The majority of the literature is devoted to the analysis of the formation of fog and haze and to the chemical composition of $PM_{2.5}$ or its influencing factors at the national scale, whereas studies on the distribution pattern of $PM_{2.5}$ concentrations on the urban scale are relatively rare. Compared with previous research, this study has two main contributions. Thematically, through the $PM_{2.5}$ pollution index, we analyze the spillover effect of haze between different cities in the BTH urban agglomeration.

The results will allow the general public to clearly understand the spatiotemporal rules of haze pollution in this urban agglomeration and provide new evidence for haze management. Methodologically speaking, abandoning the geospatial homogeneity hypothesis in spatial econometrics would better fit the real situation of the $PM_{2.5}$ concentration and facilitate the examination of the propagation path of haze pollution.

Specifically, we collected data on daily $PM_{2.5}$ concentrations in 2016 in 13 cities located within the BTH region and studied the spatial autocorrelations and aggregation patterns of $PM_{2.5}$ concentrations during different seasons. The results were intended to provide a basis for simulating $PM_{2.5}$ concentrations in the region, elucidating the underlying factors contributing to this pollution, and devising an effective monitoring point layout.

The remaining parts of this paper are organized as follows. The second section describes the data source and model description. In the third section, we present the temporal and spatial variation of $PM_{2.5}$ concentrations in the BTH region. The fourth section describes and discusses the results of spatial autocorrelation, which is followed by the final conclusions and policy implications in the fifth section.

2 Data and methodology

2.1 Overview of the BTH region and data sources

In 2013, China established a total of 612 $PM_{2.5}$ concentration monitoring sites in 74 cities, and the number of monitoring sites in each city was different. By 2015, the monitoring sites increased to 1436 (Wang et al., 2015). Presently, the sites that monitor the $PM_{2.5}$

concentration in China are mainly concentrated in the Pearl River Delta, Yangtze River Delta and BTH region. Among these areas, the urban agglomeration of the BTH region suffers from relatively more severe haze pollution than other cities, and some studies have suggested that this was because several coal - based industries, such as coal - fired powerplants and steel manufacturing, are stationed in the region (Zhao et al., 2012). This area is adjacent to Bohai Bay and is composed of 13 cities, namely, Beijing, Tianjin, Shijiazhuang, Xingtai, Handan, Hengshui, Baoding, Cangzhou, Langfang, Tangshan, Zhangjiakou, Chengde and Qinhuangdao. The land area of this region is 218000 square kilometers, and the resident population is approximately 110 million people. Industrialization has greatly bolstered the BTH region's economic development, and the main industrial types are steelmaking, petroleum processing and coking, nuclear fuel processing, building materials manufacturing and chemical manufacturing. The common feature of these industries is high energy consumption; thus, a large amount of atmospheric emissions is produced. In 2015, the total industrial emissions from this region was 4.138 million tons, including dust, sulfur dioxide and nitrogen dioxide. These chemical substances are precursors of sulfate and nitrate of $PM_{2.5}$.

In this study, the hourly $PM_{2.5}$ concentration data of the urban air quality monitoring sites in 13 cities in 2016 were obtained from the China Environmental Monitoring Station as raw data. According to the arithmetic average method, these data from cities or monitoring sites can be calculated as daily averages, monthly averages, quarterly averages and as an annual average within 2016. In accordance with the requirements of the Ambient Air Quality Standard for the effectiveness of air pollutant concentration data, the $PM_{2.5}$ data were pretreated. First, weremoved missing values. Second, when the monthly mean was calculated, the monitoring site was eliminated if the number of daily $PM_{2.5}$ concentration measurements was less than 27 days in a month. Finally, a small number of abnormal monitoring points were eliminated (such as hourly $PM_{2.5}$ concentrations greater than 900 $\mu g \cdot m^{-3}$ or hourly $PM_{2.5}$ concentrations less than 0).

2.2 Spatial interpolation method

The spatial interpolation method is often used to convert point scale data into a continuous surface scale. Additionally, this method helps us better understand the full spatial distribution of variables in the region. The accuracy ofthe spatial interpolation method is greater than that of remote sensing inversion (Lee S J et al., 2012). Many specific methods are commonly used for spatial interpolation, such as the Inverse Distance Weighted (IDW) method and the Ordinary Kriging method (OK). The accuracy of the estimates using the former method is influenced by the distance to a known point, and the requirements for the dispersion and uniformity ofthe interpolation points are relatively higher. The latter method will generate the best estimation algorithm for the output surface and will create a comprehensive calculation of the spatial behavior of the interpolation point attribute prior to the generation of the algorithm; thus, the results from the second method are preferable (Zhang et al., 2013). Generally, the OK method can reflect the spatial distribution of $PM_{2.5}$ more scientifically and the result has second-order stationarity.

2.3 Spatial clustering analysis of the $PM_{2.5}$ concentration based on Moran's Index

According to the first law of geography, entities with geographical attributes are related with each other. Clustering, random and regular distribution exist, and correlation decreases with distance (Tober, 1970). This phenomenon is called spatial autocorrelation. Spatial autocorrelation statistics can describe the potential interdependence or compactness of variables within the same area. This approach is often used to analyze the spatial agglomeration and trend of geographical elements to provide evidence for exploring spatial and temporal clustering and evolution. The spatial correlation of atmospheric activity may reveal that the values of $PM_{2.5}$ concentrations in similar areas are statistically close (Wu et al., 2015). Spatial autocorrelation analysis includes global spatial autocorrelation and local spatial autocorrelation, in which the global Moran Index is calculated as follows:

$$I = \frac{n \sum_{i=1}^{n} \sum_{j=1}^{n} w_{ij} Z_i Z_j}{S_0 \sum_{i=1}^{n} Z_i^2} \tag{1}$$

where $S_0 = \sum_{i=1}^{n} \sum_{j=1}^{n} w_{ij}$, and w_{ij} is the spatial weight matrix; in this paper, the adjacent unit is 1, and the remaining units are zero. The Moran Index is in the range [−1, 1]. A value less than 0, equal to 0, or greater than 0 indicates negative correlation, no correlation or positive correlation, respectively. For the Moran Index, the standardized statistic Z can be used to test the existence of spatial autocorrelation. Theformula for the standardized statistic Z is as follows:

$$Z_i = \frac{I - E[I]}{\sqrt{V[I]}} \tag{2}$$

$$E[I] = -1 / (n - 1) \tag{3}$$

$$V[I] = E[I^2] - E[I]^2 \tag{4}$$

Among these formulas, at the 0.05 significance level, $Z(I) > 1.96$ indicates positive spatial autocorrelation between $PM_{2.5}$ spatial units, and $-1.96 < Z(I) < 1.96$ indicates that the spatial correlation of $PM_{2.5}$ concentrations is not obvious. If $Z(I) < -1.96$, then there is a negative correlation between $PM_{2.5}$ spatial units, and the attribute value tends to be distributed. Local spatial autocorrelation is used to determine the specific location of spatial agglomeration, and the local Moran Index is calculated as follows:

$$I_i = \frac{X_i - \overline{X}}{S_i^2} \sum_{j=1, j \neq i}^{n} w_{ij}(X_i - \overline{X}) \tag{5}$$

where X_i is the attribute value of i, $\overline{X}$ is the average value, and w_{ij} is the spatial weight matrix; then,

$$S_i^2 = \frac{\sum_{j=1, j \neq i}^{n} w_{ij}}{n-1} - \overline{X}^2 \tag{6}$$

The standardized statistic of local Moran Index test is $Z[I]$:

$$Z_{Ii} = \frac{I_i - E[I_i]}{\sqrt{V[I_i]}} \tag{7}$$

$$E[I_i] = -1/(n-1) \tag{8}$$

$$V[I_i] = E[I_i^2] - E[I_i]^2 \tag{9}$$

At the 0.05 significance level, $Z>1.96$ indicates that cities with high $PM_{2.5}$ concentrations are surrounded by other cities with high $PM_{2.5}$ concentrations and that cities with low $PM_{2.5}$ concentrations are surrounded by other cities with low $PM_{2.5}$ concentrations. $Z<-1.96$ indicates that cities with high $PM_{2.5}$ concentrations are surrounded by cities with low $PM_{2.5}$ concentrations and that cities with low $PM_{2.5}$ concentrations are surrounded by cities with high $PM_{2.5}$ concentrations. When $Z=0$, the observations demonstrate an independent random distribution.

3 The variation and regularity of $PM_{2.5}$ concentrations in the BTH region

3.1 The monthly variation characteristics of $PM_{2.5}$ concentrations

An investigation of the $PM_{2.5}$ concentration changes from the cities in the BTH region over the 12 months in 2016 (see Fig. 1) reveals that the median monthly $PM_{2.5}$ concentrations in the 13 cities demonstrate U-shaped oscillations over the entire time period. Specifically, the $PM_{2.5}$ concentration demonstrated a downward trend from January to May, was overall stable from June to August, and finally increased from October to December. Among the latter months, the

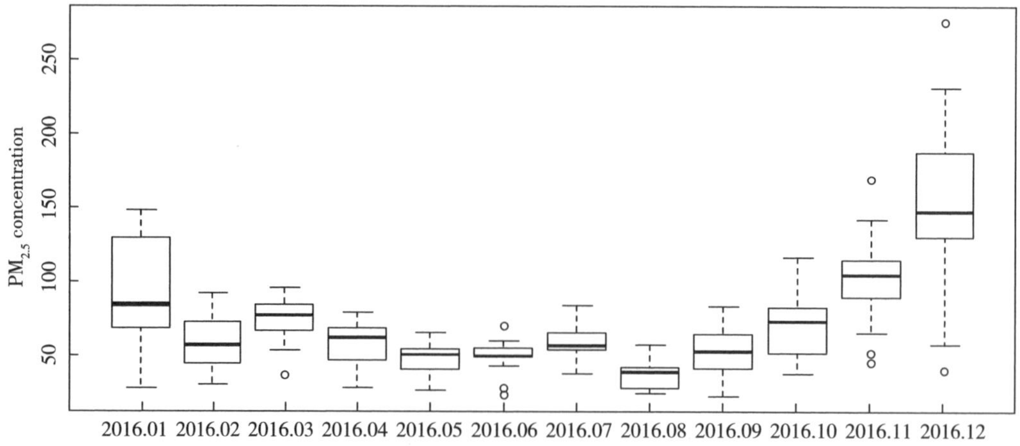

Fig. 1 The monthly average $PM_{2.5}$ concentrations of cities in the BTH region

highest peak occurred in December and was 147.34 μg · m^{-3}; meanwhile, the lowest peak appeared in August and was 38.95 μg · m^{-3}. From May to September, the median $PM_{2.5}$ concentration was under 60 μg · m^{-3}; thus, this time represents the highest air quality period in the BTH region within the entire year. In general, $PM_{2.5}$ concentrations showed significant differences by month. In the winter, $PM_{2.5}$ pollution is the most severe, but during spring, it begins to decrease and eventually maintains a stable state in the late spring. By the summer, $PM_{2.5}$ pollution decreases to the lowest level and then again begins to increase during late autumn. It can be reasonably speculated that the $PM_{2.5}$ concentration value varies with the seasonal weather and forms a cyclical change pattern.

3.2 The seasonal variation in characteristics of $PM_{2.5}$ concentrations

Hourly data of $PM_{2.5}$ concentrations were collected during spring (from March to May), summer (from June to August), autumn (from September to November) and winter (from December to February), and the quarterly averages and annual averages of $PM_{2.5}$ concentrations in each city were obtained (see Table 1). The average annual $PM_{2.5}$ concentration in the BTH region was 69.38 μg · m^{-3}, which is far above the secondary standard limit (35 μg · m^{-3}) of the Ambient Air Quality Standard. In the same period, the average annual $PM_{2.5}$ concentration in 338 cities was 47 μg · m^{-3}.

Table 1 Seasonal and annual means of the $PM_{2.5}$ concentration in cities of the BTH region during 2016

City	P ($PM_{2.5}$)/(μg · m^{-3})				
	Spring	Summer	Autumn	Winter	Annual mean
Beijing	71.35	58.16	79.55	87.72	74.2
Tianjin	64.95	49.01	73.68	83.16	67.7
Shijiazhuang	65.88	45.27	122.82	121.56	88.88
Qinhuangdao	45.89	36.63	46.99	50.02	44.88
Xingtai	68.31	52.99	91.25	134.67	86.81
Handan	65.17	48.24	71.69	118.13	75.81
Baoding	70.13	54.42	102.58	148.5	93.91
Chengde	40.32	30.07	40.11	52.86	40.84
Langfang	53.08	49.45	61.87	101.68	66.52
Zhangjiakou	30.02	28.88	34.82	30.81	31.13
Hengshui	74.4	66.09	78.93	143.67	90.77
Cangzhou	56.41	48.06	73.15	92.83	67.61
Tangshan	65.4	53.03	81.78	91.13	72.84
BTH	59.33	47.72	73.79	96.67	69.38

Furthermore, in the Yangtze River Delta and Pearl River Delta, the average annual $PM_{2.5}$ concentrations were 46 μg · m^{-3} and 32 μg · m^{-3}, respectively, indicating that the $PM_{2.5}$ pollution in the BTH region is still the most serious among all the studied areas. From the perspective of seasonal changes, the seasonal differences in ambient air pollution in the BTH region were obvious. Specifically, the $PM_{2.5}$ concentration in winter was approximately twice that in summer, and heavy pollution weather appeared primarily in winter, especially during the heating period. Among the winter months, the $PM_{2.5}$ concentration in the BTH region was 135 $\mu g \cdot m^{-3}$ during the heating period from November 15th to December 31st, 2016, which was 2.4 times that of the values during the non-heating period. Five large-scale air pollution processes occurred only in December. From the perspective of the city, $PM_{2.5}$ concentrations in all cities in the BTH region, except Zhangjiakou, were above the national secondary standards. The annual average $PM_{2.5}$ concentration in Baoding was the highest at 93.91 $\mu g \cdot m^{-3}$, which exceeds the national secondary standard limit by 168%.

3.3 The spatial distribution characteristics of $PM_{2.5}$ concentration

By using ArcGIS software and the Kriging interpolation method, we evaluated the spatial interpolation of $PM_{2.5}$ concentrations in the BTH region by month in 2016.

The results are shown in Fig. 2. The spatial interpolation of the $PM_{2.5}$ concentration was characterized by severe haze in the southern region, relatively light haze in the northern region and a slightly prominent pattern in some areas.

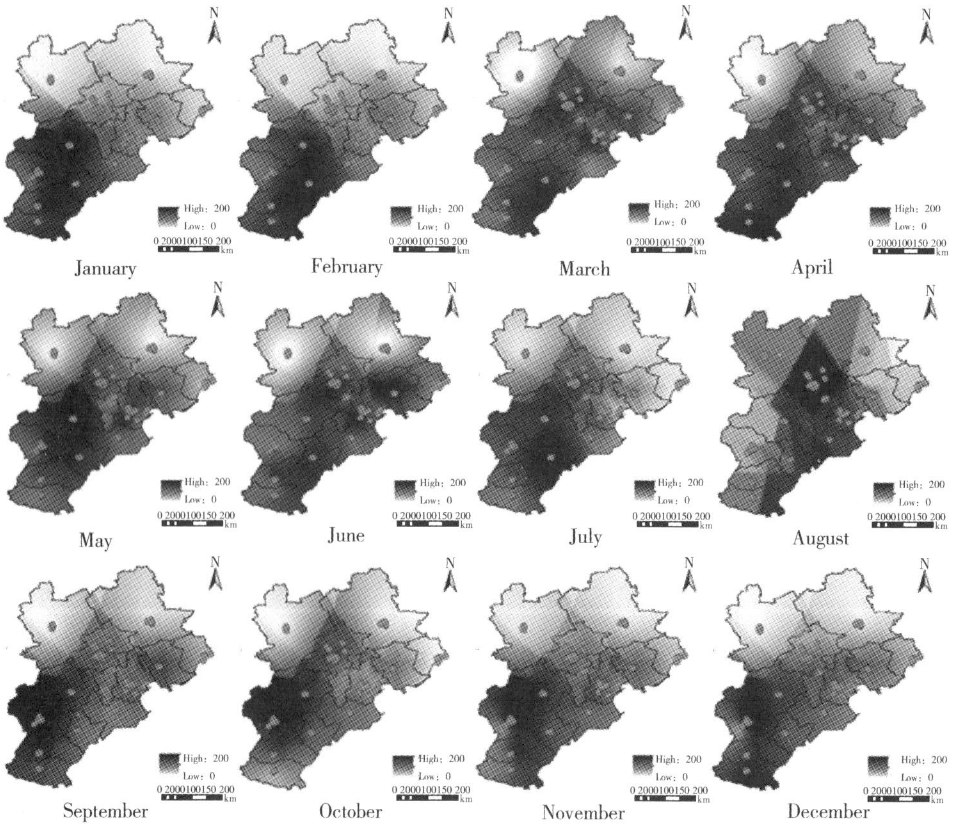

Fig. 2 Spatial distribution of $PM_{2.5}$ concentrations of cities in the BTH region in 2016

Furthermore, the difference between north and south was large. Southern cities, including Baoding, Shijiazhuang, Xingtai and Handan, within Hebei Province exhibited the highest concentrations. Through the combination of temporal and spatial analysis, we demonstrate that the $PM_{2.5}$ concentration in the BTH region began to increase from December to February in the following year; haze first appeared in the southern region but expanded to cover the whole area by February. From March to May, the scope of fog and haze narrowed from northwest to the southeast and then remained stable until September. The haze pollution suddenly increased in October and demonstrated a growing trend.

4 Results and discussion

4.1 Spatial autocorrelation test of the $PM_{2.5}$ concentration

The global spatial autocorrelation analysis can be used to compare the spatial spillover effect of $PM_{2.5}$ concentrations within different months. As shown in Table 2, the global Moran Index in terms of the $PM_{2.5}$ concentration in the BTH region during each of the twelve months from January to December was 2.971, 3.036, 0.732, 1.325, 0.797, 0.181, 2.186, 0.746, 1.190, 1.380, 1.380, 2.482, and 2.372, respectively. These data were obtained using GeoDa software. The Z (I) values of the $PM_{2.5}$ concentrations were greater than 1.96 in January, February, July, November and December, and the significance test indicated that the $PM_{2.5}$ concentrations in these months were spatially homogeneous. As a function of season, winter (from December to February of the following year) and summer (from June to August) respectively represent the highest and lowest spatial autocorrelation seasons of the $PM_{2.5}$ concentration within one year. In other words, the spatial spillover effect is higher in these two seasons than in the other seasons, and $PM_{2.5}$ pollution is more homogenous during these two seasons.

Table 2 Spatial autocorrelation index of $PM_{2.5}$ concentrations in the BTH region of China in 2016

Month	Global Moran's Index	Std-err	P-value	Z-stat
1	0.493	0.193	0.001	2.971
2	0.495	0.194	0.001	3.036
3	0.039	0.176	0.213	0.732
4	0.144	0.179	0.103	1.325
5	0.056	0.181	0.208	0.797
6	0.117	0.184	0.305	0.181
7	0.321	0.183	0.014	2.186
8	0.049	0.182	0.218	0.746
9	0.126	0.186	0.126	1.190
10	0.155	0.177	0.084	1.380
11	0.322	0.167	0.013	2.482
12	0.354	0.186	0.012	2.372

4.2 The spatial pattern evolution characteristics of $PM_{2.5}$ concentrations in the BTH region

Local spatial autocorrelation analysis was used to analyze the spatial pattern evolution characteristics of $PM_{2.5}$ pollution and included Moran's Index scatter plot and local indicators of spatial association (LISA) agglomeration analysis. Figure 3 shows the global Moran Index scatter plot for each season of the BTH region in 2016.

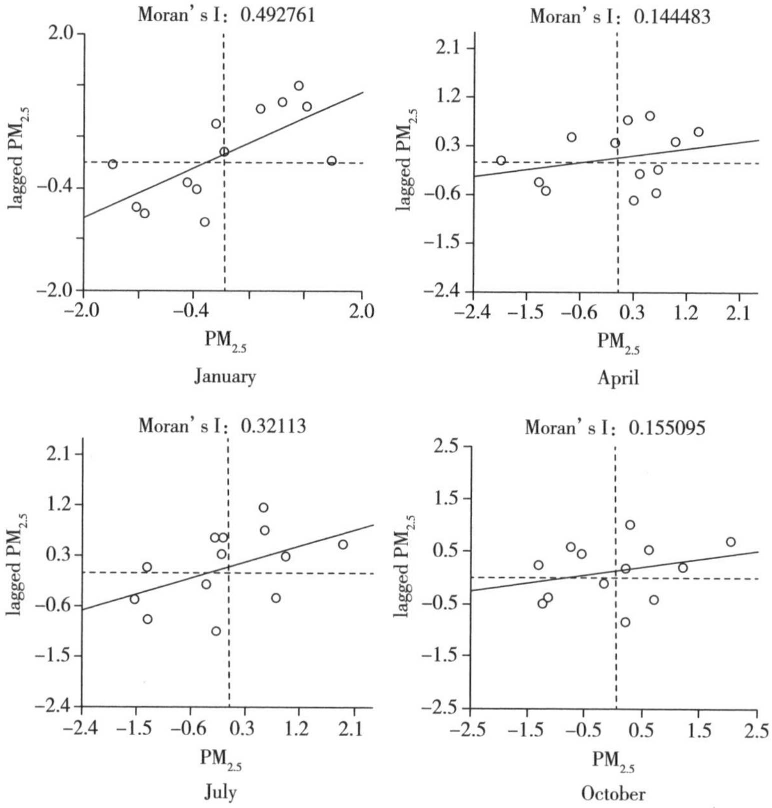

Fig. 3 Global Moran Index scatter plot of $PM_{2.5}$ concentrations in different seasons

The abscissa represents standardized $PM_{2.5}$ concentration in cities, and the ordinate is the neighboring $PM_{2.5}$ concentration value as determined by the spatial weight matrix based on the Euclidean distance, also known as the space lag vector. The four quadrants of the Moran Index scatter plot represent different agglomeration types. The first quadrant indicates the high-high (HH) agglomeration zone, which means that the $PM_{2.5}$ concentrations in a city and in its surrounding cities are high. The second quadrant indicates the low-high (LH) aggregation zone, which means that the $PM_{2.5}$ concentration in a city is low, but values in the surrounding cities are high. The third and fourth quadrants indicate low-low (LL) and high-low (HL) aggregation zones, respectively. The HH and LL agglomeration zones reflect the homogeneity of $PM_{2.5}$ pollution, which indicates positive spatial autocorrelation. The HL and LH aggregation zones reflect the heterogeneity of $PM_{2.5}$ pollution, which indicates negative spatial autocorrelation.

The spatial correlation of $PM_{2.5}$ concentration in the BTH region varies with seasons, and the spatial spillover effect in winter is the most significant. The cities with the highest $PM_{2.5}$ concentration are clustered near the origin point in spring and show strong spatial heterogeneity.

Fig. 4 shows that overall, Shijiazhuang, Hengshui and other cities that are located far from the Bohai Bay are HH agglomeration centers. This finding maybe due to favorable conditions for air diffusion in coastal cities and to the fact that atmospheric pollutants are easily spread. Meanwhile, Chengde, Qinhuangdao and other coastal cities showed LL agglomeration characteristics, which maybe attributed to the low and flat terrain of the southern BTH region, which is not conducive to $PM_{2.5}$ diffusion. Furthermore, the spatial dependence of the $PM_{2.5}$ concentration in the urban agglomeration of the BTH region shows a periodic change. During the period from November to February of the following year, the Z-value index was the highest, which indicated that the agglomeration was obvious. This phenomenon is mainly caused by the initiation of coal-fired heating in the northern parts, leading to the spread of $PM_{2.5}$ from Shijiazhuang and Hengshui to neighboring cities and triggering an increasing area of fog and haze. From March to June, the Z-value index was reduced to the lowest level in the whole year. Because no feature points dominated and the spatial homogeneity was weakened, the range of $PM_{2.5}$ pollution tended to disappear. From July to October, the scope of the HH agglomeration zone expanded again, and the fine air quality of

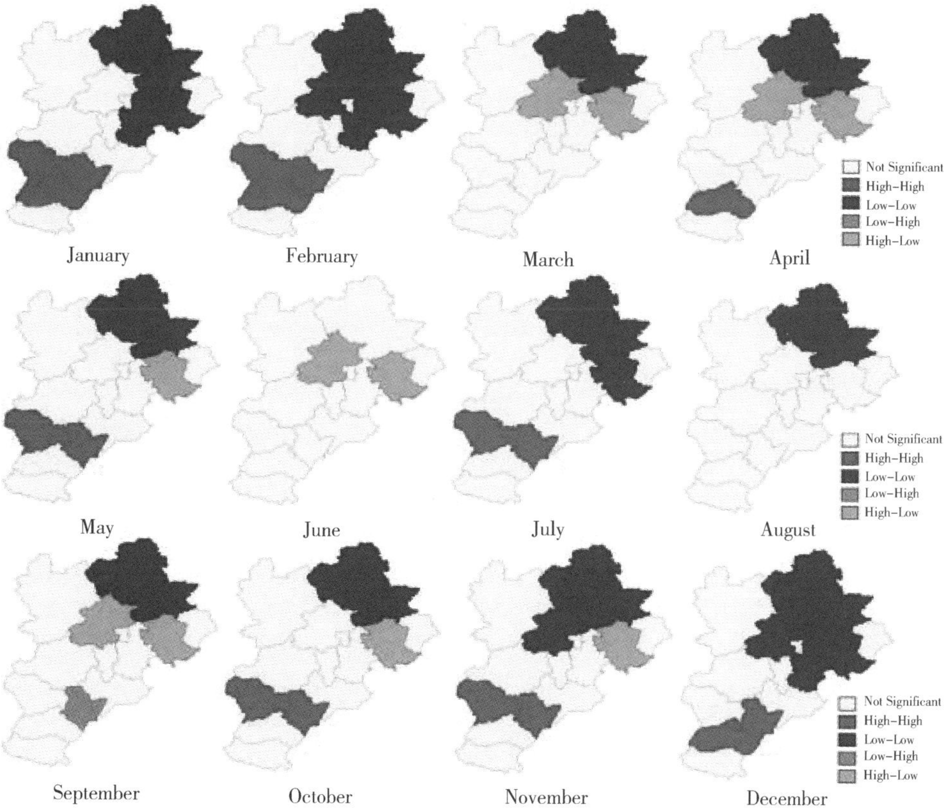

Fig. 4 Monthly spatial agglomeration diagram of $PM_{2.5}$ concentrations in the BTH region in 2016

the northern city tended to be stable. This stability maybe caused by the increase in rainfall in summer, which would enhance the purification effect on $PM_{2.5}$.

Interestingly, a regional haze pollution community has been formed, as the $PM_{2.5}$ concentration in the BTH region has shown obvious convergence characteristics. This phenomenon may exist because the BTH region is the most concentrated area of China's steel industry, and this industry consumes a great deal of fossil fuel and produces a large amount of industrial emissions. During the autumn and winter of the heating period in the BTH region, coal smoke pollutants caused by industrial boilers and heating boilers increased significantly. Although the atmosphere is stable, the frequency and intensity of inversion is high and thus prone to the agglomeration of pollutants. However, the weather is dry, windy and rainy in the spring, and when summer arrives, the atmospheric stability decreases with concentrated rainfall. Under these conditions, these seasons are not conducive to the formation of concentrated pollutants. Air is a public good without property rights, and despite the serious pollution of fog and haze, regional governments have introduced a large number of high energy-consuming industries in order to develop the economy. Thus, the internal costs have been externalized. There is a linkage between the haze pollution in the cities of the BTH region. If governance measures are implemented only for a single city, the elevated $PM_{2.5}$ concentration in the surrounding area will still cause the local haze concentration to increase. Therefore, regional joint governance is needed.

5 Conclusions and policy implications

The urban agglomeration in the BTH region is atypical haze-prone area, and studying the spatial pattern evolution characteristics of $PM_{2.5}$ concentrations in the BTH region is important for understanding the mechanism of $PM_{2.5}$ pollution and the prevention of haze phenomenon. Based on the $PM_{2.5}$ data released by the China Environmental Monitoring Station, this research analyzed the spatial autocorrelation degree and spatial clustering pattern of $PM_{2.5}$ concentrations in different seasons of the BTH region based on spatial dependence theory. The primary conclusions are follows.

(1) The distribution of $PM_{2.5}$ concentrations in this area varied greatly in 2016. On one hand, it increased from late autumn to early winter, and the spatial range expanded from southeast to northwest. On the other hand, the $PM_{2.5}$ concentration decreased rapidly from late winter to early spring, and the spatial range was narrowed from northwest to southeast.

(2) The degree of spatial dependence by season was in the following order from highest to lowest: winter, autumn, spring, summer. Winter (from December to February in the following year) and summer (from June to August) were, respectively, the highest and lowest seasons with regard to the spatial homogeneity of $PM_{2.5}$ concentrations.

(3) The agglomeration pattern of $PM_{2.5}$ concentrations in the BTH region is significant. Generally, cities such as Shijiazhuang and Hengshui, which are located far from Bohai Bay,

exhibited a high-high concentration of $PM_{2.5}$ pollution, whereas coastal cities, such as Chengde and Qinhuangdao, demonstrated a low-low concentration.

In 2013, the State Council issued the Air Pollution Prevention and Control Action Plan. By 2017, the concentration of fine particles in the BTH region, the Yangtze River Delta and the Pearl River Delta decreased by 25%, 20% and 15% respectively. Among these areas, the BTH region had the most stringent reduction targets because of its severe air pollution. The $PM_{2.5}$ concentration is the result of the atmospheric reaction of air pollutants, which is affected by atmospheric diffusion conditions. Reducing coal consumption, adjusting energy structure and implementing other initiatives can reduce pollutant emissions, but it is difficult to determine the specific air quality objectives that can be achieved. There are uncertainties in the policy effects of air pollution control programs. Based on the above empirical results, the following suggestions are put forward for haze control in the BTH urban agglomeration. Haze pollution exhibits a linkage relationship in the BTH region. If only a single city is controlled, the high $PM_{2.5}$ concentration in the surrounding cities will increase the $PM_{2.5}$ concentration in this city. Therefore, a regional joint governance approach should be adopted. For example, an air pollution joint defense mechanism could be setup for early warning of $PM_{2.5}$ pollution and rational distribution of pollution control costs among cities in the BTH region through an ecological compensation mechanism. In addition, the haze pollution in the BTH region has two main features. First is the significant spatial difference. The haze pollution in the southern region is relatively more serious, while that in the northern region is relatively light. Second is the significant seasonal difference. The haze during the heating season is serious, while that during the non-heating season is relatively light. Therefore, the policy focus for haze pollution control should be shifted to the control of pollutant discharge in the southern part of the BTH region and should reduce the frequency of severe fog and haze occurrences during the heating season. Finally, monitoring points in the southern region of the BTH region should be increased, and cities far from Bohai Bay, such as Shijiazhuang and Hengshui, should be listed as key cities for prevention and control. Generally, regional joint management is of great value in controlling pollutant emissions and improving the quality of the atmospheric environment. It is necessary to improve the BTH urban agglomeration's collaborative facilities and to promote the integration of the BTH region in order to minimize governance cost.

Presently, existing studies on the $PM_{2.5}$ concentration in China still concentrate on $PM_{2.5}$ components, source and space-time phenomena. The factors influencing $PM_{2.5}$ and its interaction with the urbanization rate based on long-term data are important topics for future research.

Acknowledgments The authors express their sincere thanks for the support from the National Natural Science Foundation of China under Grant No. 71173200, the Development and Research Center of China Geological Survey under Grant No. 12120114056601 and No. 12120113093200, the National Science and Technology Major Project under Grant No. 2016ZX05016005-003 and the Fundamental Research Funds for the Central Universities under Grant No. 53200859633.

Appendix A

Table A1 A Nomenclature List

Full name	Abbreviation
Data Envelopment Analysis	DEA
Beijing-Tianjin-Hebei region	BTH region
Environmental Kuznets Curve	EKC
Air Quality Index	AQI
Gross Domestic Product	GDP
Inverse Distance Weighted method	IDW method
Ordinary Kriging method	OK method
High-High agglomeration	HH agglomeration
Low-High aggregation	LH agglomeration
Low-Low aggregation	LL agglomeration
High-Low aggregation	HL agglomeration
Local Indicators of Spatial Association	LISA

Table A2 $PM_{2.5}$ concentrations in 13 cities in the BTH region by month

City	2016.01	2016.02	2016.03	2016.04	2016.05	2016.06	2016.07	2016.08	2016.09	2016.10	2016.11	2016.12
Beijing	68.15	43.48	92.67	67.54	53.85	59.28	68.6	46.6	54.63	84.73	99.3	130.55
Tianjin	73.52	50.29	80.84	63.35	50.65	53.32	52.97	40.75	52.47	64.07	104.49	134.9
Shijiazhuang	129.63	71.61	83.16	60.33	54.15	44.91	65.72	25.18	83.02	116.4	169.03	276.06
Qinhuangdao	40.2	38.09	63.57	39.36	34.73	42.53	39.75	27.62	36.49	39.65	64.83	84.3
Xingtai	129.09	81.34	88.2	66.1	50.64	54.17	65.86	38.95	67.35	82.23	124.17	188.04
Handan	108.58	72.3	75.09	78.67	41.76	48.75	54.73	41.23	56.12	50.87	108.07	231.65
Baoding	148.28	83.37	76.78	68.23	65.37	54.46	70.95	37.85	69.35	96.25	142.13	189.99
Chengde	44.79	40.86	53.25	37.63	30.07	27.56	37.13	25.52	27.59	41.66	51.07	56.73
Langfang	88.45	50.74	73.08	46.09	40.07	48.9	56.38	43.08	41.07	55.29	89.25	151.21
Zhangjiakou	27.52	30.14	36.49	27.45	26.11	22.87	39.78	24	22.08	37.59	44.8	39.25
Hengshui	133.78	91.38	95.28	72.41	55.52	58.11	83.57	56.59	47.27	75.19	114.34	159.27
Cangzhou	84.17	56.59	66.63	57.25	45.36	46.79	56.25	41.13	43.93	72.87	102.64	142.25
Tangshan	78.32	58.92	84.6	61.53	50.07	69.52	55.2	34.38	64.04	72.96	108.33	147.34

References

Alves C, Vicente A, Pio C, et al. Organic compounds in aerosols from selected European sites - Biogenic versus, anthropogenic sources [J]. Atmospheric Environment, 2012, 59 (59): 243-255.

Bates D V, Sizto R. Air pollution and hospital admissions in Southern Ontario: the acid summer haze effect [J]. Environmental research, 1987, 43 (2): 317-331.

Cai S, Wang Y, Zhao B, et al. The impact of the "Air Pollution Prevention and Control Action Plan" on $PM_{2.5}$ concentrations in Jing-Jin-Ji region during 2012-479 2020 [J]. Science of the Total Environment, 2017, 580: 197-209.

Chalbot M C G, Jones T A, Kavouras I G. Trends of non-accidental, cardiovascular, stroke and lung cancer mortality in Arkansas are associated withambient $PM_{2.5}$ reductions [J].

Civan M Y, Elbir T, Seyfioglu R, et al. Spatial and temporal variations in atmospheric VOCs, NO_2, SO_2, and O_3 concentrations at a heavily industrialized region in Western Turkey, and assessment of the carcinogenic risk levels of benzene [J]. Atmospheric Environment, 2015, 103: 102-113.

Dockery D W, Pope C A, Xu X, et al. An association between air pollution and mortality in six US cities [J]. New England journal of medicine, 1993, 329 (24): 1753-1759.

Guan D, Su X, Zhang Q, et al. The socioeconomic drivers of China's primary $PM_{2.5}$ emissions [J]. Environmental Research Letters, 2014, 9 (2): 024010.

Hao Y, Liu Y M. The influential factors of urban $PM_{2.5}$ concentrations in China: a spatial econometric analysis [J]. Journal of Cleaner Production, 2016, 112: 1443-1453.

Hsu C Y, Chiang H C, Chen M J, et al. Ambient $PM_{2.5}$ in the residential area near industrial complexes: Spatiotemporal variation, source apportionment, and health impact [J]. Science of the Total Environment, 2017.

Hu W, Downward G S, Reiss B, et al. Personal and indoor $PM_{2.5}$ exposure from burning solid fuels invented and unvented stoves in a rural region of China with a high incidence of lung cancer [J]. Environmental science & technology, 2014, 48 (15): 501 8456-8464.

Hussain S Q, Ahn S H, Park H, et al. Light trapping scheme of ICP-RIE glass texturing by SF 6/Ar plasma for high haze ratio [J]. Vacuum, 2013, 94: 87-91.

Jansen R C, Shi Y, Chen J, et al. Using hourly measurements to explore the role of secondary inorganic aerosol in $PM_{2.5}$ during haze and fog in Hangzhou, China [J]. Advances in Atmospheric Sciences, 2014, 31 (6): 1427-1434.

Lee S J, Serre M L, van Donkelaar A, et al. Comparison of geostatistical interpolation and remote sensing techniques for estimating long-term exposure to ambient $PM_{2.5}$ concentrations across the continental United States [J]. Environmental health perspectives, 2012, 120 (12): 1727.

Li L, Yang J, Song Y F, et al. The burden of COPD mortality due to ambient air pollution in Guangzhou, China [J]. Scientific reports, 2016, 6.

Liao T, Wang S, Ai J, et al. Heavy pollution episodes, transport pathways and potential sources of $PM_{2.5}$ during the winter of 2013 in Chengdu (China) [J]. Science of the Total Environment, 2017, 584-585: 1056.

Liu C, Henderson B H, Wang D, et al. A land use regression application into assessing spatial variation of intra-urban fine particulate matter ($PM_{2.5}$) and nitrogen dioxide (NO_2) concentrations in City of Shanghai, China [J]. Science of the Total Environment, 2016, 565: 607-615.

Liu Q, Wang Q. Sources and flows of China's virtual SO_2, emission transfers embodied in interprovincial trade: A multiregional input-output analysis [J]. Journal of Cleaner Production, 2017, 161.

Liu Z, Cheng K, Li H, et al. Exploring the potential relationship between indoor air quality and the concentration of airborne culturable fungi: a combined experimental and neural network modeling study [J]. Environ Sci Pollut Res Int, 526 2017b (6): 1-8.

Liu Z, Hu B, Wang L, et al. Seasonal and diurnal variation in particulate matter 528 (PM_{10}, and $PM_{2.5}$) at an urban site of Beijing: analyses from a 9-year study [J]. Environmental Science and Pollution Research, 2015a, 22 (1): 627-642.

Liu Z, Li H, Cao G. Quick estimation model for the concentration of indoor airborne culturable bacteria: an application of machine learning [J]. International Journal of Environmental Research and Public Health, 2017a, 14 (8): 857.

Liu Z, Zhu Z, Zhu Y, et al. Investigation of dust loading and culturable microorganisms of HVAC systems in 24 office buildings in Beijing [J]. Energy and Buildings, 2015b, 103: 166-174.

Luo J, Du P, Samat A, et al. Spatiotemporal Pattern of $PM_{2.5}$ Concentrations in Mainland China and Analysis of Its Influencing Factors using Geographically Weighted Regression [J]. Scientific Reports, 2017, 7.

Ma M, Shen L, Ren H, et al. How to Measure Carbon Emission Reduction in China's Public Building Sector: Retrospective Decomposition Analysis Based on STIRPAT Model in 2000—2015 [J]. Sustainability, 2017c, 9 (10): 1744.

Ma M, Yan R, Cai W. An extended STIRPAT model-based methodology for evaluating the driving forces affecting carbon emissions in existing public building sector: evidence from China in 2000—2015 [J]. Natural Hazards, 2017b, 89 (2): 741-545 756.

Ma M, Yan R, Du Y, et al. A methodology to assess China's building energy savings at the national level: An IPAT-LMDI model approach [J]. Journal of cleaner production, 2017a, 143: 784-793.

Moroń W, Rybak W. NO_x and SO_2 emissions of coals, biomass and their blends under different oxy-fuel atmospheres [J]. Atmospheric Environment, 2015, 116: 65-71.

National Bureau of Statistics of the People's Republic of China. Tianjin statistical yearbook 2014 [M]; China Statistical Press: Beijing, China, 2014.

Shuai C, Shen L, Jiao L, et al. Identifying key impact factors on carbon emission: evidences from panel and time-series data of 125 countries from 1990 to 2011 [J]. Applied Energy, 2017, 187: 310-325.

Su B, Thomson E. China's carbon emissions embodied in (normal and processing) exports and their driving forces, 2006—2012 [J]. Energy Economics, 2016, 59: 414-422.

Tobler W R. A computer movie simulating urban growth in the Detroit region [J]. Economic geography, 1970, 46 (supl): 234-240.

Wang X, Ding X, Fu X, et al. Aerosol scattering coefficients and major chemical compositions of fine particles observed at a rural site in the central Pearl River Delta, South China [J]. Journal of Environmental Sciences, 2012, 24 (1): 72-77.

Wang Y S, Yao L, Wang L L, et al. Mechanism for the formation of the January 2013 heavy haze pollution episode over central and eastern China [J]. Science China Earth Sciences, 2014, 57 (1): 14-25.

Wang Z, Fang C, Guang X U, et al. Spatial-temporal characteristics of the PM_ (2.5) in China in 2014 [J]. Acta Geographica Sinica, 2015 (in Chinese).

Wu J, Li J, Peng J, et al. Applying land use regression model to estimate spatial variation of $PM_{2.5}$ in Beijing, China [J]. Environmental Science and Pollution Research, 2015, 22 (9): 7045-7061.

Wu X, Chen Y, Guo J, et al. Spatial concentration, impact factors and prevention-control measures of $PM_{2.5}$ pollution in China [J]. Natural Hazards, 1-18.

Xu B, Lin B. Regional differences of pollution emissions in China: contributing factors and mitigation strategies [J]. Journal of Cleaner Production, 2015, 579 112 (4): 1454-1463.

Yan D, Lei Y, Li L, et al. Carbon emission efficiency and spatial clustering analyses in China's thermal power industry: Evidence from the provincial level [J]. Journal of Cleaner Production, 2017.

Yan R, Ma M, Pan T. Estimating energy savings in Chinese residential buildings from 2001 to 2015: A decomposition analysis [J]. Journal of Engineering Science & Technology Review, 2017, 10 (1).

Zhang A, Qi Q, Jiang L, et al. Population exposure to $PM_{2.5}$ in the urban area of Beijing [J]. PloS one, 2013, 8 (5): e63486.

Zhang Y J, Hao J F, Song J. The CO_2 emission efficiency, reduction potential and spatial clustering in China's industry: evidence from the regional level [J]. Applied Energy, 2016, 174: 213-223.

Zhao B, Wang P, Ma J Z, et al. A high-resolution emission inventory of primary pollutants for the Huabei region, China [J]. Atmospheric Chemistry & Physics, 2012, 593 11 (1): 20331-20374.

Zhao X, Wang X, Ding X, et al. Compositions and sources of organic acids in fine particles ($PM_{2.5}$) over the Pearl River Delta region, south China [J]. Journal of Environmental Sciences, 2014, 26 (1): 110-121.

Zhao Y, Liu Y, Ma J, et al. Heterogeneous reaction of SO_2 with soot: The roles of relative humidity and surface composition of soot in surface sulfate formation [J]. Atmospheric Environment, 2017, 152: 465-476.

Zou B, Wang M, Wan N, et al. Spatial modeling of $PM_{2.5}$, concentrations with a multifactoral radial basis function neural network [J]. Environmental Science and Pollution Research, 2015, 22 (14): 10395-404.

我国煤炭消费产生的温室气体和大气污染排放研究

李莉[1,2]，雷涯邻[1,2,3,*]，吴三忙[1,2]，江勇[1,2]，杨谨[1,2]

1. 中国地质大学（北京），北京 100083
2. 自然资源部资源环境承载力评价重点实验室，北京 100083
3. 北京化工大学，北京 100029

摘　要　人与自然环境和谐发展是生态文明的必然要求。当前，生态环境质量总体改善，但是与经济社会发展不平衡、不协调、不可持续的问题仍然存在。能源消费，尤其是化石能源消费对资源、环境和生态文明建设具有至关重要的影响。基于生态文明视角，首先构建了资源、环境、经济协调发展的系统。其次，以煤炭资源为例，基于 CEADs 开发的 CO_2 排放清单编纂模型核算了 1980—2016 年我国煤炭消费产生的 CO_2 排放量，并基于前人文献中煤炭的 $PM_{2.5}$ 排放因子，核算了 1980—2016 年由煤炭消费产生的 $PM_{2.5}$ 排放量。同时，对 2020—2035 年煤炭消费产生的 CO_2 排放量和 $PM_{2.5}$ 排放量进行了预测。最后，基于核算和预测结果，提出应加强煤炭消费总量控制，提高煤炭净化比重，开发利用清洁能源，继续加大 CO_2 驱油与埋存技术的大规模应用等治理措施。

关键词　温室气体排放，大气污染排放，煤炭消费

Study on greenhouse gas and air pollution emissions from coal consumption in China

LI Li[1,2]，LEI Yalin[1,2,3]，WU Sanmang[1,2]，JIANG Yong[1,2]，YANG Jin[1,2]

1. *China University of Geosciences*，*Beijing* 100083，*China*
2. *Key Laboratory of Carrying Capacity Assessment for Resource and Environment*，*Ministry of Natural Resources of the People's Republic of China*，*Beijing* 100083，*China*
3. *Beijing University of Chemical Technology*，*Beijing* 100029，*China*

Abstract　It is an inevitable requirement of ecological civilization that human and natural environment develop harmoniously. At present，the quality of the ecological environment has improved on the whole，but the problems of imbalance，incoordination，and unsustainability with economic and social development

---

原文刊于《华北水利水电大学学报（自然科学版）》，2021 年第 42 卷第 6 期。

基金项目　国家科技重大专项项目（2016ZX05016005-003）；国家自然科学基金项目（71803182，71773118）；北京市社会科学基金项目（19YJC023）；中国地质大学（北京）2020 年度本科教育质量提升计划建设项目（JGYB202028）。

* 为通信作者。

still exist. Energy consumption, especially fossil energy consumption, has a crucial impact on resources, the environment and the construction of ecological civilization. Based on the perspective of ecological civilization, firstly, a system of coordinated development of resources, environment and economy is constructed. Secondly, taking coal resources as an example, the CO_2 emission inventory compilation model developed based on CEADs has calculated the CO_2 emissions from its consumption from 1980 to 2016. Based on the $PM_{2.5}$ emission factors of coal in previous literature, the $PM_{2.5}$ emissions from coal consumption from 1980 to 2016 are calculated. At the same time, CO_2 emissions and $PM_{2.5}$ emissions from coal consumption from 2020 to 2035 are predicted. Based on the calculation and prediction results, it is proposed to strengthen the control of total coal consumption, improve the proportion of coal purification, develop and utilize clean energy, and continue to increase the large-scale application of CO_2 oil displacement and storage technology.

Keywords Resource environmental accounting, Ecological environmental management, Coal resources, Ecological civilization

能源资源是经济社会持续发展的基础性战略物资。随着工业文明的不断进步以及人们对生活品质、舒适度需求的提高，人类社会对于能源资源的需求和依赖程度也在不断加强。但同时，不可再生资源加速枯竭、区域性大气污染、自然生态环境破坏、全球气候变暖等压力依然存在。能源资源的有限性和片面追求经济增长而导致的资源浪费和环境污染等问题日益引起人们重视。由于我国富含煤炭资源，在今后较长时期内，以煤炭为主的化石能源仍将是我国的主体能源。到 2030 年，煤炭消费比重仍然会在 50%左右[1]，煤炭燃烧产生的 CO_2 仍然不容忽视。我国煤炭消费量大，能源消费的高碳特征比较明显，因而在节约能源和提高单位能源消费的产出效益上仍有较大空间。还有一些大气污染物的产生与消费尤其是煤炭消费相关，在煤炭的开采和利用中产生的 $PM_{2.5}$、SO_2、烟尘、粉尘等是我国大气污染的重要来源。就全国来说，煤炭消费对 $PM_{2.5}$ 年均浓度的贡献大约占 56%[2]。大气污染尤其 $PM_{2.5}$ 污染是影响人类健康的主要危害因素之一[3]。大气污染不仅严重影响了公众的健康[4-6]，而且也在一定程度上影响了城市形象甚至国家形象。因此，定量核算煤炭消费产生的大气污染物排放量，尤其是 $PM_{2.5}$ 排放量，是实施污染控制的前提和依据。

目前，能源消费结构中煤炭消费占比在降低，但煤炭在能源消费结构中的比例与欧美国家的相比仍然差距较大，清洁可再生能源的消费占比还需要提高。据此，本文以煤炭资源为研究对象，构建了资源-环境-经济系统，并在此基础上，研究其对 CO_2 排放量和大气污染物尤其是 $PM_{2.5}$ 排放量的影响，有利于为未来的能源消费结构转变以及 CO_2 减排和大气污染减排提供建议。

1　资源-环境-经济系统构建

在经济社会发展中，资源、环境和经济三大系统之间的关系密切。环境系统是煤炭开采规模和速度的基础，但是煤炭开采的规模和速度又会影响区域的经济系统。反之，经济系统又为改善环境系统提供物质和资金保障。资源、环境、经济形成了一个相互作用的系统，此系统的有序联动，可以为生态文明建设提供前提和基础。

第一，资源-环境系统的联动。在煤炭开采地区，资源-环境系统的联动包括以下过程：在开采和利用当地煤炭资源的同时，对环境造成了破坏；根据煤炭开采区所独有的自然系统，以及对生态环境的破坏程度，规划生态治理的线路，最终使得煤炭开采区域的生态环境得到恢复。

第二，资源-经济系统的联动。依托所在煤炭开采地区的资源，在保障资源安全和生态环境的前提下，对煤炭行业的发展进行升级改造，可以将煤炭资源开发利用、火力发电、煤化工等过程进行结合。通过在煤炭矿区周围建设设施齐全的能源资源基地，推动所在区域经济社会的可持续发展。

第三，资源-环境-经济系统的和谐联动。资源、环境、经济三大系统之间的联动意味着煤炭开采地区实现三者可持续循环发展的模式，也即当地政企之间实现发展理念一致、动作一致、目标一致，对资源的适当开采和对生态环境的保护进行合理规划，以实现煤炭资源循环渐进有序开采、经济社会可持续发展、人民生活不断前进的多重目标。

分析上述三种联动模式可知，资源、环境、经济的协调发展，指在资源开发中，应该采用清洁高效的开采技术，使资源的利用效率最大化，资源浪费最小化，同时对煤炭开采时产生的伴生资源进行综合合理利用；资源开发对生态环境的污染破坏需要达到最小化，将产生的 CO_2 和大气污染物的量降到最低，对矿区及周边生态环境及时治理并恢复，实现资源-环境-经济系统和谐发展、协调发展；在保证煤炭区域内资本存量满足当地需要的同时，又不影响后代未来发展的需要。资源、环境、经济三大系统间相互作用的机理如图 1 所示。

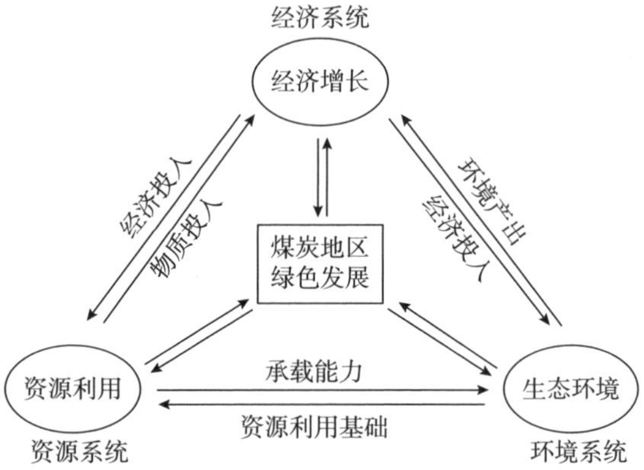

图 1　资源、环境、经济三大系统作用机理

2　资源环境核算

2.1　煤炭资源消费产生的 CO_2 排放量核算

2.1.1　核算方法

CO_2 排放量核算是气候模型构建、各国减排政策制定及国际谈判的基础。当前，国家和全球的 CO_2 排放量主要由国际机构依据 IPCC 方法和能源统计数据进行估算。CO_2 排放量的核算参数及方法主要来自于发达国家的经验值，可能会错误估计中国 CO_2 排放量。CEADs 研究团队基于物质守恒理论，以能源平衡表为基础，开发出一套适用于中国多尺度（国家-省区-地级市）的 CO_2 排放清单编纂模型，并采用更符合中

国能源品质的实测排放因子，系统编纂中国多尺度 CO_2 排放清单[7]，目的在于准确衡量中国 CO_2 排放量。一般来说，CO_2 排放的核算包括能源消费相关的 CO_2 排放和工业生产过程的 CO_2 排放。

由于能源消费中以煤炭消费为主，且其产生的 CO_2 排放占绝大部分。因此，本部分只核算煤炭消费产生的 CO_2 排放量，即煤炭资源消费量乘以 CO_2 排放因子。数据来自 2017 年的中国能源统计年鉴[8] 和实测排放因子[7]。计算公式如下：

$$E = CF \tag{1}$$

式中：E 为煤炭消费产生的 CO_2 排放量；C 为煤炭消费量；F 为煤炭消费的 CO_2 排放因子。

2.1.2　核算结果

科学合理地核算 CO_2 排放量是制定碳减排策略的基础。但是 CO_2 排放量的估算受很多因素的影响，如能源消费量等基础数据的统计口径与统计方法、排放因子的数值选取、不合理的统计数据等都会对 CO_2 排放量的估算造成一定影响。根据 IPCC（2006）国家温室气体清单指南推荐的计算方法，本文选取 1980—2016 年煤炭消费量[8] 和实测排放因子，粗略估算了煤炭消费产生的 CO_2 排放量，如图 2 所示。

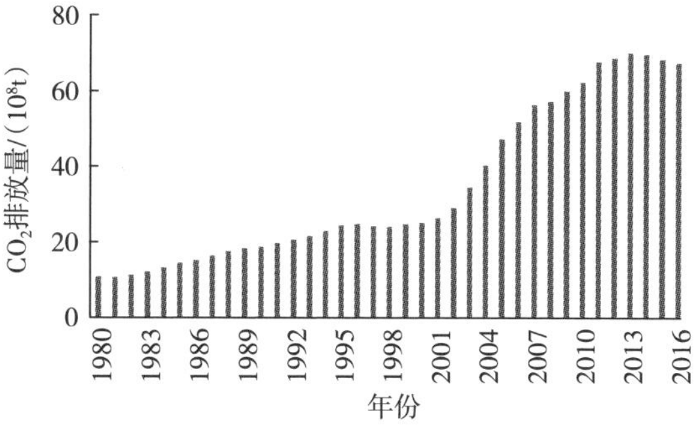

图 2　1980—2016 年煤炭消费产生的 CO_2 排放量

从图 2 中可以看出：在 1980—2002 年间，CO_2 排放量呈现平稳增长；在 2003 年后，随着经济增长和煤炭消费量的大幅增加，CO_2 排放量也随之大幅增加；在 2013 年达到此期间的最高点（70.05×10^8t），此后逐年降低；自 2014 年起 CO_2 排放量呈下降趋势，这与煤炭消费量下降、国家积极采取能源结构转变政策相关。

2.2　煤炭资源消费产生的 $PM_{2.5}$ 排放量核算

2.2.1　核算方法

各种能源的成分不同，其 $PM_{2.5}$ 排放因子也不同，参照北京市污染源普查数据和文献［9-10］，总结出了煤炭消费产生的 $PM_{2.5}$ 排放因子取值，得煤炭资源消费产生的 $PM_{2.5}$ 排放量计算公式，见表 1。

表 1　煤炭资源消费产生的 $PM_{2.5}$ 排放量计算公式

核算公式	来源
E=0.740CE=0.710C	文献［9］文献［10］

2.2.2　核算结果

本文选取 1980—2016 年煤炭消费量[8] 和文献［9］中煤炭消费的 $PM_{2.5}$ 排放因子，

粗略估算了煤炭消费产生的 $PM_{2.5}$ 排放量，如图 3 所示。

从图 3 可以看出：在 1980—2002 年间，$PM_{2.5}$ 排放量从 45.08×10^4t 增长到 120.34×10^4t；自 2003 年后，$PM_{2.5}$ 排放量随着煤炭消费量的增加而增加，在 2013 年达到最高点（291.1×10^4t）。2013 年 1 月，全国多个省市出现了不同程度的雾霾污染，由此，政府开始对 $PM_{2.5}$ 的质量浓度进行监测。随后，随着煤炭消费总量控制政策的实施，煤炭消费量下降，$PM_{2.5}$ 排放量也随之缓慢下降。2016 年，煤炭消费产生的 $PM_{2.5}$ 排放量已下降到 279.9×10^4t。

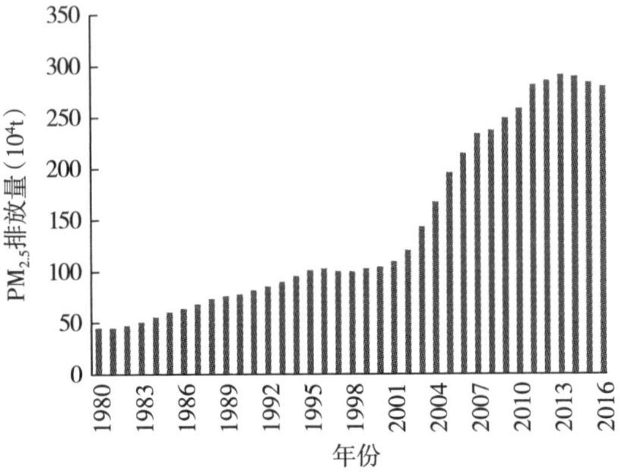

图 3　1980—2016 年煤炭消费产生的 $PM_{2.5}$ 排放量

2.3　结果分析与预测

1）煤炭消费产生的 CO_2 排放量结果分析和预测。按照中国历年实际公布的 CO_2 排放量数据（Wind 数据库中可获得）和煤炭消费产生的 CO_2 占比为 80%的背景核算出来的煤炭消费产生的 CO_2 排放量，与本文核算的图 2 中显示的 1980—2016 年煤炭消费产生的 CO_2 排放量对比，发现误差在 4%以内（2004 年除外），说明本文核算结果可靠。

根据此方法，设定中国未来存在 3 种发展模式，即低、中、高排放模式。低排放模式下，能源消费结构中高碳燃料占比小；中排放模式下能源消费结构中高碳燃料占比适中；高排放模式假定能源消费结构中高碳燃料占比较高。

基于本文所采用的核算方法和 ZHOU Nan 等[11] 关于 2020—2035 年低排放、中排放和高排放三种不同的发展模式和煤炭消费结构的预测结果，预测出 2020—2035 年煤炭消费产生的 CO_2 排放量，如图 4 所示。

由图 4 可知：在低排放模式下，煤炭消费产生的 CO_2 排放量在 2024 年达到峰值（66.16×10^8t）；在中排放模式下，煤炭消费产生的 CO_2 排放量在 2027 年达到峰值（70.68×10^8t）；在高排放模式下，煤炭消费产生的 CO_2 排放量在 2030 年达到峰值（74.99×10^8t）。

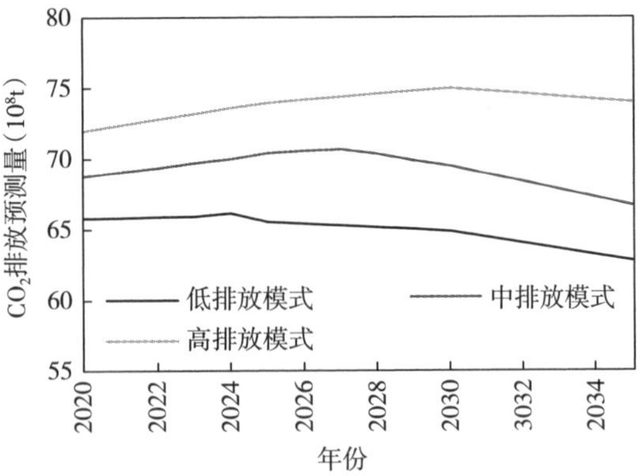

图 4　2020—2035 年煤炭消费产生的 CO_2 排放量预测

2）煤炭消费产生的 $PM_{2.5}$ 排放量结果分析和预测。根据本文关于煤炭消费产生的 $PM_{2.5}$ 排放量的计算方法，以及《国家能源革命战略（2016）》提到的 2020 年将能源消费总量控制在 50×10^8t 标煤以内，到 2030

年控制在60 10^8t 的发展目标[12]，以及对年均增速的设定[13]，本文预测出2020—2035年煤炭消费产生的 $PM_{2.5}$ 排放量，如图5所示。

由图5可以看出，在低、中、高排放模式下，2035年煤炭消费产生的 $PM_{2.5}$ 排放量分别为 341.4×10^4t、348.1×10^4t、354.8×10^4t。

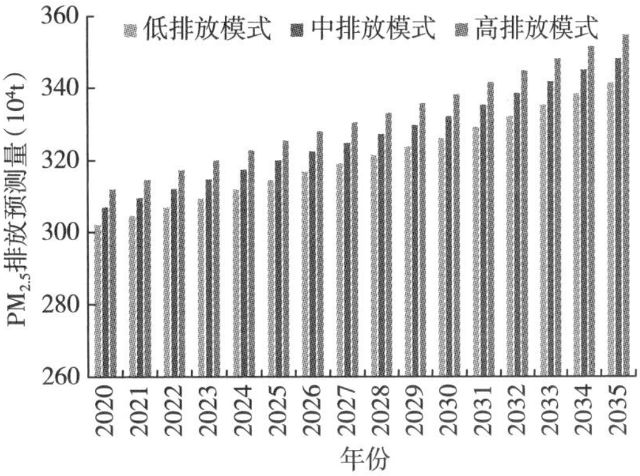

图5　2020—2035年煤炭消费产生的 $PM_{2.5}$ 排放量预测

3　结语

为降低煤炭消费产生的 CO_2 排放量和 $PM_{2.5}$ 污染量，可以从以下方面制定政策，推动环境有效治理，实现绿色可持续发展。

1）加强煤炭消费总量控制，提高煤炭净化比重。煤炭消费产生的碳排放量和 $PM_{2.5}$ 排放量较大，因此，需要继续实行煤炭消费总量控制的措施，同时，提高煤炭净化比重，降低能源碳含量，实现能源低碳化。大力发展煤净化技术是煤炭实现低碳化的一个切实可行的策略。从能源消费比例上看，煤炭占比尽管在下降，但是仍然高于世界平均水平，清洁高效能源的占比不足。正确把握能源消费结构的基本方向，引导能源消费结构向新能源转变。石油和天然气的单位消费量产生的碳排放量较煤炭的低了10%~30%。煤炭资源消费也应向煤炭、石油、天然气并重的现代型能源消费结构转变，利用好“西气东输”工程，做好能源消费结构转型。

2）开发利用清洁能源。诸如太阳能、地热能等新能源可以有效降低化石能源消费带来的二氧化碳和大气污染物，对于解决环境污染问题、实现环境治理具有重要的实际意义。比如地热能作为一种洁净无污染的可再生能源，具有巨大的能源开发和利用优势，相比风能、太阳能，它不受气候、季节、昼夜等等因素的干扰，并且储量大、分布广泛、利用率高，在节能减排、环保净化等方面，有着巨大的作用和竞争优势。我国地热资源十分丰富，远远高于很多地热能利用大国，高效、可持续地开发利用地热资源能够实现经济效益、社会效益和环境效益的协调统一。清洁能源的利用对构建资源节约型和环境友好型社会、保障国家能源安全、改善我国现有能源结构、促进国家节能减排战略目标的实现具有非常重要的意义。

3）CO_2 驱油与埋存。CO_2 驱油与埋存项目可以通过埋存，达到油气增产和 CO_2 排放量减少的双重目标，或者通过驱油提高清洁能源的资源利用率和清洁能源供给量以实现减排。要实现大规模的 CO_2 减排量，技术是最好的备选措施。虽然目前主要集中在技术研究开发和部分规模的示范阶段，减排技术的先进性和效益性也未得到全社会的认可，但 CO_2 驱油与埋存项目进程却能加速该技术的进展，并激励促进低碳领域相关技术的全面快速发展，这就为大规模 CO_2 减排提供了现实基础。同时，减排技术市场化的成功运作和示范，也将推动更多低碳技术进入项目应用和市场化运作。我国已将 CO_2 驱油与埋存纳入了国家战略规划中，此项规划将在降低温室气体排放上起到重要的作用。由于煤炭、电力、化工等部门的碳排放较高，因此需要加大 CO_2 驱油和埋存项目与煤炭、电力和化工行业的结合，为未来我国乃至全球的 CO_2 减排作出贡献。

参考文献

［1］郝宇，魏一鸣．“十三五”及 2030 年能源经济展望［EB/OL］．［2021-04-10］．http：//www.bit.edu.cn/docs/2016-01/20160108195712135846.pdf.

［2］《中国煤炭消费总量控制方案和政策研究项目》课题组．煤炭使用对中国大气污染的贡献［R］.2014.

［3］The World Bank．Cost of pollution in China［R］．USA，Washington D C：World Bank，2007.

［4］LI Li，LEI Yalin，WU Sanmang，et al. The health economic loss of fine particulate matter（$PM_{2.5}$）in Beijing［J］．Journal of cleaner production，2017，161：1153-1161.

［5］DOCKERY D W，POPE C A，XU X，et al. An association between air pollution and mortality in six US cities［J］．New England journal of medicine，1993，329（24）：1753-1759.

［6］World Health Organization．A global assessment of the burden of disease from environmental risks［R］．Geneva：World health organization，2016.

［7］SHAN Yuli，GUAN Dabo，LIU Janghua，et al. Methodology and applications of city level CO_2 emission accounts in China［J］．Journal of cleaner production，2017，161：1215-1225.

［8］国家统计局能源司．中国能源统计年鉴 2017［M］．北京：中国统计出版社，2017.

［9］郑明，马宪国．上海能源消耗与 $PM_{2.5}$ 排放量分析［J］．能源研究与信息，2015，31（1）：1-3.

［10］LIU Zhu，GUAN Dabo，WEI Wei，et al. Reduced carbon emission estimates from fossil fuel combustion and cement production in China［J］．Nature，2015，524：335-338.

［11］ZHOU Nan，FRIDLEY David，KHANNA Nina Zheng，et al. China's energy and emissions outlook to 2050：perspectives from bottom-up energy end-use model［J］．Energy policy，2013，53：51-62.

［12］中华人民共和国国家发展和改革委员会．国家发展改革委和国家能源局关于印发《能源生产和消费革命战略（2016—2030）》的通知［EB/OL］．［2021-04-10］．http：//www.ndrc.gov.cn/zcfb/zcfbtz/201704/t20170425_ 845284.html.

［13］刘卫东，仲伟周，石清.2020 年中国能源消费总量预测：基于定基能源消费弹性系数法［J］．资源科学，2016，38（4）：658-664.

区域可持续发展

Study on the coordinated development of economy, environment and resource in coal-based areas in Shanxi Province in China: Based on the multi-objective optimization model

LI Li[1,2], LEI Yalin[1,2,*], WU Sanmang[1,2], HE Chunyan[3], YAN Dan[1,2]

1. *School of Humanities and Economic Management*, *China University of Geosciences*, *Beijing* 100083, *China*
2. *Key Laboratory of Carrying Capacity Assessment for Resource and Environment*, *Ministry of Land and Resources*, *Beijing* 100083, *China*
3. *Foreign Language Department*, *China University of Geosciences*, *Beijing* 100083, *China*

Abstract Coal-based areas face resource depletion, environmental pollution, ecological damage, and other coordinated economic and social pressures. Based on the theory of policy network and the interest interaction in the policy network of the stakeholders, the paper puts forward the development goals of economic growth maximization, environmental pollution minimization and coal resource consumption minimization. Utilizing the data of the input-output table in Shanxi in 2007 and the China Environment Statistics Yearbook in 2008 and the multi-objective optimization model, this paper designs and assesses the green development policies effects, taking Shanxi Province as the case. The results are that: (1) Three stakeholders of coordinated development in coal-based regions are the government, coal enterprises and the residents of the mine area. (2) The common goals of the three stakeholders are the maximization of industry output value, the minimization of pollutant discharge, and the minimization of coal consumption. (3) The research shows the coal industry will still be the main industry in Shanxi in 2015—2020. According to the conclusions, the paper puts forward to pay attention to the government interests, coal enterprises and residents of the mining area. Furthermore, coal exploitation and utilization should be planned scientifically, the trend of the coal market and price fluctuations should be analyzed, and the appropriate tax rate should be formulated.

Keywords Coordinated development, Policy network theory, Multi-objective optimization, Shanxi Province

1 Introduction

1.1 Background

China's economy has been maintaining at a high speed since its reform and opening.

原文刊于 *Resources Policy*，2018 年第 55 卷。原文发表时为彩色。

* 为通信作者。

However, China's economy has entered a shift from high-speed growth to intermediate-speed growth in the schedule change, namely the economy has entered a new normal. The average annual growth rate from 2012 to 2015 was much lower than that from former years (National Bureau of Statistics of China, 2016). At the same time, the huge pressure of resource depletion and environmental damage which China faces are unprecedented. Contradictions among the environment, resource, ecology, human health, and social and economic development have become severe situations in China's current economic and social development. In the process of economic and social environment coordinated development, environment was deteriorated and then has been improved. Not all countries or regions have to go through the process of the treatment after pollution. Some countries control pollution emissions and resource utilization in the economic take-off phase. Since environmental degradation is irreversible, it is important to focus on the environment in coordinated development. Meanwhile, with the economic development, a certain degree of deterioration of the ecological environment is inevitable. However, economic growth, social and environment-friendly simultaneous development should be achieved under the premise of coordinated development.

Coal has been regarded as the most reliable and abundant resource in China's strategy, which occupies an important position in the energy industry and national economy. However, China's coal industry has changed since 2011 and coal-based regions are under pressure now. Coal-based regions face various problems, such as excessive rapid consumption of resources, serious ecological destruction, low-level environmental governance, unreasonable industrial structure, in-suffcient coordinated development power and so on. Shanxi Province is the first biggest in coal production and typical coal-based province in China. Under the guidance of "only paying attention on the GDP growth rate", Shanxi Province had extensive development, where many serious waste problems existed. Coal mining in Shanxi Province has made great contributions to economic and social development, but it has also brought about ecological damage, environmental damage and geological disasters. Thus, taking Shanxi Province as an example, the paper studies the coordinated development of coal resources. Studying on the policies of the coordinated development for Shanxi Province can give advice on the coordinated economic growth, environment and resources for China's coal-based regions.

According to the data of Shanxi statistical yearbooks, it can be seen the total GDP growth rate in Shanxi Province was far bigger after entering the twenty-first century. During the period between 2008 and 2009, the GDP growth rate appeared to be a relatively smooth trend, which was due to the transient stagnation caused by the financial crisis in 2008. In recent years, due to the terrible coal market and China's New Normal, the GDP growth rate has been gradually slowing down. The output of raw coal in Shanxi Province had been showing an increasing trend from 1978 to 1998. However, it showed a substantial decline in 2000 and 2001, which was due to the adjustment and rectification of the coal industrial structure. Similarly, due to the financial crisis in 2008, the output of raw coal decreased in 2009, while it increased during the following years.

1.2 Literature review

1.2.1 Research on the coordinated development of economy, environment and resource in the resource-based regions

Research on the resource - based regions has mainly concentrated on coordinated development, but the focuses are different. Former research generally constructed the development index to evaluate the coordinated development degree (Huang et al., 2008; Zhang et al., 2010; Guo, 2012; Yan et al., 2012; Zheng and Zhang, 2017). Foreign research paid more attention to the environmental protection, ecological restoration and other aspects, while domestic research tended to pay more attention to the mode choice of the development.

At the beginning of the 1950s or 1960s, resource-based regions began to appear with the phenomenon of Dutch disease and the resource curse, and the development of some resource-based regions began to slowdown. Since then, experts have begun to explore how to transform and develop the resource-based regions. Sachs and Warner (1995, 1997, 1999, 2001) studied the relationship between economic development and the resources of mining countries and introduced the concept of the "Resource Curse". Its means were that the countries with great natural resource wealth tended to grow more slowly than resource-poor countries. Ross (1999) reviewed a wide range of attempts in both economics and political science to explain the "resource curse". The results indicated that less was known about their political problems. The disparity between strong findings on economic matters and weak findings on political ones partly reflected the failure of political scientists to carefully test their own theories. Mehlum et al. (2006) claimed that countries rich in natural resources constituted both growth and growth loss, which differed according to the quality of the institutions. The paper tested this theory based on the hypothesis that the institutions were decisive for the resource curse, which was confirmed. The results claimed that the institutions did play a role for the countries rich in natural resources. Boyce and Emery (2011) demonstrated that resource abundance was negatively correlated with growth rates but positively correlated with income levels using panel data for the USA from 1970 to 2001. Morrison (2012) examined the purported causal mechanisms underlying this "curse" and showed that they all focused on the revenue that these resources generated for the government. Zaman et al. (2016) recognized that environmental conservation and human activities should be paid more close attention to in the coordinated development. Gonzúlezet al. (2017) deemed that coordinated development except the economics, society and environment Pepended on policy and culture. Now, coordinated development in the abroad is increasingly concerned with the development and utilization of re-newable resources or renewable energy, which is conducive to the reduction of environmental pollution.

Domestic research on the development of resource-based regions was conducted later than abroad research. China began to attach importance to the problem and the transformation of resource - based regional development after 1990s, and the resource - based regions were dominated by coal-based regions or cities, which mainly focused on the development modes, industrial restructuring, the development stage and related policies. Fan (1993) recognized that the industrial structure should be adjusted urgently to develop resource - based regions. Based on the law of the urban pyramid, Shen and Wan (2003) proposed that there were few middle-and small-scale cities of resource-based cities in China and that the development should be accelerated. Based on the review of the resource curse, Shi and Huang (2009) constructed a dynamic model of three sectors, including resources, manufacturing and service sectors, from the economic perspective. The results showed that resource-based regions must make a strict mining plan, prevent excessive exploitation to maintain coordinated development, and plan to continue to afford strengthening support and assistance for the manufacturing industry. Jing and Zhang (2006) recognized that the differences in resource exploitation and urbanization disrupted the power and transformation mechanism between industrialization and urbanization during development. According to the characteristics of the resource economy, this paper suggested the idea of integrating industrialization and urbanization. Zhang (2001), Wang and Li (2005), Zang et al. (2006), Duo et al. (2007) conducted research on the industrial transformation of resource-based cities from different angles. Liu and Li (2009) recognized that resource-based cities could be divided into four layers, involving resources endowment layer, industry pattern layer, the logistics industry layer and the policy system layer. Industry expanded development pattern also needed the support of science and technology, talent and market. Zhou (2014) established the evaluation index system of the coordinated development level of coal industry in Shanxi, and made a comprehensive evaluation for the coordinated de-velopment level of coal industry in Shanxi. According to the analysis, environment and resource efficiency were the main negative factors affecting the coordinated development of Shanxi's coal industry. Chen (2017) argued that in recent years, the single management pattern of coal enterprises in Anhui Province couldn't satisfy the requirement of coordinated development with the influence of the excess capacity, the falling coal prices and other factors.

From the research on the development of resource-based regions, it can be seen that the focus of this study mainly concentrates on the adjustment of the industry structure and economic transformation. Studies on the interaction of stakeholders are rare, and fewer studies have been conducted on quantitative research policy.

1.2.2 Research on policy network theory

Since the beginning of the late 1970s, policy network theory has become an important subject in the public policy field in the western countries. When the states promote public policies, they must coordinate the interests of the relevant groups, integrate of various re-sources to promote public governance capacity. The policy network is the interactive form of this stability and sustained relationship. It has formed different genres in its development process.

From the micro level, scholars in the USA mainly study the relationships among the interest group, bureaucracy and government individuals in the process of policy formulation. Wright (1988) explored the ways in which the relationship between the 'government' and 'industry' in the formation and implementation of policy could be compared cross-sectorally and cross-nationally. Thatcher (1998) considered that determining the appropriate scope and ambitions for network approaches depended on the choice of their fundamental purpose and role, together with their relationship with substantive hypotheses and models.

Most British scholars regarded the policy network as a type of interest model and carried out the research from the intermediate level. Helco (1978) recognized that the majority of decision making was a type of formal and informal network. Rhodes (1990) considered that the policy network was a type of interest community, which was formed by the mutual dependence of each organization. Its internal operation process was a type of exchange process of resources.

From the macro level of analysis, European scholars utilized the policy network to describe and analyze the relationship among the state, the citizens and society. Most European scholars regarded the policy network as a governance mode (Hood, 1986; Lester, 2002). Using the policy network, Zhang (2016a, 2016b) analyzed the policy network elements and system of China's coal mine safety production, and found that there was imbalance in the structure, interaction and rules of China's coal mine safety production policy network.

From the research on the development of the policy network, policy network theory is still at the initial stage, but it provides a new perspective for policy analysis. Additionally, policy network theory used in the coordinated development of the resource-based regions is scarce. Therefore, this paper uses policy network theory to analyze the stakeholders and their targets in the coordinated development of the resource-based regions.

2 Models and methods

This paper constructs a policy network model which consists of interactions between the stakeholders and a multi-objective function with the interest orientations of the stakeholders in the policy network model.

2.1 The policy network chart of coordinated development in coal-based regions

In policy network theory, the formulation of public policy is the result of the interaction of multiple stakeholders. Combined with the Rhodes policy network model, according to the current situation and the policy needs of coal-based regions, the policy network chart of coordinated development in coal-based regions is constructed as shown in Fig. 1.

From Fig. 1, it can be seen in the following.

(1) Policy community is a group or individual involved in the process of a particular

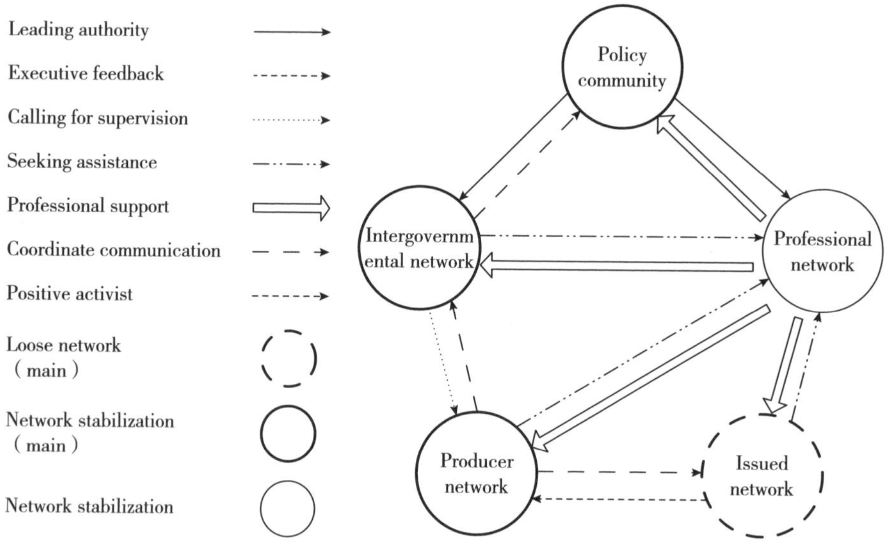

Fig. 1 The policy network chart of coordinated development in coal-based regions

policy. Policy community in the coordinated development in coal-based regions includes the state administrative agencies and legislative bodies.

(2) Professional network is mainly composed by the relevant technical persons. Professional network in the coordinated development in coal-based regions refers to coalscientific research institutions and universities, as well as the experts and scholars in the field of coal development and utilization. The stakeholders in the professional network can provide technical support for the policy community.

(3) Intergovernmental network refers to the close connection and interaction among the governments at all levels. Intergovernmental network in the coordinated development in coal-based regions refers to local governments. Local government is not only the actual executives of the policies, but also the beneficiaries of coordinated development.

(4) The main members of the producer network are front-line workers. Producer network in the coordinated development in coal-based regions refers to coal enterprises. Producer network is closely related to the intergovernmental network, and interact frequently.

(5) Issued network mainly refers to the social organizations and individuals. Issued network in the coordinated development in coal-based regions refers to the residents in the local mining area, coal industry association, environmental protection organization and media workers et al.

2.2 Stakeholder interaction analysis in coal-based regions

According to the analysis results of the stakeholders and their interaction in the policy network, three stakeholders of coordinated development in coal-based regions are the government (central government and local government), coal enterprises and residents of the mining areas.

At the same time, having surveyed typical coal-based regions for 12 times, such as Shaanxi, Shanxi, Shandong, Henan, Guizhou, Hubei, Heilongjiang, Anhui, Gansu, Inner Mongolia, and Jiangsu provinces; having studied more than 20 projects, such as opencast coal mine, underground coal mining, fly ash resource comprehensive utilization, coal gangue power generation projects, ecological reclamation area, coal chemical projects and so on; and having participated in the symposium and expert consultation 14 times, this paper obtains the result that coordinated development in coal-based regions mainly refers to the interests of the government (central government and local government), coal enterprises and residents of the mining area.

Resources here just refer to the natural resources and the paper doesn't include anything else such as the human resource, the dividend of reform, the capital and the knowledge of science and technologies etc, though these aspects are also important for the coordinated development of Shanxi Province, while these index and their goal planning aren't quantified inconveniently and the data can't be found easily. So the paper only focuses on the common goals of the three stakeholders in coal-based regions which are the maximization of industry output values, the minimization of pollutant discharge, and the mini-mization of coal consumption.

2.3 Objective function construction of coordinated development in coal-based regions

The multi-objective optimization model is constructed according to the stakeholders and the common goals of the three stakeholders in coal-based regions. As this paper needs to utilize the output value of industries in Shanxi input-output table in 2007, and the data of the waste water, waste gas and waste residue discharge coefficients in Wang's book (2013) are also the data in 2007, thus this paper selects 2007 as the base year to calculate the model parameters 2.3.1 The multi-objective function

Based on the objective function above and the constraint conditions, the basic model is constructed as follows. And x_i is the output of the i-th industry, where $1 \leqslant i \leqslant 5$.

(1) This paper uses the maximization of the output of the industries to represent the maximization of economic growth, and the max-imization of industry output value is shown in the equation (1):

$$\text{Max} \sum_{i=1}^{n} t_i x_i \tag{1}$$

t_i is the value coefficient vector and it is shown in the equation (2).

$$t_i = \frac{\text{the added value of industry } i \text{ in 2007}}{\text{the total output of industry } i \text{ in 2007}} \tag{2}$$

(2) Pollutant discharge in the paper contains waste water (w_w), waste gas (w_g) and waste residue (w_r). The minimization of pollutant discharge is in the following:

$$\text{Min}(w_w + w_g + w_r) \tag{3}$$

Where waste water discharge is shown in the equation (4), and b_i is the emission factor of waste water in per output value (ton/ten thousand yuan) (Wang, 2013);

$$w_w = \sum_{i=1}^{n} b_i x_i \tag{4}$$

waste gas discharge (waste gas mainly refers to SO_2, CO_2 and dustin this paper) is shown in the equation (5):

$$w_g = \sum_{i=1}^{n} c_i x_i \tag{5}$$

c_i is the emission factor of waste gas in per output value (ton/ten thousand yuan) (Wang, 2013); The calculation of ci refers to Wang's book (2013) and China's Environment Statistics Yearbook of 2008; Where waste residue discharge is shown in the equation (6):

$$w_r = \sum_{i=1}^{n} d_i x_i \tag{6}$$

where, d_i is the emission factor of waste residue in per output value (ton/ten thousand yuan). The calculation of d_i refers to Wang's book (2013) and China's Environment Statistics Yearbook of 2008.

(3) The minimization of coal consumption is shown in the equation (7):

$$\text{Min} \sum_{i=1}^{n} r_i x_i \tag{7}$$

where, r_i is the direct consumption coefficient vector of coal resource.

2.3.1 Constraints setting

The multi-objective optimization model is constructed according to the stakeholders and the common goals of the three stakeholders in coal-based regions. As this paper needs to utilize the output value of industries in the input-output table in Shanxi in 2007 and because the waste water, waste gas and waste residue discharge coefficient data in Wang's book (2013) are also the data in 2007, so this paper selects 2007 as the base year to calculate the model parameters.

The paper sets three constraints in the following.

(1) Size constraint of the industry output value is set in the equation (8), and R_1 is the total output target of the industrial output value (billion yuan).

$$\sum_{i=1}^{n} x_i \geqslant R_1$$
$$x_i - 5\% \leqslant x_i \leqslant x_i - 3\%,\ (1 \leqslant i \leqslant 5) \tag{8}$$

(2) Constraints of the pollutant discharge:

Constraint of waste water discharge, waste gas discharge and waste residue discharge are shown in the equations (9)-(11) in the following:

$$\sum_{i=1}^{n} b_i x_i \leqslant R_2 \tag{9}$$

$$\sum_{i=1}^{n} c_i x_i \leqslant R_3 \tag{10}$$

$$\sum_{i=1}^{n} d_i x_i \leqslant R_4 \quad (11)$$

where R_2, R_3 and R_4 are the total current emissions of wastewater, waste gas and waste residue (ten thousand tons), b_i, c_i and d_i are the emission coefficients of wastewater, waste gas, waste residue in the unit output value of industry i (ton/ten thousand yuan).

(3) Constraint of coal consumption is shown in the equation. (12), and ei is the coal consumption of industry i (ton/ten thousand yuan), R_5 is the available coal resource (ten thousand tons).

$$\sum_{i=1}^{n} e_i x_i \leqslant R_5 \quad (12)$$

3 Results and discussion

3.1 Multi-objective optimization results

Using the data from 2007 to 2013, this paper calculated the average proportion of the output value of the five industries (coal mining and dressing, petroleum processing coking and nuclear fuel processing, raw chemical materials medical and chemical fiber manufacturing, metal smelting and processing, and electricity power/heating supply), accounting for a total output value of 43.1%. Thus, these five industries played an important role in Shanxi Province's economic and social development. The development of the five industries directly affected the coordinated development in Shanxi Province. Therefore, this paper chooses the five industries to probe into the coordinated development in Shanxi Province.

As the year 2015 and 2020 are the last years of the "12th Five-Year Plan" and the "13th Five-Year Plan", these two years are important for judging whether the planning objectives of the "12th Five-Year Plan" and the "13th Five-Year Plan" will be achieved or not. Therefore, this paper selects 2015 and 2020 as the optimal years.

When selecting the model parameters, the values of R_1, R_2, R_3, R_4 and R_5 in 2015 are chosen on the basis of the "12th Five-Year Plan" in Shanxi Province (Shanxi development and reform commission, 2013), the "12th Five-Year Plan" of the coal industry in Shanxi Province (Shanxi development and reform commission, 2013), and the "12th Five-Year Plan" of environmental protection in Shanxi Province (The People's Government of Shanxi Province, 2012). As the "13th Five-Year Plan" has not been issued, the values of R_1, R_2, R_3, R_4 and R_5 in 2015 are chosen according to the "12th Five-Year Plan" in Shanxi Province; the "12th Five-Year Plan" of the coal industry in Shanxi Province; the "12th Five-Year Plan" of environmental protection in Shanxi Province; and the growth rates of R_1, R_2, R_3, R_4 and R_5 in 2015 (Fan, 2011). The object functions and constraints in 2015 and 2020 are shown in the following analysis.

On the basis of considering the questionnaire and the results of the survey, we have consulted some experts and then determined the weight coefficients of the three objective functions in this paper, which include maximizing the output growth of the industry, minimizing the discharge of pollutants and minimizing the consumption of coal resources. Taking into account the development of various regions of Shanxi and the development of various industries, industrial output growth maximization, the discharge of pollutants minimization and the consumption of coal resources minimization are all important measures to achieve green development for Shanxi Province' various regions and various industries. Hence, the three weight coefficients are set by 1/3, 1/3 and 1/3 respectively. And it is reasonable to set same weight coefficients for the three objective functions.

x_1, x_2, x_3, x_4, x_5 are used to represent the five industries of coal mining and dressing, petroleum processing coking and nuclear fuel processing, raw chemical materials medical and chemical fiber manufacturing, metalsmelting and processing, and electricity power/heating supply, respectively.

3.1.1 Multi-objective optimization in 2015

Using Lingo software, this paper can obtain the output values of the five industries, pollutant discharge and coal consumption as follows.

$$\mathrm{Max}\sum_{i=1}^{5} t_i x_i = \mathrm{Max}(0.6x_1 + 0.3x_2 + 0.2x_3 + 0.22x_4 + 0.32x_5) \tag{13}$$

$$\mathrm{Min}(w_w + w_g + w_r) = \mathrm{Min}(\sum_{i=1}^{5} b_i x_i + \sum_{i=1}^{5} c_i x_i + \sum_{i=1}^{5} d_i x_i) \tag{14}$$

$$\mathrm{Min}\sum_{i=1}^{5} r_i x_i = \mathrm{Max}(0.87x_1 + 11.74x_2 + 0.02x_3 + 0.99x_4 + 8.91x_5) \tag{15}$$

3.1.2 Multi-objective optimization in 2020

Similarly, this paper can also obtain the output values of the five industries, pollutant discharge and coal consumption as follows using Lingo software.

$$\mathrm{Max}\sum_{i=1}^{5} t_i x_i = \mathrm{Max}(0.6x_1 + 0.3x_2 + 0.2x_3 + 0.22x_4 + 0.32x_5) \tag{16}$$

$$\mathrm{Min}(w_w + w_g + w_r) = \mathrm{Min}(\sum_{i=1}^{5} b_i x_i + \sum_{i=1}^{5} c_i x_i + \sum_{i=1}^{5} d_i x_i) \tag{17}$$

$$\mathrm{Min}\sum_{i=1}^{5} r_i x_i = \mathrm{Max}(0.87x_1 + 11.74x_2 + 0.02x_3 + 0.99x_4 + 8.91x_5) \tag{18}$$

3.2 Discussion

3.2.1 Economic growth

The output values of the coal industry in 2015 and 2020 accounting for the five industries are approximately 50%, which indicates that the coal industry is still the main industry in Shanxi Province. This fact is also in line with the plan of "coal-based, diversified development, becoming bigger and stronger" in Shanxi Province. Of course, the coal industry needs green development. The value of petroleum processing coking and nuclear fuel processing in 2015 and

2020 accounting for the five industries, appears to have slightly increased, which is similar to the proportion of the coal industry output of the five industries. However, the output of raw chemical materials for medical and chemical fiber manufacturing among the five industries changes substantially from 5% in 2007 to 17.8% in 2015. This finding may be related to the future development pattern in Shanxi Province, as Shanxi Province has proposed to strengthen the growth of the chemical industry, formatting an energy-saving, emission-reducing, and efficient new development model. There was a significant change in the ratio of the output values of metal smelting and processing, decreasing from 25.3% in 2007 to 13.1% in 2015, which was relevant to Shanxi Province's accelerating the adjustment of industrial structure and eliminating outdated capacity. The ratio of the output values of electricity power/heating supply decreased from 10.9% in 2007 to 2.2% in 2015, and the first reason is that Shanxi Province had speeded up the development of "close small factory, open large factory" and old factory technical reform projects; secondly, there was a rise in coal price in the early period from 2007 to 2015, which limited the development of the electricity power/heating supply industry; thirdly, The electricity delivery to the outside of Shanxi Province was down. All these factors have contributed to a decline in the proportion of output in the electricity power/heating supply industry.

3.2.2 Environmental pollution

For the results of multi-objective optimization in 2015, the pollutant discharge of the five industries was 1.35 billion tons. For the results of multi-objective optimization in 2020, the pollutant discharge of the five industries will bes 1.56 billion tons, which increased by 20% from 2015 to 2020. It can be seen from the results, the five industries are still high in energy consumption, high emission and high pollution. The industry adjustment policies have not been implemented well. The adjustment of the five industries should be speed up and the scientific and technological innovation is needed to be accelerated. The government should supervise and urge the five industries to use the technology and equipment with low energy consumption, low emission, and promote the transformation and upgrading of industry to the direction of the recycle, low carbon. At the same time, the application and issue of pollution discharge should be levied and controlled for the five industries.

3.2.3 Coal resource utilization

The coal consumption of the five industries will be 32599.5 ten thousand tons according to the results of multi-objective optimization in 2020, which increases by approximately 20% from 2015 to 2020. This may occur because Shanxi Province will vigorously develop the coal chemical industry and make every effort to enhance power consumption to have a substantial increase in coal consumption in the "13th Five-Year Plan".

4 Conclusions and policy implications

4.1 Conclusions

For a long time, there are a large-scale development and utilization of coal resource in Shanxi Province. But the problems such as single industrial structure, extensive utilization of coal resource, excessive consumption of resources, ecological destruction and environmental pollution have seriously restricted the coordinated development of Shanxi Province. The interests in the economic, resources and environmental values of different stakeholders were different, which directly affected the implementation of the coordinated development policy. Therefore, based on the policy network theory, this paper constructs the multi-objective optimization model of Shanxi Province and draw the following conclusions:

(1) Coordinated development is the best choice of coal resource, environment, economy, and society coordinated development. In terms of its influencing factors, economic factors, resource factors and environmental factors are three main factors affecting the coordinated development in Shanxi Province. Through the regulation of these three elements, coordinated development policy is to achieve the objectives of the economic maximization, coal resource consumption minimization, environmental pollution minimization.

(2) The government (central government and local government), coal enterprises and mining residents (including coal-related practitioners) are the most important stakeholders in the coordinated development policy network of coal areas. These three key stakeholders have many interest conflicts in the policy network. In order to maximize their own goals and interests, they often carry out frequent games and interactions with another aspect. Therefore, in the process of formulating the coordinated development policy in Shanxi Province, the interests of the main stakeholders and policy needs should be given full consideration to improve the policy formulation and implementation and to make up for market failure and government failure.

(3) By constructing the multi-objective optimization model of coordinated development in Shanxi Province, the multiple objectives of maximizing industrial output growth, minimizing environmental pollution and minimizing consumption of coal resource are taken as the quantitative criteria of policy evaluation, and by setting multi-objective optimization principles and the objective function of the constraints, the paper quantitatively evaluates the implementation of the policy effects, which further makes up for the deficiencies that policy network theory can't explain, and enriches and innovates the methods and pathways of policy effect evaluation. This is of great significance for promoting coordinated development in Shanxi Province.

(4) From the multi-objective optimization results, Shanxi Province should plan the scale of coal exploitation and utilization, vigorously promote the coordinated mining of coal resource, develop circular economy and explore differentiated tax policies to achieve the coordinated goals

of maximizing economic growth, minimizing resource consumption environmental pollution.

4.2 Policy implications

(1) Attention needs to be paid to government interests (central government and local government), coal enterprises and residents of the mining area in the coordinated development of coal-based regions, because they are the three stakeholders of coordinated development in coal-based regions.

(2) In the current economic new normal, economic development is a necessary guarantee for coordinated development of the resource and environment. With regard to economic growth, first Shanxi Province should plan for coal exploitation and utilization scientifically, determine the reasonable scale of coal production, and resolutely push ahead with coal production capacity. Second, the coal industry should be transformed and upgraded, such as coal-bed methane, coal gangue, and coal mine water, to improve the comprehensive utilization of resources and achieve economic benefits. Third, the trend of the coal market will be comprehensively analyzed and price fluctuations should be rationally handled.

(3) On the one hand, referring to coal resource utilization, many coal enterprises have a large-scale loss. Taking into account the neutral principle of tax, if the tax rate is too high, it will undoubtedly increase the burden on coal enterprises. On the other hand, it cannot greatly affect the local financial income. Therefore, Shanxi Province must formulate the appropriate tax rate, which can fully reflect the adjustment role of the resource tax, improve the formation mechanism of the price of resource products, and promote regional, coordinated development.

Acknowledgements The authors express their sincere thanks for the support from the National Natural Science Foundation of China under Grant No. 71173200, the Development and Research Center of China Geological Survey under Grant No. 12120114056601 and No. 12120113093200, National Science and Technology Major Project under Grant No. 2016ZX05016005 - 003, Research Institute of Petroleum Exploration & Development: Research on the benefit evaluation index system of overseas single well under Grant No. H0719603, the Fundamental Research Funds for the Central Universities under Grant No. 53200759032.

References

Boyce, J. R., Emery, J. C. H., 2011. Is a negative correlation between resource abundance and growth sufficient evidence that there is a resource curse? Resour. Policy 36, 1-13.

Chen, L., 2017. Study on the development ideas of the coal industry in Anhui Province. Low Carbon World 44 (5).

Duo, S., Li, Z., Li, B., et al., 2007. The problem and strategy exploration of the economic transformation in the resource-based cities in China. China Popolution Resour. Environ. 17, 12-17.

Fan, J., 1993. Research on the industrial structure transformation of coal-based citied in China. Geogr. J. 48, 218-226.

Fan, Y., 2011. Study on the Cost, Path and Policy of Greenhouse Gas Emission Reduction. Science Press, Beijing.

González, M. O. A., Gonçalves, J. S., Vasconcelos, R. M., 2017. Sustainable development: case study in the implementation of renewable energy in Brazil. J. Clean. Prod. 142, 461–475.

Guo, J., 2012. Study on the technical evaluation index in green coal mine construction. Bord. Econ. Cult. 2, 27–28.

Helco, H., 1978. Issue networks and the executive establishment. Public Adm.: Concepts Cases 87–124.

Hood, Christopher, C., 1986. The Tools of Government. Chatham House Publishers, inc, Chatham, New Jersey.

Huang, J., Ni, J., Zhao, Y., et al., 2008. Study on the creation standards and evaluation indices of green mines. China Min. Mag. 17 (7), 36–39.

Jing, P., Zhang, F., 2006. Deviation and integration of industrialization and urbanization in resource-based regions-taking Shanxi Province as an example. Human Geogr. 20, 38–41.

Lester, M. Salamon, 2002. The Tools if Government: A Guide ti the New Governance. Oxford University Press, New York.

Liu, Y., Li, H., 2009. Study on the action mechanism of the logistics industry to the industry transformation of resource-based cities. Logist. Technol. 28 (12), 19–21.

Mehlum, H., Moene, K., Torvik, R., 2006. Institutions and the resource curse. Econ. J. 116, 1–20.

Morrison, K. M., 2012. What can we learn about the "resource curse" from foreign aid? World Bank Res. Obs. 27, 52–73.

Rhodes, R. A. W., 1990. Policy networks a British perspective. J. Theor. Polit. 2, 293–317. Ross, M. L., 1999. The political economy of the resource curse. World Polit. 51, 297–322.

Sachs, J. D., Warner, A. M., 1995. Natural Resource Abundance and Economic Growth. Harvard Institute of International Development, Cambrige, MA (Development Discussion Paper No. 517).

Sachs, J. D., Warner, A. M., 1997. A fundamental source of ong-run growth. Am. Econ. Rev. 87, 184–188.

Sachs, J. D., Warner, A. M., 1999. The big push, natural resource booms and growth. J. Dev. Econ. 59, 43–76.

Sachs, J. D., Warner, A. M., 2001. The curse of natural resources. Eur. Econ. Rev. 45, 827–838.

Shanxi Development and Reform Commission 12th Five-Year Plan of coal industry development in Shanxi Province. <http://www. sxdrc. gov. cn/xxlm/fzgh2/zxgh/201309/t20130927_ 69865. htm> (Accessed on September 272013).

Shen, L., Wan, H., 2003. Discuss on the reurbanization and transformation of resource-based cities. Resour. Ind. 5, 116–119.

Shi, Z., Huang, Z., 2009. Resource curse and sustainable development of resource-based regions. China Popul. Resour. Environ. 19, 33–36.

Thatcher, M., 1998. The development of policy network analyses from modest origins to overarching frameworks. J. Theor. Polit. 10, 389–416.

The People's Government of Shanxi Province Notice of the people's Government of Shanxi Province on the issuance of the 12th Five-Year Plan of the environmental protection in Shanxi Province. <http://www. shanxigov. cn/n16/n1203/n1866/n5130/n14752920/16529523. html> (Accessed on October 2012).

Wang, J., 2013. Technical Guidelines for China National Environmental Planning in Twelveth-five Years. China Environment Press, Beijing.

Wang, S., Li, Y., 2005. Research on the problems and countermeasures of the development of the resource-based cities in Northeast China. Jilin Univ. J. Sci. Ed. 45, 97–102.

Wright, M., 1988. Policy community, policy network and comparative industrial policies. Polit. Stud. 36, 593–612.

Yan, Z., Liu, Y., Wang, X., 2012. Evaluation indexes and methods of green mine construction. China

Coal 38 (2), 113-120.

Zaman, K., bin, Abdullah, A., Khan, A., et al., 2016. Dynamic linkages among energy consumption, environment, health and wealth in BRICS countries: green growth key to sustainable development. Renew. Sustain. Energy Rev. 56, 1263-1271.

Zang, S., Li, D., Han, D., 2006. Transformation and circular economy development in the resource-based cities—Taking Yichun city of Heilongjiang Province as an example. Econ. Geogr. 26, 78-82.

Zhang, D., Jia, X., Qiao, F., et al., 2010. Study on the index evaluation systems of green mines. Renew. Resour. Circ. Econ. 3 (12), 11-13.

Zhang, M., 2001. Research on industrial transformation of resource-based cities in Western China. China Soft Sci. 8, 102-105.

Zhang, X., 2016. A study based on the sustainable development of coal resources in Shanxi Province. J. Cap. Norm. Univ. (Nat. Sci. Ed.) 5 (1), 72-76.

Zhang, Z., 2016. Research on the Policy Network of Coal Mine Safety Production in China. Guangxi University, Guangxi.

Zheng, J., Zhang, Y., 2017. Study on Evaluation system about green mine Construction of non-ferrous metal. Shenhua Sci. Technol. 01, 14-17.

Zhou, M., 2014. Evaluation And Counter Measures Research On The Sustainable Development Of Shanxi's Coal Industry. Shanxi University of Finance and Economics, Shanxi.

Virtual water export and import in China's foreign trade: A quantification using input-output tables of China from 2000 to 2012

CHEN Weiming[1,2], WU Sanmang[1,2,*], LEI Yalin[1,2], LI Shantong[3]

1. *School of Humanities and Economic Management, China University of Geosciences, Beijing* 100083, *China*

2. *Key Laboratory of Carrying Capacity Assessment for Resource and Environment, Ministry of Land and Resource, Beijing* 100083, *China*

3. *Development Research Center of State Council, Beijing* 100010, *China*

Abstract Import virtual water is an effective way to alleviate water shortage. Since China is a country that suffers severely from the lack of water resources, it is badly needed to research in virtual water export and import in China's foreign trade. This paper conducted thorough research on these issues and lucubrated on the detailed destinations and origins of China's virtual water export and import using the input-output model. Moreover, the factors that affect China's virtual water trade have also been studied in this paper. The results indicated that: (1) from 2000 to 2012, the volume of virtual water export increased from 65.2 billion tons to 114.1 billion tons, and its import volume increased from 62.4 billion tons to 108.7 billion tons. China was a net exporter of virtual water; (2) the export of virtual water was mainly exported by sectors such as Textile & Garment and Leather Products, General Machinery and Equipment Manufacturing Industry, the import of virtual water was mainly imported by Agriculture, Petrochemical industry; (3) China has net exported large amounts of virtual water to developed areas such as the US, the European Union, and Japan, and has net imported from emerging economies and resource-oriented regions such as Taiwan (China), ASEAN, Brazil, Korea, Australia and Russia. Based on these conclusions, this paper proposed that China's foreign trade structure needed to be adjusted to relieve the water shortage by the following methods the first is appropriately lower the export proportion of products from sectors such as Food, Beverage & Tobacco Products, Petrochemical Industry, which exported a large proportion of virtual water; the second is to enlarge the export proportion of products with low water consumption, such as service industry; the third is to make full use of the water saving effect of import.

Keywords Water scarcity, Net exporter, Virtual water, Input-output model, Developed country

1 Introduction

Water is a crucial natural resource for the survival of human beings, and also a strategic

原文刊于 *Resources, Conservation and Recycling*, 2018 年第 132 卷。原文发表时为彩色。

* 为通信作者。

resource for the sustainable development of economy of a country (region). At present, many countries have suffered from water scarcity. It is estimated by the United Nations that there will be a population of 2-2.7 billion facing the problem of water shortage till 2050, and the per capita water resource will decrease by more than 1/3 globally in the next 20 years (UnitedNations, 2003). Among these countries, China suffered most from water shortage in that its per capita available water resource is only 2300 m^3, less than 1/4 of the world per capita, ranking the 110th globally, and is one of the 13 countries that most lack of water resources in the world. There are also considerable regional diversities between South and North China, the per capita available water resource in South China is about 3600 m^3, much more than 720 m^3 in North China (He et al., 2011). Besides, among 699 cities of China, 400 have insufficient supply of water perennially, and 110 are threatened by water scarcity, which have significantly restricted the sustainable development of China's national economy (Huang and Qin, 2009).

On the other hand, China's foreign trade value has been booming since China joined the WTO in 2001. The import and export volume increased from RMB1863.8 billion and RMB2063.4 billion in 2000 to RMB14388.3 billion and RMB12035.8 billion in 2014, and China has become the No. 1 trading country in the world (China's National Bureau of Statistics, 2015). Owning to the water consumption during the manufacturing process of the commodity, there is hidden water in the commodity, which is called "virtual water" (Allan, 1993). Such massive imports and exports of commodities have inevitably led to the fact that China has exported and imported a great amount of virtual water.

In 1999, Prof. Allan proposed the point that a country (region) could implement the virtual water trade strategy to import water resource intensive products from another country (region), and to reduce the exports of products with high water consumption, in order to import (virtual) water resources and save its own water resources. It is expensive to import or export real water, so the trade of virtual water has overcome this drawback, and is an effective way to deal with water shortage for countries with scarce water resources (Allan, 1999; Hoekstra, 2003). In thc past 20 ycars, the trade flow and trade volume of virtual water among all countries has at least doubled (Dalin et al., 2012). With the rapid development of China's international trade, the water shortage problem will be alleviated to a large extent through the virtual water trade strategy to adjust China's foreign trade structure, cut down virtual water export and increase virtual water import.

However, the first problem for implementing virtual water trade strategy is to grasp relevant information such as the import and export volume, the destination and source of virtual water trade and its changes with time goes by. From present literature, researchers have not investigated the abovementioned problem quantitatively. Therefore, this paper based on previous studies on virtual water trade, adopted the input-output tables of China from 2000 to 2012 and detailed trade data of subsectors between China and its trade partners to investigate these problems, and put forward corresponding suggestions on the adjustment of the foreign trade structure.

2 Literature reviews

Initially, the object of quantitative study on virtual water trade is mainly agricultural products. Firstly, this is because that crops contain abundant water resources and thus water consumption in agriculture accounts for a great part of global water consumption, and therefore, agriculture sector bears the largest water consumption in the world. Secondly, the calculation of the virtual water content in industrial and service products is complicated, and the actual water consumption of industrial products is less than that of agriculture products, and thus often being neglected. The calculation methods for the virtual water content in crops in most studies origin from the Penman formula proposed by FAO (the UN Food and Agriculture Organization).

There are now mainly two specific methods to measure the virtual water content in products: one is to investigate the production tree of different products proposed by Chapagain and Hoekstra (2003); the other is to categorize and differentiate products into agricultural products, industrial products and animal products and etc. (Zimmer and Renault, 2003). Many studies have focused on agriculture, and even directly adopted the trade volume of virtual water in agriculture among countries (regions) to measure the whole status of the virtual water trade among relevant areas. For instance, Novo et al. (2009), Affuso (2010) and El-Sadek (2010) took Spain, North America and Egypt respectively as examples to study the effects of international trade of agricultural products on the water resource and food security in these areas with virtual water embedded in food as measuring object. Roson and Sartori (2010) and Antonelli et al. (2012) concentrated on the Mediterranean regions, illustrated the virtual water flow among these countries, and pointed out that the virtual water trade in agriculture among these countries had achieved the reasonable allocation of water resources and alleviated the water scarcity in relevant countries. Some other researchers focused on a global level. They calculated the global flow of virtual water based on the food trade among all countries. According to their findings, in respect of the international trade of virtual water in agriculture, countries such as America, Canada, Thailand, and India are major net exporters of virtual water, while countries such as Holland, Japan and Korea are major net importers of virtual water (Aldaya et al., 2010; Hoekstra and Chapagain, 2007; Hoekstra and Hung, 2005; Schwarzet al., 2015; Biewald et al., 2014).

Some researchers have looked at the water use or virtual water trade in agriculture in China. Cao et al. (2015) and Liu et al. (2015) calculated the water consumption of grain cultivation in China; the result was approximately 689.04 billion m^3. Shi et al. (2014) selected 27 main crops, measured the virtual water trade among China and other trade partners from 1986 to 2009, and found that China had net exports of virtual water to Asia, Africa and Europe, but had net imports from America, and overall it tended to be an net importer of virtual water. Zhang et al. (2016) found that China's virtual water export in agriculture had been declining while the import had been increasing, which is related to the booming ofthe imports of agriculture products.

However, the virtual water is not only essential for crops, but also deeply rooted in the whole social economy and in all walks of life. With further exploration on virtual water trade, researchers have increasingly realized the limitation of using virtual water in agriculture as a substitute for the virtual water trade among regions. Zhu and Gao (2009) calculated the virtual water volume in China's foreign trade in all industries, and the results manifested that the exports of virtual water surpassed the imports and the gap was expanding. Zhu (2014) calculated the virtual water volume in China's foreign trade in all industries in 2010, and the results illus-trated that some industrial products with high water consumption accounted for a large proportion in China's export trade, and this was the vital reason that China became a net exporter of virtual water. Compared with the findings of Zhang et al. (2016), it can be seen that the status of China's foreign trade of virtual water goes quite opposite when the research object transferred from agriculture to all industries, which also proved the limitation of studying only virtual water trade in agriculture. With the continuous global industrialization, and the rapid expansion of the trade flow in industries and service industries among various countries, it is necessary to measure the trading status of virtual water in all industries among various countries (regions), so as to evaluate the effects of international trade on water resources in relevant areas. For this purpose, VanOel et al. (2009) conducted a quantitative study on the external water footprint (EWFP) of Holland, and found that up to 89% of the water footprint came from external regions, and 60% of which were from industrial products. Islam et al. (2006) and Orlowsky et al. (2014) launched a qualitative study on the water resources and virtual water trade in various countries, and indicated the need for some countries to conduct virtual water trade. However, restricted by their research method, they failed to calculate the virtual water trade volume in all industries among various regions.

According to Su and Ang (2012), the SDA (structural decomposition analysis) can be used to analyze the driving forces to the changes of virtual water and time-series virtual water estimate. In addition, the SDA can also be applied on the multi-region comparisons of water use or carbon emission performance (Su andAng, 2016). For instance, Yang et al. (2016) analyzed on the growing water use in China during 1997—2007 by applying the dynamic SDA model. Zhi et al. (2014) studied the water footprint changes in the Haihe River basin in China with the help of IO tables and SDA model.

Since the manufacturing process of industrial products is intricate, and relevant to various sectors in the whole economic system, the quantitative measurement of the virtual water in industrial or service products has always been the difficulty in this field. The input-output model with water consumption coefficient belongs to the environmental input-output model, which covers the manufacturing process and usage of the products in the whole economic system, and is able to track the water resources effects behind the products trade with the universal link among all industries. What's more, it is conducive to differentiating direct and indirect consumption of virtual water, and making the calculation more intuitive and more precise. Therefore, this method has become a main research method for studies on virtual water trade. A recent study conducted by Hubacek and Feng (2016) have showed that the input-output model has great advantage in tracking the water consumption and pollution, land displacement or embodied land and other environmental factors. Lenzen et al. (2013) adopted the international input-output model to study the global flow of virtual water, and found that developed countries had imported an increasing amount of virtual water, which to some extent alleviated their own water scarcity problem. Chen and Chen (2013) calculated the water footprints of 112 countries (regions) and the virtual water trade among these countries with the input-output model. Mubako et al. (2013) used the input-output model to evaluate the utilization of water resources, and quantified the virtual water transfer between the state of California and Illinois. In addition, the input-output model can also be used to trace the energy or carbon flows. Chen and Chen (2015) calculated the embodied energy consumption used to produce goods and services in the city of Beijing, including the direct and indirect energy consumption. In 2016, by combining network analysis and input-output analysis, they also tracked the inter-regional carbon flows in Beijing (Chen and Chen, 2016).

Many scholars have conducted quantitative studies on China's virtual water problems with the input-output model, and their research can be generally divided into two kinds. The first is to study virtual water trade among different regions in China. For instance, Zhang et al. (2011) and Wang et al. (2013) took Beijing as their research object, while Wang et al. (2014) and Dong et al. (2013) took Shandong Province and Liaoning Province as their research object, calculated the virtual water trade volume among the above mentioned areas and other regions within China using the input-output method. Feng et al. (2012) and White et al. (2015) investigated the virtual water trade in the Yellow River Basin and Hai River Basin. More studies focused on the virtual water trade among all regions within China, and the results indicated that the virtual water transferred from the Midwest to the eastern developed areas (Zhang andAnadon, 2014; Deng et al., 2016; Jiang et al., 2015; Dong et al., 2014). For instance, Feng et al. (2014) assessed the virtual water flows among 30 provinces in China based on the multiregional input-output analysis and found that some water-scarce regions are facing a serious water crisis driven by rapid economic growth of China. The second is to study virtual water trade among China and other countries. Jiang (2012), Zhao et al. (2009) calculated the total volume of the imports and exports of the virtual water in all industries in China, given a specific year, and from the empirical results, it can be seen that China in general was a net exporter of virtual water, and had exported high proportion of products with high water consumption. At present, China's foreign trade structure is aggravating China's water scarcity. However, their research did not traced back to the export destination and import source of virtual water in various sectors, and thus could not put forward detailed suggestions for the adjustment of the trade structure.

In conclusion, although numerous studies have been conducted by scholars in terms of virtual water trade, its research fields keep expanding from agriculture to all sectors of national economy, and its research methods are under continuous perfection with the input - output method being widely used. In addition, many scholars did massive studies on China's virtual water trade, but there is no scholar who has split China's imports and exports of virtual water to different countries and observed the changes over a long time, and analyzed the factors that affect virtual water trade between China and each country. Therefore, this paper conducted a quantitative study on the above mentioned problems with the input - output tables of China from 2000 from 2012, and proposed relevant policy suggestions.

3 Research method and data sources

3.1 Input-output model of virtual water trade

In 1936, Leontief published the article which named Quantitative input and output relations in the economic systems of the United States and first put forward the input - output analysis (Leontief, 1936). The input-output model can reflect the interdependence between the various sectors of the economic system and reflect the relationship between the production and distribution of the various sectors of the national economy. Subsequently, the input-output method has been widely used in new areas including analysis of industrial structure, accounting for environmental pollution, income distribution, wealth and capital flows, and international trade forecasting.

3.1.1 Basic structure of the input-output model

Suppose a country (region) has n industrial sectors, then the basic mathematic structure of the input-output model involves n linear equations, which depicts the dependency relationship between the output of economic sectors and other correlated sectors and the final demand. The equation is as follows:

$$x_i = \sum_{j=1}^{n} z_{ij} + f_i \tag{1}$$

In this equation, n reflects the number of sectors; x_i means the total output of sector i; z_{ij} means the intermediate input of sector i to sector j; f_i represents the final demand of sector i.

Define the direct input coefficient as a_{ij} to express the demand of sector i's input for a unit output of sector j. The formula is as follows:

$$a_{ij} = \frac{z_{ij}}{x_j} \tag{2}$$

Hence, the Eq. (1) can be expressed as:

$$x_i = \sum_{j=1}^{n} a_{ij}x_j + f_i \quad (3)$$

Expressed in matrix form:

$$X = AX + F \quad (4)$$

X, A and F represent total output matrix, direct consumption coefficient matrix and final demand matrix respectively. Based on A, it is known that $(I-A)$ is a invertible non-singular matrix. The Formula (4) can be rewritten as the following demand-oriented form:

$$X = (I - A)^{-1}F,\ \text{make } B = (I - A)^{-1} = [b_{ij}] \quad (5)$$

In this formula, $(I-A)^{-1}$ is a Leontief inverse matrix; b_{ij} shows the whole (indirect and direct) demand of the output of sector i for the unit final products of sector j. Hence, the relationship between the final demand and output has been established through Leontief coefficient.

3.1.2 Eliminating the imports in the intermediate and final demands

The input-output model can incorporate either the noncompetitive or competitive imports assumption. The study by Su and Ang (2013) firstly analyze the imports assumption issues and find out the relationship between the embodiment estimates using different assumptions (Su and Ang, 2013). The original I-O tables used in this study are the China's input-output tables compiled by China National Bureau of Statistics, which are tables using the competitive imports assumption. This means that there is a certain amount of imported products in the intermediate and final demands of all sectors in these tables, the imports need to be eliminated when calculating the import and export of virtual water separately. After the imports of each sector have been eliminated, those I-O tables are no longer the tables using competitive imports assumption. We adopted the noncompetitive imports assumption in this study. It is unknown that among the inputs of various sectors in other sectors, how many are from imported products, and thus, we assume the import ratio is the same as the proportion of imported products in the total usage (total out-put+import-export) in the sector (Weber et al., 2008). For sector i, the ratio of imported products to the total usage can be expressed as follows:

$$s_i = \frac{m_i}{x_i + m_i - e_i} \quad (6)$$

In this formula, m_i represents the import of sector i, e_i means the export of sector i. The intermediate usage matrix and final demand matrix within the country excluding the import can be expressed as follows:

$$A_d = diag(1 - s)A,\ F_d = diag(1 - s)F \quad (7)$$

Thus, in this paper, the complete consumption coefficient matrix is:

$$B = (I - A_d)^{-1} = [b_{ij}] \quad (8)$$

3.1.3 Direct water consumption coefficient and complete water consumption coefficient

The direct water consumption coefficient and complete water consumption coefficient need to be determined in order to introduce the water consumption volume into the input - output structure. The direct water consumption coefficient reflects the direct water consumption volume to manufacture unit product, and its mathematical expression is as follows:

$$W = [\omega_i], \ \omega_i = \frac{w_i}{x_i} \tag{9}$$

Define the complete water consumption coefficient as δ_j and the complete water consumption coefficient matrix as D. δ_j can be calculated by multiplying the direct water consumption coefficient by Leontiefcoefficient, and it reflects the water consumption volume of all sectors in social economy during the whole manufacturing process of the product that meets the final demand of the unit sector j.

$$\delta_j = \sum_{i=1}^{n} \omega_i b_{ij}, \ D = [\delta_j] \tag{10}$$

3. 1. 4 Calculation of the virtual water export and import

From the above deduction, with export being a component of the final demand (consumption, investment, export), the virtual water content in the exported products can be calculated by multiplying the complete water consumption coefficient matrix by the export matrix excluding the import; hence the export of virtual water can be expressed as:

$$W_E = W(I - A_d)^{-1} E_d = DE_d \tag{11}$$

In this formula, $E_d = \text{diag}(1-s)\ E$ means the export matrix excluding the import.

As for the calculation of virtual water import, when a country (region) imports a commodity, it saves water resources that are needed for producing the commodity domestically, and the saved-amount of water is equal to the water consumption for producing the commodity in the importing country. In this way, the water-saving volume of imported products (or the virtual water avoided by the imports) is only related to the level of production technology (input-output relationship among all sectors) of the importing country. When importing a product from certain sector, not only the direct water consumption of that sector is saved, but also the indirect water consumption contained in the intermediate input products during the manufacturing process. Therefore, the calculation of the water-saving volume of imported products should be in accordance with the complete water consumption method for domestic final output. Although the virtual water content in the imported products calculated by this method is not the actual water consumption during the manufacturing process aboard, when focusing on the importing country, this method can better reflect the influence of the import on the water resources in the country. However, when the research scale has expanded to all countries in the world, and the research target is the global water flow under the international trade, this method cannot be applied, and the technological differences of various countries ought to be considered (Weber et al., 2008; Zhu and Gao, 2009).

This paper used the input-output coefficient of the importing country (China) to calculate the virtual water import:

$$W_M = W(I - A_d)^{-1}M = DE_d \tag{12}$$

M is the import matrix.

Thus, the net exports of virtual water of a country (region) can be expressed as:

$$NW = W_E - W_M \tag{13}$$

Owning to the productive relations among all sectors, the production of a commodity of certain sector may be in need of the product input from other sectors, which leads to the transfer of virtual water among various sectors. For a certain sector, the virtual water transferred from other sectors can be calculated with the input-output formula and the direct water consumption coefficient. W_{ij} represents the transferred amount of virtual water of sector i to sector j for its unit final product. It can be achieved by Formulas (8) and (9):

$$W_{ij} = b_{ij}\omega_i \tag{14}$$

Then the virtual water in a certain sector's exports came from all sectors. In other words, the virtual water export that came from a certain sector was exported by all sectors. So, when studying the sector characters of virtual water export, for any sector i, the "virtual water of sector i that was exported by all sectors" (shortly defined as "virtual water export of sector i"), and the "virtual water exported by sector i" and make the former W_i, according to Formula (14), it can be shown as:

$$W_i = W_{i1} + W_{i2} + \cdots + W_{ij} + \cdots + W_{in} = \sum_{j=1}^{n} W_{ij} \tag{15}$$

The "virtual water exported by sector i" means the virtual water embodied in the exported products of sector i (it may come from all sectors). Define it as W'_i, and according to Formula (14), it can be shown as:

$$W'_i = W_{i1} + W_{i2} + \cdots + W_{ij} + \cdots + W_{in} = \sum_{j=1}^{n} W_{ij} \tag{16}$$

For better comprehension, we assume there are only two sectors, i and j, and each sector exports only 1 unit of product, then the virtual water transfer among sectors resulted from the export can be shown as Fig. 1. For sector i, the virtual water export of sector i is $W_{ii}+W_{ij}$, and the virtual water exported by sector i is: $W_{ii}+W_{ji}$.

3.2 Data sources

This paper used China's input-output tables of six years from 2002 to 2012 (2000, 2002, 2005, 2007, 2010, 2012), which were compiled and released by China's National Bureau of Statistics (National Bureau of Statistics of China, 2015). This paper merged 33 sectors into 14 sectors in the input-output tables, and the specific classification is

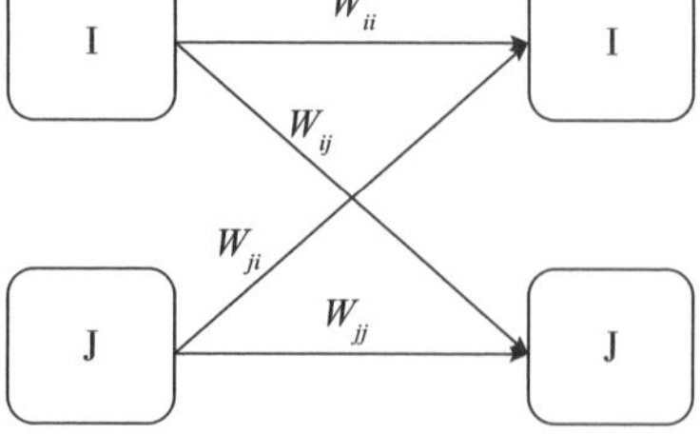

Fig. 1 Virtual Water Transfer among Sectors

exhibited in Table 1.

Table 1 Sector Classification

Sector code	Sectors
A01	Agriculture
A02	Mining Industry
A03	Food, Beverage & Tobacco Products
A04	Textile, Garment, Leather Products
A05	Wood and Furniture Manufacturing Industry
A06	Paper and Printing Industry
A07	Petrochemical Industry
A08	Other Nonmetallic Mineral Products Industry
A09	Metal Products Industry
A10	General Machinery and Equipment Manufacturing Industry
A11	Transportation Equipment Manufacturing Industry
A12	Other Manufacturing industry
A13	Electric and Water Industry
A14	Architecture and Service Industry

For the water consumption data of various sectors in China, the data of agriculture sector is from China Statistical Yearbook of relevant years (China's National Bureau of Statistics, 2013, 2001, 2003, 2006, 2008, 2011); and data of other sectors are from China Statistical Yearbook on Environment (China's National Bureau of Statistics, 2013). Eventually, the trading data between China's sectors and their counterparts of other countries are from OECD international trade database.

In fact, sector aggregation may have an impact on the results, and different combing methods may lead to different results, especially at the sectorial level results. Su et al. (2010) firstly analyzed the sector aggregation issues on the embodiment studies, and calculated the results of energy-related CO_2 emissions embodied in China's exports under different levels of sector aggregation. The main reason why we merged the 33 sectors into the 14 sectors in Table 1 is this aggregation can make the sectors cover all the sectors in the input-output table, the Environmental Statistical Yearbook and the OECD database, and farthest ensures the reliability of the results.

4 Research results and discussion

4.1 Analysis of China's export and import

There was a vigorous growth of China's exports after China joined WTO in 2001. The export volume increased from RMB 2063.4 billion in 2000 to RMB 12935.9 billion in 2012, by 6.2 times (China's National Bureau of Statistics, 2015). Although China has established trading relationships with most countries and areas in the world, it mainly exports to regions such as the EU, the US, Japan, ASEAN, Hong Kong, Korea, Taiwan, Australia, Russia and Brazil (Fig. 2).

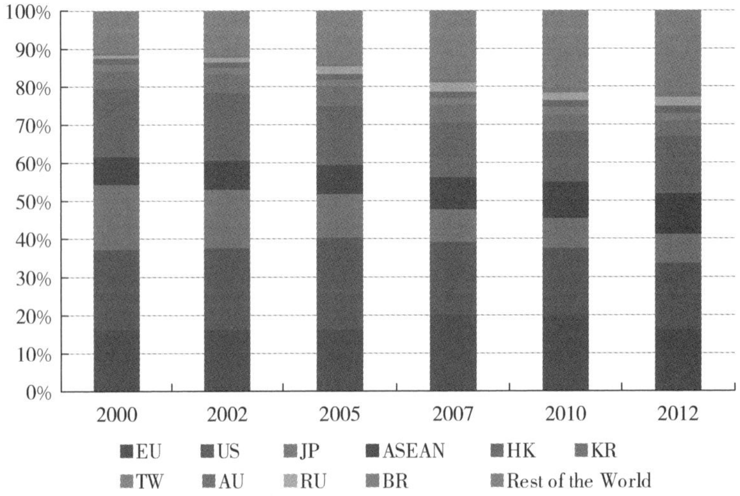

Fig. 2 Export Ratio of China to Other Regions

Pursuant to Fig. 2, the export volume of China to the ten regions such as EU and so on (except "Rest of the world") is basically accounting for 80% and above in its total export volume, and the EU, the US, Japan, ASEAN and Hong Kong are China's dominating export destinations of commodities. These five regions are traditionally developed economies except ASEAN, and possess substantial economic aggregate except HongKong.

From the perspective of the export ratio of various sectors in China (Fig. 3), A04 (Textile, Garment, Leather Products), A07 (Petrochemical Industry) and A10 (General Machinery and Equipment Manufacturing Industry) are China's major export sectors, while Agriculture and Service Industry have few export volume. From the aspect of the variation trend of exports in various sectors, the export ratio of A10 (General Machinery and Equipment Manufacturing Industry) keeps rising, and has achieved above 40% of the total export since 2007 while the export ratio of A04 (Textile, Garment, Leather Products) keeps declining.

China's imports have been booming with the accession to WTO. The import volume increased from RMB1863.8 billion in 2000 to RMB11480 billion in 2012, by 5.16 times. The EU, the US, Japan, ASEAN, Hong Kong, Korea, Taiwan, Australia, Russia and Brazil are also the key import sources of China, and the import ratio of China from the above ten regions kept about 80% of its total import volume from 2000 to 2012 (China's National Bureau of Statistics, 2015).

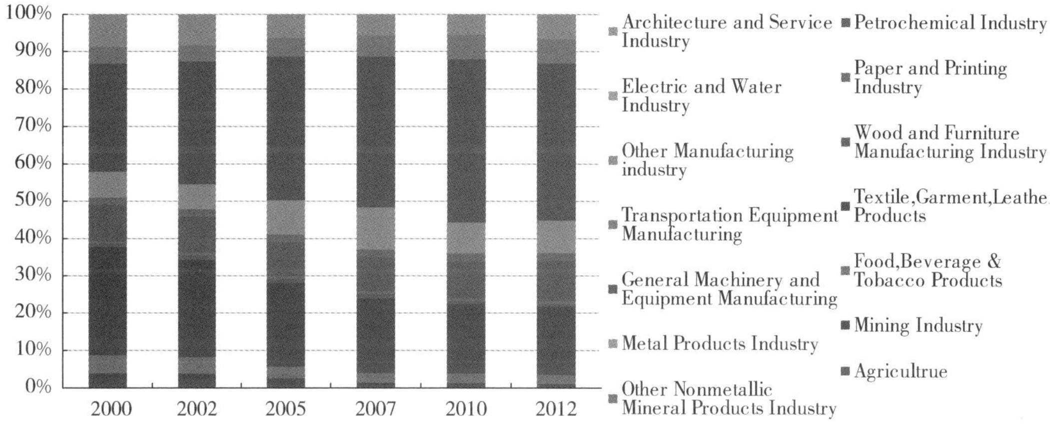

Fig. 3　Export Ratio of Various Sectors in China

4.2　Analysis of China's virtual water import and export from the perspective of sectors

4.2.1　The variation of China's total virtual water import and export

Based on our calculation, there was in total 65.3 billion tons of virtual water export in China's foreign trade in 2000. Due to the rapid development of China's export, the export volume of virtual water increased by 75% to 114.2 billion tons in 2012 compared with that of2000 (Fig. 4). China had exported accumulatively 549 billion tons of virtual water in the sixyears, namely 2000, 2002, 2005, 2007, 2010, and 2012 (the six years hereafter all refers to these years).

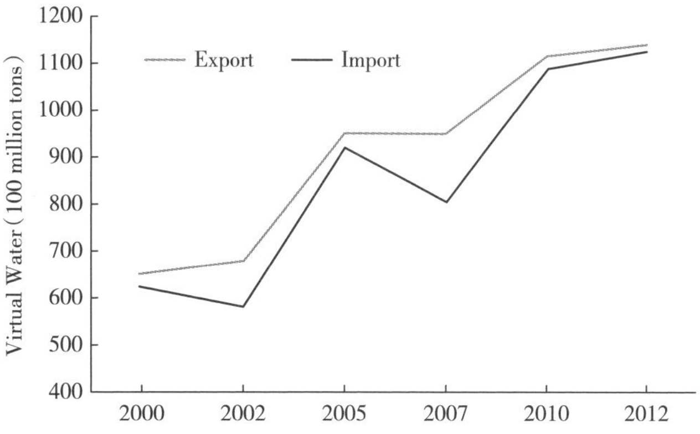

Fig. 4　Variation of Virtual Water Import and Export Volume in China's Foreign Trade

Meanwhile, China's virtual water import increased from 62.3 billion tons in 2000 to 112.6 billion tons in 2012, and China had accumulatively imported 514.4 billion tons of virtual water in the six years. China was mainly a net exporter of virtual water, and had a net export of 34.6 billion tons of virtual water in the six years, and thus China's foreign trade had exacerbated its water scarcity, which further confirmed the research conclusion of Zhu and Gao (2009) and

Zhu (2014), but was distinctively different from the research results of Shi et al. (2014), who studied China's virtual water import and export from the perspective of crops.

As we have mentioned in the part of Literature Review, there are already many scholars who have studied the virtual water export and import in China's international trade. So after we got the results of China's virtual water export and import in the period of 2000—2012, we draw a comparison between our results and those of them (Table 2). It can be seen that the results between the different studies are similar. But this paper studies the change of China's virtual water import and export in a long period of time.

Table 2 Studies about the virtual water export (VWE) and virtual water import (VWI) in China's international trade (100 million tons)

	2000		2002		2005		2007		2010		2012	
	VWE	VWI	VWE	VWI	VWE	VWI	VWE	VWI	VWE	VWI	VWE	VWI
Zhao et al. (2009)	×	×	789	636	×	×	×	×	×	×	×	×
Zhu and Gao (2009)	×	×	575	474	×	×	×	×	×	×	×	×
Jiang (2012)	775	627	736	613	1123	982	1208	951	×	×	×	×
Zhu (2014)	×	×	×	×	×	×	×	×	1397	1394	×	×
This Study	653	623	677	582	951	921	951	805	1116	1087	1142	1126

4.2.2 Sector characters of virtual water export

We first analyzed the virtual water export of each sector, to explore which sectors are China's virtual water exports mainly from, this paper adopted Formula (15) to calculate them (Fig. 5). As is reflected in Fig. 5, the virtual water exports of A01 (Agriculture) and A13 (Electric and Water Industry) are far more than other sectors. During the six years, the total of the virtual water exports of these two sectors has been occupied about 80% of China's overall virtual water exports. There are two reasons for the larger virtual water exports in A01 and A13. The first is from the view of direct water consumption coefficient (Fig. 6). The water consumption is massive in the manufacturing process of the products of the two sectors, and thus the two sectors belong to high water consumption sectors; the second is from the angle of industrial chain. These two sectors are basic industries that are at the end of the industrial chain, and the exports of other sectors will stimulate the exports of these two sectors.

We also analyzed the virtual water exported by each sector's export by Formula (16) (Fig. 7). As is illustrated in Fig. 7, the sectors that can greatly pull China's virtual water exports are A03 (Food, Beverage & Tobacco Products), A04 (Textile, Garment, Leather Products), A07 (Petrochemical Industry) and A10 (General Machinery and Equipment Manufacturing Industry) and etc. For 2000, 2002, 2005, 2007, 2010 and 2012, the summation of China'svirtual water exported by the above - mentioned 4 sectors' exports were respectively 51.8 billion tons, 181.5 billion tons, 51.1 billion tons and 102.9 billion tons.

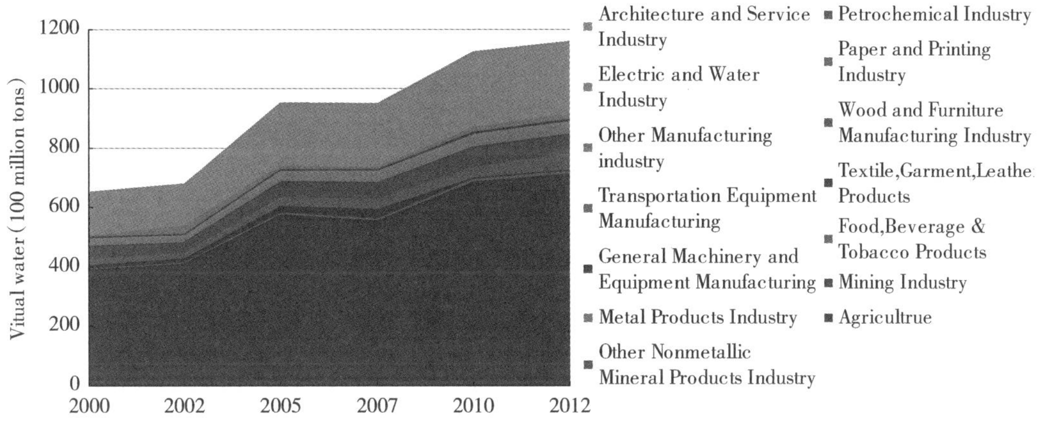

Fig. 5 Virtual Water Export of China's Various Sectors

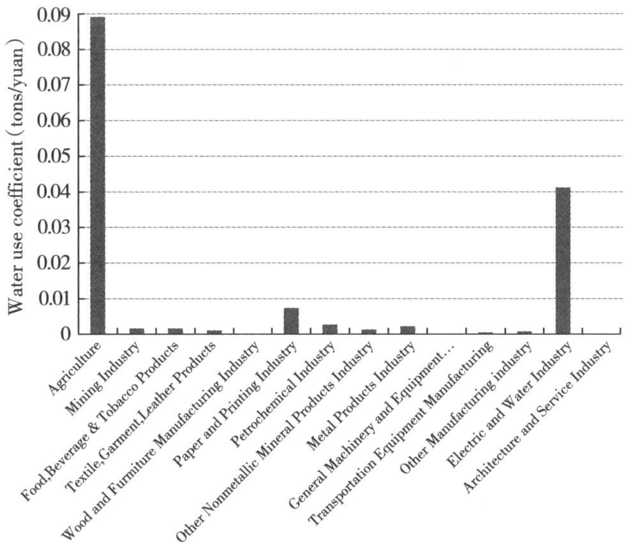

Fig. 6 Direct Water Consumption coefficient of China's Various Sectors

The proportion of the virtual water exported by these 4 sectors in China's total virtual water exports had been kept above 70%. These 4 sectors all belong to manufacturing industry, which indicates that exports in manufacturing industry are the major force for pulling the exports of virtual water in China, while agriculture and service industry's promoting effects are not significant. On the one hand, there is a rapid development of China's manufacturing industry since the reform and opening up policy. China has become the No. 1 manufacturing country in the world, and the manufacturing sector is a dominating sector for China's export. In fact, it can be found with Fig. 3 that A03 (Food, Beverage & Tobacco Products), A04 (Textile, Garment, Leather Products), A07 (Textile, Garment, Leather Products) and A10 (General Machinery and Equipment Manufacturing Industry) are also large sectors for China's export. On the other hand, China has a large population, and the per capita cultivated landis limited, and hence the export of agriculture products is few. The development of service industry has long been lagged behind in China. In 2014, the proportion of China's service export in the trade export is 25%, while the proportion of the world's service export in the trade export is 26.4% (China's National Bureau of Statistics, 2015).

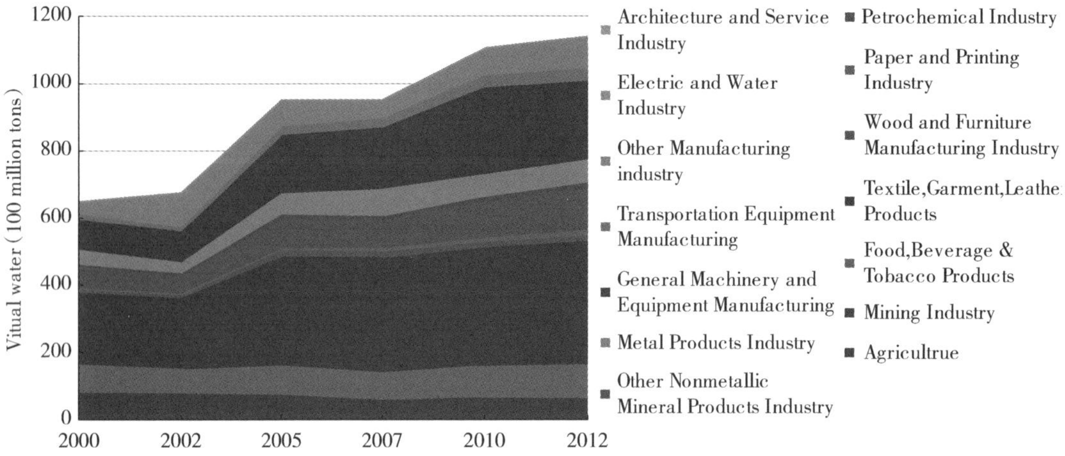

Fig. 7 Virtual Water Exported by Various Sectors' Export of China

Comparing the virtual water export of various sectors with the virtual water exported by various sectors' exports in China, it can be found that: firstly, there is a huge difference, and little correlation between these two indicators. For instance, there is a large amount of virtual water export in Agriculture and Electric and Water Industry, nevertheless, there is few virtual water exported by these two sectors' exports. This is because the amount of virtual water exported is highly correlated with the export volume of the sector, while the amount of virtual water export is correlated with the sector's direct water consumption coefficient and its position in the industry chain. Secondly, the export ratio of virtual water in various sectors has been stable, while there is a great variation of the ratio of virtual water exported by various sectors' exports. For example, the ratio of virtual water exported by agriculture exports declined from 16% in 2000 to 5% in 2012, and this was closed related to the variation of the export volume of all sectors. In accordance to Fig. 2, the proportion ofexport volume in agriculture descended from 2% in 2000 to 0.7% in 2012.

4.2.3 Virtual water imported by China's various sectors' imports

Fig. 8 demonstrates the virtual water imported by China's various sectors' imports. It is clear that China's sectors that have significantly pulled the virtual water import are A01 (Agriculture), A10 (General Machinery and Equipment Manufacturing Industry), A03 (Food, Beverage & Tobacco Products), A02 (Mining Industry), A07 (Petrochemical Industry) and etc. The summation of virtual water imported by the above-mentioned five sectors' imports in 2000, 2002, 2005, 2007, 2010 and 2012 were respectively 127.4 billion tons, 116.4 billion tons, 84.6

billion tons, 50 billion tons and 39.6 billion tons. The virtual water imported by these 5 sectors' imports has occupied about 80% in China's total virtual water import volume.

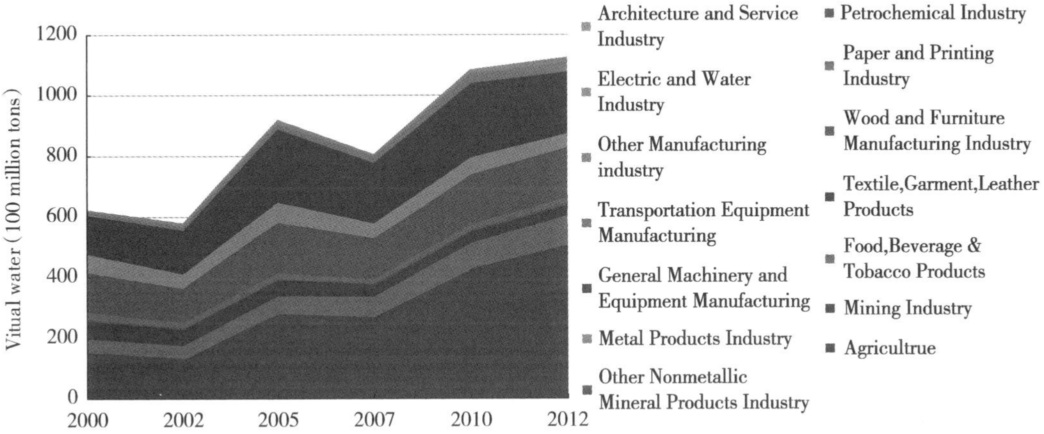

Fig. 8 Virtual Water Imported by China's Various Sectors' Imports

4.2.4 The net export of virtual water driven by China's various sectors

A04 (Textile, Garment, Leather Products) and A12 (Other Manufacturing industry) are two sectors that net export most of virtualwater in China (Fig. 9), and their aggregate in the six years can achieve 154.4 billion tons and 38.8 billion tons. Therefore, these two sectors deserve substantial attention in order to reduce China's virtual water export. A01 (Agriculture), A02 (Mining Industry), and A07 (Petrochemical Industry) had a net import of 87.8 billion tons, 47 billion tons and 33.5 billion tons of virtual water. China has to some degree saved its water resources through the imports of these sectors.

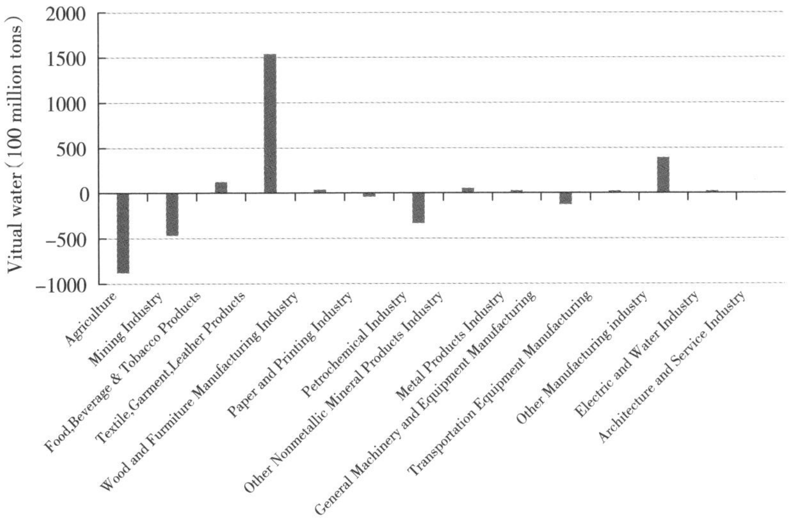

Fig. 9 Virtual Water Net Exported by Various Sectors' Foreign Trade

4.3 Analysis of virtual water import and export between China and other countries (regions) in the world

Based on the above analysis, the ten areas such as the EU, the US and etc. are the main export destinations as well as import sources of China's products. Thus this paper analyzed the virtual water import and export between China and the above-mentioned areas (Table 3). Please note that this paper calculated the virtual water import using the water consumption coefficient of importer (China). Therefore, the virtual water import of China in this paper is the virtual water avoided by importing, not the actual virtual water embodied in imported goods. But the "virtual water avoided by importing" is more accurate to measure the impact of import activities on China's water resources. So the following analyses of the bilateral virtual water balance are used to evaluate the impact of trade between China and other regions on China's domestic water resources. This article does not study the actual water balance in international trade, and also does not assess the gains and losses of China in bilateral trade with each region in Table 3.

Table 3 Virtual Water Trade between China and Other Major Regions in the World (100 million tons)

		2000	2002	2005	2007	2010	2012	Total
	Export	112	112	129	113	120	139	725
Hong Kong	Import	21	18	14	9	5	4	71
	Net Export	91	94	115	104	115	135	654
	Export	94	99	169	180	217	189	948
European Union	Import	73	61	83	72	100	104	493
	Net Export	21	38	86	108	117	85	455
	Export	112	136	188	168	194	187	985
The United States	Import	70	60	100	86	140	163	619
	Net Export	42	76	88	82	54	24	366
	Export	135	119	134	103	105	105	701
Japan	Import	88	79	105	81	97	75	525
	Net Export	47	40	29	22	8	30	176
	Export	8	10	25	32	27	30	132
Russia	Import	22	30	38	34	29	31	184
	Net Export	-14	-20	-13	-2	-2	-1	-52
	Export	9	9	14	14	19	21	86
Australia	Import	27	23	35	29	47	52	213
	Net Export	-18	-14	-21	-15	-28	-31	-127
	Export	41	43	53	50	48	46	281
Korea	Import	61	50	85	66	78	74	414
	Net Export	-20	-7	-32	-16	-30	-28	-133

Continued

		2000	2002	2005	2007	2010	2012	Total
Brazil	Export	3	3	5	8	16	17	52
	Import	11	17	35	38	69	87	257
	Net Export	-8	-14	-30	-30	-53	-70	-205
ASEAN	Export	52	53	72	82	117	137	513
	Import	79	78	124	119	161	162	723
	Net Export	-27	-25	-52	-37	-44	-25	-210
Taiwan	Export	12	11	16	14	18	18	89
	Import	61	62	80	62	65	51	381
	Net Export	-49	-51	-64	-48	-47	-33	-292
Rest of the World	Export	75	82	146	187	235	253	978
	Import	110	104	222	209	296	323	1264
	Net Export	-35	-22	-76	-22	-61	-70	-286
Total	Export	653	677	951	951	1116	1142	5490
	Import	623	582	921	805	1087	1126	5144
	Net Export	30	95	30	146	29	16	346

4.3.1 The gross volume of virtual water import and export between China and other countries regions in the world

Developed areas such as The US, the EU, Hong Kong (China), Japan and etc. are main areas for China's virtual water export. In the six years, China had accumulatively exported 98.5 and 94.8 billion tons of virtual water to the US and the EU, accounting for 17.9% and 17.3% of China's virtual water export. The main reason that China had exported such massive virtual water to the US, the EU, Hong Kong and Japan was that there was a large amount of virtual water in the manufactured goods exported from China to those areas. For instance, during the six years, China had implicitly exported 43.1 and 17.8 billion tons of virtual water to the US through exports of products in A10 (General Machinery and Equipment Manufacturing Industry), and A04 (Textile, Garment, Leather Products), accounting for 46.1% and 18.9% in China's total virtual water export to the US. Moreover, during the six years, China had implicitly exported 38.5 and 19.7 billion tons of virtual water to the EU through exports of products in A10 (General Machinery and Equipment Manufacturing Industry), and A04 (Textile, Garment, Leather Products), accounting for 42.8% and 21.9% in China's total virtual water export to the EU. Meanwhile, China has also imported substantial virtual water from those regions. During the six years, China had accumulatively imported 72.3 and 61.9 billion tons of virtual water from the EU and the US, accounting for 14.1% and 11.9% of China's virtual water import. The primary reason that China had imported such massive virtual water from these regions was that there was a large amountof virtual water hidden in the products

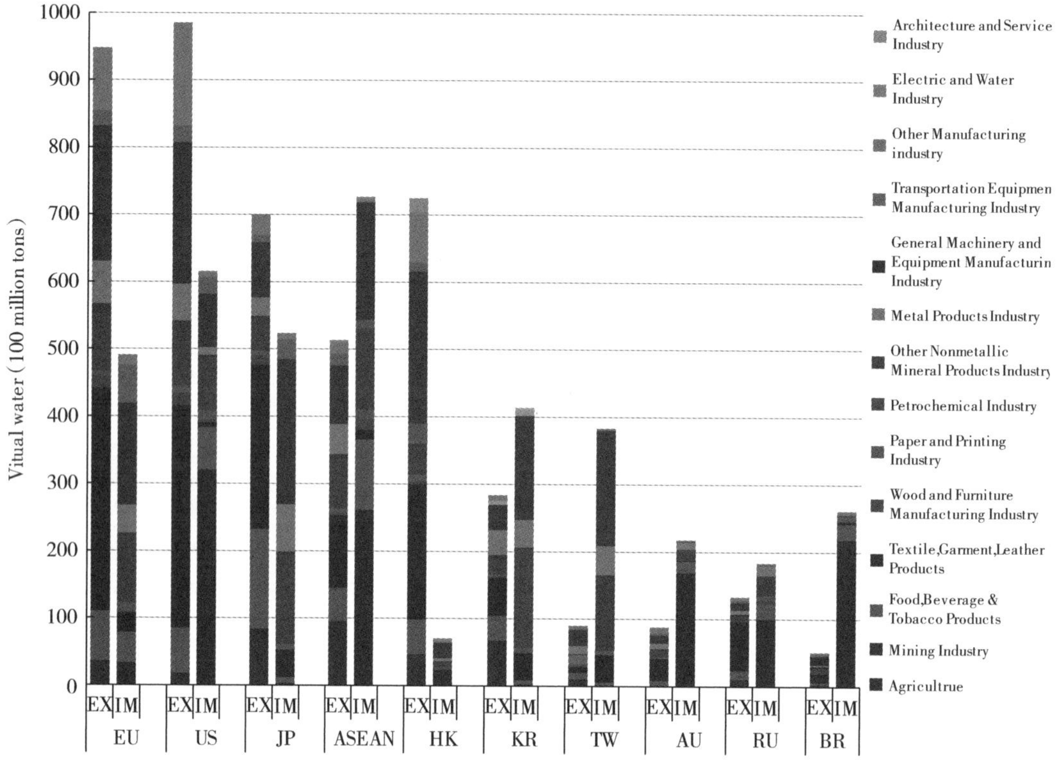

Fig. 10　Sector Composition of the Virtual Water Trade between China and Other Major Regions

of sectors such as A01 (Agriculture) imported from these areas to China (Fig. 10). For instance, during the six years, China had accumulatively imported 22.6 and 31.6 billion tons of virtual water from ASEAN and the US through imports of products in A01 (Agriculture), accounting for 31.2% and 51.0% in China's total virtual water import from ASEAN and the US respectively.

4.3.2　Net virtual water trade between China and other regions in the world

Hong Kong (China), the EU, the US and Japan are primary areas for China's net export of virtual water. In the six years, China had accumulatively exported 65.4 billion tons, 45.5 billion tons, 36.6 billion tons and 17.6 billion tons of virtual water to Hong Kong (China), the EU, the US and Japan respectively. The reason for this net export is as above-mentioned: there is a large amount of virtual water hidden in the manufactured goods exported from China to those areas through A10 (General Machinery and Equipment Manufacturing Industry), and A04 (Textile, Garment, Leather Products) (Fig. 11)❶.

Taiwan (China), ASEAN, Brazil, Korea, Australia, and Russia are major areas for China's net import of virtual water. In the six years, China has accumulatively imported 29.2

❶ 图 11 见原文，刊于 *Resources, Conservation and Recycling*，2018 年第 132 卷。原图发表时为彩色。

billion tons, 21 billion tons, 20.5 billion tons, 13.3 billion tons and 12.7 billion tons of virtual water from Taiwan (China), ASEAN, Brazil, Korea, Australia, and Russia respectively. In respect of type, these regions belong to emerging economies or resource-oriented areas. There are two aspects accounting for China's net import of virtual water from these regions: the first is that there is a great deal of virtual water hidden in the products of agriculture and extractive industry imported from resource-oriented regions. For example, the A01 (Agriculture) products from the ASEAN and Brazil contain a lot of virtual water, and so is the A01 (Agriculture) and A02 (Mining Industry) products from Australia. The second is that the manufactured goods imported from the emerging economies contain vast virtual water, for instance, there is a lot of virtual water in the A10 (General Machinery and Equipment Manufacturing Industry) products imported from Taiwan (China) and Korea.

4.4 The virtual water intensity of China's import and export

In order to grasp the main sectors and areas for the structural adjustment in China's foreign trade, the virtual water intensity of China's import and export trade from 2000 to 2012 were calculated from the perspective of sectors, that is the virtual water volume pulled by various sectors' unit import and export. (Since this paper used the input-output coefficient of the importing country to calculate the virtual water import, the virtual water export and import are equal in the unit import and export of various sectors, and hence, this paper only analyzed the virtual water volume pulled by the unit export of all sectors) (Fig. 12) In addition, this paper evaluated the virtual water intensity of the import and export trade between China and other major regions, that is, the unit virtual water volume contained in products China exported to other regions and products imported from other regions (Fig. 13).

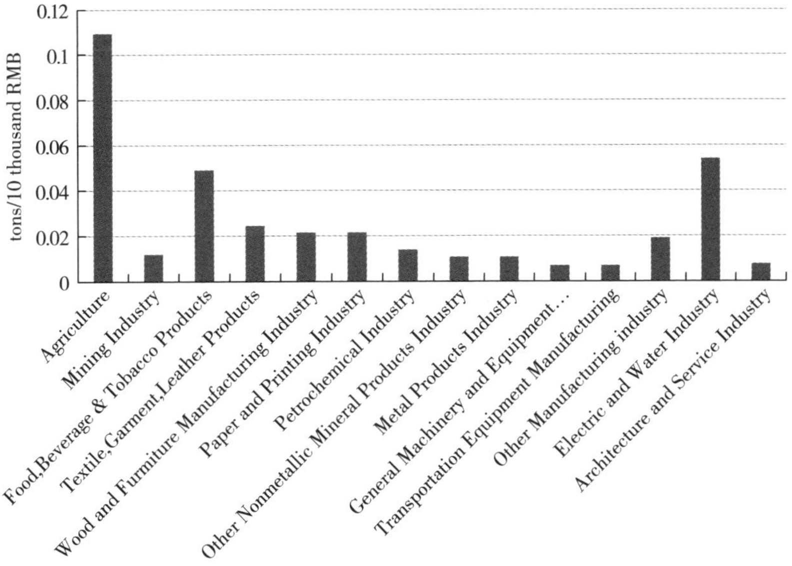

Fig. 12 Virtual Water Intensity of the Export Trade of China's Various Sectors

It can be seen from Fig. 13 that the virtual water intensity of the products imported from Brazil, Australia, and Russia to China has greatly surpassed that of products from Korea, Japan, Taiwan (China) and the EU, with Brazil being the highest, 0.046 t/10 thousand RMB, and Japan being the lowest, 0.013 t/10 thousand RMB. Comparing the virtual water intensity of import and export in the same region, the virtual water intensity of the products imported from Brazil and Australia to China is far exceeding that of the products exported to those two countries, which is on the contrary to the trade with Korea and Japan, indicating that in general, China's trade with Brazil and Australia is more conducive for saving its water resources.

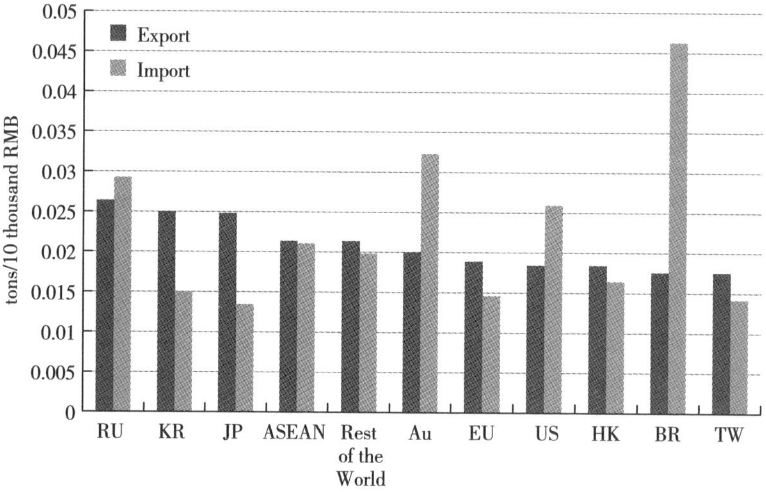

Fig. 13 Virtual Water Intensity of the Trade between China and other Major Regions

According to Fig. 12, although the virtual water intensity of the exports in A01 (Agriculture) and A13 (Electric and Water Industry) are the biggest, compared with the direct water consumption cofficient, there is little difference between the virtual water intensity of the exports in these two sectors and the other sectors. Compared with the direct water consumption coefficient, the virtual water intensity is higher in the export trade of A03 (Food, Beverage & Tobacco Products), A04 (Textile, Garment, Leather Products) and A06 (Paper and Printing Industry), indicating that they have more intermediate demands for sectors with high water consumption in the manufacturing process.

5 Conclusions and recommendations

This paper conducted a quantitative study on the virtual water import and export in China's foreign trade from 2002 to 2012 with China's input-output tables from 2002 to 2012 with a view to providing scientific foundation for the trade structural adjustment to alleviate China's water scarcity. The main conclusions are as follows:

(1) Since 2000, China's rapid development of its foreign trade has led to massive virtual water export, which had doubled during 2000—2012, that is China's virtual water export

volume increased from 63.5 billion tons in 2000 to 114.2 billion tons in 2012. Meanwhile, China had imported a great deal of virtual water, and the imported volume ascended from 62.3 billion tons in 2000 to 112.6 billion tons in 1012. Overall, China was a net exporter of virtual water, and its foreign trade aggravated its water scarcity.

(2) Manufacturing industry is the major sector for pulling China's virtual water export. Manufacturing industries such as Food, Beverage & Tobacco Products, Textile, Garment, Leather Products, Petrochemical Industry, General Machinery and Equipment Manufacturing Industry had accumulatively exported 387.3 billion tons of virtual water in China for the six years, accounting for 72% of its total virtual water export volume. There was no significant pulling effect of agriculture and service industry on China's virtual water export. Nevertheless, Agriculture and Electric and Water Industry are the two sectors that bear the most virtual water export. During the six years, the summation of virtual water exports of these two sectors accounted for about 80% of China's total virtual water export. China imported virtual water through three sectors: Agriculture; Petrochemical Industry; General Machinery and Equipment Manufacturing Industry. During the six years, the summation of virtual water import of these two sectors accounted for about 64% of China's total virtual water import.

(3) China has net exports of virtual water to development regions such as the EU, the US, Japan and Hong Kong (China), while at the same time has net imports from emerging economies or resource-oriented regions such as Taiwan (China), ASEAN, Brazil, Korea, Australia, Russia and etc. The reason for the net exports is because there is substantial virtual water contained in the products that China has exported to the developed regions through General Machinery and Equipment Manufacturing Industry, Textile, Garment, Leather Products. While the reason for the net imports from emerging economies and resource-oriented regions is because there is a great amount of virtual water contained in the products that China has imported from those regions through Agriculture and manufacturing industry.

(4) The virtual water intensity is pretty strong in the exports of sectors such as Agriculture; Electric and Water Industry; Food, Beverage & Tobacco Products; Textile, Garment, Leather Product industry of China. From the regional perspective, the virtual water intensity ofthe products imported from Brazil and Australia to China is far exceeding that ofthe products exported to those two countries from China.

Based on the above conclusion and analysis, the author believes there is a great potential to alleviate China's water scarcity through the adjustment of China's foreign trade structure.

First and foremost, properly lower the exports of the industries that play important roles in promoting China's virtual water export. For example, reducing the export ratio of sectors such as Food, Beverage & Tobacco Products; Textile, Garment, Leather Products; Petrochemical Industry, will help save water consumption in China's export trade. On the one hand, exports can be abated by levying export tariff and resource tax on these industries; on the other hand, these industries should be promoted to upgrade from the manufacturing stage to R&D (research and development) and sales stage in the industrial chain. Compared with the manufacturing stage, R&D and sales stage in the industrial chain will obtain more profits with less water consumption.

What's more, the export ratio of products with low water consumption needs to be expanded, especially for service industry. China's Service industry not only bears the merit of low water consumption, its export ratio is lower than the world average. Therefore, service export will be the crucial part in China's foreign trade in the future. The water consumption in China's foreign export trade will be diminished through enlarging the export ratio of service industry.

Last but not least, the water scarcity of China can be relieved by making full use of the water-saving effect from the import trade. Firstly, expand imports of sectors such as Agriculture; General Machinery and Equipment Manufacturing Industry; Petrochemical Industry. These sectors have higher virtual water intensity in China's import and export trade. Enhancing the import of commodities in these sectors can better ease the tension of China's water consumption. Secondly, attach great importance to the trade with resource-oriented regions. This is not only because China is a net importer of virtual water in the trade with resource-oriented regions, but also in that the virtual water intensity is higher in the products imported from resource-oriented regions such as Brazil, Australia, and Russia.

It is worth reminding that this study merged the 33 sectors in the input-output table into 14 sectors. So the virtual water export and import results at the sector level may lack of precision. When the sector classification is refined, the results of some sectors may even change dramatically. On the other hand, the "same technology assumption of imports" used in this study is particularly important, as single-region input-output model is notable to correctly calcu-late the virtual water in China's imports, only water avoided by importing. So the above results can only be used to measure the impact of China's international trade activities on China's domestic water resources and cannot be used to evaluate the results of virtual water balance of bilateral trade. For example, if the manufacturing procedure of product in some countries is more water-saving than in China, the actual virtual water in China's imports from these countries will be less than the virtual water calculated using the Chinese water use coefficient. So the results derived in this paper tend to underestimate China's virtual water net export to developed countries, and also tend to underestimate China's virtual water net import from some of the less developed regions.

Acknowledgements The authors are grateful for financial support from the Funda-mental Research Funds for the Central Universities under Grant No. 52300859631, the National Natural Science Foundation of China under grant nos. 71003066 and 71133003.

References

Affuso, E., 2010. Virtual Water Flows in North America, Holl-Aubum: Sectorof Agricultural Economics and

Rural Sociology. 301A Comer Hall Auburn University, pp. 1-25.

Aldaya, M. M., Allan, J. A., Hoekstra, A. Y., 2010. Strategic importance of green water in international crop trade. Ecol. Econ. 69 (4), 887-894.

Allan, J. A., 1993. Fortunately There Are Substitutes for Water Otherwise Our Hydro-Political Futures Would Be Impossible, Priorities for Water Resources Allocation and Management. Overseas Development Administration, London UK, pp. 13-26.

Allan, J. A., 1999. Water Stress and Global Mitigation: Water Food and Trade. Arid Lands Newsletter, pp. 45.

Antonelli, M., Roson, R., Sartori, M., 2012. Systemic input-output computation of green and blue virtual water 'flows' with an illustration for the Mediterranean region. Water Resour. Manag. 26(14), 4133-4146.

Biewald, A., Rolinski, S., Lotze-Campen, H., et al., 2014. Valuing the impact of trade on local blue water. Ecol. Econ. 101, 43-53.

Cao, X., Wang, Y., Wu, P., Zhao, X., Wang, J., 2015. An evaluation of the water utilization and grain production of irrigated and rainfed croplands in China. Sci. Total Environ. 529, 10-20.

Chapagain, A. K., Hoekstra, A. Y., 2003. Virtual water trade: a quantification of virtual water flows between nations in relation to international trade of livestock and livestock products, Virtual water trade. In: Proceedings of the International Expert Meeting on Virtual Water Trade, UNESCO-IHE (United Nations Educational, Scientific and Cultural Organization-Institute for Water Education), Delft, The Netherlands.

Chen, Z. M., Chen, G. Q., 2013. Virtual water accounting for the globalized world economy: national water footprint and international virtual water trade. Ecol. Indic. 28, 142-149.

Chen, S., Chen, B., 2015. Urban energy consumption: different insights from energy flow analysis, input-output analysis and ecological network analysis. Appl. Energy 138 (C), 99-107.

Chen, S., Chen, B., 2016. Tracking interregional carbon flows: a hybrid network model. Environ. Sci. Technol.

Dalin, C., Konar, M., et al., 2012. Evolution of the global virtual water trade network. Proc. Natl. Acad. Sci. 109 (16), 5989-5994.

Deng, G., Ma, Y., Li, X., 2016. Regional water footprint evaluation and trend analysis of China-based on interregional input-output model. J. Clean Prod. 112, 4674-4682.

Dong, H., Geng, Y., Sarkis, J., Fujita, T., Okadera, T., Xue, B., 2013. Regional water footprint evaluation in China: a case of Liaoning. Sci. Total Environ. 442, 215-224.

Dong, H., Geng, Y., Fujita, T., Fujii, M., Hao, D., Yu, X., 2014. Uncovering regional disparity of China's water footprint and inter-provincial virtual water flows. Sci. Total Environ. 500-501, 120-130.

El-Sadek, A., 2010. Water desalination: an imperative measure for water security in Egypt. Desalination 250 (3), 876-884.

Feng, K., et al., 2012. Assessing regional virtual water flows and water footprints in the Yellow River Basin, China: a consumption based approach. Appl. Geogr. 32, 691-701.

Feng, K., Hubacek, K., Pfister, S., Yu, Y., Sun, L., 2014. Virtual scarce water in china. Environ. Sci. Technol. 48 (14), 7704.

He, X. B., et al., 2011. Analysis of chinese international virtual water trade in an input-output method. Environ. Sci. Manage. 36 (3), 7-10 (in Chinese).

Hoekstra, A. Y., Chapagain, A. K., 2007. The water footprints of Morocco and the Netherlands: global water use as a result of domestic consumption of agricultural commodities. Ecol. Econ. 64 (1), 143-151.

Hoekstra, A. Y., Hung, P. Q., 2005. Globalisation of water resources: international virtual water flows in relation to crop trade. Global Environ. Chaange 15 (1), 45-56.

Hoekstra, A. Y., 2003. Virtual water: an introduction. In: Hoekstra, A. Y. (Ed.), Virtual Water

Trade: Proceedings of the International Expert Meeting on Virtual Water Trade. Value of Water Research Report Series No. 12. IHE, Delft, the Netherlands.

Huang, D., Qin, J., 2009. Inspiration ofJapan's Water Resources Compensation Mechanism to China. CLS-AERL, pp. 365-369 (in Chinese).

Hubacek, K., Feng, K., 2016. Comparing apples and oranges: some confusion about using and interpreting physical trade matrices versus multi-regional input-output analysis. Land Use Policy 50, 194-201.

Islam, M. S., Oki, T., Kanae, S., Hanasaki, N., Agata, Y., Yoshimura, K., 2006. A grid-based assessment of global water scarcity including virtual water trading. In: Integrated Assessment ofWater Resources and Global Change. Springer, Netherlands, pp. 19-33.

Jiang, Y., Cai, W., Du, P., et al., 2015. Virtual water in interprovincial trade with implications for China's water policy. J. Clean. Prod. 87, 655-665.

Jiang, L., 2012. The Study of Chinese Virtual Water Trade Based on Input-Output Analysis Guangdong. South China University ofTechnology (in Chinese).

Lenzen, M., Moran, D., Bhaduri, A., et al., 2013. International trade of scarce water. Ecol. Econ. 94, 78-85.

Leontief, W. W., 1936. Quantitative input and output relations in the economic systems of the united states. Rev. Econ. Stat. 18 (3), 105-125.

Liu, J., Sun, S., Wu, P., Wang, Y., Zhao, X., 2015. Evaluation of crop production, trade, and consumption from the perspective ofwater resources: a case study of the Hetao irrigation district China, for 1960—2010. Sci. Total Environ. 505, 1174-1181.

Mubako, S., Lahiri, S., Lant, C., 2013. Input-output analysis of virtual water transfers: case study of California and Illinois. Ecol. Econ. 93, 230-238.

National Bureau of Statistics of China, 2013. China Environment Statistics Yearbook 2012. China Statistics Press, Beijing, China (in Chinese).

National Bureau of Statistics of China, 2015. China Input-Output Table 2012. China Statistics Press, Beijing, China (in Chinese).

Novo, P., Garrido, A., Varela-Ortega, C., 2009. Are virtual water flows in Spanish grain trade consistent with relative water scarcity? Ecol. Econ. 68 (5), 1454-1464.

Orlowsky, B., Hoekstra, A. Y., Gudmundsson, L., Seneviratne, S. I., 2014. Today's virtual water consumption and trade under future water scarcity. Environ. Res. Lett. 9 (7), 074007.

Roson, R., Sartori, M., 2010. Water Scarcity and Virtual Water Trade in the Mediterranean. University Ca'Foscari ofVenice, Dept. of Economics Research Paper Series, (08-10).

Schwarz, J., Mathijs, E., Maertens, M., 2015. Changing patterns of global agri-food trade and the economic efficiency ofvirtual water flows. Sustainability 7 (5), 5542-5563.

Shi, J., Liu, J., Pinter, L., 2014. Recent evolution of China's virtual water trade: analysis of selected crops and considerations for policy. Hydrol. Earth Syst. Sci. 18 (4), 1349-1357.

Su, B., Ang, B. W., 2012. Structural decomposition analysis applied to energy and emissions: some methodological developments. Energy Econ. 34 (1), 177-188.

Su, B., Ang, B. W., 2013. Input-output analysis of CO_2 emissions embodied in trade: competitive versus non-competitive imports. Energy Policy 56, 83-87.

Su, B., Ang, B. W., 2016. Multi-region comparisons of emission performance: the structural decomposition analysis approach. Ecol. Indic. 67, 78-87.

Su, B., Huang, H. C., Ang, B. W., Zhou, P., 2010. Input-output analysis of CO_2 emissions embodied in trade: the effects of sector aggregation. Energy Econ. 32 (1), 166-175.

United Nations, 2003. The 1st UN world water development report. In: Water for People, Water for Life., pp. 1-595, Kyoto, Japan.

Van Oel, P. R., Mekonnen, M. M., Hoekstra, A. Y., 2009. The external water footprint of the Netherlands: geographically-explicit quantification and impact assessment. Ecol. Econ. 69 (1), 82-92.

Wang, Z., Huang, K., Yang, S., et al., 2013. An input-output approach to evaluate the water footprint and virtual water trade of Beijing, China. J. Clean. Prod. 42, 172-179.

Wang, W., et al., 2014. Relationships between regional economic sectors and water use in a water-scarce area in China: a quantitative analysis. J. Hydrol. 515, 180-190.

Weber, C. L., Peters, G. P., Guan, D., et al., 2008. The contribution of Chinese exports to climate change. Energ Policy 36 (9), 3572-3577.

White, D. J., et al., 2015. A hydro-economic MRIO analysis of the Haihe River Basin's water footprint and water stress. Ecol. Model. 318, 157-167.

Yang, Z., Liu, H., Xu, X., Yang, T., 2016. Applying the Water Footprint and dynamic Structural Decomposition Analysis on the growing water use in China during 1997—2007. Ecol. Indic. 60, 634-643.

Zhang, C., Anadon, L. D., 2014. A multi-regional input-output analysis of domestic virtual water trade and provincial water footprint in China. Ecol. Econ. 100, 159-172.

Zhang, Z., et al., 2011. Analyses of water footprint of Beijing in an interregional input-output framework. Ecol. Econ. 70, 2494-2502.

Zhang, Y., Zhang, J., Tang, G., et al., 2016. Virtual water flows in the international trade of agricultural products of China. Sci. Total Environ. 557, 1-11.

Zhao, X., Yang, Z. F., Chen, B., 2009. Study on chinese virtual water trade and consumption in an input-output framework. J Nat Resour. 24 (2), 286-294 (in Chinese).

Zhi, Y., Yang, Z. F., Yin, X. A., 2014. Decomposition analysis of water footprint changes in a water-limited river basin: a case study of the Haihe River basin, China. Hydrol. Earth Syst. Sci. 18(5), 1549-1559.

Zhu, Q. R., Gao, J. F., 2009. Study on the issue of foreign trade and virtual water in China by input-output model. China Soft Sci. 5, 40-45 (in Chinese).

Zhu, Q. R., 2014. Study on virtual water of chinese foreign trade and adjusting of trade structure. Chin. Ind. Econ. 2, 58-70 (in Chinese).

Zimmer, D., Renault, D., 2003. Virtual water in food production and global trade: review of methodological issues and preliminary results. Virtual Water Trade: Proceedings of the International Expert Meeting on Virtual Water Trade, Value of Water Research Report Series 12 (1), 1-19.

Virtual water embodied in the export from various provinces of China using multi-regional input-output analysis

HUANG Yue[1,2], LEI Yalin[1,2,*], WU Sanmang[1,2]

1. *School of Humanities and Economic Management, China University of Geosciences, Beijing* 100083, *China*
2. *Key Laboratory of Carrying Capacity Assessment for Resource and Environment, Ministry of Land and Resources, Beijing* 100083, *China*

Abstract China is a large exporter and consumer of water embodied in export products. In addition, there is much trade among regions within China, and exports from various provinces consume not only local water but national non-local water. However, former calculations of the water consumption embodied in the exports of various provinces do not reflect the sources of the water consumption. Previous studies on virtual water in China's exports have focused on national aggregate analyses and have paid little attention to the inter-regional water consumption transfer driven by exports. We used a multi-regional input-output model of thirty provinces to examine the virtual water export. The results show that the total virtual water in China's exports was 106.3 billion m^3, and 77.0% of the embodiment was from exports from the eastern provinces, where over 90% of China's exports occur. However, the virtual water driven by per unit of export in the eastern regions is far less than that in the central and western regions. Moreover, the central and western provinces, whose exports are small, indirectly export much virtual water by supporting eastern China's exports via inter-regional economic linkages. The results yield important implications for China's export and virtual water export control policy.

Keywords Inter-provincial trade, Multi-regional input-output, Provinces of China, Regional water transfer, Virtual water export

1 Introduction

According to studies on agricultural water management by Mohammad Valipour et al. (2015; and with colleagues, 2015), we believe that it is essential for China to have a water policy due to the unpromising state of water resources. China is plagued with severe water

原文刊于 *Water Policy*，2017 年第 19 卷。原文发表时为彩色。

* 为通信作者。

shortages, and its fresh water resources comprise 2800 billion m^3, accounting for 6% of global freshwater resources. However, China contains 21% of the world's population, resulting in a per capita water availability of only 2200 m^3. This availability represents a quarter of the world's average, one-fifth of the United States availability, and ranks 121 in the world. China is one of the thirteen countries that lack per capita water resources. Regional water shortages will not only lead to pressure for water use, but will also negatively affect the water supply sector, resulting in a deterioration of the quality of the water supplied to the residents (Sadiq et al., 2007; Petkovic et al., 2011; Lindhe et al., 2012; Boryczko & Tchórzewska-Cies'lak, 2014). In addition, China is a major exporter according to figures released by the General Administration of Customs. China's national total export reached ¥14.39 trillion in 2014, continuing as the largest exporter in the world. Such large commodity exports will drive significant virtual water outflow. Virtual water refers to the water resources that the production of goods and services require, an innovative concept put forward by Allan (1993) in the early 1990s. Much virtual water export undoubtedly exacerbates the shortage of water resources and the deterioration of the quality of the water supplied to the residents. However, China remains a country with large regional differences, which not only manifest in its level of economic development but also reflect its endowment of water resources. Due to the extensive division of labor in manufacturing processes and trade contacts among provinces, provinces produce different quantities of export products at different stages. Any given province indirectly exports the water resources of other relevant provinces when exporting commodities, which leads to the redistribution of water resources in various regions with exportation from various provinces.

Previous studies have generally only considered the trade flow of virtual water between China and other countries or the virtual water export in various industries across the country. They have not used appropriate methods and data to embody the export of virtual water in various industries in various provinces to assess their differences. Furthermore, previous studies on virtual water export in China have failed to evaluate the virtual water export in each province that is indirectly driven by the exports of the other domestic provinces. Addressing these issues will provide reasonable policy basis recommendations and suggestions to relieve the shortage of water resources in China, and in particular, ease the issue of unevenly distributed water resources.

2 Literature review

Many scholars have studied virtual water, although the depth and methods of research are different.

For example, Sun et al. (2013) studied water footprint intensity spatial distribution patterns and evolution trends over 15 years via an exploratory spatial data analysis based on the water footprint calculation of China's thirty-one provinces (cities, districts) from 1995—

2009. Liu et al. (2011) developed a general model and an improved model of agricultural virtual water domestic trade gravity of time and space based on the classical gravity model and the time gravity model. They analyzed the panel data of China's twenty-nine provinces (cities, districts) and the influencing factors on China's agricultural virtual water net domestic outflow. Xu et al. (2011) developed mathematical analysis models from the technology level, the industry level and the district level and demonstrated how to promote a water shortage region's economic growth via virtual water trade. They also illustrated the contribution of virtual water trade on the economic growth in China based on China's international agricultural virtual water flow patterns from 2005—2009. However, among the previously reported studies on virtual water, the input-output method is the most common and most effective method used in China and abroad. When analyzing the problems related to virtual water using the input-output method, scholars have primarily used the methods and levels discussed below.

Studies on the water footprint and virtual water problems in some regions use a single regional input-output model to measure and analyze the total water footprint or industry strength of virtual water and occasionally involves the virtual water trade with other regions. Cai et al. (2012) calculated and analyzed the intensity of virtual water and the water footprint of primary industries, secondary industries and tertiary industries of the Gansu Province in 1997, 2002 and 2007 based on the single regional input-output method. They found that the intensity of virtual water of primary industries was the highest and decreased year-by-year. They also analyzed the virtual water trade of the Gansu Province and showed that Gansu is a net exporter of virtual water. Wang et al. (2013) used an adjusted single regional input-output method to measure the total water use of all departments in Beijing in 2002 and 2007 to analyze the total water consumption and its change in primary water consumption departments. The results showed that the intensity of the water for agriculture and industries was decreasing. Yu et al. (2010) also used a single regional input-output method to explore the virtual water trade patterns in England. The authors found that the water-poor southeast area's water footprint was 22% higher than the water-rich northeast area. The total water footprint of the southeast area was two times that of the northeast. Mubako et al. (2013) used this method to study the state of virtual water of California and Illinois and found that the two states are net exporters of virtual water and that most of the virtual water outflow was due to the high water consumption intensity and low value-added departments.

The multi-regional input-output (MRIO) model primarily focused on virtual water trade and transfer between a region and the remaining areas. This method creates a new category that differs from the method described above. When studies involve multiple regions, it is necessary to introduce a MRIO model, which is the primary method to evaluate inter-regional virtual water trade and transfer. To estimate the virtual water trade between Shanxi and other provinces and to reveal the spatial pattern of Shanxi's virtual water trade, Li et al. (2012) applied a MRIO model in their study. By combining the trade data of other provinces in China, both the degrees of agriculture and industry virtual water trade were calculated. Shanxi primarily imported virtual water from western and northeast China, and exported virtual water to eastern coastal provinces. This is contrary to the traditional virtual water strategy. Based on MRIO model and the socio-economic data of 2007, Tan et al. (2013) studied the water footprint and the composition of its origin in the Guangdong Province, revealing how each province contributed to the water footprint of the Guangdong Province. In addition, using the MRIO model, Zhang et al. (2011) studied the water footprint composition and the virtual water transfer of Beijing. Using the MRIO model, scholars have studied the water footprint of each province in China and have measured the virtual water transfer due to regional trade. They have also utilized the MRIO model in the same manners as described above. The difference is the research object, which extends from a single region to all parts of the country. Wang (2014) studied regional virtual water trade space distribution law and change tendency in China from 2002 to 2007 based on the regional input-output model to identify whether regional trade in China truly played a role in relieving the pressure of water deficiencies. According to the results, the current trade adjusting tend among the various regional departments in China and the changes in regional virtual water trade patterns were both not conducive to relieve the water resources shortage problem caused by uneven distributions. Zhou and Shi (2008) pre-sented an improved input-output table calculation model of water resources, bringing fresh water usage and wastewater discharge into the national economic industry value input-output table to construct a mixed input-output table of value-entity type. Then, the volume of inter-regional virtual water trade was obtained intuitively and accurately using the method and the authors put forward various problems that should be urgently addressed when China implements a regional virtual water strategy. Guan and Hubacek (2007) divided China into eight regions and designed a corresponding MRIO model to measure and evaluate China's inter-regional trade structure and virtual water transfer. The results showed that the current structure of China's trade is not conducive for the rational allocation of water resources. In the north, water scarcity outflowed 5% of their available water, whereas the south is rich in water resources, but instead imported water from other regions. Dong et al. (2014) used the model to study the water footprint and virtual water trade of each province in China. The results showed that the water footprint was closely related to the GDP of the province, as developed provinces tended to import virtual water and the north and west provinces with poor water exported the most virtual water.

Using a single regional input-output model, we can study virtual water net exports of various industries in the development of China's external trade. Because China has become the largest exporting country in the world, many merchandise exports cause loads on the virtual water outlets. This type of research utilizes a single regional input-output model, computing the coefficient of water usage, calculating the virtual water net export quantity based on foreign trade data, and then providing ideas for foreign trade strategies from water-saving point of view. He et al. (2011) computed virtual water input and output quantities of seventeen industries across the entire country of China for 2002, 2005, and 2007. The results showed that China was a virtual water net export country, and the quantity was still increasing. They also provided a detailed instruction of the primary regions and industries of virtual water export. Zhao et al. (2009) reported a computational model for the water footprint by summarizing the virtual trade based on input-output analysis technics and the consumption calculation model. They computed the direct water usage coefficient, the final consumed virtual water quantity, the water foot-print and the import and export of virtual water using the input-output table of 2002. The result showed that the direct water consumption coefficient and the volume of virtual water in agriculture and power, heat, and water were bigger than that in other departments in China, and other departments consumed much water indirectly. China is a net exporter of virtual water, which is primarily concentrated in the light industries and services. Zhu and Gao (2009) also used the single regional input-output method and calculated the foreign trade of virtual water from 2002 to 2007. They performed an empirical analysis of the problems in the current structure of China's foreign trade in light of water resource conservation.

In summary, previous studies generally include the following two deficiencies in their input-output method for studying the issue of China's virtual water:

(1) When studying the issue of China's virtual water, previous studies have adopted MRIO model but only studied the situation of transferring virtual water among various regions of China due to inter-provincial trade. They failed to extend their research to foreign export and did not calculate and analyze the export quantity of virtual water to foreign countries.

(2) When combining studies on the issue of virtual water with foreign export, previous studies applied the single regional input-output model and usually considered the entire nation of China as the research object. Thus, they did not evaluate virtual water export from various provinces in detail. As a result, the direct export of local virtual water and the indirect export of non-local virtual water due to the export products of the each province are neglected. In addition, studying the foreign export of China's virtual water with the single regional input-output model utilized the national average water use coefficients of various industries to measure the export quantity of virtual water. Such studies neglect different industrial structures and the different water-use efficiency of products among various provinces of China. Therefore, the virtual water export from China cannot be accurately measured and calculated to show differences in the virtual water export among various regions of China.

Hence, the MRIO model is more appropriate for comprehensively evaluating the source of actual water consumption of export products among various provinces of China. Its advantages lie in combining the model with the input-output data, water-use data, and foreign trade data of various provinces and calculating the export quantity of virtual water from various provinces due to the exports from each industry in each province.

3 Methodology and data

3.1 MRIO model

The input-output method was first proposed in the 1930s by the economist Leontief and was used to reflect the relationship between the various sectors (industry) of the economy. As a MRIO table is an extension of single region input-output tables in that it reflects the inter-regional trade of commodities and services. It is more comprehensive and systematically reflects the economic relations among multiple regions in contrast to the single regional input-output table. Chenery (1953) and Moses (1955) put forward the MRIO model. The model assumes that the composition and sources of regional goods that have different uses (intermediate input, final consumption and investment of the different departments) are indiscriminate. Therefore, we only need to obtain the data for each department product flow in each region and do not need to develop a regional product flow matrix. Compiling a MRIO table that contains the volume of water consumption can aid in finding the physical and technical connection of water consumption among different departments in different regions. The specific methods and calculation steps are as follows.

Assume that there are n regions, and each region is divided into m sectors (each sector produces a homogeneous product). The balance of production activities in region r can be expressed as equation (1), that is, the total output in each sector equals the output used as the intermediate input plus the output used as the final consumption:

$$x_i^r = \sum_{s=1}^{n}\sum_{j=1}^{m} x_{ij}^{rs} + \sum_{s=1}^{n} f_i^{rs} + e_i^r \tag{1}$$

where x_{ij}^{rs} is the intermediate sector monetary flow from sector i in region r to sector j in region s (r, s=1, 2⋯ m; i, j=1, 2 ⋯ n), f_i^{rs} is the final demand of region s for products in sector i from region r, e_i^r is the export of sector i in region r, and x_i^r is the total output of sector i in region r.

a_{ij}^r is the direct consumption coefficients of region r, which indicates that sector j in region r produces per unit of product in demand of sector i in the national regions. We define it as:

$$a_{ij}^r = \frac{x_{ij}^r}{x_j^r} \tag{2}$$

According to the product-flow origin and destination (O-D) matrix, we can calculate the proportion of product i used in region r that comes from region s and the proportion of product i used in region r that comes from region r, as follows:

$$c_i^{sr} = \frac{z_i^{sr}}{d_i^r},\quad c_i^{rr} = \frac{z_i^{rr}}{d_i^r} \tag{3}$$

where z_i^{sr} is the monetary flow of product i from region s (the origin) to region r (the destination) and d_i^r is the total shipments of product i into region r from all regions (see Table 1).

Table 1 O-D matrix

		Destination		
		1	…	n
Origin	1	$z_i^{1,1}$	$z_i^{1,s}$	$z_i^{1,1}$
	…	…	…	…
	n	$z_i^{n,1}$	$z_i^{n,s}$	$z_i^{1,n}$
	Total	d_i^1	d_i^s	d_i^n

Substituting equation (2) and equation (3) into equation (1) yields the following equation:

$$x_i^r = \sum_{s=1}^{n}\sum_{j=1}^{m} c_i^{rs} a_{ij}^s x_j^s + \sum_{s=1}^{n} c_i^{rs} f_i^s + e_i^r \tag{4}$$

Converting the above formula to matrix form yields:

$$X = CAX + CF + E \tag{5}$$

where X is the output matrix, A is the technical coefficient matrix, and C is the trade coefficient matrix, F is the final demand matrix, and E is the export matrix.

x_i^r is the output of sector i in region r.

Let

$$C = \begin{bmatrix} C^{11} & \cdots & C^{1n} \\ \cdots & \cdots & \cdots \\ C^{n1} & \cdots & C^{nn} \end{bmatrix}, \quad C^{rs} = \begin{bmatrix} c_1^{rs} & 0 & 0 & 0 \\ 0 & c_2^{rs} & 0 & 0 \\ 0 & 0 & \vdots & 0 \\ 0 & 0 & 0 & c_m^{rs} \end{bmatrix}$$

in which the elements show the proportion of the total amount of each product used in region s that comes from region r.

Let

$$A = \begin{bmatrix} A^1 & 0 & 0 & 0 \\ 0 & A^2 & 0 & 0 \\ 0 & 0 & \vdots & 0 \\ 0 & 0 & 0 & A^n \end{bmatrix}, \quad A^r = \begin{bmatrix} a_{11}^r & \cdots & a_{1m}^r \\ \cdots & \cdots & \cdots \\ a_{m1}^r & \cdots & a_{mm}^r \end{bmatrix},$$

where A^r is the technical coefficient matrix indicating the intermediate inputs of region r from both within the region and outside regions (including imports).

$$F = \begin{bmatrix} F^1 \\ F^2 \\ \vdots \\ F^n \end{bmatrix} \text{ and } F^r = \begin{bmatrix} f_1^r \\ f_2^r \\ \vdots \\ f_m^r \end{bmatrix},$$ where f_i^r is the final consumption demand of sector i in region r.

$$E=\begin{bmatrix}E^1\\E^2\\\vdots\\E^n\end{bmatrix} \text{ and } E^r=\begin{bmatrix}e_1^r\\e_2^r\\\vdots\\e_m^r\end{bmatrix}$$, where e_i^r is the export demand of sector i in region r.

Rewriting equation (5), we can obtain equation (6),

$$CAX + CF + E \Rightarrow (I - CA)X = CF + E \Rightarrow X = (I - CA)^{-1}(CF + E) \tag{6}$$

To remove CF from equation (6), we obtain equation (7),

$$XE = (I - CA)^{-1}E \tag{7}$$

where XE is the output driven by the exports of the entire nation. We can further obtain the production in region s driven by the exports of region r:

$$X_{EX^r} = (I - CA)^{-1}E^r \tag{8}$$

$$X_{EX^r} = \begin{bmatrix}x_{EX^r}^1\\x_{EX^r}^2\\\vdots\\x_{EX^r}^s\\\vdots\\x_{EX^r}^n\end{bmatrix}$$

in which the element represents the production in region s driven by the exports of region r.

Thus, we can further obtain the virtual water in region s induced by the exports of region r:

$$W_{EX^r} = \widehat{W}X_{EX^r} = \widehat{W}(I - CA)^{-1}E^r \tag{9}$$

$$W_{EX^r} = \begin{bmatrix}w_{EX^r}^1\\w_{EX^r}^2\\\vdots\\w_{EX^r}^s\\\vdots\\w_{EX^r}^n\end{bmatrix},\ \widehat{W} = \begin{bmatrix}\widehat{W}^1 & & & & \\ & \ddots & & & \\ & & \widehat{W}^s & & \\ & & & \ddots & \\ & & & & \widehat{W}^n\end{bmatrix} \text{ and}$$

$$\widehat{W}^s = \begin{bmatrix}w_1^s & & & & \\ & \ddots & & & \\ & & w_i^s & & \\ & & & \ddots & \\ & & & & w_m^s\end{bmatrix},$$

where w_i^s is the direct water use coefficient of sector i in region s, $\widehat{W}^s$ is the diagonal matrix of the direct water use coefficient in region s, $\widehat{W}$ is the diagonal matrix of the direct water use coefficient in the national various regions, and $w^{s}_{EX^{r}}$ is the virtual water in region s induced by the exports of region r.

Thus, we can calculate the total virtual water induced by exports from region r using the following equation:

$$w^{\text{tot}}_{EX^{r}} = \sum_{s=1}^{n} w^{s}_{EX^{r}} \tag{10}$$

In addition, we can calculate the virtual water in region s induced by the national exports:

$$w^{s}_{EX^{\text{tot}}} = \sum_{r=1}^{n} w^{s}_{EX^{r}} \tag{11}$$

3.2 Data sources and processing methods

In this article, we used the 2007 MRIO tables (Xu and Li, 2008) for China developed by the State Development Research Center and the State Statistical Bureau. Until now, the 2007 MRIO table for China has been the latest regional input-output table for China. These MRIO tables include thirty provinces (all provinces of China except Xizang, Taiwan, Hong Kong and Macao), and each province includes forty-two sectors. For convenience, this study merged forty-two sectors into thirteen sectors. Moreover, the flows of bulk commodities supplied by service sectors was poorer compared to other industrial sectors, and the data of water use was difficult to obtain. Furthermore, the water use coefficient was very low, and the quantity of virtual water in exports is very small, this study only analyzed the cases of nine sectors (except for four service sectors) among the thirteen sectors.

The water consumption of each province and that of the relevant industrial sectors of China in 2007 were obtained from the China Environment Yearbook (National Bureau of Statistics, 2008) and China Water Resources Bulletin (Ministry of Water Resources People 's Republic of China, 2008). The data of agricultural water use in various regions were obtained directly from the water resources bulletin. To calculate the industrial water use coefficient of various departments of each province, first, we calculated the national average water use coefficient of each industrial department w_i, and then made $W = [w_i]$. For province r, the output matrix of industrial departments was $X^r = [x_i^r]'$. Second, we made $S^{r'} = WX^r$, which means the pending revision of total water consumption of all industrial departments in province r, it was calculated by the national average water use coefficient. After that, we made S^r the real water consumption of all industrial departments in province r, and made $\delta^r = S^r / S^{r'}$, which means the correc-tion factor of province r. Finally, the revised water use coefficient matrix of industrial departments in province r was $W^r = \delta^r W = [\delta^r W_i]$.

4 Results and discussion

Obvious differences among the eastern, central, and western regions of China exist in the level of development and water resource potential (the eastern region includes eleven provinces-

Beijing, Tianjin, Hebei, Liaoning, Shanghai, Jiangsu, Zhejiang, Fujian, Shandong, Guangdong, and Hainan; the central region includes eight provinces – Shanxi, Jilin, Heilongjiang, Anhui, Jiangxi, Henan, Hubei, and Hunan; the western region includes eleven provinces – Inner Mongolia, Guangxi, Chongqing, Sichuan, Guizhou, Yunnan, Shaanxi, Gansu, Qinghai, Ningxia, and Xinjiang). Therefore, in this paper we analyzed the state of virtual water export of thirty provinces and the three macro-regions (that is, the eastern, central, and western regions) in China.

4.1 An analysis of the pulling effect of export from various regions on the export of national virtual water

For each province, the total national virtual water induced by the province's exports includes the water consumption of export commodities locally and in other provinces. This indicator can reflect the impact of each region's exports on the water consumption of the entire country. Table 2 not only presents data from all provinces but also the commodity export data of the provinces, so the relationship between the national virtual water driven by exports from each province and local exports can be analyzed.

Table 2 Export quantity of various regions and the export situation of the national virtual water pulled by exports from various regions in 2007

Region (r)	$W^{tot}_{EX^r}$ ($10^9 m^3$) *	EX^r (10^9 ¥) **	100 [$W^{tot}_{EX^r}/\sum_r W^{tot}_{EX^r}$] (%) ***	100 [$EX^r/\sum_r EX^r$] (%) ****
Eastern provinces	81.9	8, 420.0	77.0	92.2
Western provinces	15.7	312.8	14.8	3.4
Central provinces	8.7	400.9	8.2	4.4
Guangdong	23.1	2, 870.0	21.7	31.4
Jiangsu	18.7	1, 350.0	17.6	14.8
Zhejiang	9.8	959.0	9.2	10.5
Shandong	9.7	676.5	9.1	7.4
Xinjiang	9.4	27.8	8.9	0.3
Shanghai	7.0	1, 120.0	6.6	12.3
Fujian	5.8	382.9	5.4	4.2
Liaoning	2.4	246.6	2.3	2.7
Heilongjiang	2.2	38.3	2.1	0.4
Beijing	2.1	436.3	2	4.8
Hebei	1.5	138.8	1.4	1.5
Tianjing	1.4	224.8	1.3	2.5
Anhui	1.3	65.0	1.2	0.7
Hubei	1.2	59.9	1.2	0.7
Guangxi	1.2	43.5	1.1	0.5

Continued

Region (r)	$W^{tot}_{EX^r}$ (10^9m^3) *	EX^r (10^9 ¥) **	100 [$W^{tot}_{EX^r}/\sum_r W^{tot}_{EX^r}$] (%) ***	100 [$EX^r/\sum_r EX^r$] (%) ****
Ningxia	1.2	12.2	1.1	0.1
Jiangxi	1.1	40.8	1	0.4
Gansu	1.0	44.2	0.9	0.5
Jilin	0.9	27.4	0.8	0.3
Hunan	0.9	44.8	0.8	0.5
Henan	0.8	67.9	0.7	0.7
Sichuan	0.7	49.7	0.6	0.5
Inner Mongolia	0.5	26.8	0.5	0.3
Hainan	0.4	18.4	0.4	0.2
Chongqing	0.4	30.4	0.3	0.3
Yunnan	0.4	18.2	0.4	0.2
Shaanxi	0.4	38.6	0.4	0.4
Qinghai	0.4	6.8	0.3	0.1
Shanxi	0.3	56.9	0.3	0.6
Guizhou	0.2	14.6	0.2	0.2

* $W^{tot}_{EX^r}$ represents the total national virtual water induced by exports from region r.

** EX^r represents the quantity of production exports from region r.

*** 100 [$W^{tot}_{EX^r}/\sum_r W^{tot}_{EX^r}$] represents the proportion of the total national virtual water induced by exports from region r.

**** 100 [$EX^r/\sum_r EX^r$] represents the proportion of quantity of production exports from region r.

First, the national virtual water exports are mainly driven by exports in eastern China due to the high volume of exports from the eastern region, but the virtual water export driven by each export in the eastern region is less. According to Table 2 and from the perspective of the eastern, central and western parts, the eastern region exported commodities worth ¥8420 billion, which comprised 92.2% of the total national export. However, the export quantity of the national virtual water driven by the exports from the eastern region only occupied 77.0% of the total export quantity of national virtual water. The exports from the western region accounted for 3.4% of the national export, promoting 14.8% of the export of national virtual water. The central region, whose exports accounted for 4.4% of national exports, pulled 8.2% of the export of national virtual water. Accordingly, the unit export commodity of the eastern region consumed one-fifth of the water used by the unit export commodity of the western part and half of the water used by the unit export commodity of the central part. There were two main reasons for this result. First, the eastern region used water more efficiently when producing the same export commodity compared to the central and western regions. Second, the respective export commodities of the eastern, central and western regions presented huge differences. Specifically, the eastern region focused on exporting commodities characterized by low water consumption, whereas the central and western regions primarily exported commodities characterized by high water consumption.

Second, provinces with small exports are likely to pull a large amount of the national virtual water exports. The Guangdong Province ranked highest according to the array of the export quantity of national virtual water driven by export from various provinces. In 2007, the exports from the Guangdong Province boosted export to 23.1 billion m^3 of virtual water across the country, accounting for 21.7% of the total export of national virtual water. The major cause was that the Guangdong Province had a large amount of exports and exported a total of ¥2870 billion in 2007, which accounted for 31.4% of the national exports. Ranking second in export quantity, the Jiangsu Province, with 14.8% of export quantity, propelled 17.6% of the export of national virtual water. The export quantity of virtual water pulled by the export of the Jiangsu Province was also second in the country. The eight top-ten provinces in pulling the export quantity of the national virtual water were in the eastern region, and the export quantity of these provinces ranked top throughout the country. Xinjiang and Heilongjiang accounted for 8.9% and 2.1%, respectively, of the export quantity of national virtual water, but their export quantities only held a small proportion of the national export quantity, namely 0.3% and 0.4%, respectively. This is because the two provinces are typical provinces exporting agricultural products, and agricultural sectors belong to typical sectors with low prices and high water consumption. The virtual water export driven by agricultural from Xinjiang and Heilongjiang respectively reached 12.95 billion m^3 and 3.72 billion m^3. Moreover, a great deal of virtual water was exported by the mining industry, the petrochemical industry and power, heat, and water industries in the Heilongjiang Province. These sectors were also characterized by high water consumption and exported 1.06 billion m^3 virtual water.

4.2 An analysis on the water consumption intensity of unit export commodities of various provinces

We divided the national virtual water in exports from each region by the volume of the exports of each region to produce the water consumption per unit of export commodities. This value can reflect th e differences in the efficiency of water use in the production processes of exports in the various regions of China and it can aid in the analysis of the different types of exports and different water technologies in the various regions.

As shown in Fig. 1, the export quantity of the national virtual water driven by the unit export of Xinjiang reached 0.339 m^3/¥, which exceeded all other provinces. Xinjiang was followed closely by Ningxia, with a unit export value of 0.098 m^3/¥. Guangdong, Tianjin, Shanghai, Shanxi, and Beijing ranked the lowest. Except for Shanxi, these provinces made up regions with relatively developed economies and that focused on exporting low water consumption and high value-added commodities. Furthermore, the proportion of tertiary industries in these provinces far exceeded the proportion of secondary industries, whose production links primarily consumed little water. Among the top ten-ranking provinces in pulling the export quantity of national virtual water by virtue of their unit export, only the Hainan Province with a relatively underdeveloped economy is in the eastern part; the other provinces are in central and western parts. Moreover, several traditional agricultural provinces, such as Xinjiang, Heilongjiang, Jilin, Jiangxi and so on, have unit exports with high water consumption, which indicated the primary cause for the large differences in the export quantity of national virtual water pulled by the unit export of various provinces was that the export commodities among the various provinces were highly different.

4.3 An analysis on the pulling effect of national exports on the virtual water export of various provinces

For each province, the pulling effect of national exports on the local virtual water export is the export volume of local virtual water that is driven by local exports and exports from other regions; specifically, the local virtual water embodied in China's total exports. It can reflect the impact of China's exports on the water consumption of each region. Table 3 presents the results of this analysis. We analyzed the data together with the export volume of each region to take the water consumption into account when evaluating the benefits created by the exports of various provinces. Each region is presented in descending order of the ratio of the local virtual water embodied in China's total exports to the local export volume.

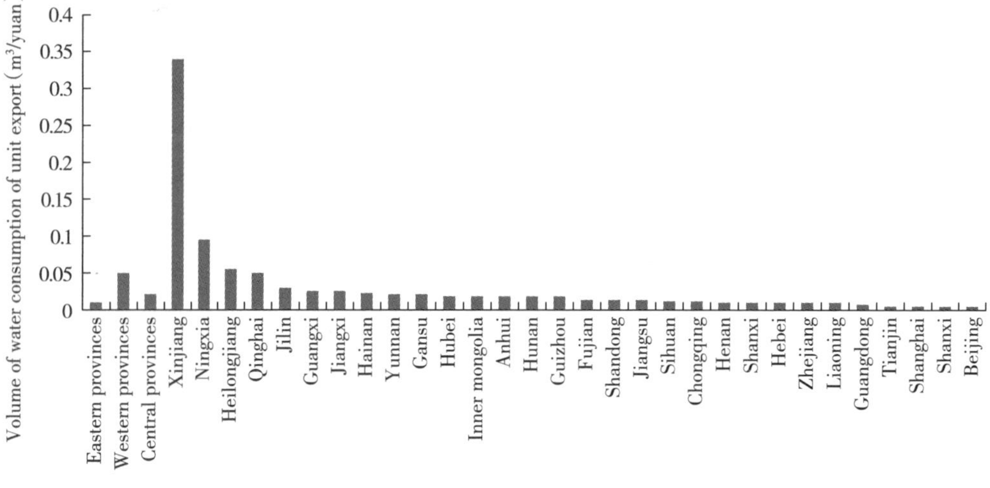

Fig. 1 Export quantity of the national virtual water pulled by the unit exports of various provinces

Table 3 Export quantity of various regions and the export situation of the virtual water pulled by national exports in 2007

Region (r)	$W^r_{EX^{tot}}$ (109 m^3)*	EX^r (10^9 ¥)	100 [$W^r_{EX^{tot}}/\sum_r W^r_{EX^{tot}}$] (%)**	100 [$EX^r/\sum_r EX^r$] (%)
Western provinces	27.4	312.8	25.8	3.4
Central provinces	20.4	400.9	19.2	4.4

Continued

Region (r)	$W^r_{EX^{tot}}$ (109 m^3) *	EX^r (10^9 ¥)	100 [$W^r_{EX^{tot}}/\sum_r W^r_{EX^{tot}}$] (%) **	100 [$EX^r/\sum_r EX^r$] (%)
Eastern provinces	58. 5	8, 420. 0	55. 0	92. 2
Xinjiang	13. 1	27. 8	12. 4	0. 3
Ningxia	1. 4	12. 2	1. 3	0. 1
Heilongjiang	4. 9	38. 3	4. 6	0. 4
Inner Mongolia	2. 6	26. 8	2. 4	0. 3
Guangxi	3. 5	43. 5	3. 3	0. 5
Hunan	3. 2	44. 8	3	0. 5
Yunnan	1. 2	18. 2	1. 1	0. 2
Jilin	1. 6	27. 4	1. 5	0. 3
Guizhou	1. 1	14. 6	1	0. 2
Jiangxi	2	40. 8	1. 9	0. 4
Anhui	3. 4	65	3. 2	0. 7
Qinghai	0. 4	6. 8	0. 4	0. 1
Hubei	2. 8	59. 9	2. 6	0. 7
Hainan	0. 6	18. 4	0. 6	0. 2
Henan	1. 9	67. 9	1. 8	0. 7
Sichuan	1. 3	49. 7	1. 2	0. 5
Gansu	1. 3	44. 2	1. 2	0. 5
Shaanxi	0. 9	38. 6	0. 9	0. 4
Hebei	3. 3	138. 8	3. 1	1. 5
Chongqing	0. 6	30. 4	0. 6	0. 3
Fujian	5. 3	382. 9	5	4. 2
Jiangsu	17. 6	1350	16. 6	14. 8
Shanxi	0. 6	56. 9	0. 6	0. 6
Liaoning	2. 4	246. 6	2. 2	2. 7
Shangdong	5. 2	676. 5	4. 9	7. 4
Zhejiang	5. 4	959	5. 1	10. 5
Guangdong	13. 5	2870	12. 7	31. 4
Shanghai	4. 3	1120	4	12. 3
Tianjing	0. 4	224. 8	0. 4	2. 5
Beijing	0. 5	436. 3	0. 5	4. 8

* $W^r_{EX^{tot}}$ represents the virtual water export in region r induced by the total national exports.

** 100 [$W^r_{EX^{tot}}/\sum_r W^r_{EX^{tot}}$] represents the proportion of the virtual water export in region r induced by the total national exports.

First, the central and western regions have contributed a large amount of water to the exports of eastern region. As shown in Table 3, the pulled export quantity of virtual water in various provinces of the eastern part is the largest, namely 58. 5 billion m^3, followed by 27. 4

billion m^3 in the western regions, and 20.4 billion m^3 in the central regions. The pulled exports of the virtual water of the three regions correspond to 55.0%, 25.8% and 19.2%, respectively. When these data are compared with the proportion of total national virtual water induced by exports from these three regions (see Table 2), the following findings are quite striking Namely, the eastern region, with 92.2% of national export quantity, boosted 77.0% of the total export of national virtual water (as in Table 2), but the quantity of the virtual water export pulled by the national exports in this region only made up 55.0% of the total export quantity of the national virtual water. In other words, the export quantity of the virtual water pulled by the exports of the eastern region in other regions made up at least 22% of the total export quantity of the national virtual water (55.0% of the export quantity of the virtual water in the eastern region was partly driven by other regions). On the contrary, the western region exported only 3.4% of the national commodities, but the pulled export of the virtual water in the western provinces accounted for 25.8% of the total export quantity of the national virtual water. The central region exported only 4.4% of the national export quantity, but its export quantity of virtual water amounts to 19.2%.

The central and western regions had small export quantities. However, the central and western regions contributed a good deal of water to national (mainly eastern region) exports. Without doubt, this situation will intensify the problem of the shortage of water resources in the central and western regions. In addition to the above-mentioned differences in the export commodities of the various regions in water consumption intensity, there were two additional reasons for this situation. First, the central and western regions primarily undertook the upstream links of producing export commodities due to geographic and technical differences. These links usually consume a large amount of water. The eastern region primarily undertook technology-intensive links and consumed less water. Second, the central and western provinces primarily exported high water consumption and low value-added products to the eastern provinces in the process of provincial import and export. Inversely, the eastern provinces exported low water consumption and high value-added products to the central and western provinces.

Second, most central and western provinces export a large amount of local virtual water but less products. The distinct confines of 'west, central, and east' among various provinces run from the top to the bottom of Table 3. Six of the top ten provinces are from the western region, and four from the central region. However, the ten provinces at the bottom of the list are all in eastern region except for Hebei and Hainan, which indicates that the exports of most of the central and western provinces play a lesser role in promoting their economies, but they consumed a relatively large amount of local water resources due to the national exports, and the opposite occurred in the eastern provinces. For example, Xinjiang consumed 13.1 billion m^3 of virtual water due to China's total exports while receiving an economic benefits of only ¥2.78 billion from local exports. The ratio of these two quantities was up to 0.5, while this ratio for Beijing was only 0.0011, indicating that while the same benefit was received from exports, the cost of water consumption in Xinjiang was much higher than that in Beijing by supporting China's total exports. However, in the past, the evaluation of the export activities of the provinces only encompassed the economic benefits without considering the differences in the cost of water consumption in the provinces, or the need for relevant support and compensation in the central and western provinces.

4.4 A comparative analysis on pulling and pulled export quantities of virtual water of various provinces

Respective discussions were held regarding the national virtual water driven by the export of various provinces and the virtual water export in various provinces driven by national exports, and an analysis was conducted by combining the export quantity of the various provinces above. However, the above two datasets of various provinces were not displayed and analyzed comparatively. For each province, if the former is greater than the latter, the province has a net import of virtual water from other provinces, if the former is less than the latter, the province has a net export of virtual water to other provinces. In addition, we added the indicator of local embodied virtual water to the local exports, and compared it with the above two indicators to reflect the local virtual water from non-local exports and the non-local virtual water from local exports (Fig. 2).

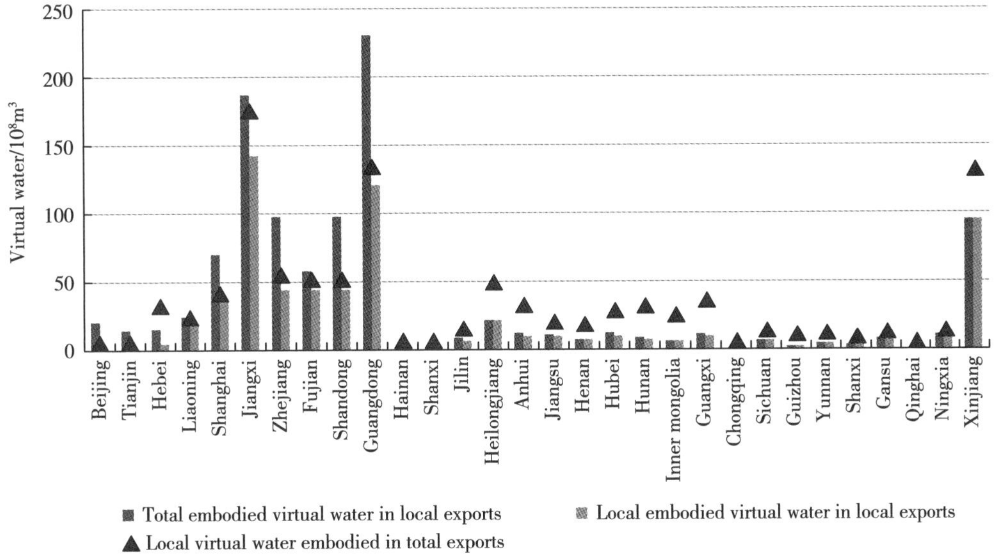

Fig. 2 Detailed information of the quantity of virtual water export in various provinces

First, individual eastern provinces provide large amounts of water to support the export of the other regions, whereas the export of the eastern provinces pulls a large export of the virtual water of other provinces. According to Fig. 2, the national virtual water embodied in the local exports in the eastern region, including Beijing, Tianjin, Shanghai, Zhejiang, Shandong and Guangdong, was much higher than the local virtual water embodied in the national exports in these provinces. Thus, the export quantity of the virtual water in these provinces was pulled by their own exports, which indicated that the export of these provinces consumed much of the water resources of other regions in addition to using their own water resources. Except for consuming the plentiful water resources of other provinces for its export, Jiangsu also expended significant local water resources due to the influence of national exports. Hebei is a water-shortage region, and the local virtual water embodied in national exports was much larger than in local exports, as Hebei provided abundant water resources to support the economic development of Beijing and Tianjin.

Second, the export quantity of national virtual water pulled by the export of the central and western provinces is derived locally and these provinces provide a certain amount of water consumption for the exportation of products in other regions. For each province in central and western regions, the total export quantity of national virtual water in local exports was less than that of the local virtual water in total national exports, which was in contrast to the state of the eastern provinces. In addition, this finding was more obvious in Heilongjiang, Anhui, Hunan, Inner Mongolia, Guangxi and other provinces, which further illustrated that the central and western regions offered plenty of water to the exports of the eastern region. Additionally, in terms of the central and western provinces, the local virtual water embodied in local exports was roughly equal to the national virtual water embodied in local exports. Namely, the exports of these provinces consumed almost no water resources of other provinces.

4.5 An analysis of virtual water export of each sector in the provinces

We calculated and analyzed the proportion of virtual water exports in various industrial sectors in each region driven by China's exports (Fig. 3), to comprehensively identify the key sectors of virtual water export in each province and explore the internal reasons for the disparity between the provinces.

First, the export of virtual water in each province mainly arises from agriculture and power, heat and water. The proportion of virtual water export from these two sectors accounted for 61.4% and 21.9% of the nation's total virtual water exports. Especially in Xinjiang and Ningxia, where the contribution of agriculture was greater than 90%. On the other hand, the construction industry and building material production are sectors with less virtual water export, and their proportion of virtual water export accounted for only 0.26% and 0.03% of the national's total virtual water export. These phenomena result from the high water consumption by agriculture and power, heat and water and the low water consumption and lower export volume by the construction and building material industries.

Second, a huge difference in the proportion of virtual water export exists in each sector

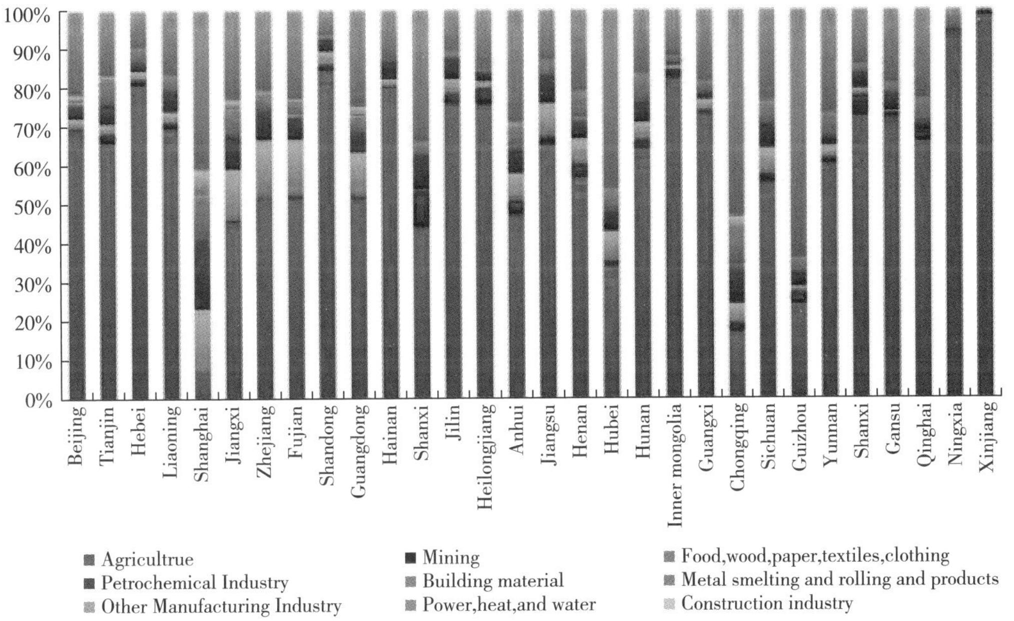

Fig. 3 Proportion of virtual water exports of each sector in the provinces

among the provinces. For example, agriculture and power, heat and water in Shanghai are responsible for only 48% of Shanghai's total virtual water exports, while the percentage for other provinces was greater than 68%. In addition, the proportion of virtual water exports from the food, wood, paper, textiles and clothing industries in several of the eastern provinces (that is, Shanghai, Jiangsu, Zhejiang, Fujian and Guangzhou) is higher than in the central and western regions. However, the mining industry in the central and western provinces was responsible for a higher percentage of virtual water exports than in the eastern provinces. This is closely related to the marked differences in the industrial structure of the provinces.

5 Conclusions and recommendations

According to previous studies, exports from China contain significant virtual water export. However, in this study, using MRIO, we found that our exports not only caused significant virtual water outflow, but that exports from eastern China contributed more than 90% of the total exports and generated around 77% of China's virtual water outflow, the central and western provinces accounted for less than 10% of exported goods, nonetheless, pulled approximately 23% of China's virtual water outflow.

In addition, this study also found that, due to the differences in water use efficiency of production, export commodity groups, and production division of labor, economically developed regions, such as the eastern provinces, generally export commodities of low water consumption intensity, whereas the central and western provinces generally export commodities of high water consumption intensity.

Finally, by comparing the volume of exports and virtual water export, we found that, although the eastern region exports many products, its local virtual water export driven by its own exports is low. In contrast, the central and western regions contributed to significant virtual water exports for the eastern region export. In addition, virtual water induced by China's exports in the central and western regions is greater than that in the eastern regions but the same economic benefit is derived from the exports. Regarding the central and western regions, the national virtual water export driven by their own exports are derived almost completely from local sources; that is, the product exports of the central and western regions consume scarce water resources from other regions. This shows that in the provincial trade of high water consumption products, primarily central and western regions export to the eastern region, and most of products of the eastern region export to the central and western regions for the export-oriented products. Furthermore, the volume of the virtual water exports from agriculture and power, heat and water in the various provinces were usually large, while the construction and building materials industries exported less virtual water. Because there are differences in various provinces in terms of industrial structure and the division of labor, the volumes of virtual water export in each sector from each province are distinctly different.

In terms of policy recommendations, firstly, we should prompt the central and western regions to improve the production process and technical level and to improve water use efficiency, especially for high water consumption industries in the water-shortage provinces.

Secondly, when evaluating the benefits created by the exports of various provinces, the water consumption caused by export should also be taken into account. The central and western regions contributed a large amount of virtual water for the eastern regions ' exports; therefore, because they received great economic benefits from these exports, the eastern regions should provide appropriate financial compensation and technical support to the appropriate water supply regions, so as to alleviate water shortages in these regions. According to the analysis above, the current industrial structure in China will worsen the condition of water shortages in the water-short regions, and the water-rich regions consume little water at present, which means that the current industrial structure in China is not conducive to balance the unevenly distributed water resources in China. Promoting nationwide industrial restructuring by transferring the high water-consuming export industries from the western region to the eastern region will effectively ease the water shortage in China's central and western regions. In addition, the water-short regions should reduce the production and export of high water-consuming products, especially for those with low added value that are useless for stimulating the local economy.

Acknowledgements The authors express their sincere thanks for the support from the National Natural Science Foundation of China under Grant No. 71173200 and the support from the Development and Research Center of China Geological Survey under Grant No. 12120114056601.

References

Allan, J. A. (1993). Fortunately there are substitutes for water otherwise our hydro-political futures would be impossible. In Priorities for Water Resources Allocation and Management, Vol. 13, p. 26.

Boryczko, K. & Tchórzewska-Cieślak, B. (2014). Analysis of risk of failure in water main pipe network and of delivering poor quality water. Environment Protection Engineering 40 (4), 77-92.

Cai, Z. H., Shen, L. X., Liu, J. G. & Zhao, X. (2012). Applying input-output analysis method for calculation of water footprint and virtual water trade in Gansu Province. Acta Ecologica Sinica 32 (20), 6481-6488 (in Chinese).

Chenery, H. B. (1953). Regional analysis. In: The Structure and Growth of the Italian Economy. Chenery, H. B., Clark, P. G. & Pinna, V. C. (ed.). US Mutual Security Agency, Rome, pp. 97-129.

Dong, H. J., Geng, Y., Fujita, T., Fujii, M., Hao, D. & Yu, X. M. (2014). Uncovering regional disparity of China's water foot-print and inter-provincial virtual water flows. Science of the Total Environment 500-501, 120-130.

Guan, D. & Hubacek, K. (2007). Assessment of regional trade and virtual water flows in China. Ecological Economics 61, 159-170.

He, X. B., Zhang, H. W., Wang, Y. & Ma, S. Y. (2011). Analysis of Chinese international virtual water trade in an input-output method. Environmental Science and Management 36 (3), 7-10 (in Chinese).

Li, F. Y., Liu, W. D. & Liu, H. G. (2012). Inter-regional virtual water trade model and its application in Shanxi Province. Resources Science 34 (5), 802-810 (in Chinese).

Lindhe, A., Norberg, T. & Rosén, L. (2012). Approximate dynamic fault tree calculations for modelling water supply risks. Reliability Engineering & System Safety 106, 61-71.

Liu, H. M., Li, G. J. & Wang, K. Q. (2011). Analysis on the influencing factors of China's agricultural virtual water domestic trade based on gravity model. Chinese Rural Economy (5), 21-32 (in Chinese).

Ministry of Water Resources People's Republic of China (2008). China Water Resources Bulletin 2008. China Water & Power Press, Beijing (in Chinese).

Moses, L. N. (1955). The stability of interregional trading patterns and input-output analysis. American Economic Review 45, 803-832.

Mubako, S., Lahiri, S. & Lant, C. (2013). Input-output analysis of virtual water transfers: case study of California and Illinois. Ecological Economics 93, 230-238.

National Bureau of Statistics of China (2008). China Environment Yearbook 2008. China Statistical Publishing House, Beijing (in Chinese).

Petkovic, S., Gregoric, E., Slepcevic, V., Blagojevic, S., Gajic, B., Kljujev, I., Žarković, B., Djurovic, N. & Draskovic, R. (2011). Contamination of local water supply systems in suburban Belgrade. Urban Water Journal 8 (2), 79-92.

Sadiq, R., Rodriguez, M. J., Imran, S. A. & Najjaran, H. (2007). Communicating human health risks associated with disinfection by-products in drinking water supplies: a fuzzy-based approach. Stochastic Environmental Research and Risk Assessment 21 (4), 341-353.

Sun, C. Z., Chen, S. & Zhao, L. S. (2013). Spatial correlation pattern analysis of water footprint intensity based on ESDA model at provincial scale in China. Journal of Natural Resources 28 (4), 571-582 (in Chinese).

Tan, S. L., Liu, Z. F., Xiong, Y. J., Liao, C. & Chen, Z. H. (2013). Evaluation of water footprint for Guangdong Province in an interregional input-output framework. Ecology and Environmental Sciences 22 (9), 1564-1570 (in Chinese).

Valipour, M. (2015). A comprehensive study on irrigation management in Asia and Oceania. Archives of Agronomy and Soil Science 61 (9), 1247–1271.

Valipour, M., Ziatabar Ahmadi, M., Raeini–Sarjaz, M., Gholami Sefidkouhi, M. A., Shahnazari, A., Fazlola, R. & Darzi–Naftchali, A. (2015). Agricultural water management in the world during past half century. Archives of Agronomy and Soil Science 61 (5), 657–678.

Wang, X. N. (2014). Research on distribution and tendency of virtual water trade in China based on inter–regional input–output model. Management Review 26 (7), 46–54 (in Chinese).

Wang, Z. Y., Huang, K., Yang, S. S. & Yu, Y. J. (2013). An input–output approach to evaluate the water footprint and virtual water trade of Beijing, China. Journal of Cleaner Production 42, 172–179.

Xu, X. C. & Li, S. T. (2008). Preparation and Analysis of Regional Input – Output Table of China. Journal of Tsinghua University, Beijing (in Chinese).

Xu, C. X., Ma, C., Tian, G. L. & Xie, W. X. (2011). Study on mechanisms of virtual water trade and its contribution to regional economic growth. China Soft Science (12), 110–119 (in Chinese).

Yu, P. P., Zhang, J. Z. & Lin, C. G. (2006). China agriculture development strategy in twenty–first Century. Resources Environment & Development 7 (2), 21–27 (in Chinese).

Yu, Y., Hubacek, K., Feng, K. & Guan, D. (2010). Assessing regional and global water footprints for the UK. Ecological Economics 69, 1140–1147.

Zhang, Z. Y., Yang, H. & Shi, M. J. (2011). Analyses of water footprint of Beijing in an interregional input–output framework. Ecological Economics 70, 2494–2502.

Zhao, X., Yang, Z. F. & Chen, B. (2009). Study on Chinese virtual water trade and consumption in an input–output framework. Journal of Natural Resources 24 (2), 286–294 (in Chinese).

Zhou, J. & Shi, A. N. (2008). Method for calculating virtual water trade and demonstration. China Population Resources and Environment 18 (4), 185–188 (in Chinese).

Zhu, Q. R. & Gao, J. F. (2009). Study on the issue of foreign trade and virtual water in China by input–output model. China Soft Science (5), 40–46 (in Chinese).

Received 29 December 2015; accepted in revised form 16 July 2016. Available online 23 September 2016

Opportunities for low-carbon socioeconomic transition during the revitalization of Northeast China: Insights from Heilongjiang Province

CHEN Weiming[1,2], LEI Yalin[1,2,*], WU Sanmang[1,2], LI Li[1,2]

1. *School of Economics and Management*, *China University of Geosciences*, *Beijing* 100083, *China*

2. *Key Laboratory of Carrying Capacity Assessment for Resource and Environment*, *Ministry of Natural Resources of the People's Republic of China*, *Beijing* 100083, *China*

Abstract The Strategies of Reviving the Old Industrial Bases provide opportunities for low-carbon transition in Northeast China, which is one of the earliest regions to industrialize and the largest rustbelt in China, but study on the impacts of its socioeconomic factors on CO_2 emissions is still in short, though it is essential for guiding the pathways to achieve low-carbon socioeconomic transition. We adopted the structural decomposition analysis (SDA) to identify the main contributors to emissions increase in Heilongjiang Province during 2002—2012, which is the heartland of Northeast revitalization. The results show that the increase in CO_2 emissions was mainly driven by growth in per-capita final demand, which generated 203.8 Mt (153.6%) upstream CO_2 emissions between 2002 and 2012. Changes in production structure and final demand structure had smaller impacts on CO_2 emissions increase (36.1 Mt and 27.0 Mt). However, the positive influences were largely overwhelmed by change in emission intensity, which avoided 135.4 Mt (−102%) CO_2 emissions. Therefore, appropriate measures related to energy structure optimization and efficiency improvement should be implemented. Especially, increasing the proportion of wind, solar and biomass energy in Heilongjiang, where renewable energy is abundant, would reduce the CO_2 emissions significantly. In addition, domestic export took the lead position in driving the CO_2 emissions in Heilongjiang, accounting for 37.6%−43.1% annual emissions between 2002 and 2012. Thus, some financial instrument, such as tax relief for less carbon intensive exports could be adopted to prompt upstream suppliers to decarbonize their production processes.

Keywords Structural decomposition analysis (SDA), Low-carbon transiton, Input-outpont model, CO_2 emissions, Heilongjiang Provience

1 Introduction

The CO_2 emissions from fossil fuel consumption are considered to be the primary cause of

原文刊于 *Science of the Total Environment*，2019 年第 683 卷。原文发表时为彩色。

* 为通信作者。

climate change, posing great threat to global sustainability (Canadell et al., 2007; Pachauri et al., 2014; Peters et al., 2012; Wang et al., 2016). As the major contributor to global CO_2 emissions (Feng et al., 2009; Liu et al., 2015b), China has promised to reach emission peak before 2030, and to reduce the carbon intensity by 60%-65% in 2030 relative to 2005. It's widely acknowledged that, even after the US withdrawal from the Paris Agreement, China can and will lead on climate change mitigation (Dai et al., 2017; Wang and Wang, 2017; Wang et al., 2019a; Zhang et al., 2017a, b). Comprehensive emission reduction targets were set in National Plan on Climate Change (2014—2020) and the 12th/13th Five-Year Plan for Economic and Social Development of the People's Republic of China to fulfill the terms of Paris Agreement. Numerous efforts have been devoted to reducing the carbon emissions in China (Chen et al., 2017a; Feng et al., 2013; Liu et al., 2013, 2015a), such as implementing coal consumption cap plan, investing in renewable energy and establishing carbon trading market.

The government's 13th Five-Year Plan (2016—2020) aims at a 15% reduction in energy intensity and a 18% reduction in carbon intensity. The national targets should be further broken down to smaller ones to be accomplished by individual provinces, considering the heterogeneity of China's regional socioeconomic conditions (Shan et al., 2016; Zhang et al., 2016). More importantly, special attentions should be paid to some typical industrial provinces, especially those in Northeast China (i. e., Heilongjiang, Jilin and Liaoning Province) (Li et al., 2012; Sun et al., 2013). The first two five-year plans were designed to build the foundations of China's industries, of which most heavy industries were located at Northeast China, resulting in the resource-based and emission-intensive economic development mode. In 2012, the indus-trial sector accounted for around 70% of the total energy consumption in Northeast China, of which 97% was dominated by fossil energy (National Bureau of Statistics of China, 2013). Northeast China has witnessed a rapid economic growth until the central government shifted the focus to coastal regions through the Chinese economic reform in the late 1970s and early 1980s. However, the high proportion of heavy industry and neglect of protecting the environment has made it difficult to transform the economy in this region (Li, 2004; Wang et al., 2014a, b). As a result, Northeast China has gradually declined into the largest rustbelt in China, if not in the world, tinged with resource depletion, environment degradation, population loss, and economic recession (Acs et al., 2009; Li, 2004; Li et al., 2016). In addition, the carbon productivity (the ratio of GDP to carbon emissions) in this region is also much lower than that of east China, leading to serious environmental pollution in this area (Li and Wang, 2019). To tackle with the issue, a revitalization strategy was implemented in 2003 and further reinforced in a series of policy instruments since then. Though there may be a conflict between economic development and climate change mitigation (Hubacek et al., 2017), it also provides opportunities for low-carbon socioeconomic transition during the revitalization of Northeast China.

Heilongjiang Province, accounting for 35% and 30% of Northeast China's population and

GDP, respectively, is in the heartland of revitalization and carbon reduction. Under the revitalization strategy, Heilongjiang Province has more than quadrupled its value added of the industrial sector during 2002—2012. The industry upgrading has also resulted in the decrease of secondary industry's contribution to GDP, from 51% in 2002 to 44% in 2012 (Heilongjiang Statistical Bureau, 2013). However, the carbon emissions induced by energy consumption in Heilongjiang Province also doubled (Shan et al., 2016), while the population has stagnated during this period. Further revitalization will continue to change socioeconomic conditions significantly in Heilongjiang Province (e. g. production technological, final demand, emission intensity), and then change the total carbon emissions (Wang et al., 2019c). A prerequisite to integrate low-carbon development into the revitalization strategy is to uncover the relative contributions of these socioeconomic factors to the total emission change (Chen et al., 2018a; Feng et al., 2016; Hammond and Norman, 2012; Li et al., 2018; Mi et al., 2017; Sun et al., 2010; Wang et al., 2019b).

Many efforts have been devoted to shedding light on the driving forces of emission change by using decomposition analysis, of which index decomposition (IDA) and structural decomposition analysis (SDA) are the two mostly used methods (Cai et al., 2019; Chen et al., 2018; Wang et al., 2017). With advantage of simplicity, transparency and lower data requirements, IDA can provide less detailed decomposition for economic system (Ang, 2004, 2005, 2015; Su and Ang, 2012). The SDA usually requires a complete input-output table which provides all the sectoral information and their interchange relationships (Chen et al., 2018b; Chen et al., 2019; Lenzen et al., 2014; Zhang et al., 2014). Therefore, SDA has the unique strength that it can take the production structure change into consideration (Dietzenbacher and Los, 1998; Feng et al., 2017; Hoekstra and Van den Bergh, 2003; Wang et al., 2017). Given this, SDA has been widely applied to evaluate the driving factors for historical changes in resources depletion and pollutants emission, such as energy consumption (Lan et al., 2016; Zhang and Lahr, 2014, p. 200), water consumption (Cazcarro et al., 2013; Zhang et al., 2012), carbon emissions (Guan et al., 2008; Su and Ang, 2016) and other air pollutants (Guan et al., 2014; Liang et al., 2013). The SDA analysis has also been conducted at a wide spectrum of geographical regions, including national level (Feng et al., 2012; Liu and Wang, 2017; Mi et al., 2017), regional level (Chang and Lin, 1998; Zhang et al., 2017a, b), city level (Chang et al., 2008) and sectoral level (Kopidou and Diakoulaki, 2017). At regional level, decomposition analyses of CO_2 emissions have highlighted the role of some socioeconomic factors in driving emissions in many parts of China (Geng et al., 2013; Wang et al., 2014a, b; Wang and Yang, 2015). But few studies gave a full treatment of emissions related to all demand categories (i. e. private consumption, government consumption, gross fixed capital formation, changes in inventory and exports) and drivers of emission changes, and carbon decomposition studies for Heilongjiang are also very few.

Given this, decomposition analysis is urgently needed to support effective and efficient policy design for low - carbon transitions when implementing the revitalization strategies in Northeast China. This paper aims to fill such a research gap by using structural decomposition analysis (SDA) to identify and quantify the impact of socioeconomic factors on carbon emissions increase during 2002—2012 in Heilongjiang Province. According to the input-output model and the process of carbon emissions generation, carbon emission intensity (emissions per unite of output) and total output are the two most important factors of carbon emission (Dong et al., 2018; Wang et al., 2017). Among them, the total output can be calculated by Leontief inverse matrix and total final demand, the former is determined by the direct input coefficient matrix, which can be seen as the production input structure, and the latter can be viewed as a product of the final demand structure (the matrix with the share of different kinds of final demand in total final demand), per capita final demand and the population. Thus, we can get five socioeconomic factors for our decomposition analysis, including emission intensity, production input structure, final demand structure, per capita final demand, population (Ma et al., 2019). These factors cover all major aspects of the social and economic of Heilong-jiang. The findings will be useful for evaluating the impacts of recent energy efficiency and emission reduction initiatives as well as for developing future policy.

The revitalization strategy was originally proposed in 2003 by the Chinese government and further emphasized in many national and provincial five-year plans. Therefore, the 2002 is chosen as the beginning year of the period under research. The 2012 is selected as the final year, as 2012 in the latest year that the input-output table for Heilongjiang Province is available.

The rest of this study is organized as follows: Section 2 introduces the methodology and data; the detailed results are illustrated in Section 3; some discussions and policy implications are made in Section 4.

2 Methodology and data sources

2.1 CO_2 emissions inventory

Following the approach provided in the Intergovernmental Panel on Climate Change (IPCC), the total CO_2 emissions for Heilongjiang Province in a certain year is calculated based on Heilongjiang provincial energy statistics, which is typically called "production-based CO_2 emissions" (Marques et al., 2012; Peters and Hertwich, 2008). By summing the carbon emission from different types of fossil energy (including the coal and its products, the oil and its products, and natural gas and liquefied natural gas), the formula of production-based carbon emission accounting can be written as below:

$$C = \sum^{i} E_i \times N_i \times F_i \times O_i \times M \tag{1}$$

where C is the fossil-fuel-related CO_2 emission, the subscript i represents fuel type, E_i

denotes the amount of energy consumption from different fuel types (in physical unit), N_i is the average net calorific value of different fuel types, F_i is the carbon content representing CO_2 emissions when unit heat is released, O_i is the oxygenation efficiency of the i type fuel, and Mis the molecular weight ratio of CO_2 and carbon (44/12). The value of each coefficient from different fuel types are listed in Table 1.

Table 1 Compilation of sectors in the urban system

Fossil-fuel energy type	Average net calorific value (kJ/kg or kJ/m^3)	Carbon content of unit heat (kg C/GJ)	Carbon oxygenation efficiency	The coefficient to transformstandard coal (kg/kg or kg/m^3)
Row coal	20908	26.37	94%	0.7143
Clean coal	26344	25.41	93%	0.9
Other washed coal	8363	25.41	93%	0.4643
Coke	28435	29.42	93%	0.9714
Briquette	20908	33.56	90%	0.6
Gangue	8363	25.8	93%	0.2857
Coke oven gas	16726	13.58	99%	0.5714
Blast furnace gas	3763	70.8	99%	0.1286
Converter gas	7945	49.6	99%	0.2714
Other coal gas	5227	12.2	99%	0.3571
Other coking products	28435	25.8	93%	1.3
Raw oil	41816	20.08	98%	1.4286
Fuel oil	41816	21.1	98%	1.4286
Gasoline	43070	18.9	98%	1.4714
Diesel oil	42652	20.2	98%	1.4571
Petroleum coke	31947	26.6	98%	1.0918
Other petroleum products	41031	20	98%	1.4
Liquefied petroleum	50179	17.2	98%	1.7143
refinery gas	45998	18.2	98%	1.5714
Natural gas	38931	15.32	99%	1.33
Liquefied natural gas	51434	15.3	98%	1.7572

Data source: Eggleston et al., 2006.

2.2 Structural decomposition analysis

Based on the environmentally extended input-output analysis, the consumption-based CO_2 emissions can be calculated as:

$$C = E(I - A)^{-1}F \tag{2}$$

which are total emissions required by thefinal demand of an economy C (Feng et al., 2014; Kan et al., 2019; Mi et al., 2016; Peters, 2008). Where E is the sectoral direct emissions intensity matrix, $(I-A)^{-1}$ is the Leontief inverse matrix, F is the final demand matrix.

Make $L=(I-A)^{-1}$ represents the Leontief inverse matrix, the final demand F can be calculated with final demand structure (F_s), per-capita final demand (y_p) and population (p): $F=F_s y_p p$. The Eq. (2) can there-fore be written as:

$$C = ELF_s y_p p \tag{3}$$

Then a total difference of the Eq. (2) results in the decomposition form:

$$\Delta C = \Delta ELF_s y_p p + E\Delta LF_s y_p p + EL\Delta F_s y_p p + ELF_s \Delta y_p p + ELF_s y_p \Delta p + \sigma \tag{4}$$

where ΔC is the emission change during a certain period and the right 5 terms stand for the emission change induced by the change of emission intensity ($\Delta ELF_s y_p p$), production input structure ($E\Delta LF_s y_p p$), final demand structure ($EL\Delta F_s y_p p$), per capita final demand ($ELF_s \Delta y_p p$) and population ($ELF_s y_p \Delta p$), σ accounts for the fact the decomposition is not exact. Theoretically, there are n types of decomposition results when decomposing the total emission change into n factors.

In this study, the average of two polar decompositions is applied to obtain the relatively accurate results without complicated calculations (Dietzenbacher and Los, 1998). For the period t_0-t_1, the change in carbon emission can be written as:

$$\begin{aligned}\Delta C &= \Delta E(L^0 F_s^0 y_p^0 p^0 + L^1 F_s^1 y_p^1 p^1)/2 \\ &+ \Delta L(E^1 F_s^0 y_p^0 p^0 + E^0 F_s^1 y_p^1 p^1)/2 \\ &+ \Delta F_s(E^1 L^1 y_p^0 p^0 + E^0 L^0 y_p^1 p^1)/2 \\ &+ \Delta y_p(E^1 L^1 F_s^1 p^0 + E^0 L^0 F_s^0 p^1)/2 \\ &+ \Delta p(E^1 L^1 F_s^1 y_p^1 + E^0 L^0 F_s^0 y_p^0)/2\end{aligned} \tag{5}$$

2.3 Data sources

The monetary input-output tables for Heilongjiang Province are derived from the Heilongjiang Bureau of Statistics, imports from foreign countries and other provinces are excluded from intermediate use and final demand based on their share in total output. Average net calorific value and the coefficient to conversion to standard coal equivalent are collected from China Energy Statistics Yearbook (National Bureau of Statistics of China, 2013), carbon emissions coefficient and carbon oxygenation efficiency are collected from IPCC (Eggleston et al., 2006). To make the classification of sectors in the input-output table for Heilongjiang Province consistent with the sectoral emission satellite, the sectors are integrated into a uniform 28-sector format (see Table 2). It's noted that Heilongjiang Bureau of Statistics provides the competitive input-output table, assuming that the imports are produced with local technology (Miller and Blair, 2009). Thus, the domestic and foreign imports are removed from the Heilongjiang provincial input-output table to isolate the influence of local factors on the carbon emissions change, by using the method introduced in many previous studies (Guan et al., 2014; Weber et al., 2008).

Table 2 Sectors and classification.

Code	Sector	Classification
1	Agriculture	Agriculture (1)
2	Coal mining	Mining (2-4)
3	Petroleum and gas	
4	Metal mining	
5	Nonmetal mining	Manufacturing (5-21)
6	Food processing and tobaccos	
7	Textile	
8	Clothing, leather, fur, etc.	
9	Wood processing and furnishing	
10	Paper making, printing, stationery, etc.	
11	Petroleum refining, coking, etc.	
12	Chemical industry	
13	Nonmetal products	
14	Metallurgy	
15	Metal products	
16	General and specialist machinery	
17	Transport equipment	
18	Electrical equipment	
19	Electronic equipment	
20	Instrument and meter	
21	Other manufacturing	
22	Electricity and hot water production and supply	Electricity, gas and water production and supply
23	Gas and water production and supply	(22-23)
24	Construction	Construction (24)
25	Transport and storage	Services (25-28)
26	Wholesale and retailing	
27	Scientific research	
28	Other services	

3 Results and discussions

3.1 Analysis of production-based CO_2 emissions

Fig. 1a presents the energy-related CO_2 emissions (hereinafter referred to as CO_2 emissions) trajectory of Heilongjiang Province during 2002—2012. CO_2 emissions in Heilongjiang Province have witnessed a rapid increase from 113.2 Mt in 2002 to 246.0 Mt in 2012, with an average annual growth rate of 8.1%. Emissions induced by coal use, the major source of electricity generation, grew from 77.5 Mt in 2002 to 169.0 Mt in 2012, accounting for around 70% of Heilongjiang's total CO_2 emissions during the accounting period. In specific, more than 80% of the coal-related CO_2 emissions were contributed by raw coal, which generates more pollutions averagely than other types of coals (Cheng et al., 2017; Tian et al., 2011; Xia et al., 2017). Though emissions due to oil consumption (i.e., crude oil, gasoline, kerosene, diesel oil and fuel oil) increased moderately from 23.8 Mt to 41.3 Mt, their share in Heilongjiang's total emissions decreased slightly from 21.0% to 16.8%, which were largely offset by emission increase from coke consumption. Moreover, emission intensity declined significantly from 311.4 t/million RMB in 2002 to 179.5 t/million RMB in 2012, which is still much higher than that in other provinces/cities (Li et al., 2018; Shan et al., 2016).

Sectoral compositions of Heilongjiang's production-based CO_2 emissions are depicted in Fig. 1b. The 28 sectors are classified into 6 industries (Table 2). In general, there was an obvious increase trend in Manufacturing, Services and Electricity, gas and hot water production and supply between 2002 and 2012. The Electricity, gas and hot water production and supply contributed most to the Heilongjiang's annual CO_2 emissions (more than half), followed by Manufacturing, whose share roared from 18.1% in 2002 to 26.1% in 2012. Notably, Services has gradually replaced Mining as the third largest contributor. Direct emissions from Mining have stagnant during 2007—2012, due to the downturn of coal and oil industry in Heilongjiang. The production-based emissions of Construction could be ignored.

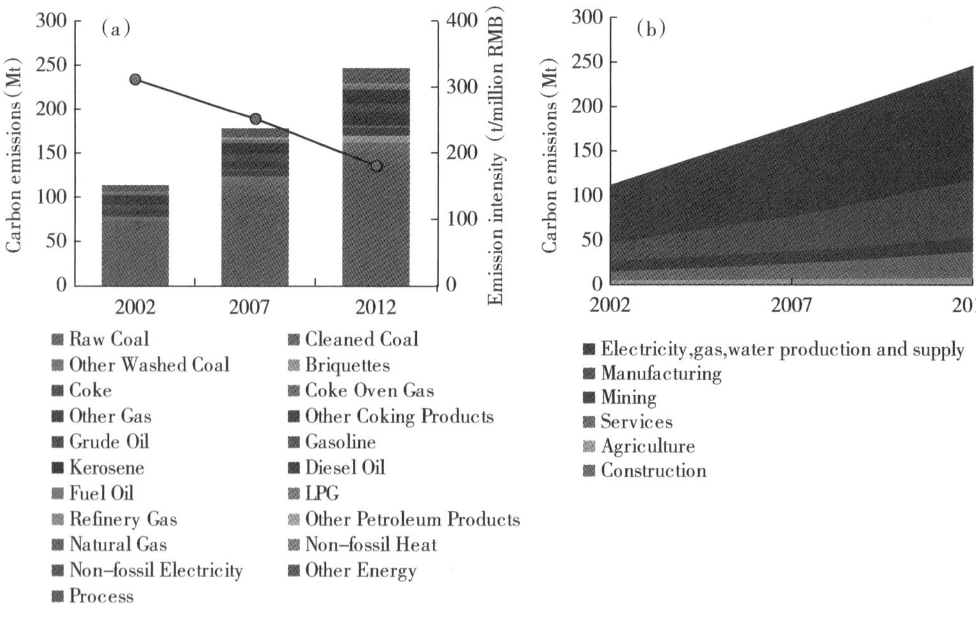

Fig. 1 Heilongjiang's production-based CO_2 emissions by different energy types (a) and industries (b) between 2002 and 2012

3.2 Analysis of consumption-based CO_2 emissions

Fig. 2a presents the reallocation of production-based CO_2 emissions to different final demand categories, which distinguishes underlying driving forces through the supply chains. An increasing trend can be identified for domestic export, household consumption, fixed capital formation and government consumption, while inventory change and foreign export remain relatively stable during the accounting period. Domestic export took the lead position in driving the CO_2 emissions in Heilongjiang, accounting for 37.6%–43.1% annual emissions between 2002 and 2012. More than three quarters of the 92.4 Mt emissions induced by domestic export in 2012 were contributed by Manufacturing, Mining and Agriculture (see Table 3), in which Food processing and tobaccos, Petroleum refining, coking, etc., Agriculture and Petroleum and gas were in the forefront (Fig. 3). In other words, other provinces in China have outsourced their emissions-intensive manufacturing to Heilongjiang through the domestic supply chains. Household consumption ranked as the second largest driver, which were responsible for 27.0%–30.5% annual emissions between 2002 and 2012. In 2012, 71.4 Mt emissions could be attributed to household consumption, mainly due to the increasing electricity consumption. Fixed capital formation contributed 16.2%–19.7% annual emissions between 2002 and 2012, 70% of which were occupied by Construction alone.

Table 3 Heilongjiang's CO_2 emissions induced by final demand in 2012 (Unit: Mt)

Final demand	Agriculture	Mining	Manufacturing	Electricity, gas and water production and supply	Construction	Services
Household consumption	3.0	1.2	8.3	40.1	0.3	18.5
Government consumption	0.2	0.0	0.0	0.0	0.0	24.7
Fixed capital formation	1.0	0.0	9.9	0.0	32.8	4.7
Inventory change	0.6	0.4	1.7	0.0	0.0	0.4
Foreign export	0.4	0.0	4.0	0.0	0.0	1.3
Domestic export	11.5	12.9	46.9	9.9	0.8	10.3

Sectoral contributions of Heilongjiang's consumption-based CO_2 emissions are shown in Fig. 2b, which is quite different from production-based sectoral contributions. From a consumption-based perspective, emissions of Manufacturing more than doubled since 2002, to 70.8 Mt, making it the largest driver of Heilongjiang's CO_2 emissions. As shown in Table 3, consumption-based emissions from Manufacturing were mainly for domestic export. In additions, Services increased from 27.7 Mt in 2002 to 59.9 Mt in 2012, accounting for around one quarter of the total annual emissions. Consumption-based emissions induced by Electricity, gas and hot water production and supply has gradually slowed its annual growth rate, soaring form 14.9 Mt in 2002 to 45.3 Mt in 2007, followed by a sluggish growth to 50.0 Mt in 2012. It can be seen

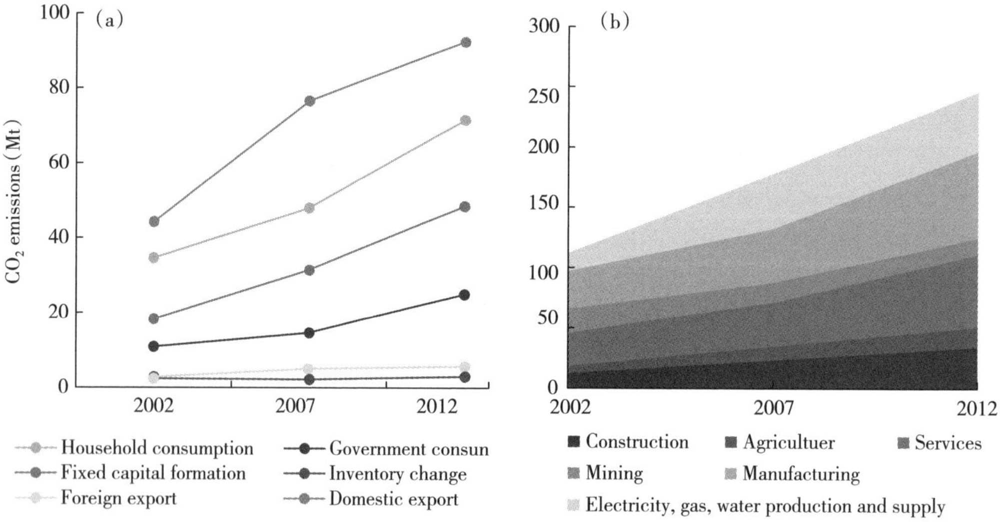

Fig. 2 Heilongjiang's consumption-based CO_2 emissions by different final demand categories (a) and industries (b) between 2002 and 2012

form Table 3 that household consumption and domestic export were the two main drivers. Due to the decline of Heilongjiang's electricity transferred to other regions in China,[1] consumption-based emissions of Electricity, gas and hot water production and supply by domestic export encountered a large decrease from 23.8 Mt to 9.4 Mt between 2007 and 2012, partly offsetting the emission increase from 21.3 Mt to 40.1 Mt induced by household consumption (see Fig. 3), leading to the slow growth during 2007—2012. Similar to the variation trend from production-based perspective, Mining sector also reduced its consumption-based emissions, largely due to the falling demand for coal and oil exports. As one of the agriculturally developed regions in China, consumption-based emissions of Heilongjiang's Agriculture continued to grow from 6.8 Mt in 2002 to 16.7 Mt in 2012, mainly stimulated by domestic export. Though Construction sector barely emitted CO_2 emissions directly, fixed capital formation of Construction caused 12.1-34.0 Mt emissions accumulated along the supply chains between 2002 and 2012.

3.3 Socioeconomic drivers of CO_2 emission change

The overall CO_2 emission change can be decomposed into 5 different socioeconomic factors, namely emission intensity, production input structure, final demand structure, per-capita demand and population. The decomposition results are shown in Fig. 4. From an aggregated perspective, the sole curbing factor was the emission intensity, which decreased nearly half between 2002 and 2012. The decline of emission intensity avoided 135.4 Mt (-102%) CO_2 emissions if other factors had remained constant. Such efforts have been totally offset and even reversed by population growth, per-capita demand increase, final demand structure and

[1] http://www.xinhuanet.com/politics/2015lh/2015-03/14/c_1114640094.htm.

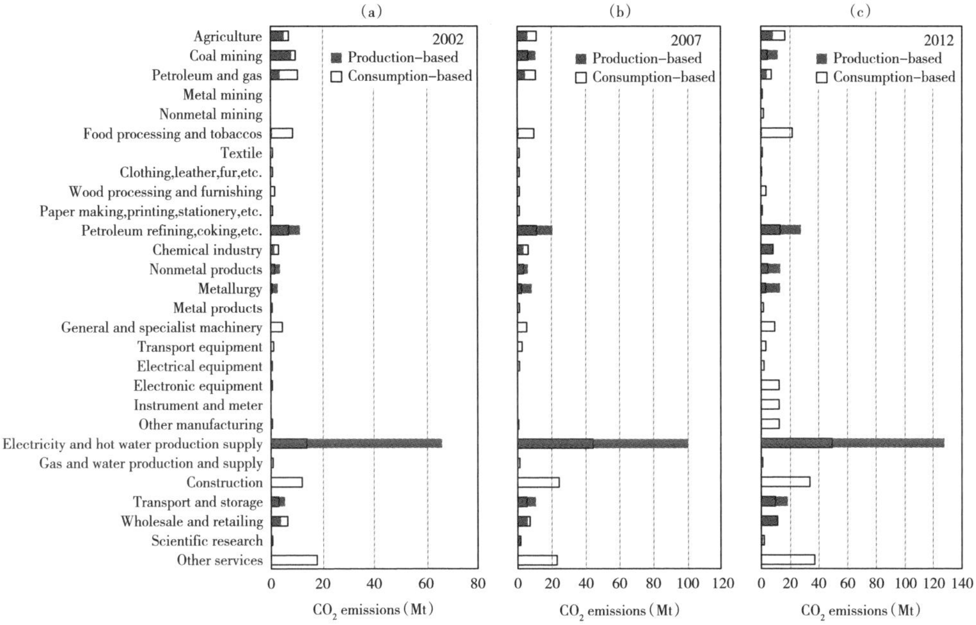

Fig. 3 Production-based and consumption-based sectoral emissions of Heilongjiang in 2002 (a), 2007 (b), 2012 (c)

production input structure change, leading to the overall emission raise between 2002 and 2012. Per-capita demand level was the most predominant factor driving the CO_2 emission increase, led to another 203.8 Mt (153.6%) upstream CO_2 emissions if other factors had remained constant. Production input structure and final demand structure change had smaller driving force on CO_2 emissions increase, contributing 36.1 Mt (27.2%) and 27.0 Mt (20.3%) to Heilongjiang's total emission increase, respectively. The contribution from population change was negligible, as Heilongjiang's population barely rose during these 10 years. In order to reveal the relative contributions of these 5 factors in shorter periods, we further decompose the results during 2002—2007 and 2007—2012, respectively. As presented in Fig. 4, all the factors, except production input structure, exerted the same effects (i. e., negative or positive) on CO_2 emissions change, although the influencing degree varied during these two periods. For example, emission intensity has attenuated its impacts, whose contributions to CO_2 emissions reduction decreased from 94.9 Mt during 2002—2007 to 40.5 Mt during 2007—2012. The de-carbonization of the final demand structure led to the weakening impacts of final demand structure on CO_2 emissions increase. On the contrary, the effects of per-capita final demand have strengthened due to the ever-increasing consumer demand in Heilongjiang, contributing 90.8 Mt during 2002—2007 and an additional 113.0 Mt during 2007—2012. The production input structure has undergone a striking turnaround, whose influencing shifted from positive to negative.

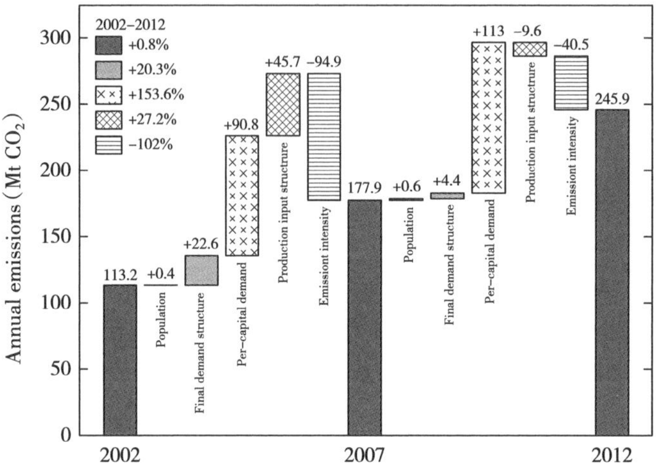

Fig. 4 Contributions of different socioeconomic factors to Heilongjiang's CO_2 emission changes between 2002 and 2012

To investigate the sectoral details of the decomposition results, the contributions of emissions intensity change and per-capita final demand change, the largest driving force of CO_2 emissions decrease and increase, were further broken down to 28 sectoral resolution (shown in Fig. 5). Generally, Heilongjiang's CO_2 emission reduction driven by emission intensity change was mostly generated in energy-related sectors. The carbon productivity improvement in Electricity and hot water production and supply avoided 90.1 Mt CO_2 emissions between 2002 and 2012, accounting for 66.5% of the total contributions by emission intensity change. The Coal mining ranked second (−15.9 Mt), followed by Petroleum refining, coking, etc. (−12.4 Mt). These efforts were slightly offset by the carbon intensity decline in Chemical industry, resulting in a 2.1 Mt emissions increase. In terms of per-capita final demand, the growth in demand for Construction and Other services has triggered 40.9 Mt and 33.4 Mt upstream emissions between 2002 and 2012, respectively. Moreover, the increasing food demand, mostly driven by local household consumption and domestic export, led to the emission increase in Agriculture and Food processing and tobaccos by 21.6 Mt and 22.0 Mt, respectively.

4 Discussions and policy implications

In 2003, the central government initiated a large project called "Revitalizing Northeast Old Industrial Base", aiming to upgrade the local industrial structure. Subsequently, large investment poured in to renovate conventional industries (i.e., petrochemical industry) as well as develop emerging industries (i.e., modern service industry). For example, many large infrastructure projects were implemented in Heilongjiang Province, such as high-speed railway, large petrochemical engineering project and highway construction, etc., resulting in large direct

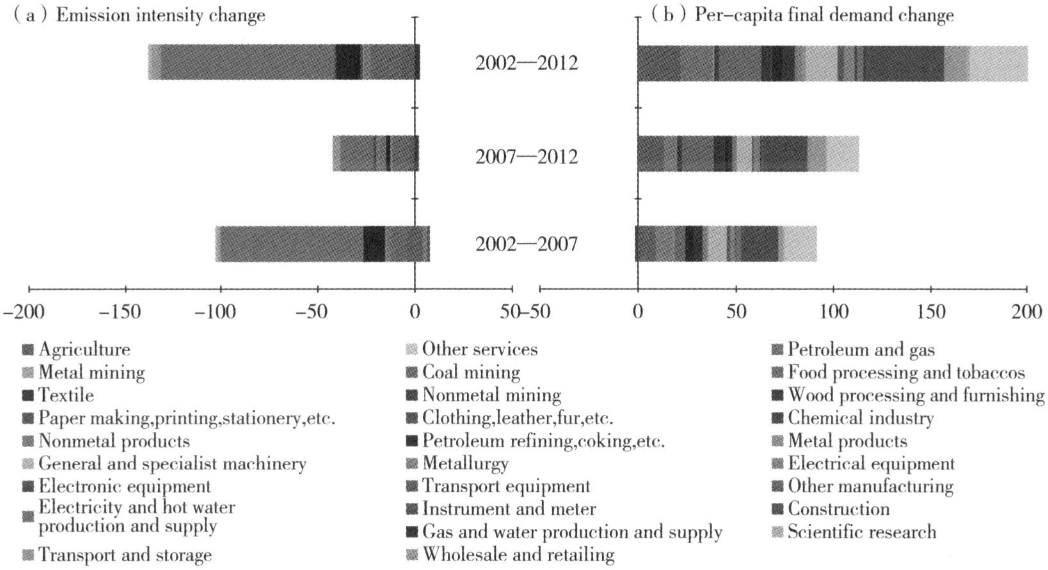

Fig. 5 Sectoral details of the decomposition results of emissions intensity change (a) and per-capita final demand change (b)

CO_2 emissions. Therefore, energy-related CO_2 emissions in Heilongjiang Province more than doubled during 2002—2012. Since the revitalization strategy will continue in the next decades, Heilongjiang Province may face a great challenge to mitigate its carbon emission while revitalizing the old industrial base. Thus, concrete mitigation policies should be formulated by considering the local realities to achieve low-carbon transition when implementing the revitalization strategies. We employ the input-output model to quantify the production-based and consumption-based CO_2 during 2002—2012 in Heilongjiang Province, and the structural decomposition analysis (SDA) to identify the main contributors to emissions increase during this period from five socio-economic factors. The results can provide some detailed policy implica-tions for low-carbon transition in Heilongjiang, which are summarized as follows:

(1) First, the decomposition results have shown that emission intensity change was the largest factor curbing CO_2 emission increase, making it a priority to lower emission intensity in Heilongjiang. Emission intensity depends on energy structure and efficiency. Therefore, appropriate measures related to energy structure optimization and efficiency improvement should be implemented, especially for these sectors whose emission intensity change contributing most to the total emission reduction, such as Electricity and hot water production and supply, Coal mining and Petroleum refining, coking, etc. (Fig. 5). In other parts of the northeast, such as Liaoning and Jilin, heavy industries with high carbon intensity also account for a high proportion. Since it is difficult to make significant adjustments to the industrial structure in a short period of time, it is also critical for these regions to improve energy efficiency and reduce the intensity of carbon emissions.

(2) Second, with China's biggest oil field (i. e., Daqing Oil Field) and many large coal mines, Heilongjiang's economy relies heavily on fossil fuels. Thus, more renewable energy should be introduced to optimize the energy structure in Heilongjiang. As illustrated in Heilongjiang 13th Five Year Plan for Energy Development, wind and solar energy are abundant in Heilongjiang. Moreover, as the leading agriculture and forestry producing province in China, Heilongjiang has rich biomass resources left untapped. Increasing the proportion of wind, solar and biomass energy in Electricity and hot water production and supply sector would reduce the CO_2 emissions significantly in the future. As the culprit of Heilongjiang's CO_2 emissions, coal and oil were responsible for more than 90% of the total emissions. On one hand, Heilongjiang has implemented a coal consumption cap plan to control the total coal use. On the other hand, more advanced technology should be developed to reduce the pollutants and improve the efficiency in Coal mining and Petroleum refining, coking, etc. Besides, Heilongjiang should also support high value-added manufacturing and service industries to upgrade the industrial structure gradually, but not too radically.

(3) In addition, from a consumption-based perspective, per-capita final demand contributed most to the CO_2 emission increase between 2002 and 2012. Specifically, domestic export has been playing an ever-increasing important role in driving upstream emissions in Heilongjiang, accounting for around 40% of total emissions between 2002 and 2012. The revitalization strategies further encourage export to be one of the pillars of Heilongjiang's economy. Heilongjiang Province has committed to actively undertake manufacturing outsourcing from other provinces/cities. Many previous studies have highlighted the role of exports in redistributing the environmental impacts (Chen et al., 2017b, 2018c; Chen et al., 2017; Mi et al., 2017). Thus, low-carbon transition for Heilongjiang Province or Northeast China, should not only focus on local CO_2 mitigation, but also consider the domestic and foreign supply chains. Some financial instrument, such as tax relief for less carbon intensive exports in Food processing and tobaccos and Petroleum refining, coking, could be adopted. This will encourage exporters to produce low carbon products and can, further, prompt upstream suppliers to decarbonize their production processes.

(4) Another important demand-side driver was the household consumption, whose embodied CO_2 emissions more than doubled during 2002—2012. The revitalization has stimulated local economic growth and raised residents'income significantly, leading to the ever-increasing life quality and consumption demand. On one hand, the government of Heilongjiang Province should establish an incentive mechanism to promote low carbon consumption, such as carbon pricing. On the other hand, major enterprises in Heilongjiang Province are encouraged to add the carbon footprint label on their products. It's verified that integrating carbon footprint into supply chain management will win more profits for enterprises (Kumaret al., 2012). The SDA results also show that production input structure became the inhibitive factor to CO_2

emissions during 2007—2012 (Fig. 4). Thus, optimizing production input structure by using low-carbon inputs from upstream suppliers is advocated.

(5) Finally, it's interesting to note that sectors like Manufacturing, Construction and Services had larger consumption-based emissions than production-based emissions, indicating that these sectors accumulate more emissions along the supply chains. As shown in Fig. 5 (b), per-capita final demand increases for Manufacturing, Construction and Services contributed most to the CO_2 emission increase between 2002 and 2012. The revitalization strategies encourage high value-added manufacturing and service industries. Unless the emission along the supply chains of these sectors can be regulated, other such strategy will only obtain moderate emission reduction gains in Heilongjiang.

Acknowledgements The authors express their sincere thanks for the support from Beijing Propaganda Culture High-level Talent Training Subsidy Program under Grant No. 2017XCB031, the National Natural Science Foundation of China under Grant No. 71773118.

References

Acs, Z., Plummer, L. A., Sutter, R., 2009. Penetrating the knowledge ? lter in "rust belt" economies. Ann. Reg. Sci. 43, 989.

Ang, B. W., 2004. Decomposition analysis for policymaking in energy: which is the preferred method? Energy Policy 32, 1131-1139.

Ang, B. W., 2005. The LMDI approach to decomposition analysis: a practical guide. Energy Policy 33, 867-871.

Ang, B. W., 2015. LMDI decomposition approach: a guide for implementation. Energy Policy 86, 233-238.

Cai, B., Zhang, W., Hubacek, K., Feng, K., Li, Z., Yawen, Liu, Liu, Yu, 2019. Drivers of virtual water flows on regional water scarcity in China. J. Clean. Prod. 207, 1112-1122.

Canadell, J. G., Le Quéré, C., Raupach, M. R., Field, C. B., Buitenhuis, E. T., Ciais, P., Conway, T. J., Gillett, N. P., Houghton, R. A., Marland, G., 2007. Contributions to accelerating atmospheric CO_2 growth from economic activity, carbon intensity, and efficiency of natural sinks. Proc. Natl. Acad. Sci. 104, 18866-18870.

Cazcarro, I., Duarte, R., Sánchez - Chóliz, J., 2013. Economic growth and the evolution of water consumption in Spain: a structural decomposition analysis. Ecol. Econ. 96, 51-61.

Chang, Y. F., Lin, S. J., 1998. Structural decomposition of industrial CO_2 emission in Taiwan: an input-output approach. Energy Policy 26, 5-12.

Chang, Y. F., Lewis, C., Lin, S. J., 2008. Comprehensive evaluation of industrial CO_2 emission (1989—2004) in Taiwan by input-output structural decomposition. Energy Policy 36, 2471-2480. https://doi.org/10.1016/j.enpol.2008.01.043.

Chen, B., Yang, Q., Li, J. S., Chen, G. Q., 2017a. Decoupling analysis on energy consumption, embodied GHG emissions and economic growth—the case study of Macao. Renew. Sust. Energ. Rev. 67, 662-672.

Chen, B., Yang, Q., Zhou, S., Li, J. S., Chen, G. Q., 2017b. Urban economy's carbon flow through external trade: spatial-temporal evolution for Macao. Energy Policy 110, 69-78.

Chen, B., Li, J. S., Zhou, S. L., Yang, Q., Chen, G. Q., 2018a. GHG emissions embodied in Macao's internal energy consumption and external trade: Ddriving forces via decomposition analysis. Renew.

Sust. Energ. Rev. 82, 4100-4106.

Chen, B., Han, M. Y., Peng, K., Zhou, S. L., Shao, L., Wu, X. F., Wei, W. D., Liu, S. Y., Li, Z., Li, J. S., 2018b. Global land-water nexus: agricultural land and freshwater use embodied in worldwide supply chains. Sci. Total Environ. 613, 931-943.

Chen, B., Li, J. S., Wu, X. F., Han, M. Y., Zeng, L., Li, Z., Chen, G. Q., 2018c. Global energy flows embodied in international trade: a combination of environmentally extended input-output analysis and complex network analysis. Appl. Energy 210, 98-107.

Chen, B., Wang, X. B., Li, Y. L., Yang, Q., Li, J. S., 2019. Energy-induced mercury emissions in global supply chain networks: structural characteristics and policy implications. Sci. Total Environ. 670, 87-97.

Chen, G., Hou, F., Chang, K., Zhai, Y., Du, Y., 2018. Driving factors of electric carbon productivity change based on regional and sectoral dimensions in China. J. Clean. Prod. 205, 477-487.

Chen, W., Wu, S., Lei, Y., Li, S., 2017. China's water footprint by province, and inter-provincial transfer of virtual water. Ecol. Indic. 74, 321-333.

Cheng, M., Zhi, G., Tang, W., Liu, S., Dang, H., Guo, Z., Du, J., Du, X., Zhang, W., Zhang, Y., 2017. Air pollutant emission from the underestimated households′ coal consumption source in China. Sci. Total Environ. 580, 641-650.

Dai, H. C., Zhang, H. B., Wang, W. T., 2017. The impacts of US withdrawal from the Paris Agreement on the carbon emission space and mitigation cost of China, EU, and Japan under the constraints of the global carbon emission space. Adv. Clim. Chang. Res. 8, 226-234.

Dietzenbacher, E., Los, B., 1998. Structural decomposition techniques: sense and sensitivity. Econ. Syst. Res. 10, 307-324.

Dong, F., Yu, B., Hadachin, T., Dai, Y., Wang, Y., Zhang, S., Long, R., 2018. Drivers of carbon emission intensity change in China. Resour. Conserv. Recycl. 129, 187-201.

Eggleston, S., Buendia, L., Miwa, K., Ngara, T., Tanabe, K., 2006. 2006 IPCC Guidelines for National Greenhouse Gas Inventories. Institute for Global Environmental Strategies Hayama, Japan.

Feng, K., Hubacek, K., Guan, D., 2009. Lifestyles, technology and CO_2 emissions in China: a regional comparative analysis. Ecol. Econ. 69, 145-154.

Feng, K., Siu, Y. L., Guan, D., Hubacek, K., 2012. Analyzing drivers of regional carbon dioxide emissions for China: a structural decomposition analysis. J. Ind. Ecol. 16, 600-611.

Feng, K., Davis, S. J., Sun, L., Li, X., Guan, D., Liu, W., Liu, Z., Hubacek, K., 2013. Outsourcing CO_2 within China. Proc. Natl. Acad. Sci. 110, 11654-11659.

Feng, K., Hubacek, K., Sun, L., Liu, Z., 2014. Consumption-based CO_2 accounting of China's megacities: the case of Beijing, Tianjin, Shanghai and Chongqing. Ecol. Indic. 47, 26-31.

Feng, K., Davis, S. J., Sun, L., Hubacek, K., 2016. Correspondence: reply to reassessing the contribution of natural gas to US CO_2 emission reductions since 2007. Nat. Commun. 7, 10693.

Feng, L., Chen, B., Hayat, T., Alsaedi, A., Ahmad, B., 2017. The driving force of water foot-print under the rapid urbanization process: a structural decomposition analysis for Zhangye city in China. J. Clean. Prod. 163, S322-S328.

Geng, Y., Zhao, H., Liu, Z., Xue, B., Fujita, T., Xi, F., 2013. Exploring driving factors of energy-related CO_2 emissions in Chinese provinces: a case of Liaoning. Energy Policy 60, 820-826.

Guan, D., Hubacek, K., Weber, C. L., Peters, G. P., Reiner, D. M., 2008. The drivers of Chinese CO_2 emissions from 1980 to 2030. Glob. Environ. Chang. 18, 626-634.

Guan, D., Su, X., Zhang, Q., Peters, G. P., Liu, Z., Lei, Y., He, K., 2014. The socioeconomic drivers of China's primary $PM_{2.5}$ emissions. Environ. Res. Lett. 9, 024010.

Hammond, G. P., Norman, J. B., 2012. Decomposition analysis of energy-related carbon emissions from UK manufacturing. Energy 41, 220-227.

Heilongjiang Statistical Bureau, 2013. Heilongjiang Statistical Yearbook 2013. China Statistics Press, Beijing, China.

Hoekstra, R., Van den Bergh, J. C., 2003. Comparing structural decomposition analysis and index. Energy Econ. 25, 39-64.

Hubacek, K., Baiocchi, G., Feng, K., Patwardhan, A., 2017. Poverty eradication in a carbon constrained world. Nat. Commun. 8, 912.

Kan, S. Y., Chen, B., Wu, X. F., Chen, Z. M., Chen, G. Q., 2019. Natural gas overview for world economy: from primary supply to final demand via global supply chains. Energy Policy 124, 215-225. https://doi.org/10.1016/j.enpol.2018.10.002.

Kopidou, D., Diakoulaki, D., 2017. Decomposing industrial CO_2 emissions of Southern European countries into production- and consumption-based driving factors. J. Clean. Prod. 167, 1325-1334.

Kumar, S., Teichman, S., Timpernagel, T., 2012. A green supply chain is a requirement for profitability. Int. J. Prod. Res. 50, 1278-1296.

Lan, J., Malik, A., Lenzen, M., McBain, D., Kanemoto, K., 2016. A structural decomposition analysis of global energy footprints. Appl. Energy 163, 436-451.

Lenzen, M., Geschke, A., Wiedmann, T., Lane, J., Anderson, N., Baynes, T., Boland, J., Daniels, P., Dey, C., Fry, J., 2014. Compiling and using input-output frameworks through col-laborative virtual laboratories. Sci. Total Environ. 485, 241-251.

Li, C., 2004. China's northeast: from largest rust belt to fourth economic engine? China Leadership Monitor 9, 1-15.

Li, H., Mu, H., Zhang, M., Gui, S., 2012. Analysis of regional difference on impact factors of China's energy-related CO_2 emissions. Energy 39, 319-326.

Li, H., Lo, K., Wang, M., Zhang, P., Xue, L., 2016. Industrial energy consumption in North-east China under the revitalisation strategy: a decomposition and policy analysis. Energies 9, 549.

Li, J. S., Zhou, H. W., Meng, J., Yang, Q., Chen, B., Zhang, Y. Y., 2018. Carbon emissions and their drivers for a typical urban economy from multiple perspectives: a case analysis for Beijing city. Appl. Energy 226, 1076-1086.

Li, S., Wang, S., 2019. Examining the effects of socioeconomic development on China's carbon productivity: a panel data analysis. Sci. Total Environ. 659, 681-690. https://doi.org/10.1016/j.scitotenv.2018.12.409.

Liang, S., Xu, M., Liu, Z., Suh, S., Zhang, T., 2013. Socioeconomic drivers of mercury emissions in China from 1992 to 2007. Environmental science & technology 47, 3234-3240.

Liu, Q., Wang, Q., 2017. How China achieved its 11th Five-Year Plan emissions reduction target: a structural decomposition analysis of industrial SO_2 and chemical oxygen demand. Sci. Total Environ. 574, 1104-1116.

Liu, Z., Guan, D., Crawford-Brown, D., Zhang, Q., He, K., Liu, J., 2013. Energy policy: a low-carbon road map for China. Nature 500, 143-145. https://doi.org/10.1038/500143a.

Liu, Z., Guan, D., Moore, S., Lee, H., Su, J., Zhang, Q., 2015a. Climate policy: steps to China's carbon peak. Nature 522.

Liu, Z., Guan, D., Wei, W., Davis, S. J., Ciais, P., Bai, J., Peng, S., Zhang, Q., Hubacek, K., Marland, G., 2015b. Reduced carbon emission estimates from fossil fuel combustion and cement production in China. Nature 524, 335.

Ma, J. J., Du, G., Xie, B. C., 2019. CO_2 emission changes of China's power generation system: input-output subsystem analysis. Energy Policy 124, 1-12.

Marques, A., Rodrigues, J., Lenzen, M., Domingos, T., 2012. Income - based environmental responsibility. Ecol. Econ. 84, 57-65.

Mi, Z., Zhang, Y., Guan, D., Shan, Y., Liu, Z., Cong, R., Yuan, X. -C., Wei, Y. -M., 2016. Consumption-based emission accounting for Chinese cities. Appl. Energy 184, 1073-1081.

Mi, Z., Meng, J., Guan, D., Shan, Y., Song, M., Wei, Y. -M., Liu, Z., Hubacek, K., 2017. Chinese CO_2 emission flows have reversed since the global financial crisis. Nat. Commun. 8, 1712.

Miller, R. E., Blair, P. D., 2009. Input - Output Analysis: Foundations and Extensions. Cambridge university press.

National Bureau of Statistics of China, 2013. China Energy Statistical Yearbook 2013. China Statistics Publishi.

Pachauri, R. K., Allen, M. R., Barros, V. R., Broome, J., Cramer, W., Christ, R., Church, J. A., Clarke, L., Dahe, Q., Dasgupta, P., 2014. Climate Change 2014: Synthesis Report. Con-tribution of Working Groups I, II and III to the Fifth Assessment Report of the Inter-governmental Panel on Climate Change. IPCC.

Peters, G. P., 2008. From production-based to consumption-based national emission inventories. Ecol. Econ. 65, 13-23.

Peters, G. P., Hertwich, E. G., 2008. Post - Kyoto greenhouse gas inventories: production versus consumption. Clim. Chang. 86, 51-66.

Peters, G. P., Andrew, R. M., Boden, T., Canadell, J. G., Ciais, P., Le Quéré, C., Marland, G., Raupach, M. R., Wilson, C., 2012. The challenge to keep global warming below 2 C. Nat. Clim. Chang. 3, 4.

Shan, Y., Liu, J., Liu, Z., Xu, X., Shao, S., Wang, P., Guan, D., 2016. New provincial CO_2 emission inventories in China based on apparent energy consumption data and updated emission factors. Appl. Energy 184, 742-750.

Su, B., Ang, B. W., 2012. Structural decomposition analysis applied to energy and emissions: some methodological developments. Energy Econ. 34, 177-188.

Su, B., Ang, B. W., 2016. Multi - region comparisons of emission performance: the structural decomposition analysis approach. Ecol. Indic. 67, 78-87.

Sun, L., Hong, E., Li, T., 2010. Incorporating technology diffusion, factor mobility and structural change into cross-region growth regression: an application to China. J. Reg. Sci. 50, 734-755.

Sun, Q. Q., Shi, K., Damerell, P., Whitham, C., Yu, G. H., Zou, C. L., 2013. Carbon dioxide and methane fluxes: seasonal dynamics from inland riparian ecosystems, northeast China. Sci. Total Environ. 465, 48-55.

Tian, H., Wang, Y., Xue, Z., Qu, Y., Chai, F., Hao, J., 2011. Atmospheric emissions estimation of Hg, As, and Se from coal-fired power plants in China, 2007. Sci. Total Environ. 409, 3078-3081.

Wang, C., Wang, F., 2017. China can lead on climate change. Science 357 (764-764).

Wang, C., Wang, F., Zhang, H., Ye, Y., Wu, Q., Su, Y., 2014b. Carbon emissions decomposi-tion and environmental mitigation policy recommendations for sustainable development in Shandong Province. Sustainability 6, 8164-8179.

Wang, D., Wei, S., Luo, H., Yue, C., Grunder, O., 2017. A novel hybrid model for air quality index forecasting based on two-phase decomposition technique and modified extreme learning machine. Sci. Total Environ. 580, 719-733.

Wang, H., Ang, B. W., Su, B., 2017. A multi-region structural decomposition analysis of global CO_2 emission intensity. Ecol. Econ. 142, 163-176.

Wang, M., Chen, Z., Zhang, P., Tong, L., Ma, Y., 2014a. Old Industrial Cities Seeking New Road of Industrialization: Models of Revitalizing Northeast China. World Scientific.

Wang, S., Li, Q., Fang, C., Zhou, C., 2016. The relationship between economic growth, energy consumption, and CO_2 emissions: empirical evidence from China. Sci. Total Environ. 542, 360-371.

Wang, S., Fang, C., Sun, L., Su, Y., Chen, X., Zhou, C., Feng, K., Hubacek, K., 2019a. Decarbonizing China's urban agglomerations. Annals of the American Association of Geographers 109, 266-285. https://doi.org/10.1080/24694452.2018.1484683.

Wang, S., Shi, C., Fang, C., Feng, K., 2019b. Examining the spatial variations of determinants of energy-related CO_2 emissions in China at the city level using Geographically Weighted Regression Model. Appl. Energy 235, 95-105. https://doi.org/10.1016/j.apenergy.2018.10.083.

Wang, S., Zeng, J., Liu, X., 2019c. Examining the multiple impacts of technological progress on CO_2 emissions in China: a panel quantile regression approach. Renew. Sust. Energ. Rev. 103, 140-150. https://doi.org/10.1016/j.rser.2018.12.046.

Wang, Z., Yang, L., 2015. Delinking indicators on regional industry development and carbon emissions: Beijing-Tianjin-Hebei economic band case. Ecol. Indic. 48, 41-48.

Weber, C. L., Peters, G. P., Guan, D., Hubacek, K., 2008. The contribution of Chinese exports to climate change. Energy Policy 36, 3572-3577.

Xia, X. H., Chen, B., Wu, X. D., Hu, Y., Liu, D. H., Hu, C. Y., 2017. Coal use for world economy: provision and transfer network by multi-region input-output analysis. J. Clean. Prod. 143, 125-144.

Zhang, C., Zhang, M., Zhang, N., 2017b. CO_2 emissions from the power industry in the China's Beijing-Tianjin-Hebei Region: decomposition and policy analysis. Pol. J. Environ. Stud. 26, 903-916. https://doi. org/10.15244/pjoes/66718.

Zhang, H., Lahr, M. L., 2014. China's energy consumption change from 1987 to 2007: a multi-regional structural decomposition analysis. Energy Policy 67, 682-693.

Zhang, H. B., Dai, H. C., Lai, H. X., Wang, W. T., 2017a. US withdrawal from the Paris Agreement: reasons, impacts, and China's response. Adv. Clim. Chang. Res. 8, 220-225.

Zhang, Wei, Li, K., Zhou, D., Zhang, Wenrui, Gao, H., 2016. Decomposition of intensity of energy-related CO_2 emission in Chinese provinces using the LMDI method. Energy Policy 92, 369-381.

Zhang, Y., Zheng, H., Fath, B. D., Liu, H., Yang, Z., Liu, G., Su, M., 2014. Ecological network analysis of an urban metabolic system based on input-output tables: model development and case study for Beijing. Sci. Total Environ. 468, 642-653.

Zhang, Z., Shi, M., Yang, H., 2012. Understanding Beijing's water challenge: a decomposition analysis of changes in Beijing's water footprint between 1997 and 2007. Environmental science & technology 46, 12373-12380.

The impact of climate change on urban resilience in the Beijing-Tianjin-Hebei region

LIU Lingna[1,2,4], LEI Yalin[1,2,3,*], ZHUANG Minghao[5], DING Shuang[6]

1. *School of Economics and Management, China University of Geosciences, Beijing* 100083, *China*

2. *Beijing University of Chemical Technology, Beijing* 100029, *China*

3. *Key Laboratory of Carrying Capacity Assessment for Resource and Environment, Ministry of Land and Resources, Beijing* 100083, *China*

4. *Risk and Resilience Program, International Institute for Applied System Analysis, Laxenburg* 1190, *Austria*

5. *College of Resources and Environmental Sciences, National Academy of Agriculture Green Development, Key Laboratory of Plant-Soil Interactions, Ministry of Education, China Agricultural University, Beijing* 100193, *China*

6. *China National Institute of Standardization, Beijing* 100191, *China*

Abstract The increasing uncertainty related to disaster risk under climate change brings about new challenges for sustainable urban management. The emergence of the urban resilience concept can improve the ability and extent to which cities can absorb and resolve risks, providing insight into the sustainable development of cities and regions. Yet, to date, the impact of climate change on regional urban resilience is not well understood. This paper measures the changes in urban resilience of the Beijing-Tianjin-Hebei (BTH) region from 1998 to 2019, and then explores the contribution of climate influencing factors such as temperature, precipitation and wind speed to urban resilience using econometric models. Results demonstrate the following: (1) Urban resilience shows a large spatial heterogeneity in the BTH region. Overall, Beijing and Tianjin have better and more stable resilience than Hebei Province. (2) Regarding the static impact of climate change on urban resilience, a 1 unit increase in Ln temperature and Ln precipitation will respectively increase Ln resilience by 1.01 units and 0.54 units, indicating that it has a significant positive impact on urban resilience. Each 1 unit increase in Ln wind speed will decrease resilience by 1.65 units, representing a significant negative effect. (3) Regarding the dynamic impact of climate change on urban resilience, a positive 1 unit impact of climatic factors indicates that an increase in temperature will first increase and then decrease urban resilience, and an increase in precipitation and wind speed will initially support improvement in urban resilience. Based on these findings, this article offers policy recommendations to improve urban resilience.

Keywords Climate change, Urban resilience, Beijing-Tianjin-Hebei region, Temperature

---

原文刊于 *Science of the Total Environment*，2022 年第 827 卷。原文发表时为彩色。

* 为通信作者。

1 Introduction

In 2020, the global average temperature was approximately 14.9 ℃, which is 1.2 ℃ higher than preindustrial levels (WMO, 2020). All aspects of global ecology, resources, the environment, economies and human life are seriously threatened by climate change (IPCC, 2014). As an area with a highly dense population and a thriving economy, city is a major battle-field in fighting the risks of climate change, and urban issues have received extensive attention from the international community. Over the past 30 years, global and Chinese urbanization has progressed rapidly, bringing profound changes in production and lifestyle and dynamic changes in urban and rural economic, social and spatial structures. China's National Development and Reform Commission and its Ministry of Housing and Urban-Rural Development issued the "City Adaptation Action Plan" in February 2016, which included a proposal to help 30 pilot cities adapt to climate change and thereby actively promote urban adaptation. The ultimate goal of climate change adaptation is to improve urban resilience, that is, to improve the ability and extent of cities to absorb and resolve climate change risks (Alberti and Marzluff, 2004).

As the political and cultural centre of China, the Beijing-Tianjin-Hebei (BTH) region is the largest and most dynamic region in northern China, and it is also highly sensitive to climate change risks (Kejun et al., 2016). The BTH region has the most intensive innovation resources and the greatest potential for innovation in China, and it shoulders an important historical mission in China's progress toward global influence in science and technology. However, the BTH region still confronts many challenges, such as uneven regional development and environmental problems. In particular, as a typical urban region with high energy consumption and high emissions, a large portion of China's greenhouse gas emissions come from Hebei Province. According to the 2012 greenhouse gas emissions inventory released by the National Development and Reform Commission's Department of Climate Change in 2017, China emitted 9.53 billion tons of carbon dioxide in 2012, accounting for 83.7% of total global greenhouse gas emissions, and the Beijing-Tianjin-Hebei region accounted for approximately 15% of the national total.

For China, actions to address climate change can bring synergies in many areas (Qin, 2014), and so measures to address climate change will essentially offer economic-environmental and green-low-carbon win-win effects (Du, 2010). However, the issue of climate change is also one of the major challenges to China's sustainable development (Zhuang, 2004); in particular, the challenges of climate change and the vulnerabilities of large cities in the urbanization process should not be underestimated (Hunt and Watkiss, 2011; Fleurbaey et al., 2014). Therefore, some scholars believe that the "2030 Agenda for Sustainable

Development" issued by the United Nations provides a powerful new impetus for China's economic transformation and upgrading and its sustainable development. In cities, climate change has and will continue to affect the operation of urban life-line systems, the quality of human settlements and the safety of residents' lives and property (Cui et al., 2015). In terms of regions, meteorological risks increase property and economic losses, and the national economy in East China is fairly sensitive to meteorological conditions (Wang et al., 2008). At the same time, the restrictions on the use of fossil energy required to mitigate climate change will profoundly change the global energy trade model and have an asymmetrical impact on regional development (Johansson et al., 2016; McGlade et al., 2015), although some studies argue that climate change can bring new opportunities for economic development.

The term "resilience" originated from the concept of physics, which refers to the variability of an object around its inherent benchmark while maintaining its essential characteristics. It is the reaction force of a substance in response to external forces. Ecologist Holling believes that the basic meaning of resilience is the ability of the system to anticipate and resolve external shocks while maintaining its main functions when a crisis occurs (Holling, 1973). At the same time, resilience is also the system's ability to respond to or mitigate disasters (Perrings, 2001). With further development of the concept, different scholars have noted that resilience can not only resist interference for an extended time, but can also provide opportunities and means for cities to deal with interference in the form of system updates and new trajectories (Folke, 2006). This provides a valuable back-ground for the analysis of cities and is also a transition to sustainable development (Folke et al., 2002). Subsequently, the proposal of engineering resilience, ecological resilience and evolutionary resilience further enriched the connotation of resilience theory. Engineering resilience is the original cognitive view of resilience theory, that is, the ability to restore it to its original state. Berkes et al. (2003) pointed out that engineering resilience emphasizes the stability around the current equilibrium state, which can be evaluated by the speed at which the system returns to its equilibrium state. With the further understanding of the structure of the system and the mechanism of change, Walker et al. (2004) proposed that resilience is not only the ability of the system to restore to its original state, but also the ability of the complex social-ecological system to change and adapt in response to shocks, which is the evolution of resilience. In recent years, resilience theory has been widely used in the fields of psychology (Snyderand McCullough, 2000), economics (Perrings, 2001), human development (Masten and Obradovic, 2008), urban planning and environmental security (Walker and Cooper, 2011). In 2002, the International Council for Regional Sustainable Development (ICLEI) introduced "resilience" into the field of urban construction and disaster prevention and mitigation at the United Nations Global Summit on Sustainable Development. From that point forward, related research on urban resilience came into being.

Under the United Nations Millennium Sustainable Development Goals, urban resilience is seen not only as essential to relieve external shocks and pressures but also as a foundation for economic growth and prosperity; a resilient socioeconomic ecosystem is central to effectively promoting sustainable development. The resilience of the socioeconomic ecosystem needs to be "strengthened", "constructed" and "developed". Urban resilience can provide a framework for the theoretical analysis of a complex, cross-scale, nonlinear, and indefinitely evolving complex system such as a city or region; such analysis will provide a clearer understanding of cyclical evolution, development status, and processes of the system mechanism (Cutter et al., 2008; Tanner et al., 2009). As a comprehensive, systematic and complex concept, resilience results from the interaction of social, economic, ecological and governance aspects. Clarifying the factors influencing urban resilience presents a challenge: first, there is a lack of systematic behavioural data; second, the city itself is complex; and finally, the size and nature of external shocks or pressures are difficult to predict (Kharrazi et al., 2015).

At present, most resilience measures are constructed by inductive methods, and factors are determined according to a conceptual framework or based on the previous literature. Related variables and weights are usually identified through principal component analysis and factor analysis (Briguglio et al., 2016). Indexes are very useful for measuring the level of resilience in various situations and over time. However, the shortcomings of the index method are obvious. The indicators considered maybe incomplete or irrelevant, and intangibles in the socioeconomic system cannot be considered. A few scholars proposed applying the threshold method to focus on a small number of participants and their interactions, while considering the minimum applicability of future interference and loss (Bennett et al., 2005), but this approach relies on too many assumptions, has high requirements for the researcher's computing power, and offers lower applicability for the evaluation object (Zhang and Yang, 2014). Most existing related research offering in-depth investigation of the influencing factors of regional urban resilience focuses on the qualitative level (Pu, 2016; Man, 2019). A few scholars have adopted quantitative research to explore the relationship between influencing factors and urban resilience (Zheng et al., 2018; Zhu and Sun, 2020), but research rarely involves the impact of climate change on the resilience of regional cities nor does it discuss cop-ing strategies.

Overall, the existing literature has performed more detailed research, but in our opinion, the following shortcomings still exist. First, the resilience-related literature gradually transitioned to the field of social sciences in the 1990s (Adger, 2000). As the urban system includes social, economic, ecological and other intangible processes, urban resilience as a new theory and method still faces unresolved challenges in developing quantitative methods and identifying influencing factors. Second, the literature focused on regions more often investigates the coordinated development of the regional economy, society, and the environment. There is almost no literature focused on how climate change impacts the resilience of regional cities. Third, the existing literature offers general analyses; discussions of the factors influencing urban development, whether at the national level or the city level, rarely involved climate change; in particular, little consideration is given to the additional challenges that high climate vulnerability brings to regional sustainable development. These deficiencies motivate the direction of this article.

This paper takes the BTH region as the research object, applies a quantitative method to measure the urban resilience of this region, and analyses and explores differences in the development of resilience levels between regions and the impact of climate change factors on urban resilience. The conclusions of this paper can provide more targeted countermeasures for improving regional urban resilience. The specific sections of this article are organized as follows: Section 2 introduces the construction of the urban resilience index system, data sources, data description and statistical testing. Section 3 introduces and discusses the results of the calculated urban resilience for the BTH region and the static and dynamic impacts of climate change. Section 4 provides the conclusions and policy implications of this article.

2 Methods and data

2.1 Construction of urban resilience evaluation indicators

The evaluation system developed in this article is based on an idea proposed by the Rockefeller Foundation in New York in 2013 to guide the evaluation of urban resilience, which it describes as "a measurable and data-based way to describe urban resilience. This method can be used for urban planning, practice and investment (Rockefeller Foundation, 2014). Our quantitative study of urban resilience in China is to a certain extent based on this model. The definition of urban resilience in this article follows Leichenko's definition of urban resilience in 2011: urban resilience is the ability of cities to deal with various disturbances and pressures (Leichenko, 2011). This ability is used to face and resolve various existing or potentially uncertain interference, regardless of whether such interference has occurred (Thornbush et al., 2013). In practical research, the comprehensive ability of cities to deal with complex disturbances is considered, rather than specific events and disasters.

Social resilience should focus on human development. The level of education and medical care can reflect the ability of cities to cope with risks (Walker et al., 2004; WHO, 2016); higher urbanization rates often reflect higher levels of urbanization and stronger urban resilience (Lester and Nguyen, 2015). Considering the important impact of an increase in the number of scientific and technological personnel on changes in the social elements of urban resilience, it has become increasingly important to increase the educational level of human capital in urban society to impart the appropriate form of knowledge and skills (Wall and Marzall, 2006).

Inter-city diversification is fostered by the social activities of residents, and the bonds between residents and the government, which indirectly enable the government and residents to trust and support each other, also increase social integration (Rockefeller Foundation, 2014). The main evaluation subject in social resilience is composed of people, and the development of people drives overall social development. In addition, a city's social governance capabilities cannot be ignored. When a city is resilient, it can continuously absorb experience from the changes and disturbances it has faced in the past, and take appropriate actions based on the experience. This shows that cities must have effective leadership and urban management capabilities, decisionmaking capabilities, and at the same time, they must also provide information and educational channels to enhance the capabilities of their stakeholders so that individuals and organizations can take appropriate actions. Communities can better realize the construction of bottom-up urban resilience through effective interactions with the government.

Turning to the indicators related to economic resilience, foreign trade dependence is the ratio of total imports and exports to the region's GDP; this can reflect the degree of external dependence of the city's economy and is negatively correlated with economic resilience (Man, 2019). The per capita GDP and per capita disposable income of urban residents reflects the level of microeconomic development, and it also shows that urban residents have some type of economic foundation to help them bear any losses caused when they face risks. Diversified economic behaviour and economic innovation are key factors in economic resilience, and it is important not to overlook the important role of different economic entities in economic resilience (Martin, 2012). When urban residents experience a disaster and cannot carry out labour activities, savings can better maintain their living standards at the normal level (Rockefeller Foundation, 2014).

Among the indicators related to engineering resilience, urban infrastructure is often vulnerable when cities respond to change, especially power levels (Walker et al., 2004). Therefore, the electricity consumption of the total city is selected for the engineering resilience evaluation to reflect the city's ability to robustly respond to changes in infrastructure, and the number of Internet and mobile phones is used to reflect the communication capabilities at critical moments. When a city is facing a disaster, public transportation cannot quickly move all residents to a safe place, and the ownership of private cars can to some extent alleviate the problem of public transportation congestion (Zhang and Yang, 2014). Highway line mileage is used to measure residents access to routes for emergency evacuation. Since technological innovation capabilities play a key role in improving infrastructure (Rockefeller Foundation, 2014), the number of patents granted is selected to reflect the city's technical strength and ability to respond to changes.

Among the indicators related to ecological resilience, the green coverage rate can reflect the urban ecological foundation and park area, measured per capita, provides a natural barrier when responding to climate change and risks (Rockefeller Foundation, 2014). Since ecological resilience is inseparable from the urban ecosystem, the total energy consumption and total water resources are selected to reflect changes in energy consumption levels and water resource utilization, respectively. Both of these are basic resource guarantees for urban residents dealing with external disturbances. In terms of protecting the ecological environmental, indicators such as "inhalable particulate matter (PM10) concentration, sulphur dioxide (SO_2) emissions, sewage treatment rate, industrial solid waste utilization rate, and domestic garbage harmless treatment rate" and other indicators were selected to analyse environmental issues such as air, water and soil pollution. The specific index evaluation system is as follows (Table 1):

Table 1 China's urban resilience evaluation system

Domains	Criterion Level	Indicators	Weight	Unit	Direction
Urban social system	Human development level	Urbanization rate	a_1	%	Positive
	Social development level	Natural growth rate	a_2	%	Positive
		Population ageing	a_3	%	Negative
		Number of divorced persons	a_4	Person	Negative
		R&D practitioners	a_5	Person	Positive
	Social governance level	Art and cultural activities	a_6	Piece	Positive
		Unemployment rate at the end of the period	a_7	%	Negative
		Number of communities per 10000 people	a_8	Per 10000 people	Positive
		Number of hospital beds	a_9	Piece	Positive
		Number of undergraduate and junior college students per 10000 students	a_{10}	Per 10000 people	Positive
Urban engineering system	Economic endowment	Number of industrial enterprises above designated size	b_1	Piece	Positive
		Resident deposits as a proportion of total residents'income	b_2	%	Positive
		The total retail sales of social consumer goods	b_3	Ten thousand yuan	Positive
		Tourism income	b_4	Ten thousand yuan	Positive
	Economic structure	Fixed asset investment	b_5	Ten thousand yuan	Positive
		per capita GDP	b_6	Yuan	Positive
		Engel coefficient of urban residents	b_7	%	Negative
		Contribution rate of tertiary industry	b_8	%	Positive
		Foreign trade dependence	b_9	%	Negative

Continued

Domains	Criterion Level	Indicators	Weight	Unit	Direction
Urban engineering system	Infrastructure level	Electricity consumption of the whole society Highway route mileage Length of water supply pipeline Per capita refuge area Public transport passenger volume Personal car ownership Mobile phone users Internet users	c_1 c_2 c_3 c_4 c_5 c_6 c_7 c_8	Ten thousand kilowatt hours Kilometer Kilometer M2/person Per 10,000 Vehicle people Per 10000 households Per 10000 households	Positive Positive Positive Positive Positive Positive Positive Positive
	Technological innovation ability	Total power of agricultural machinery Number of patents granted	c_9 c_{10}	Ten thousand kilowatts Piece	Positive Positive
Urban ecological system	Resource utilization level	Energy efficiency Water availability per capita Harmless treatment rate of domestic garbage Green coverage	d_1 d_2 d_3 d_4	Tons of standard coal/ 10000 yuan Cubic meter/person % %	Positive Positive Positive Positive
	Environmental governance capability	Sewage treatment rate Industrial solid waste utilization rate PM10 concentration Sulfur dioxide emissions	d_5 d_6 d_7 d_8	% % Mg/m^3 Ten thousand tons	Positive Positive Negative Negative

2.2 Index weight calculation

Due to the inconsistency of the dimensions (units) of the indicators in the urban resilience measurement framework, before measuring urban resilience, it is necessary to standardize the original data and measure the weights of the indicators and social, economic, engineering and ecological subsystems, after which urban resilience can be calculated (Appendix A). CRITIC is an objective weighting method proposed by Diakoulaki. Compared with other methods, it is better able to consider the contrast strength and conflict between evaluation indicators (Zhang and Zhang, 2015)(Table 2). However, the traditional CRITIC model does not consider the degree of dispersion of the indicators and only uses the standard deviation to measure the contrast strength. This study applies the improved CRITIC model to determine the impact of indicators on the system. Its mathematical expression is:

$$C_k = \delta_k \sum_{k=1}^{n} (1 - | r_{jk} |), \quad k = 1 \cdots n$$

where C_k represents the degree of influence of the k-th evaluation indexes on the corresponding system, and δ_k shows the standard deviation of the k-th evaluation index. The larger the value of C_k is, the greater the degree of influence the k-th evaluation index has on the system, and the more obvious its importance, while a smaller value has the opposite meaning. r_{jk} represents the correlation coefficient between the j-th and the k-th evaluation index. However, the measurement results of r_{jk} in the traditional model have positive and negative values, which actually reflect the degree of correlation and are not affected by the sign of the indicator. Therefore, in actual measurement, the absolute value of the correlation coefficient can be used for greater accuracy.

To determine the weight of the indicator, that is, normalize the difference coefficient,

$$W_k = \frac{C_k}{\sum_{k=1}^{n} C_k}, \quad k = 1 \cdots n$$

where W_k is the objective weight of the k-th evaluation indicator.

Table 2 Statistical description of data

Variable	Sample size	Mean	Std. dev.	Minimum	Maximum
Resilience	66	0.459	0.115	0.249	0.680
Ln resilience (Ln R)	66	-0.812	0.265	-1.389	-0.386
Temperature	66	12.436	0.636	11.013	13.947
Ln Temperature (Ln T)	66	2.519	0.051	2.399	2.635
Precipitation	66	568.684	76.986	412.778	788.442
Ln Precipitation (Ln P)	66	6.334	0.136	6.023	6.670
Wind speed	66	2.257	0.199	1.835	2.668
Ln Wind speed (Ln W)	66	0.810	0.089	0.607	0.982

2.3 Urban resilience measurement model

In the urban resilience measurement model, A_1, A_2, A_3, and A_4 represent the weights of urban social, economic, engineering, and ecological resilience, respectively. *UR* represents the overall level of urban resilience, *URS* represents the urban social resilience level, *URE* represents the urban economic resilience level, *URT* represents the urban engineering resilience level, and *URC* represents the urban ecological resilience level (Pu, 2016). Therefore, the final calculation of the overall urban resilience is expressed as:

$$UR = A_1 * URS + A_2 * URE + A_3 * URT + A_4 * URC$$

Among them, the expression of social resilience is:

$$URS = A_1 * \sum_{k=1}^{n} (W_k L_{kj}), \quad k = 1 \cdots n$$

The expressions of economic resilience, engineering resilience and ecological resilience are the same. Where W_k represents the weight of the index, and L_{kj} represents the quantitative index of the k-th index in the j-th layer.

2.4 Econometric model

The static impact focuses on the study of whether climate change has an impact on the resilience of BTH region, and the dynamic impact focuses on further study of the duration and dynamic changes of the climate change impact on the resilience of Beijing-Tianjin-Hebei cities. Based on the data of the BTH region from 1998 to 2019, VAR models were used to study the relationship between static influence and dynamic influence.

2.4.1 Panel VAR model

The panel VAR model was first proposed by Douglas et al. (1988). Compared with other models, the panel VAR offers two advantages. In terms of data characteristics, the panel data-based VAR model not only contains more samples but also overcomes the possible heteroscedasticity issues of cross-sectional data and the possible self-existence of time series data. From an analytical perspective, the panel data-based VAR can analyse changes in the influence of different variables over time and avoid biased estimates that occur when variables are ignored (Douglas et al., 1988). Based on panel data of urban resilience and climate change-related indicators in the Beijing-Tianjin-Hebei region from 1998 to 2019, this paper establishes a panel vector autoregressive model (panel VAR) to explore the impact of climate change on the resilience of provincial capitals. The selection of indicators related to climate change is based on Kim and Song's (2018) study of climate resilience in Korea, which revealsthe external influence of climate factors on urban resilience. Therefore, this article selects factors such as temperature, precipitation, and wind speed to construct an econometric model linking climate factors and urban resilience. The specific model is as follows:

$$y_t = a_1 y_{t-1} + \cdots a_p y_{t-p} + p_t \; t = 1, 2, \cdots, T$$

where y_t is the k-dimensional endogenous variable column vector; p is the lag order; T is the sample size; H is the coefficient matrix to be estimated; and p_t is the k-dimensional disturbance column vector.

2.4.2 Dynamic VAR model

Through two dynamic measurement methods, the generalized impulse response function and the variable parameter model, we assume that climate change has an impact on urban resilience in the short term and, based on the time series, dynamically explore the sensitivity of urban resilience to climate change and the quantitative impact of climate change on urban resilience. The impulse response function is an analysis method based on the VAR that overcomes a limitation of the VAR whereby the order of each variable has an influence on the result (Li et al., 2017a, 2017b). The specific model is as follows:

$$y_t = \sum_{i=1}^{p} A_i Y_{t-i} + \varepsilon_t$$

where y_t is a vector composed of n endogenous variables in the VAR model, P is the lag order of the VAR model, A_i is a $n \times n$ coefficient matrix, and ε_t is a $n \times 1$ residual vector that satisfies $E(\varepsilon_t) = 0$, $E(\varepsilon_t\varepsilon_t') = 0$, and $E(\varepsilon_t\varepsilon_s) = 0$; therefore, the generalized impulse influence function formula is:

$$GI_y(n,\ \delta_k,\ t-1) = E(y_{t+n} \mid \varepsilon_{kt} = \delta_k,\ t-1) - E(y_{t+n} \mid t-1)$$

where δ_k represents the impact from the last variable, n is the number of response periods of the impact, and $t-1$ represents all the available information sets when the impact occurs. The impulse function value of the n-period impulse response actually considers the difference caused by the impact of δ_k on the expected value of y_{t+n}. Assuming ε_t–N $(0,\ \Omega)$, the expected value of the impact condition is:

$$E(\varepsilon_t \mid \varepsilon_{kt} = \delta_k) = (\sigma_{1k},\ \sigma_{2k},\ \cdots,\ \sigma_{nk})'\sigma_{kk}^{-1}\delta_k = \sum e_k\,\sigma_{kk}^{-1}\,\delta_k$$

where e_k is a unit vector with the k-th element being 1 and the other elements being 0.

2.5 Data test

2.5.1 Data stationarity test

In the regression part of the static panel, to remove possible heterogeneity in the economic time series and smooth the data, we will take the logarithm of each variable, which we represent by Ln R, Ln T, Ln P and Ln W (Table 2). When establishing a model for the quantitative analysis of data, particularly when establishing a VAR model, to ensure that the established model can accurately reflect reality and has economic significance, the regression quantity must be stable or integrated to avoid false regression. This requires that the time data of the VAR model be stable or have a cointegration relationship. Therefore, to avoid spurious regression, the stability of the data needs to be tested by a unit root test before panel estimation. There are many methods for such tests, which can be roughly divided into two categories, namely, unit root tests under the same conditions and unit root tests under different root conditions. If the data are balanced, then the Levin-Lin-Chu test (LLC test) is often used for the same root, and the Im-Pesaran-Shin test (IPS test) is often used for different roots (Li et al., 2017a, 2017b). To ensure that the unit root test is comprehensive, this article uses the above two methods to test the variables at the same time, and the test results are shown in Table 3.

In the table, D represents the first-order difference of the variables. As shown in the above table, they are all single integers of the same order, without unit roots.

It can also be seen from the table that only the IPS test of the original precipitation sequence did not reject the null hypothesis at any significant level, while the original sequence of all the other variables rejected the null hypothesis with a significance level of more than 10%, and the time series was not stable. However, after the first-order difference is taken of each variable, the null hypothesis is rejected at the 1% level, indicating that each variable has no unit root in the case of the same-order difference; that is, the sequence is stable after the first-order difference, and the four variables can be cointegrated.

Table 3　Data stationarity test (panel unit root tests)

Variable	LLC	P	IPS	P
Ln R	-1.2602	0.1038	1.0858	0.8612
D. Ln R	-5.6395	0.0000	-4.1704	0.0000
Ln T	0.2233	0.5884	-1.8218	0.0342
D. Ln T	-3.562	0.0002	-4.7834	0.0000
Ln P	-2.0677	0.0193	-3.8176	0.0001
D. Ln P	-3.5261	0.0002	-5.2695	0.0000
Ln W	-0.3813	0.3515	-1.3216	0.0932
D. Ln W	-2.5385	0.0056	-4.7637	0.0000

2.5.2　Cointegration test

In the 1980s, Engle and Granger et al. proposed the concept of cointegration and pointed out that stationary or lower-order single integration maybe a linear combination of two or more nonstationary time series characteristics. Some time series may be nonstationary, whereas their linear combination is stationary. This stable linear combination reflects a cointegration relationship, that is, a long - term stable proportional relationship between variables. Cointegration analysis is a tool to analyse the quantitative relationship between nonstationary economic variables. It is based on the VAR analysis of time series and is a modelling and theoretical analysis method that combines spatial structure and time dynamics. Cointegration analysis describes the linear adjustment mechanism between economic variables through the linear error correction model (ECM).

To test the cointegration of variables, three methods are used to judge whether there is a long-term equilibrium relationship between them. The results are shown in the table. The result shows that the minority obeys the majority, and regardless of whether it contains a time trend, there is a long-term cointegration relationship between variables (Table 4).

Table 4　Results of cointegration test

Method	Unit	Statistics	P	Result
westerlund	Variance ratio	4.4894	0.0000	cointegration relationship
	Modified Phillips-Perron t	1.9264	0.027	cointegration relationship
pedroni	Phillips-Perron t	0.449	0.3267	Accept the null hypothesis
	Augmented Dickey-Fuller t	2.0964	0.018	cointegration relationship
	Modified Dickey-Fuller t	-1.366	0.086	cointegration relationship
kao	Dickey-Fuller t	-1.5718	0.058	cointegration relationship
	Augmented Dickey-Fuller t	-0.6022	0.2735	Accept the null hypothesis
	Unadjusted modified Dickey-Fuller t	-2.6648	0.0039	cointegration relationship
	Unadjusted Dickey-Fuller t	-2.0795	0.0188	cointegration relationship

2.6 Data sources

The 37 indicators selected in this article come from the National Bureau of Statistics, the China City Statistical Yearbook, the websites of provincial and municipal statistical bureaus, the statistical yearbooks of various provinces and cities, and the statistical bulletins of various provinces and cities (Beijing Bureau of Statistics, 1999—2020; Tianjin Bureau of Statistics, 1999—2020; Hebei Bureau of Statistics, 1999—2019). Environmental data come from the website of the Ministry of Ecology and Environment of the People's Republic of China and the "Environmental Protection Bulletin" of various provinces and cities; energy data come from the China Energy Statistical Yearbook; population data come from the China Demographic Year-book; patent data come from the National Intellectual Property Bureau and City Statistical Yearbook; and the total green coverage rate is mainly derived from the national land survey and its annual change survey, the geographic national conditions survey (monitoring) and remote sensing monitoring data. Very little data are missing, but imputation is used to fit the data when needed.

3 Results and discussion

3.1 Change in urban resilience in the BTH region from 1998 to 2019

From 1998 to 2019, the BTH region showed positive growth overall, but the growth rate remained volatile (Fig. 1). The highest average resilience in the Beijing-Tianjin-Hebei region was 6.9% in 2007, and the lowest was -0.8% in 2000. It is worth noting that the sharp decline in the growth rate of resilience at several time nodes coincides with the emergence of major disasters in China during that period. Among them, troughs occurred in 2000, 2005, 2008, etc. China suffered a rare nationwide drought in 2000, which was one of the most severe experienced since the

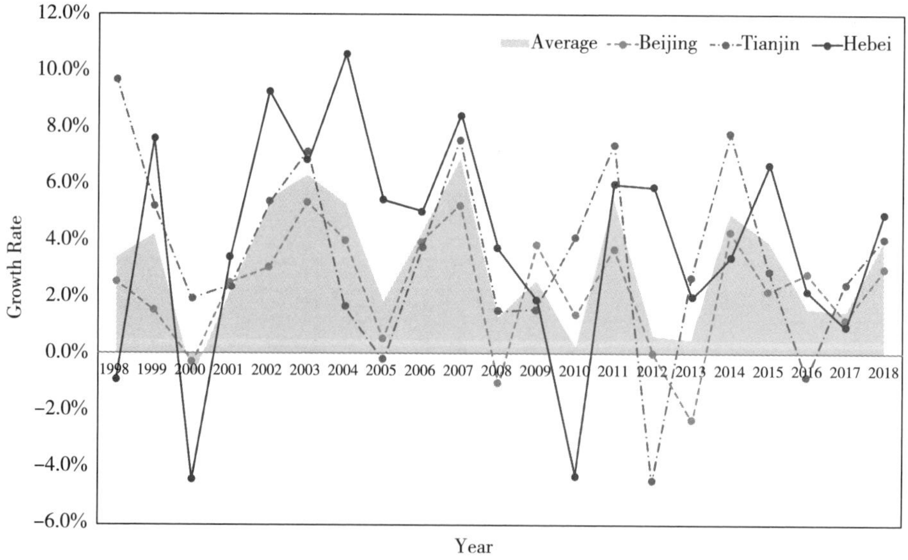

Fig. 1 Growth rate trend of urban resilience in the BTH region from 1998 to 2019

founding of the People's Republic of China, and it triggered a more severe locust plague. These combined disasters had a high impact on the growth rate of urban resilience in Hebei Province, as it dropped to −4.4%. The outbreak of the "H1N1" bird flu in 2005, the global financial crisis and the occurrence of other events in 2008 also affected the growth rate, but the provinces as a whole were able to maintain resilient growth, showing that the region has good resilience.

Urban resilience levels in the BTH region have shown an upward trend with slight fluctuation between 1998 and 2019, and the differences in levels between cities have gradually decreased (Fig. 2). The average urban resilience in the BTH region rose from 0.32 to 0.61, with a growth rate of 90.6%. The urban resilience level of Beijing in the observation years was higher than the average level of the BTH region, rising from 0.43 in 1998 to 0.68 in 2019, an increase of approximately 58.1%. As China's political and cultural centre, Beijing has set an example for other cities and has a relatively high and stable level of resilience. The urban resilience of Hebei Province and Tianjin City showed a "catch-up" phenomenon, changing from 0.25 to 0.51 and 0.28 to 0.64, respectively, with increases of 104% and 128.6%. Since 2008, the average resilience of Hebei Province has exceeded the average level of the BTH region, except in 2011 and 2012, when it was below it. This was mainly because the frequency of meteorological disasters in Hebei Province during those two years and the losses caused were higher than the average level since the 1990s, as it was a "medium-heavy" year. Changes in extreme weather have greatly affected the development of resilience in Hebei Province. However, Hebei Province subsequently narrowed the resilience gap with other cities by improving its resilience level, which largely relied on the continuous improvement of its own level and the radiating influence of coordinated development across the BTH region. For the BTH region, since 2013, as it became more integrated, regional differences have shown a gradual shrinking trend.

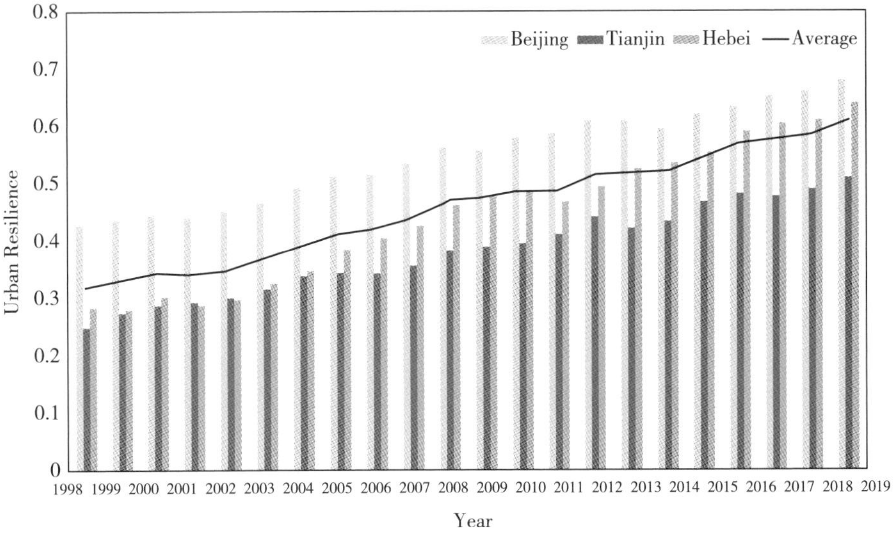

Fig. 2 Trends of urban resilience in the BTH region from 1998 to 2019

In the past two decades, social resilience in Beijing and Tianjin has been better and more stable than that in Hebei Province as a whole (Fig. 3). During the observation years, the overall differentiation of social resilience in the BTH region continued to shrink, but the effect was not significant. The level of resilience in Beijing in 2019 declined compared with that in 2013, mainly due to the continuous expansion of Beijing's urban scale and the relatively saturated level of urbanization. The increase in the level of ageing and the increase in the number of divorces caused a decline in resilience. However, Hebei Province and Tianjin City have improved their social resilience level through the development of their cities and have seized the opportunity to integrate the BTH region to promote their social resilience level. In terms of economic resilience, the BTH region has experienced different degrees of growth during the past two decades, the gap in economic resilience among regions has gradually narrowed, and development has become more balanced. The overall resilience of the engineering system in the BTH region showed differentiated development, and the development between regions had an uneven trend. In Hebei Province, engineering resilience increased from 0.04 to 0.22, giving this province the highest engineering resilience in the BTH region in 2019. This is largely due to the investment and development of infrastructure in Hebei Province in recent years. Studies have shown that the development of urban infrastructure in Hebei Province is ahead of its level of economic development. The Tianjin－Hebei region has been

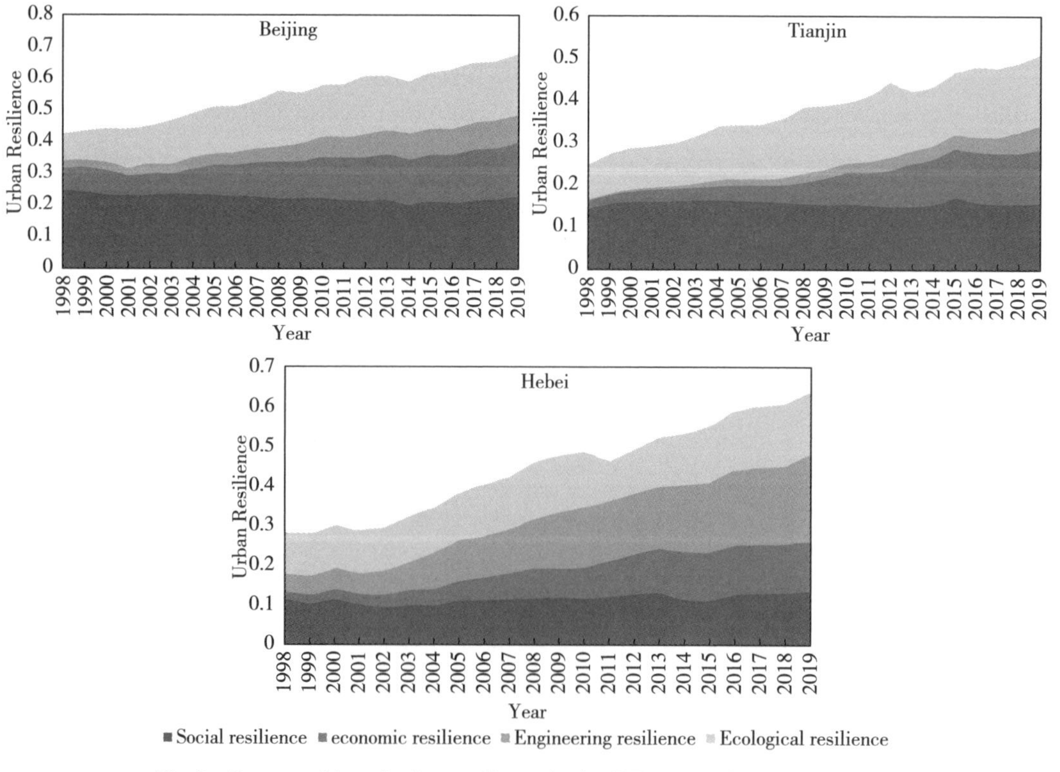

Fig. 3 Decomposition of urban resilience in the BTH region from 1998 to 2019

greatly improved, and this change shows agglomeration among regions. Compared with the regional-level changes in social resilience, economic resilience, and engineering resilience, the ecosystem has achieved a true comprehensive balance over the past 20 years. This shows that the integration strategy of the BTH region has had an impact on the ecological resilience; the promotion of this strategy has played a powerful role. Tianjin ranks last in ecological resilience among the four municipalities directly under the central government. This is mainly because the per capita water resource level of Tianjin has always been the lowest ranked of the 30 provinces. By the end of 2017, the per capita water resources were 83.4 cubic metres per person. The shortage of per capita water resources has become a key factor restricting the growth of ecological resilience in Tianjin.

3.2 The impact of climate change on urban resilience in the BTH region

By adopting the stepwise regression method and adding variables in sequence, the goodness of fit of models 1-4 are determined to be 0.104, 0.270, 0.122 and 0.389, respectively, indicating that model 4 has the best fit. The results show that the regression coefficient of Ln temperature is 1.019, and it is significant at the 90% confidence interval. Each 1 unit increase in Ln T will increase Ln resilient by 1.019 units, and Ln T has a significant positive effect on Ln resilient. Each 1 unit increase in Ln W will reduce Ln resilient by 1.652 units, which represents a significant negative impact; while each 1 unit increase in Ln P will increase Ln resilient by 0.538 units, which represents a significant positive impact (Table 5).

Table 5 Results of benchmark regression

Variable	Method 1	Method 2	Method 3	Method 4
Ln T	1.674***			1.019*
	(2.68)			(1.74)
Ln W		-2.119***		-1.652***
		(-4.79)		(-3.61)
Ln P			0.565***	0.538***
			(2.93)	(3.22)
_cons	-5.030***	0.905**	-4.394***	-5.447**
	(-3.20)	(2.52)	(-3.59)	(-2.54)
N	66	66	66	66
R-sq	0.104	0.270	0.122	0.389

t statistics in parentheses.
* $p < 0.1$.
** $p < 0.05$.
*** $p < 0.01$.

To further analyse the differences in the impact of climate change on urban resilience in the BTH region, the total sample was divided into three groups based on Beijing, Tianjin, and

Hebei, and the regional heterogeneity test was conducted. While testing the regional heterogeneity, one-period lagging explanatory variables are introduced as instrumental variables to avoid the problem of endogeneity. The results of the regional heterogeneity test are shown in Table 6.

Table 6 The heterogeneity of the impact of three climate factors on the BTH region

Variable	Beijing	Tianjin	Hebei
Ln T	-0.809	3.051***	0.828
	(-1.17)	(3.93)	(0.55)
Ln W	-2.226***	0.952	-2.900***
	(-4.81)	(0.64)	(-3.61)
Ln P	0.0798	0.595**	1.330***
	(0.55)	(2.62)	(2.91)
_cons	2.597	-13.42***	-9.197
	(1.02)	(-3.85)	(-1.67)
N	22	22	22
R-sq	0.757	0.560	0.496

t statistics in parentheses.

** $p < 0.05$.

*** $p < 0.01$.

The results show that, in terms of regional heterogeneity, in the Beijing area, wind speed has a significant negative impact on resilience. In the Tianjin area, both Ln T and Ln P have a significant positive impact on resilience. In the Hebei area, Ln W has a significant negative impact on resilience, and Ln P has a significant positive impact on resilience. Compared with Beijing, the negative effect of wind speed on the resilience of Hebei Province is more significant. Compared with Tianjin, the positive effect of precipitation on the resilience of Hebei Province is more significant. Therefore, urban resilience in Hebei Province is more sensitive to climate change, and the related impacts include both positive and negative effects. The precipitation factor has a positive impact on the urban resilience in all three regions, mainly because it has played a positive role in the continuous improvement of infrastructure in the BTH region, especially under extreme precipitation prompting urban weather disaster forecasting and emergency management capabilities. The increase in temperature has a negative impact on the resilience of Beijing city, indicating to a certain extent that the impact of temperature may first appear in megacities; for wind speed, since Tianjin is a coastal city, wind speed changes have played a positive role in improving urban resilience.

3.3 Dynamic shock

Since urban resilience is the ability of a city to respond to internal and external changes, it is particularly important to understand how changes in this ability occur and how to improve it. To further explore how climate change affects urban resilience and how this effect develops over time, this article uses the panel VAR model to study the dynamic relationship between urban resilience and climate change factors based on data from the BTH region from 1998 to 2019.

The results of the cointegration test in the second section show that there is a long-term equilibrium relationship between climate factors and urban resilience. Lag period calculation results are in Appendix B. It is necessary to further verify whether this equilibrium relationship constitutes a causal relationship through the Granger causality test. The Granger causality test in this study set the included lag order to 1, and the following results were obtained (Table 7). There is a one-way causal relationship between climate factors and urban resilience. Temperature Granger-causes urban resilience, precipitation Granger-causes resilience, and resilience Granger-causes wind speed. Therefore, changes in temperature and precipitation can cause changes in urban resilience, but changes in urban resilience are not the cause of changes in temperature and precipitation. The conclusion of the Granger causality test is only statistical causality, not necessarily true causality. Although it can be used as a support for true causality, it cannot be used as the ultimate basis for affirming or denying causality.

Table 7 Results of gran causality test

Equation	Excluded	chi2	df	Prob > chi2
(a) Temperature and Resilience				
h_d resilience	h_d temperature	6.4306	1	0.011
h_d resilience	ALL	6.4306	1	0.011
h_d temperature	h_d resilience	0.51746	1	0.472
h_d temperature	ALL	0.51746	1	0.472
(b) Precipitation and Resilience				
h_d resilience	h_d precipitation	6.4653	3	0.091
h_d resilience	ALL	6.4653	3	0.091
h_precipitation	h_d resilience	3.039	3	0.386
h_precipitation	ALL	3.039	3	0.386
(c) Wind speed and Resilience				
h_d resilience	h_d wind speed	0.28148	1	0.596
h_d resilience	ALL	0.28148	1	0.596
h_d wind speed	h_d resilience	3.588	1	0.058
h_d wind speed	ALL	3.588	1	0.058

h represents the forward difference.

To further explore the dynamic impact of climate change on urban resilience, this study introduced an impulse response function. An impulse response function is often used to study the short-term and long-term effects of shocks caused by changes in one variable on others. Based on the VAR model, one positive standard deviation in innovation impact is applied to the variable, and the specific impulse response effect can be seen through the measurement software. Thus, the impulse response function of urban resilience to temperature and precipitation is simulated, and the result is shown in Fig. 4. The horizontal axis is the number of lag periods, and the vertical axis is the response value. In general, the figure shows what happens when urban resilience is affected by a positive impact of one standard deviation in a certain influencing factor, where the X-axis is time, the period of these data is one year, the Y-axis is the response value, the solid line represents the true impulse response curve, the dashed line is deviation from the curve at plus or minus two standard deviations for the response in each period. The red line is the confidence interval, and the blue line is used to analyse the magnitude and positiveness of the shock response. If the first period is assumed at the beginning, the coefficient corresponding to the blue line is positive, and within the confidence interval, it represents a shock to the shock source variable. After that, the response variable is positive. The horizontal axis below represents the number of periods, the first period and the second period, in chronological order.

For a 95% confidence interval, Fig. 4 (a) shows the impulse response path of urban resilience to temperature. The impact of a positive one unit change in temperature has a positive

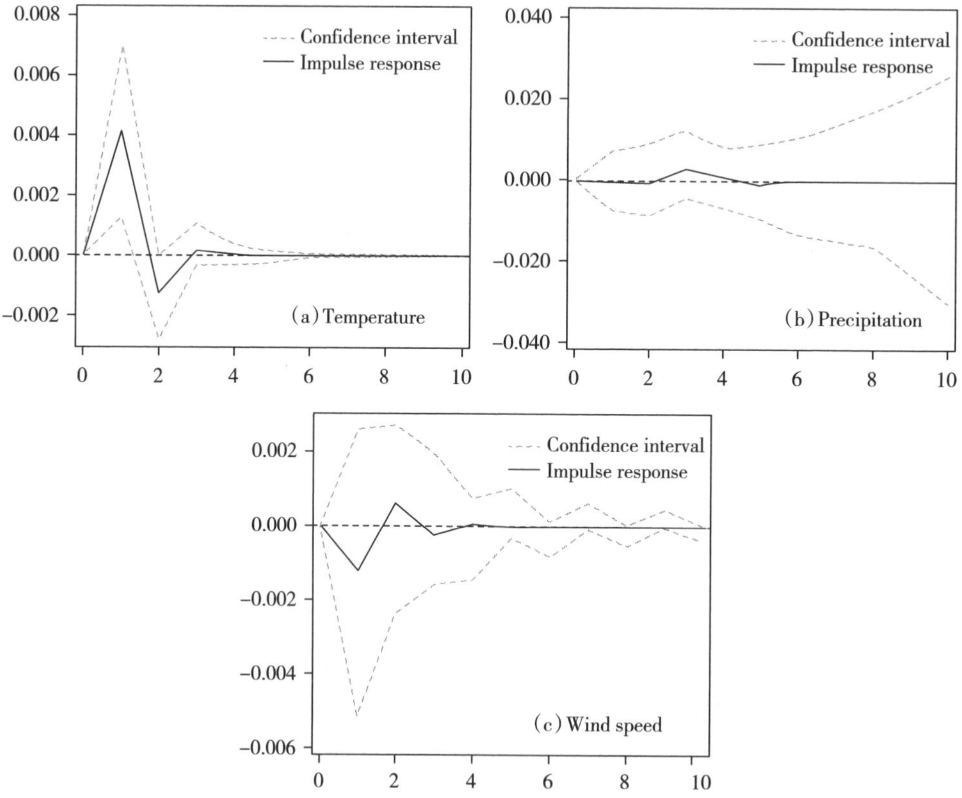

Fig. 4 Impulse response curve between various variables and urban resilience

impact on urban resilience at 1.5 years and 2 years. This shows that a temperature rising in the short term will promote the development of urban resilience in the BTH region. After the second year, there is a negative effect, which gradually disappears over the next 2.5-year period. Taking climate change governance policies as an example, a quantitative analysis of climate governance policies in the BTH region and its surrounding areas from 2008 to 2017 shows that governments at all levels issued a total of 183 policies during this period, of which 133 were "action plans" issued by the State Council in 2013. After the simultaneous issuance of the "Detailed Implementation Rules" for the implementation of policies in the BTH region and surrounding areas, local governments at alllevels issued supporting policies and specific implementation plans. The increasing attention given by the government to environmental issues, to a large extent, promoted improvements in regional economic, social, and ecological resilience and thus weakened or offset the negative impact of rising temperatures. Based on the current development level of BTH, rising temperatures will have a gradually weakening impact on urban resilience in the future.

Fig. 4 (b) shows the impulse response path of urban resilience to precipitation. Given the impact of a positive unit of precipitation, resilience decreased slightly in the first two periods, increased positively in the third period, and then stabilized. For the BTH region, the increase in precipitation will not have a major impact on urban resilience overall. The three provinces and cities will continue to improve their urban resilience with the increase in precipitation to allow them to cope with these changes. Since the "Twelfth Five-Year Plan" period, China's urban municipal infrastructure investment has continued to increase, with a cumulative investment of 95 trillion yuan, representing an increase of nearly 90% over the "Eleventh Five-Year Plan" period, and this has led to the continuous improvement of infrastructure capacity, ensuring that the city can cope with any damage caused by extreme precipitation.

Fig. 4 (c) shows the impulse response path of urban resilience to wind speed. Under the impact of a positive unit of wind speed, urban resilience will first drop significantly, then show a positive response, and ultimately stabilize. The increase in wind speed level will have a negative impact on urban resilience for a certain period of time, but this negative impact will not last long.

4 Conclusions and suggestions

Based on measurements of urban resilience levels in the BTH region from 1998 to 2019, this paper explores the temporal evolution of urban resilience in the BTH region and analyses the components of urban resilience: social resilience, economic resilience, engineering resilience and ecological resilience. Based on our results, we discuss the effects of climate change factors, temperature, precipitation and wind speed on urban resilience, and the specific conclusions are as follows:

(1) Since 1998—2019, the urban resilience level of the BTH region has been on the rise overall, and the difference in levels between cities has gradually decreased. The average urban resilience in the BTH region rose from 0.32 to 0.61, with a growth rate of 90.6%. The urban resilience level of Beijing in the observation years was higher than the average level of the BTH region, rising from 0.43 in 1998 to 0.68 in 2019, an increase of approximately 58.1%. The urban resilience of Hebei Province and Tianjin City showed a "catch-up" phenomenon, changing from 0.25 to 0.51 and 0.28 to 0.64, respectively, representing increases of 104% and 128.6%. Regarding social resilience and economic resilience, the overall differentiation within the region continued to shrink during the observation years, but the effect of social resilience was not significant. In terms of engineering resilience, the overall level of engineering system resilience developed in a differentiated manner, and regional development has shown an uneven trend. For Hebei Province, engineering resilience increased from 0.04 to 0.22, making it the province with the highest engineering resilience in the BTH region in 2019. Compared with the regional-level changes in social resilience, economic resilience, and engineering resilience, the ecosystem has achieved a true and comprehensive balance. The integration strategy of the BTH region has played a powerful role in improving the ecological resilience.

(2) To assess the static impact of climate change on urban resilience, an increase of 1 unit in Ln T will increase Ln R by 1.019 units, and Ln T has a significant positive effect on Ln R. Each 1 unit increase in Ln W will reduce Ln R by 1.652 units, demonstrating a significant negative impact; each 1 unit increase in Ln P will increase the Ln R by 0.538 units, which represents a significant positive impact. Across the BTH region, there are certain differences in the level of regional economic development and urban resilience development. Specifically, in the Beijing area, Ln W has a significant negative impact on Ln R. In the Tianjin area, both ln temperature and Ln P have a significant positive impact on Ln R. In Hebei, Ln W has a significant negative impact on Ln R, and Ln P has a significant positive impact on Ln R.

(3) Considering the dynamic impact of climate change on urban resilience, given a positive impact of 1 unit in climatic factors, an increase in temperature will first increase and then decrease urban resilience, while an increase in precipitation and wind speed will first decrease and then increase urban resilience.

In the future, the development of urban resilience should take the following directions: ensure good performance in the top-level design of urban resilience and accelerate the construction of pilot "resilient cities". Since 2015, the Chinese government has successively launched pilot projects for the construction of "sponge cities" and "climate-adapted" cities, focusing on disaster risks and improving the ability to respond to climate disasters. A total of 58 pilot cities have been launched and shown good results. This approach will enhance urban resilience and develop resilient cities. The national development plan will increase support for pilot cities in terms of policy support and capital investment, speed up the construction of pilot cities, and establish and improve the performance ratings and assessment methods of resilient cities to improve urban resilience, which in turn can provide more detailed guidance for upgrading other cities (Wang et al., 2018). Other important steps are to initiate a scientific evaluation of ecological resilience capabilities; increase the role of artificial intelligence technology in building ecologically resilient cities; establish and improve emergency systems, pilot cities, evaluation mechanisms, and assessment methods to support work related to the construction of ecologically resilient cities; and speed up incentive mechanisms and policy systems that involve the use of synergies. Climate change will impact the regional economy-society-ecosystem nexus, thus forming a transmission mechanism for climate change-regional economy-urban resilience, and eventually leading cities to fall into a "lose-lose" dilemma in coping with climate change and increasing regional coordinated development.

Acknowledgements The authors are grateful for financial support from China Postdoctoral Science Foundation under Grant No. 2021M692999, National Major Science and Technology Projects of China under Grant No. 2016ZX05016005-003.

Appendix A

A.1 Standardized processing

To standardized the indicators, we need to consider the different calculation units of the indicators. There are also positive and negative indicators, which have different positive and negative effects on the level of resilience. When the indicator is a positive indicator, a greater indicator value means a higher developed urban system, such as the number ofhealth institutions, per capita GDP and other indicators. When the indicator is a negative indicator, a smaller indicator value means a higher degree of urban system development; on the contrary, a lower value means higher unemployment rate at the end of the period and SO_2 emissions. By standardizing the indicator data, it is possible to avoid some extreme data from affecting the final result. The data standardization processing method is described as follows:

Positive indicators:

$$Y_{ij} = \frac{x_{ij} - minx_{ij}}{maxx_{ij} - x_{ij}} \tag{1}$$

Negative indicators:

$$Y_{ij} = \frac{maxx_{ij} - x_{ij}}{maxx_{ij} - minx_{ij}} \tag{2}$$

where x_{ij} represents the actual value of the j-th index in the i-th year, $maxx_{ij}$ and $minx_{ij}$ respectively represent the maximum and minimum values of the j-index in the i-th year among the 30 capital cities. It is known from the above equation that Y_{ij} is a standardized index, and a larger

$Y_{ij} \in [0,1]$ value represents the greater contribution of x_{ij} to the development of the subsystem, and vice versa. In order to ensure that the actual calculation results are comparable across cities and years, the following calculation weights will use uniform weights.

A. 2 Calculation results of urban resilience in the BTH region

Table a Urban resilience in the BTH region

Year	Urban Resilience		
	Beijing	Tianjin	Hebei
1998	0. 426	0. 249	0. 283
1999	0. 437	0. 273	0. 280
2000	0. 444	0. 288	0. 302
2001	0. 443	0. 293	0. 288
2002	0. 454	0. 300	0. 298
2003	0. 467	0. 316	0. 326
2004	0. 492	0. 339	0. 348
2005	0. 512	0. 345	0. 385
2006	0. 515	0. 344	0. 406
2007	0. 535	0. 357	0. 426
2008	0. 563	0. 384	0. 462
2009	0. 558	0. 390	0. 480
2010	0. 579	0. 396	0. 489
2011	0. 587	0. 412	0. 468
2012	0. 609	0. 443	0. 496
2013	0. 609	0. 423	0. 525
2014	0. 595	0. 434	0. 536
2015	0. 620	0. 468	0. 554
2016	0. 634	0. 482	0. 591
2017	0. 652	0. 478	0. 604
2018	0. 660	0. 490	0. 610
2019	0. 682	0. 513	0. 639

Appendix B

B. 1 Lag order selection

It is necessary to determine the optimal lag period of the VAR model in the beginning. A longer selected lag period will lower the degree offreedom of the VAR model and weaken the explanatory power. Therefore, in order to strengthen the explain ability of the variables in the model, it is necessary to select an appropriate lag period and find a better balance between the lag period and the degree of freedom. This paper chooses the method of multicriteria joint determination to determine the best lag period, adopts Akaike information criterion (AIC), Bayesian Information Criterion (BIC) and Hannan-Queen information criterion (HQIC) to estimate the lag length of the VAR model through these three indicators. The results show that the lag period of temperature and wind speed are 1 period, and the precipitation period is 3 periods. Changes in temperature and wind speed will have a greater impact on urban resilience in a short period of time, and changes in precipitation will take longer to affect urban resilience.

Table b (1)　Temperature and urban resilience

lag	AIC	BIC	HQIC
1	-4.30893 *	-3.9505 *	-4.16963 *
2	-4.20004	-3.68438	-4.00117
3	-3.75985	-3.07803	-3.49931
4	-2.956	-2.09836	-2.6319
5	-2.18988	-1.14603	-1.80075

Notes: * Indicates a significance level of 10%.

Table b (2)　Precipitation and urban resilience

lag	AIC	BIC	HQIC
1	6.88463	7.24306	7.02393
2	6.42023	6.9359	6.61911
3	5.69774	6.37956 *	5.95828 *
4	5.69554 *	6.55317	6.01964
5	6.27688	7.32073	6.66602

Notes: * Indicates a significance level of 10%.

Table b (3)　Wind speed and urban resilience

lag	AIC	BIC	HQIC
1	-7.19351 *	-6.83508 *	-7.05421 *
2	-6.89529	-6.37963	-6.69642
3	-6.18805	-5.50623	-5.9275
4	-5.32583	-4.46819	-5.00173
5	-4.87785	-3.834	-4.48872

Notes: * Indicates a significance level of 10%.

References

Adger, W. N., 2000. Social and ecological resilience: are they related. Prog. Hum. Geogr. 24(3), 347-364.

Alberti, M., Marzluff, J. M., 2004. Ecological resilience in urban ecosystems: linking urban patterns to human and ecological functions. Urban Ecosyst. 7 (3), 241-265.

Beijing Bureau of Statistics, 1999—2020. Beijing Statistical Yearbook. China Statistics Press, Beijing.

Bennett, E. M., Cumming, G. S., Peterson, G. D., 2005. A systems model approach to determining resilience surrogates for case studies. Ecosystems 8, 945-957.

Berkes, F., Colding, J., Folke, C., 2003. Navigating Social-ecological Systems: Building Resilience for Complexity and Change. CUP, UK.

Briguglio, Giusi, Libra, 2016. Occupational exposure to pesticides as a possible risk factor for the development of chronic diseases in humans. Mol. Med. Rep. 4475-4488.

Cui, S. H., Xu, L. L., Huang, Y. F., et al., 2015. Research progress and prospects of urban spatial morphology in response to climate change. Prog. Geogr. Sci. 34 (10), 1209-1218 (in Chinese).

Cutter, S. L., Barnes, L., Berry, M., et al., 2008. A place-based model for understanding community resilience to natural disasters. Glob. Environ. Chang. 18 (4), 0-606.

Douglas, Holtz-Eakin, Whitney, 1988. Estimating vector autoregressions with panel data. Econometrica 56 (6), 1371-1395.

Du, X. W., 2010. Discussion on China's green and low-carbon energy strategy. J. Taiyuan Univ. Technol. 41 (5), 453 (in Chinese).

Fleurbaey, M., Kartha, S., Bolwig, S., Chee,, Y. L., Chen, Y., et al., 2014. Sustainable development and equity. In: Edenhofer, O., Pichs-Madruga, R., Sokona, Y. (Eds.), Climate Change 2014: Mitigation of Climate Change. Contribution of Working Group III to the Fifth Assessment Report of the Intergovernmental Panel on Climate Change. Cambridge University Press, Cambridge, New York.

Folke, C., 2006. Resilience: the emergence of a perspective for social-ecological systems analyses. Glob. Environ. Chang. 16 (3), 253-267. Science of the Total Environment 827 (2022) 154157.

Folke, C., Carpenter, S., Elmqvist, T., et al., 2002. Resilience and sustainable development: building adaptive capacity in a world of transformations. Ambio 31 (5), 437-440.

Hebei Bureau of Statistics, 1999—2019. Hebei Economic Yearbook. China Statistics Press, Hebei. Holling, S. C., 1973. Resilience and stability of ecological systems. Annu. Rev. Ecol. Syst. 4 (1), 1-23.

Hunt, A., Watkiss, P., 2011. Climate change impacts and adaptation in cities: a review of the literature. Clim. Chang. 104 (1), 13-49.

IPCC, 2014. In: Edenhofer, O., Pichs-Madruga, R., Sokona, Y., Farahani, E., Kadner, S., Seyboth, K., Adler, A., Baum, I., Brunner, S., Eickemeier, P., Kriemann, B., Savolainen, J., Schlömer, S., von Stechow, C., Zwickel, T., Minx, J. C. (Eds.), Climate Change 2014: Mitigation of Climate Change. Contribution of Working Group III to the Fifth Assessment Report of the Intergovernmental Panel on Climate Change. Cambridge University Press, Cambridge, United Kingdom and New York, NY, USA.

Johansson, C., Abreu, P. C., Carstensen, J., et al., 2016. Human activities and climate variability drive fast-paced change across the world's estuarine-coastal ecosystems. Glob. Chang. Biol. 22 (2).

Kejun, Jiang, Chenmin, He, Xing, Zhuang, Jia, Liu, Gao Ji, Xu., Xiangyang, Chen Sha, 2016.

Scenario and feasibility study for peaking CO_2 emission in betwen 2020 and 2022 from energy activities in China. Adv. Clim. Chang. Res. 12 (3), 167-171.

Kharrazi, S., Karlsson, R., Sandin, J., 2015. Performance based standards for high capacity transports in Sweden. FIFFI Project 2013-03881 Report 1, Review of Existing Regulations and Literature.

Kim, D., Song, S. K., 2018. Measuring changes in urban functional capacity for climate resilience: perspectives from Korea. Futures 102 (SEP.), 89-103.

Leichenko, R. M., 2011. Climate change and urban resilience. Curr. Opin. Environ. Sustain. 3(3), 164-168.

Lester, T. W., Nguyen, M. T., 2015. The economic integration of immigrants and regional resilience. J.

Urban Aff. 38 (1), 42-60.

Li, G., Lei, Y., Yao, H., et al., 2017. The influence of land urbanization on landslides: an empirical estimation based on chinese provincial panel data. Sci. Total Environ. 595, 681.

Li, L., Liu, Q., Tang, D., et al., 2017. Media reporting, carbon information disclosure, and the cost of equity financing: evidence from China. Environ. Sci. Pollut. Res. 24 (10), 1-13.

Man, S., 2019. Study on the Measurement of Urban Economic Toughness in Northeast China [D]. Harbin Normal University (in Chinese).

Martin, R., 2012. Regional economic resilience, hysteresis and recessionary shocks. J. Econ. Geogr. 12(1), 1-32.

Masten, A., Obradovic, J., 2008. Disaster preparation and recovery: lessons from research on resilience in human development. Ecol. Soc. 13 (1), 1-9.

McGlade, Christophe, Ekins, 2015. The geographical distribution of fossil fuels unused when limiting global warming to 2 degrees C. Nature 517 (Jan. 8 TN. 7533): 187-a3.

Perrings, C., 2001. Resilience in the dynamics of economy - environment systems. Environ. Resour. Econ. 11 (3-4), 503-520.

Pu, B., 2016. Measurement and Spatiotemporal Analysis of Urban Elasticity [D]. Southwest Jiaotong University (in Chinese).

Qin, D. H., 2014. Climate change science and human sustainable development. Adv. Geogr. Sci. 33 (07), 874-883.

Rockefeller Foundation, 2014. City resilience framework [EB/OL]. http: www. rockerfellerfoundation. org/ app/uploads/City-Resilience-Framework1 pdf.

Snyder, C. R., Mccullough, M. E., 2000. A positive psychology field of dreams: "If you build it, they will Come…". J. Soc. Clin. Psychol. 19 (1), 151-160.

Tanner, T., Mitchell, T., Polack, E., Guenther, B., 2009. Urban Governance for Adaptation: Assessing Climate Change Resilience in Ten Asian Cities.

Thornbush, M., Golubchikov, O., Bouzarovski, S., 2013. Sustainable cities targeted by combined mitigation-adaptation efforts for future-proofing. Sustain. Cities Soc. 9, 1-9.

Tianjin Bureau of Statistics, 1999—2020. Tianjin Statistical Yearbook. China Statistics Press, Tianjin.

Walker, J., Cooper, M., 2011. Genealogies of resilience: from systems ecology to the political economy of crisis adaptation. Secur. Dialogue 42 (2), 143-160.

Walker, B., Holling, C. S., Carpenter, S. R., et al., 2004. Resilience, adaptability and transformability in social-ecological systems. Ecol. Soc. 9 (2), 1-9.

Wall, E., Marzall, K., 2006. Adaptive capacity for climate change in Canadian rural communities. Local Environ. 11 (4), 373-397.

Wang, K., Wang, C., Chen, J. N., 2008. The simulation of technological change and the application in climate policy models. China Pop. Res. Environ. 3, 31-37.

Wang, Z., Deng, X., Wong, C., et al., 2018. Learning urban resilience from a social - economic - ecological system perspective: a case study of Beijing from 1978—2015. J. Clean. Prod. 183, 343-357.

WHO, 2016. WHO household energy database. Retrieved from https: //www. who. int/ airpollution/data/ household-energy-database/en/.

World Meteorological Organization, 2020. The state of the global climate. https: //public. wmo. int/en/our-mandate/climate/wmo-statement-state-of-global-climate.

Zhang, L., Yang, C. L., 2014. Research on the evolution of the sustainable development level of urban agglomerations and its influencing factors: evidence from China's top ten urban agglomerations. Forum on Statistics and Information. 1, pp. 87-93.

Zhang, L. J., Zhang, X., 2015. Weighted clustering method based on improved CRITIC method. Statist.

Decisions 442 (22), 65-68.
Zheng, Y., Xie, X. L., et al., 2018. Development as adaptation: framing and measuring urban resilience in Beijing. Adv. Clim. Chang. Res. 9 (04), 234-242.
Zhu, J. H., Sun, H. X., 2020. Study on the temporal and spatial evolution and influencing factors of urban resilience of three major urban agglomerations in China. Soft Sci. 34 (02), 72-79 (in Chinese).
Zhuang, G. Y., 2004. Climate change and sustainable development. World Econ. Polit. (04).